SMALL WASTEWATER SYSTEM OPERATION AND MAINTENANCE

Volume II

Second Edition

A Field Study Training Program

prepared by

Office of Water Programs
College of Engineering and Computer Science
California State University, Sacramento

≈ ≋ ≈

Kenneth D. Kerri, Project Director

≈ ≋ ≈

2012

Cover photographs:

Background: Courtesy Naturally Wallace Consulting, Stillwater, MN

Foreground, left to right: (1) Courtesy Hoot Systems, LLC,
Lake Charles, LA; (2–4) Rich Parkhurst, OWP

Cover design: Rich Parkhurst

Funding for this operator training manual was provided by the Office of Water Programs, California State University, Sacramento. Mention of trade names or commercial products does not constitute endorsement or recommendation for use by the Office of Water Programs or California State University, Sacramento.

ISBN
978-1-59371-063-7

www.owp.csus.edu

OFFICE OF WATER PROGRAMS

The Office of Water Programs is a nonprofit organization operating under University Enterprises, Inc., California State University, Sacramento, to provide distance learning courses for persons interested in the operation and maintenance of drinking water and wastewater facilities. These training programs were developed by people who explain, through the use of our manuals, how they operate and maintain their facilities. The university, fully accredited by the Western Association of Schools and Colleges, administers and monitors these training programs, under the direction of Dr. Ramzi J. Mahmood.

Our training group develops and implements programs and publishes manuals for operators of water treatment plants, water distribution systems, wastewater collection systems, and municipal and industrial wastewater treatment and reclamation facilities. We also offer programs and materials for pretreatment facility inspectors, environmental compliance inspectors, and utility managers. All training is offered as distance learning, using correspondence, video, or computer-based formats with opportunities for continuing education and contact hours for operators, supervisors, managers, and administrators.

Materials and opportunities available from our office include manuals in print, CD, DVD, or video formats, and enrollments for courses providing CEU (Continuing Education Unit) contact hours. Here is a sample:

- Industrial Waste Treatment, 2 volumes (print, course enrollment)
- Operation of Wastewater Treatment Plants, 2 volumes (print, CD, course enrollment, online courses)
- Advanced Waste Treatment (print, course enrollment)
- Treatment of Metal Wastestreams (print, course enrollment)
- Pretreatment Facility Inspection (print, video, course enrollment)
- Small Wastewater System Operation and Maintenance, 2 volumes (print, course enrollment)
- Operation and Maintenance of Wastewater Collection Systems, 2 volumes (print, course enrollment)
- Collection System Operation and Maintenance Training Videos (video, course enrollment)
- Utility Management (print, course enrollment)
- Manage for Success (print, course enrollment)
- and more

These and other materials may be ordered from the Office of Water Programs:

Office of Water Programs
California State University, Sacramento
6000 J Street
Sacramento, CA 95819-6025
(916) 278-6142 – phone
(916) 278-5959 – FAX

or

visit us on the web at www.owp.csus.edu

ADDITIONAL VOLUMES OF INTEREST

Operation of Wastewater Treatment Plants, Volume I

The Treatment Plant Operator
Why Treat Wastes?
Wastewater Treatment Facilities
Racks, Screens, Comminutors, and Grit Removal
Sedimentation and Flotation
Trickling Filters
Rotating Biological Contactors
Activated Sludge (Package Plants and Oxidation Ditches)
Wastewater Stabilization Ponds
Disinfection Processes

Operation of Wastewater Treatment Plants, Volume II

Activated Sludge (Operation of Conventional Activated Sludge Plants)
Sludge Digestion and Solids Handling
Effluent Discharge, Reclamation, and Reuse
Plant Safety
Maintenance
Laboratory Procedures and Chemistry
Applications of Computers for Plant O&M
Analysis and Presentation of Data
Records and Report Writing
Treatment Plant Administration

Operation and Maintenance of Wastewater Collection Systems, Volume I

The Wastewater Collection System Operator
Why Collection System Operation and Maintenance?
Wastewater Collection Systems
Safe Procedures
Inspecting and Testing Collection Systems
Pipeline Cleaning and Maintenance Methods
Underground Repair

Operation and Maintenance of Wastewater Collection Systems, Volume II

Lift Stations
Equipment Maintenance
Sewer Renewal (Rehabilitation)
Safety/Survival Programs for Collection System Operators
Administration
Organization for System Operation and Maintenance
Capacity Assurance, Management, Operation, and Maintenance (CMOM)

Small Wastewater System Operation and Maintenance, Volume I

The Small Wastewater System Operator
Small Collection, Treatment, and Discharge Systems
Safety
Septic Tanks and Pumping Systems
Wastewater Treatment and Effluent Discharge Methods
Collection Systems
Maintenance and Troubleshooting
Setting Rates for Small Wastewater Utilities

Small Water System Operation and Maintenance

The Small Water System Operator
Water Sources and Treatment
Wells
Small Water Treatment Plants
Disinfection
Safety
Laboratory Procedures
Setting Water Rates and System Security for Small Water Utilities

PREFACE TO THE SECOND EDITION

This *Small Wastewater System Operation and Maintenance* field study training program serves the following purposes:

1. To develop new qualified small wastewater system operators
2. To expand the abilities of existing operators, permitting better service to both their employers and the public
3. To prepare operators for higher-paying positions by passing civil service and *CERTIFICATION EXAMINATIONS.*[1]

To provide you with the knowledge, skills, and abilities needed to safely operate and maintain small wastewater systems as efficiently and effectively as possible, experienced wastewater system operators prepared the material in each chapter of this manual with an emphasis on what operators need to know to do their jobs.

Small wastewater systems vary from town to town and from region to region. The material in this program is presented to provide you with an understanding of the basic operation and maintenance aspect of your system and with information to help you analyze and solve operation and maintenance problems. This information will help you operate and maintain your system in a safe and efficient manner.

Small wastewater system operators continuously have to learn how to operate and maintain new technologies. They are confronted with having to comply with an ever-expanding array of complex regulations. To stay current in this critical field, operators must have a continuing education program. Joining professional organizations and maintaining contact with regulatory agencies helps operators stay informed so they can perform their jobs and protect the public health.

The project directors are deeply indebted to the many operators and other persons who contributed to this operator training manual. Because of their help, this Second Edition contains the most relevant information about current technology in the field of small on-site and decentralized wastewater operation and maintenance systems.

The publications staff of the Office of Water Programs are recognized for their contributions to this training manual: Janet Burton (Publications Manager); Jan Weeks (Senior Editor); Robert Beck (Production Editor); Lea Washington (Technical Editor).

KENNETH D. KERRI

2012

---

1. *Certification Examination.* An examination administered by a state agency or professional association that operators take to indicate a level of professional competence. In the United States, certification of operators of water treatment plants, wastewater treatment plants, water distribution systems, and small water supply systems is mandatory. In many states, certification of wastewater collection system operators, industrial wastewater treatment plant operators, pretreatment facility inspectors, and small wastewater system operators is voluntary; however, current trends indicate that more states, provinces, and employers will require these operators to be certified in the future. Operator certification is mandatory in the United States for the Chief Operators of water treatment plants, water distribution systems, and wastewater treatment plants.

OBJECTIVES OF THIS MANUAL

Proper installation, inspection, operation, maintenance, repair, and management of small wastewater systems have a significant impact on the operation and maintenance costs and effectiveness of the systems. The objective of this manual is to provide small wastewater system operators with the knowledge and skills required to operate and maintain these systems effectively, thus eliminating or reducing the following problems:

1. Health hazards created by the discharge of improperly treated wastewater

2. System failures that result from the lack of proper installation, inspection, preventive maintenance, surveillance, and repair programs designed to protect the public's investment in these facilities

3. Fish kills and environmental damage caused by wastewater system problems

4. Corrosion damages to pipes, equipment, tanks, and structures in the wastewater system

5. Complaints from the public or local officials due to the unreliability or failure of the wastewater system to perform as designed

6. Odors caused by insufficient or inadequate wastewater and sludge handling and treatment procedures

SCOPE OF THIS MANUAL

Operators with the responsibility for operating and maintaining package wastewater treatment plants and wastewater treatment processes, which are used by smaller communities and treatment agencies, will find this manual very helpful. Topics covered include wastewater stabilization ponds; activated sludge; rotating biological contactors; and some alternative treatment systems. Other topics include disinfection, chlorination and dechlorination; wastewater treatment, discharge, and reuse methods; laboratory procedures; and management of a small wastewater collection and treatment agency.

Material in this manual furnishes you with information concerning situations encountered by most small wastewater system operators in most areas. These materials provide you with an understanding of the basic operational and maintenance concepts for small wastewater systems and with an ability

to analyze and solve problems when they occur. Operation and maintenance programs for small wastewater systems will vary with the age of the system, the extent and effectiveness of previous programs, and local conditions. You will have to adapt the information and procedures in this manual to your particular situation.

Technology is advancing very rapidly in the field of operation and maintenance of small wastewater systems. To keep pace with scientific advances, the material in this program must be periodically revised and updated. This means that you, the system operator, must be aware of new advances and recognize the need for continuous personal training reaching beyond this program. Training opportunities exist in your daily work experience, from your associates, and from attending meetings, workshops, conferences, and classes.

USES OF THIS MANUAL

This manual was developed to serve the needs of operators in several different situations. The format used was developed to serve as a home-study or self-paced instruction course for operators in remote areas or persons unable to attend formal classes either due to shift work, personal reasons, or the unavailability of suitable classes. This home-study training program uses the concepts of self-paced instruction where you are your own instructor and work at your own speed. In order to certify that a person has successfully completed this program, objective tests and special answer sheets for each chapter are provided when a person enrolls in this course.

Also, this manual can serve effectively as a textbook in the classroom. Many colleges and universities have used this manual as a text in formal classes (often taught by operators). In areas where colleges are not available or are unable to offer classes in the operation of small wastewater systems, operators and utility agencies can join together to offer their own courses using this manual.

Cities or utility agencies can use this manual in several types of on-the-job training programs. In one type of program, a manual is purchased for each operator. A senior operator or a group of operators are designated as instructors. These operators help answer questions when the persons in the training program have questions or need assistance.

This manual was prepared to help operators operate and maintain their small wastewater systems. Please feel free to use the manual in the manner that best fits your training needs and the needs of other operators. We will be happy to work with you to assist you in developing your training program. Please feel free to contact the Project Director:

Project Director
Office of Water Programs
California State University, Sacramento
6000 J Street
Sacramento, CA 95819-6025
(916) 278-6142 – phone
(916) 278-5959 – FAX
wateroffice@csus.edu – e-mail

TECHNICAL REVIEWERS

Russ Armstrong
Pat Conway
David Dauwalder
Steve Desmond
Dwight Lancaster
Martin Lopez

Thurlow Morrow
Stew Oakley
Kurt Ohlinger
Don Snelling
Tim Sullivan

INSTRUCTIONS TO PARTICIPANTS IN HOME-STUDY COURSE

Procedures for reading the lessons and answering the questions in this manual are contained in this section.

To progress steadily through this training program, establish a regular study schedule. Some of the chapters are longer and more difficult than others, so many of them are divided into two or more lessons. The time required to complete a lesson will depend on your background and experience. The important thing is that you understand the material in the lesson before starting the next one.

Each lesson is arranged for you to read a short section, write your answers to the questions at the end of the section, and check your answers against the suggested answers. You can then decide if you understand the material sufficiently to continue or whether you need to read the section again. You may find that this procedure is slower than reading a typical textbook, but you will probably remember much more when you have finished.

Discussion and review questions follow each lesson in the chapters. These questions are provided to help you review the important points covered in the lesson. Write your answers to the discussion and review questions to help yourself retain the material better.

In the appendix at the end of this manual, you will find comprehensive review questions and suggested answers. These questions and answers are provided for you to review how well you remember the material. You will probably need to review the entire manual before attempting to answer these questions. Some of the questions are essay-type questions, which are used by some states for higher-level certification examinations. After you have answered all the questions, compare your answers with those provided and determine the areas in which you might need additional review before your next certification or civil service examination. Please do not send your answers to the Office of Water Programs at California State University, Sacramento.

You are your own teacher in this training program. You could merely look up the suggested answers to the questions at the end of the chapters and in the comprehensive review section, but doing so will not help you understand the material. Consequently, you would not be able to apply the material to the performance of your job or recall it during an examination for certification or a civil service position. You will get out of this program what you put into it, so we encourage you to make the most of the material presented.

SUMMARY OF PROCEDURE

To complete this program, you need to work through all of the chapters. Study the material in chapter order or in a sequence that works best for you. Many manuals include an arithmetic appendix to give you a quick review of the math concepts found in the manual.

The following is a summary of the suggested procedure for studying the material in this manual:

1. Read what you are expected to learn in each chapter (the Objectives).

2. Read the sections in each lesson or chapter.

3. Write your answers to the questions at the end of each section, just as you would if they were questions on an actual test.

4. Compare your answers with the suggested answers.

5. Decide whether to review the section or to continue on to the next section.

6. Write your answers to the discussion and review questions at the end of each lesson.

NOTE: Safety is an important topic in all of the manuals in this operator training series. Because operators daily encounter situations and equipment that can cause serious disabling injury or illness, operators need to be aware of potential dangers and exercise adequate precautions when performing their jobs. As you work through the chapters, pay close attention to the safe procedures that are stressed throughout.

SMALL WASTEWATER SYSTEM OPERATION AND MAINTENANCE

Volume II

Second Edition

CHAPTER 9

WASTEWATER STABILIZATION PONDS

by

A. Hiatt

Revised by

John Brady

TABLE OF CONTENTS
Chapter 9.   WASTEWATER STABILIZATION PONDS

OBJECTIVES

Chapter 9.   WASTEWATER STABILIZATION PONDS

1. Explain how wastewater stabilization ponds work and what factors influence and control pond treatment processes.

2. Identify the different types of ponds.

3. Place a new pond into operation.

4. Schedule and conduct normal and abnormal operation and maintenance duties.

5. Collect samples, interpret lab results, and make appropriate adjustments in pond operation.

6. Recognize factors that indicate a pond is not performing properly, identify the source of the problem, and take corrective action.

7. Develop a pond operating strategy.

8. Conduct your duties in a safe fashion.

9. Determine pond loadings.

10. Keep records for a waste treatment pond facility.

11. Review plans and specifications for new ponds.

WORDS

Chapter 9.   WASTEWATER STABILIZATION PONDS

AEROBIC (air-O-bick)   AEROBIC

A condition in which atmospheric or dissolved oxygen is present in the aquatic (water) environment.

ALGAE (AL-jee)   ALGAE

Microscopic plants containing chlorophyll that live floating or suspended in water. They also may be attached to structures, rocks, or other submerged surfaces. Excess algal growths can impart tastes and odors to potable water. Algae produce oxygen during sunlight hours and use oxygen during the night hours. Their biological activities appreciably affect the pH, alkalinity, and dissolved oxygen of the water.

ALGAL (AL-gull) BLOOM   ALGAL BLOOM

Sudden, massive growths of microscopic and macroscopic plant life, such as green or blue-green algae, which can, under the proper conditions, develop in lakes, reservoirs, and ponds.

ANAEROBIC (AN-air-O-bick)   ANAEROBIC

A condition in which atmospheric or dissolved oxygen (DO) is *NOT* present in the aquatic (water) environment.

APPURTENANCE (uh-PURR-ten-nans)   APPURTENANCE

Machinery, appliances, structures, and other parts of the main structure necessary to allow it to operate as intended, but not considered part of the main structure.

BOD (pronounce as separate letters)   BOD

Biochemical Oxygen Demand. The rate at which organisms use the oxygen in water or wastewater while stabilizing decomposable organic matter under aerobic conditions. In decomposition, organic matter serves as food for the bacteria and energy results from its oxidation. BOD measurements are used as a surrogate measure of the organic strength of wastes in water.

BACTERIA (back-TEER-e-uh)   BACTERIA

Bacteria are living organisms, microscopic in size, that usually consist of a single cell. Most bacteria use organic matter for their food and produce waste products as a result of their life processes.

BIOCHEMICAL OXYGEN DEMAND (BOD)   BIOCHEMICAL OXYGEN DEMAND (BOD)

See BOD.

BIOFLOCCULATION (BUY-o-flock-yoo-LAY-shun)   BIOFLOCCULATION

The clumping together of fine, dispersed organic particles by the action of certain bacteria and algae. This results in faster and more complete settling of the organic solids in wastewater.

CHEMICAL OXYGEN DEMAND (COD)   CHEMICAL OXYGEN DEMAND (COD)

A measure of the oxygen-consuming capacity of organic matter present in wastewater. COD is expressed as the amount of oxygen consumed from a chemical oxidant in mg/L during a specific test. Results are not necessarily related to the biochemical oxygen demand (BOD) because the chemical oxidant may react with substances that bacteria do not stabilize.

COLIFORM (KOAL-i-form)

COLIFORM

A group of bacteria found in the intestines of warm-blooded animals (including humans) and also in plants, soil, air, and water. The presence of coliform bacteria is an indication that the water is polluted and may contain pathogenic (disease-causing) organisms. Fecal coliforms are those coliforms found in the feces of various warm-blooded animals, whereas the term coliform also includes other environmental sources.

COMPOSITE (PROPORTIONAL) SAMPLE

COMPOSITE (PROPORTIONAL) SAMPLE

A composite sample is a collection of individual samples obtained at regular intervals, usually every one or two hours during a 24-hour time span. Each individual sample is combined with the others in proportion to the rate of flow when the sample was collected. Equal volume individual samples also may be collected at intervals after a specific volume of flow passes the sampling point or after equal time intervals and still be referred to as a composite sample. The resulting mixture (composite sample) forms a representative sample and is analyzed to determine the average conditions during the sampling period.

DETENTION TIME

DETENTION TIME

The time required to fill a tank at a given flow or the theoretical time required for a given flow of wastewater to pass through a tank. In septic tanks, this detention time will decrease as the volumes of sludge and scum increase.

DISINFECTION (dis-in-FECT-shun)

DISINFECTION

The process designed to kill or inactivate most microorganisms in water or wastewater, including essentially all pathogenic (disease-causing) bacteria. There are several ways to disinfect, with chlorination being the most frequently used in water and wastewater treatment plants. Compare with STERILIZATION.

DISSOLVED OXYGEN

DISSOLVED OXYGEN

Molecular oxygen dissolved in water or wastewater, usually abbreviated DO.

DUCKWEED

DUCKWEED

A small, green, cloverleaf-shaped floating plant, about one-quarter inch (6 mm) across, which appears as a grainy layer on the surface of a pond.

EFFLUENT (EF-loo-ent)

EFFLUENT

Water or other liquid—raw (untreated), partially treated, or completely treated—flowing *FROM* a reservoir, basin, treatment process, or treatment plant.

FACULTATIVE (FACK-ul-tay-tive) POND

FACULTATIVE POND

The most common type of pond in current use. The upper portion (supernatant) is aerobic, while the bottom layer is anaerobic. Algae supply most of the oxygen to the supernatant.

FREE OXYGEN

FREE OXYGEN

Molecular oxygen available for respiration by organisms. Molecular oxygen is the oxygen molecule, O_2, that is not combined with another element to form a compound.

GRAB SAMPLE

GRAB SAMPLE

A single sample of water collected at a particular time and place that represents the composition of the water only at that time and place.

INFLUENT

INFLUENT

Water or other liquid—raw (untreated) or partially treated—flowing *INTO* a reservoir, basin, treatment process, or treatment plant.

MEDIAN

MEDIAN

The middle measurement or value. When several measurements are ranked by magnitude (largest to smallest), half of the measurements will be larger and half will be smaller.

MOLECULAR OXYGEN MOLECULAR OXYGEN

The oxygen molecule, O_2, that is not combined with another element to form a compound.

NPDES PERMIT NPDES PERMIT

National Pollutant Discharge Elimination System permit is the regulatory agency document issued by either a federal or state agency that is designed to control all discharges of potential pollutants from point sources and stormwater runoff into US waterways. NPDES permits regulate discharges into US waterways from all point sources of pollution, including industries, municipal wastewater treatment plants, sanitary landfills, large animal feedlots, and return irrigation flows.

OVERTURN OVERTURN

The almost spontaneous mixing of all layers of water in a reservoir or lake when the water temperature becomes similar from top to bottom. This may occur in the fall/winter when the surface waters cool to the same temperature as the bottom waters and also in the spring when the surface waters warm after the ice melts. This is also called turnover.

PARALLEL OPERATION PARALLEL OPERATION

Wastewater being treated is split and a portion flows to one treatment unit while the remainder flows to another similar treatment unit. Also see SERIES OPERATION.

PATHOGENIC (path-o-JEN-ick) ORGANISMS PATHOGENIC ORGANISMS

Bacteria, viruses, protozoa, or internal parasites that can cause disease (such as giardiasis, cryptosporidiosis, typhoid fever, cholera, or infectious hepatitis) in a host (such as a person). There are many types of organisms that do not cause disease and are not called pathogenic. Many beneficial bacteria are found in wastewater treatment processes actively cleaning up organic wastes.

PERCOLATION (purr-ko-LAY-shun) PERCOLATION

The slow passage of water through a filter medium; or, the gradual penetration of soil and rocks by water.

pH (pronounce as separate letters) pH

pH is an expression of the intensity of the basic or acidic condition of a liquid. Mathematically, pH is the logarithm (base 10) of the reciprocal of the hydrogen ion activity.

$$pH = \text{Log} \frac{1}{\{H^+\}}$$

If $\{H^+\} = 10^{-6.5}$, then pH = 6.5. The pH may range from 0 to 14, where 0 is most acidic, 14 most basic, and 7 neutral.

PHOTOSYNTHESIS (foe-toe-SIN-thuh-sis) PHOTOSYNTHESIS

A process in which organisms, with the aid of chlorophyll, convert carbon dioxide and inorganic substances into oxygen and additional plant material, using sunlight for energy. All green plants grow by this process.

POPULATION EQUIVALENT POPULATION EQUIVALENT

A means of expressing the strength of organic material in wastewater. In a domestic wastewater system, microorganisms use up about 0.2 pound (90 grams) of oxygen per day for each person using the system (as measured by the standard BOD test). May also be expressed as flow (100 gallons (378 liters)/day/person) or suspended solids (0.2 lb (90 grams) SS/day/person).

$$\text{Population Equivalent, persons} = \frac{\text{Flow, MGD} \times \text{BOD, mg/L} \times 8.34 \text{ lbs/gal}}{0.2 \text{ lb BOD/day/person}}$$

or

$$\text{Population Equivalent, persons} = \frac{\text{Flow, cu m/day} \times \text{BOD, mg/L} \times 10^6 \text{ L/cu m}}{90,000 \text{ mg BOD/day/person}}$$

RIPRAP RIPRAP

Broken stones, boulders, or other materials placed compactly or irregularly on levees or dikes for the protection of earth surfaces against the erosive action of waves.

ROTATING BIOLOGICAL CONTACTOR (RBC)

ROTATING BIOLOGICAL CONTACTOR (RBC)

A secondary biological treatment process for domestic and biodegradable industrial wastes. Biological contactors have a rotating shaft surrounded by plastic discs called the media. The shaft and media are called the drum. A biological slime grows on the media when conditions are suitable and the microorganisms that make up the slime (biomass) stabilize the waste products by using the organic material for growth and reproduction.

SEPTIC (SEP-tick)

SEPTIC

A condition produced by anaerobic bacteria. If severe, the sludge produces hydrogen sulfide, turns black, gives off foul odors, contains little or no dissolved oxygen, and the wastewater has a high oxygen demand.

SERIES OPERATION

SERIES OPERATION

Wastewater being treated flows through one treatment unit and then flows through another similar treatment unit. Also see PARALLEL OPERATION.

SHOCK LOAD (ACTIVATED SLUDGE)

SHOCK LOAD

The arrival at a plant of a waste that is toxic to organisms in sufficient quantity or strength to cause operating problems. Possible problems include odors and bulking sludge, which will result in a high loss of solids from the secondary clarifiers into the plant effluent and a biological process upset that may require several days to a week to recover. Organic or hydraulic overloads also can cause a shock load.

SHORT-CIRCUITING

SHORT-CIRCUITING

A condition that occurs in tanks or basins when some of the flowing water entering a tank or basin flows along a nearly direct pathway from the inlet to the outlet. This is usually undesirable because it may result in shorter contact, reaction, or settling times in comparison with the theoretical (calculated) or presumed detention times.

SPLASH PAD

SPLASH PAD

A structure made of concrete or other durable material to protect bare soil from erosion by splashing or falling water.

STABILIZED WASTE

STABILIZED WASTE

A waste that has been treated or decomposed to the extent that, if discharged or released, its rate and state of decomposition would be such that the waste would not cause a nuisance or odors in the receiving water.

STERILIZATION (STAIR-uh-luh-ZAY-shun)

STERILIZATION

The removal or destruction of all microorganisms, including pathogens and other bacteria, vegetative forms, and spores. Compare with DISINFECTION.

STOP LOG

STOP LOG

A log or board in an outlet box or device used to control the water level in ponds and also the flow from one pond to another pond or system.

SUPERNATANT (soo-per-NAY-tent)

SUPERNATANT

The relatively clear water layer between the sludge on the bottom and the scum on the surface of an anaerobic digester or septic tank (interceptor).

(1) From an anaerobic digester, this water is usually returned to the influent wet well or to the primary clarifier.

(2) From a septic tank, this water is discharged by gravity or by a pump to a leaching system or a wastewater collection system.

Also called clear zone.

TOXIC

TOXIC

A substance that is poisonous to a living organism. Toxic substances may be classified in terms of their physiological action, such as irritants, asphyxiants, systemic poisons, and anesthetics and narcotics. Irritants are corrosive substances that attack the mucous membrane surfaces of the body. Asphyxiants interfere with breathing. Systemic poisons are hazardous substances that injure or destroy internal organs of the body. Anesthetics and narcotics are hazardous substances that depress the central nervous system and lead to unconsciousness.

TOXICITY (tox-IS-it-tee)   TOXICITY

The relative degree of being poisonous or toxic. A condition that may exist in wastes and will inhibit or destroy the growth or function of certain organisms.

TRICKLING FILTER   TRICKLING FILTER

A treatment process in which wastewater trickling over media enables the formation of slimes or biomass, which contain organisms that feed upon and remove wastes from the water being treated.

CHAPTER 9. WASTEWATER STABILIZATION PONDS

USED FOR TREATMENT OF WASTEWATER AND OTHER WASTES

(Lesson 1 of 3 Lessons)

9.0 USE OF PONDS

Shallow ponds (3 to 6 feet or 1 to 2 meters deep) are often used to treat wastewater and other wastes instead of, or in addition to, conventional waste treatment processes. Table 9.1 lists the purpose of pond parts and Figure 9.1 shows a typical pond plant layout. When discharged into ponds, wastes are treated or *STABILIZED*[1] by several natural processes acting at the same time. Heavy solids settle to the bottom where they are decomposed by bacteria. Lighter suspended material is broken down by bacteria in suspension. Some wastewater is disposed of by evaporation from the pond surface or by *PERCOLATING*[2] into the ground.

Dissolved nutrient materials such as nitrogen and phosphorus are used by green *ALGAE*,[3] which are actually microscopic plants floating and living in the water. The algae use carbon dioxide (CO_2) and bicarbonate to build body protoplasm. To grow, they need nitrogen and phosphorus in their metabolism much as land plants do. Like land plants, they release oxygen and some carbon dioxide as waste products.

Ponds can serve as very effective treatment facilities. Extensive studies of their performance have led to a better understanding of the natural processes by which ponds treat wastes.

This chapter provides information about the natural processes and ways operators can regulate pond processes for efficient waste treatment.

TABLE 9.1 PURPOSE OF POND PARTS

Part	Purpose
Flowmeter	Measures and records flows into the pond and possibly effluent.
Bar Rack	Removes coarse material from pond influent. Also called bar screen.
Pond Inlets	Distribute influent in the pond.
Pond Depth and Outlet Control	Regulates outflow from the pond and depth of water in the pond. Allows pond to be drained for cleaning and inspection.
Outlet Baffle	Prevents scum and other surface debris from flowing to the next pond or into receiving waters. (Part of outlet structure.)
Dike or Levee	Separates ponds and holds wastewater being treated in ponds.
Transfer Line	Conveys wastewater from one pond to another.
Recirculation Line	Returns pond effluent rich in algae and oxygen from the second pond to the first pond for seeding, dilution, and process control.
Chlorination	Applies chlorine to treated wastewater for disinfection purposes.
Chlorine Contact Basin	Provides contact time for chlorine to disinfect the pond effluent.
Effluent Line	Conveys treated wastewater to receiving waters, to point of reuse (irrigation), or to land disposal site.

---

1. *Stabilized Waste.* A waste that has been treated or decomposed to the extent that, if discharged or released, its rate and state of decomposition would be such that the waste would not cause a nuisance or odors in the receiving water.

2. *Percolation* (purr-ko-LAY-shun). The slow passage of water through a filter medium; or, the gradual penetration of soil and rocks by water.

3. *Algae* (AL-jee). Microscopic plants containing chlorophyll that live floating or suspended in water. They also may be attached to structures, rocks, or other submerged surfaces. Excess algal growths can impart tastes and odors to potable water. Algae produce oxygen during sunlight hours and use oxygen during the night hours. Their biological activities appreciably affect the pH, alkalinity, and dissolved oxygen of the water.

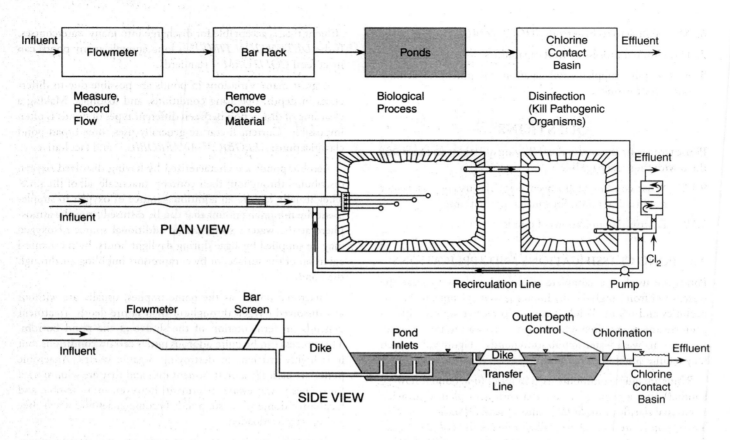

Fig. 9.1   *Typical plant; ponds only*

9.1   WASTE TREATMENT BY PONDS

Since 1958, engineers have designed and constructed a great number of ponds based on research by qualified biological consultants, current scientific knowledge of ponding, and the experience of past successes and failures. When operated in a knowledgeable and purposeful manner, these ponds have successfully performed a variety of functions.

As a complete process, wastewater stabilization ponds offer many advantages for smaller installations where land is not costly and the location is isolated from residential, commercial, and recreational areas. The following are advantages of ponds:

1. Do not require expensive equipment

2. Do not require highly trained operating personnel

3. Are economical to construct

4. Provide treatment that is equal or superior to some conventional processes

5. Are a satisfactory method of treating wastewater on a temporary basis

6. Are adaptable to changing waste loads

7. Are adaptable to land application

8. Consume little energy

9. Serve as wildlife habitats

10. Have increased potential design life

11. Have few short-term sludge handling and disposal problems

12. Are probably the most trouble-free of any treatment process when used correctly, provided a consistently high-quality effluent is not required

The following are the limitations of ponds:

1. May produce odors

2. Require a large area of land

3. Treat wastes inconsistently depending on climatic conditions

4. May contaminate groundwaters if not properly lined

5. May have high suspended solids levels in the effluent due to algae

6. May be slow to recover from *SHOCK LOADS* [4]

7. Have limited process control capabilities

8. May require supplemental aeration equipment to achieve odor-free treatment

9.2 POND CLASSIFICATIONS AND APPLICATIONS

Ponds are used as a complete treatment process to treat the wastewater from single-family homes as well as from small communities and towns. Ponds designed to receive wastes with no prior treatment are often referred to as "raw wastewater (sewage) lagoons" or "wastewater stabilization ponds" (Figure 9.2), which may require sizable areas of land.

Ponds are quite commonly used in series (one pond following another) after a primary wastewater treatment plant to provide additional clarification, *BOD* [5] removal, and *DISINFECTION.* [6] These ponds are sometimes called *FACULTATIVE PONDS.* [7] Ponds are sometimes used in series after a *TRICKLING FILTER* [8] plant or *ROTATING BIOLOGICAL CONTACTOR (RBC).* [9] These ponds are sometimes called "polishing ponds." Ponds placed in series with each other can provide a high-quality

effluent that is acceptable for discharge into many watercourses. If the *DETENTION TIME* [10] is long enough, many ponds can meet fecal *COLIFORM* [11] standards.

A great many variations in ponds are possible due to differences in depth, operating conditions, and loadings. Making a clear line of distinction between different types of ponds is often impossible. Current literature generally uses three broad pond classifications: *AEROBIC,* [12] *ANAEROBIC,* [13] and facultative.

Aerobic ponds are characterized by having dissolved oxygen distributed throughout their contents practically all of the time. They usually require an additional source of oxygen to supplement the minimal amount that can be diffused from the atmosphere at the water's surface. The additional source of oxygen may be supplied by algae during daylight hours, by mechanical agitation of the surface, or by compressors bubbling air through the pond.

Anaerobic ponds, as the name implies, usually are without any dissolved oxygen throughout their entire depth. Treatment depends on fermentation of the sludge at the pond bottom. This process can be quite odorous under certain conditions, but it is highly efficient in destroying organic wastes. Anaerobic ponds are basically a pretreatment unit and they are seldom used for domestic wastewater treatment; however, some aerobic and facultative domestic-waste ponds become anaerobic when they are severely overloaded.

Facultative ponds are the most common type of pond in use. The upper portion (supernatant) of these ponds is aerobic, while the bottom portion is anaerobic. Algae supply most of the oxygen to the supernatant. Facultative ponds are most common

---

4. *Shock Load (Activated Sludge).* The arrival at a plant of a waste that is toxic to organisms in sufficient quantity or strength to cause operating problems. Possible problems include odors and bulking sludge, which will result in a high loss of solids from the secondary clarifiers into the plant effluent and a biological process upset that may require several days to a week to recover. Organic or hydraulic overloads also can cause a shock load.

5. *BOD* (pronounce as separate letters). Biochemical Oxygen Demand. The rate at which organisms use the oxygen in water or wastewater while stabilizing decomposable organic matter under aerobic conditions. In decomposition, organic matter serves as food for the bacteria and energy results from its oxidation. BOD measurements are used as a surrogate measure of the organic strength of wastes in water.

6. *Disinfection* (dis-in-FECT-shun). The process designed to kill or inactivate most microorganisms in water or wastewater, including essentially all pathogenic (disease-causing) bacteria. There are several ways to disinfect, with chlorination being the most frequently used in water and wastewater treatment plants. Compare with STERILIZATION.

7. *Facultative* (FACK-ul-tay-tive) *Pond.* The most common type of pond in current use. The upper portion (supernatant) is aerobic, while the bottom layer is anaerobic. Algae supply most of the oxygen to the supernatant.

8. *Trickling Filter.* A treatment process in which wastewater trickling over media enables the formation of slimes or biomass, which contain organisms that feed upon and remove wastes from the water being treated.

9. *Rotating Biological Contactor (RBC).* A secondary biological treatment process for domestic and biodegradable industrial wastes. Biological contactors have a rotating shaft surrounded by plastic discs called the media. The shaft and media are called the drum. A biological slime grows on the media when conditions are suitable and the microorganisms that make up the slime (biomass) stabilize the waste products by using the organic material for growth and reproduction.

10. *Detention Time.* The time required to fill a tank at a given flow or the theoretical time required for a given flow of wastewater to pass through a tank. In septic tanks, this detention time will decrease as the volumes of sludge and scum increase.

11. *Coliform* (KOAL-i-form). A group of bacteria found in the intestines of warm-blooded animals (including humans) and also in plants, soil, air, and water. The presence of coliform bacteria is an indication that the water is polluted and may contain pathogenic (disease-causing) organisms. Fecal coliforms are those coliforms found in the feces of various warm-blooded animals, whereas the term coliform also includes other environmental sources.

12. *Aerobic* (air-O-bick). A condition in which atmospheric or dissolved oxygen is present in the aquatic (water) environment.

13. *Anaerobic* (AN-air-O-bick). A condition in which atmospheric or dissolved oxygen (DO) is NOT present in the aquatic (water) environment.

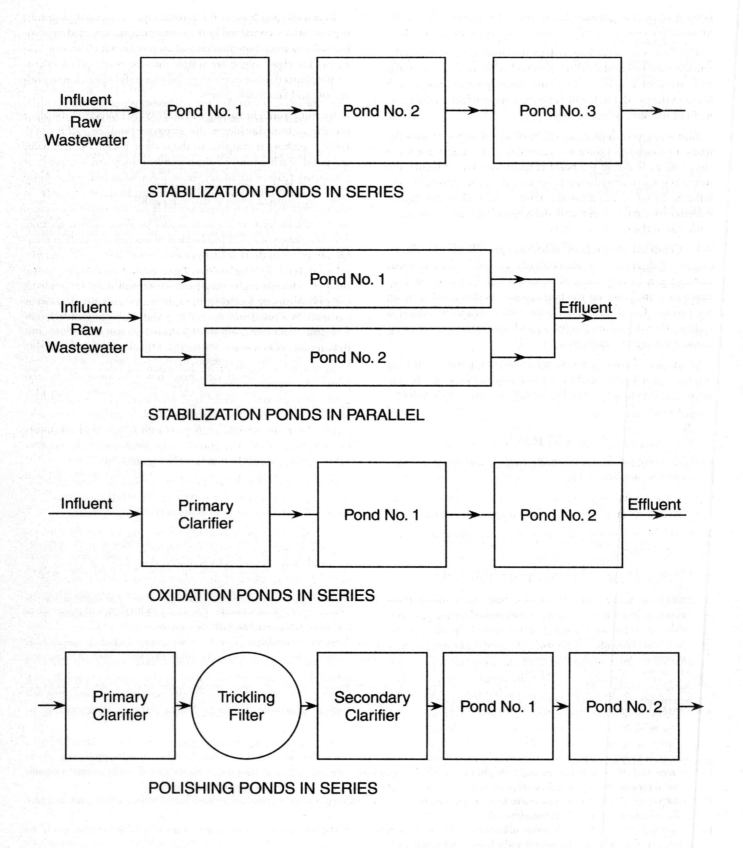

Fig. 9.2   Pond classifications

because it is almost impossible to maintain completely aerobic or anaerobic conditions all the time at all depths of the pond.

Pond uses vary according to detention time. A pond with a detention time of less than three days will perform in ways similar to a sedimentation or settling tank. Some growth of algae will occur in the pond, but it will not have a major effect on the treatment of the wastewater.

Abundant growth of algae will be observed in ponds with detention periods from three to twenty days, and large amounts of algae will be found in the pond effluent. In some effluents, the stored organic material may be greater than the amount in the influent. Detention times in this range merely allow the organic material to change form and delay problems until the algae settle out in the receiving waters.

Longer detention times in ponds may provide time for the reduction of algae due to a reduction of nutrients in the water that sustain algae. Usually, this will occur in facultative ponds having anaerobic conditions on the bottom and aerobic conditions on the surface. Combined aerobic–anaerobic treatment provided by long detention times produces good stabilization of the pond influent during treatment in the pond.

Controlled-discharge ponds are facultative ponds with long detention times of up to 180 days or longer. These ponds may discharge effluent only once (in the fall) or twice (in the fall and spring) a year.

QUESTIONS

Please write your answers to the following questions and compare them with those on page 44.

9.2A   What is the difference between the terms "aerobic," "anaerobic," and "facultative" when applied to ponds?

9.2B   How does the use of a pond vary depending on detention time?

9.3   EXPLANATION OF TREATMENT PROCESS

As mentioned in the previous section, wastewater stabilization ponds also are classified according to their dissolved oxygen content. Oxygen in an aerobic pond is distributed throughout the entire depth practically all the time. An anaerobic pond is predominantly without oxygen most of the time because oxygen requirements are much greater than the oxygen supply. In a facultative pond, the upper portion is aerobic most of the time, whereas the bottom layer is predominantly anaerobic.

In aerobic ponds or in the aerobic layer of facultative ponds, organic matter contained in the wastewater is converted by aerobic bacteria to carbon dioxide and ammonia, which serve as nutrients for algae. Algae are simple one- or many-celled microscopic plants that are essential to the successful operation of both aerobic and facultative ponds.

By using sunlight through *PHOTOSYNTHESIS*,[14] the algae use the carbon dioxide in the water to produce *FREE OXYGEN*,[15] making it available to the aerobic bacteria that inhabit the pond. Each pound of algae in a healthy pond is capable of producing 1.6 pounds of oxygen on a normal summer day. Algae live on carbon dioxide and other nutrients in the wastewater. At night, when light is no longer available for photosynthesis, algae use up the oxygen by respiration and produce carbon dioxide. The alternating use and production of oxygen and carbon dioxide can result in diurnal (daily) variations of both *pH*[16] and dissolved oxygen. During the day, algae use carbon dioxide, raising the pH, while at night they produce carbon dioxide, lowering the pH. Algae are found in the soil, water, and air; they occur naturally in a pond without seeding and multiply greatly under favorable conditions. Figure 9.3 illustrates the role of algae in treating wastes in a pond.

In anaerobic ponds or in the anaerobic layer of facultative ponds, the organic matter is first converted to organic acids by a group of organisms called "acid producers." In an established pond, at the same time, a group called "methane fermenters" breaks down the organic acids produced by the acid producers to form methane gas, carbon dioxide, ammonia, water, and alkalinity. This process is illustrated in Figure 9.3.

In a successful facultative pond, the processes characteristic of aerobic ponds occur in the surface layers, while those similar to anaerobic ponds occur in its bottom layers.

During certain periods, sludge decomposition in the anaerobic zone is interrupted, so sludge begins to accumulate. If sludge accumulation occurs and decomposition does not set in, it is probably due to a lack of the right bacteria, low pH, presence of substances that slow or stop the process, or a low temperature. Under these circumstances, the acid production will continue at a slower rate, but the rate of gas (methane) production slows down considerably.

Sludge storage in ponds is continuous, with small amounts stored during warm weather and larger amounts when it is cold. During low temperatures, the bacteria cannot multiply fast enough to handle the waste. When warm weather comes, the

---

14. *Photosynthesis* (foe-toe-SIN-thuh-sis).   A process in which organisms, with the aid of chlorophyll, convert carbon dioxide and inorganic substances into oxygen and additional plant material, using sunlight for energy. All green plants grow by this process.

15. *Free Oxygen.*   Molecular oxygen available for respiration by organisms. Molecular oxygen is the oxygen molecule, O_2, that is not combined with another element to form a compound.

16. *pH* (pronounce as separate letters).   pH is an expression of the intensity of the basic or acidic condition of a liquid. Mathematically, pH is the logarithm (base 10) of the reciprocal of the hydrogen ion activity.

$$pH = Log \frac{1}{\{H^+\}}$$

If $\{H^+\} = 10^{-6.5}$, then pH = 6.5. The pH may range from 0 to 14, where 0 is most acidic, 14 most basic, and 7 neutral.

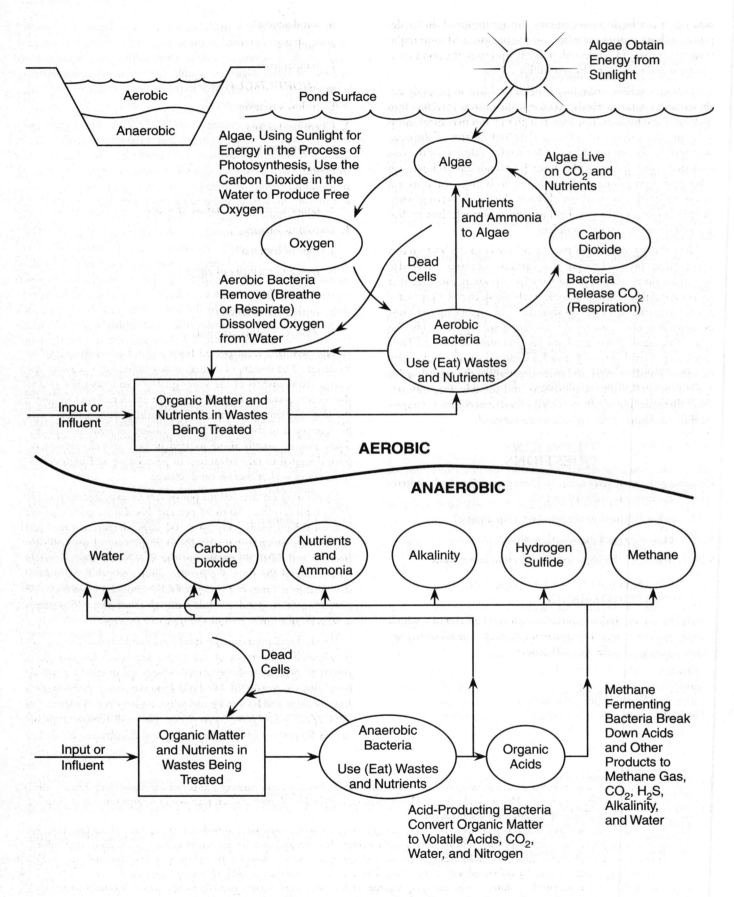

Fig. 9.3 Process of decomposition in aerobic and anaerobic layers of a pond

acid producers begin to decompose the accumulated sludge deposits built up during the winter. If the organic acid production is too great, the pH is lowered, possibly upsetting the pond environment and causing hydrogen sulfide odors.

Hydrogen sulfide ordinarily is not a problem in properly designed and operated ponds because it dissociates (divides) into hydrogen and hydrosulfide ions at high pH and may form insoluble metallic sulfides or sulfates. This high degree of dissociation and the formation of insoluble metallic sulfides are the reasons that ponds having a pH above 8.5 do not emit odors, even when hydrogen sulfide is present in relatively large amounts. An exception occurs in northern climates during the spring when the pH is low and the pond is just getting started; then hydrogen sulfide odors can be a problem.

All of the organic matter that finds its way to the bottom of a stabilization pond through the various processes of sludge decomposition is subject to methane fermentation, provided that proper conditions exist or become established. In order for methane fermentation to occur, an abundance of organic matter must be deposited and continually converted to organic acids. An abundant population of methane bacteria must be present. They require a pH level of from 6.5 to 7.5 within the sludge, alkalinity of several hundred mg/L to buffer (neutralize) the organic acids (volatile acid/alkalinity relationship), and suitable temperatures. Once methane fermentation is established, it accounts for a considerable amount of the organic solids removal.

QUESTIONS

Please write your answers to the following questions and compare them with those on pages 44 and 45.

9.3A What happens to organic matter in a pond?

9.3B How is oxygen produced by algae?

9.3C Where do the algae found in a pond come from?

9.4 POND PERFORMANCE

Pond performance is determined by treatment efficiencies, which vary more than most other treatment devices. The following are some of the factors affecting efficiency:

1. Physical Factors
   a. type of soil (permeability)
   b. surface area
   c. water depth
   d. wind action
   e. sunlight
   f. temperature
   g. *SHORT-CIRCUITING*[17]
   h. inflow variations
2. Chemical Factors
   a. organic material
   b. pH
   c. solids
   d. concentration and nature of waste
3. Biological Factors
   a. type of bacteria
   b. type and quantity of algae
   c. activity of organisms
   d. nutrient deficiencies
   e. *TOXIC*[18] concentrations

The performance expected from a pond depends largely on its design. The design, of course, is determined by the waste discharge requirements or the water quality standards to be met in the receiving waters. Overall treatment efficiency may be about the same as primary treatment (only settling of solids), or it may be equivalent to the best secondary biological treatment plants. Some ponds, usually those located in hot, arid climates, have been designed to take advantage of percolation and high evaporation rates so that there is no discharge.

Depending on the design, ponds can be expected to provide BOD removals from 50 to 90 percent. Facultative ponds, under normal design loads with 50 to 60 days' detention time, will usually remove approximately 90 to 95 percent of the coliform bacteria and 70 to 80 percent of the BOD load approximately 80 percent of the time. Controlled-discharge ponds with 180-day detention times can produce BOD removals from 85 to 95 percent, total suspended solids removals from 85 to 95 percent, and fecal coliform reductions of up to 99 percent.

Physical sedimentation by itself has been found to remove approximately 90 percent of the suspended solids in three days. About 80 percent of the dissolved organic solids can be removed by biological action in 10 days. However, in a pond with a healthy algae and bacteria population, a phenomenon known as *BIOFLOCCULATION*[19] can occur. This will remove approximately 85 percent of both suspended and dissolved solids within

---

17. *Short-Circuiting.* A condition that occurs in tanks or basins when some of the flowing water entering a tank or basin flows along a nearly direct pathway from the inlet to the outlet. This is usually undesirable because it may result in shorter contact, reaction, or settling times in comparison with the theoretical (calculated) or presumed detention times.

18. *Toxic* (TOX-ick). A substance that is poisonous to a living organism. Toxic substances may be classified in terms of their physiological action, such as irritants, asphyxiants, systemic poisons, and anesthetics and narcotics. Irritants are corrosive substances that attack the mucous membrane surfaces of the body. Asphyxiants interfere with breathing. Systemic poisons are hazardous substances that injure or destroy internal organs of the body. Anesthetics and narcotics are hazardous substances that depress the central nervous system and lead to unconsciousness.

19. *Bioflocculation* (BUY-o-flock-yoo-LAY-shun). The clumping together of fine, dispersed organic particles by the action of certain bacteria and algae. This results in faster and more complete settling of the organic solids in wastewater.

hours. Bioflocculation is accelerated by increased temperature, wave action, and high dissolved oxygen content.

Pond detention times are sometimes specified by regulatory agencies to ensure adequate treatment and removal of *PATHO-GENIC ORGANISMS*.[20] Many agencies specify effluent or receiving water quality standards in terms of *MEDIAN*[21] and maximum MPN (most probable number) values that should not be exceeded. In critical water use areas, chlorination or other means of disinfection can be used to further reduce the coliform level (see Chapter 12, "Disinfection and Chlorination"). In many areas, the pond effluent must not contain any chlorine residual, so the effluent must be dechlorinated.

A pond is generally regarded as not fulfilling its function when the discharge

• creates visual or odor nuisances;

• contains high effluent BOD levels; or

• contains high effluent solids, grease, or coliform-group bacteria concentrations.

The exception is the pond that was designed to be anaerobic in the first stages and aerobic in later ponds for final treatment.

QUESTIONS

Please write your answers to the following questions and compare them with those on page 45.

9.4A   What biological factors influence the treatment efficiency of a pond?

9.4B   What is bioflocculation?

9.4C   What factors indicate that a pond is not fulfilling its function (operating properly)?

END OF LESSON 1 OF 3 LESSONS

on

WASTEWATER STABILIZATION PONDS

Please answer the discussion and review questions next.

DISCUSSION AND REVIEW QUESTIONS
Chapter 9.   WASTEWATER STABILIZATION PONDS

(Lesson 1 of 3 Lessons)

At the end of each lesson in this chapter, you will find discussion and review questions. Please write your answers to these questions to determine how well you understand the material in the lesson.

1. Where is the chlorine contact basin located in a typical plant with ponds only?

2. When wastewater flows through different treatment processes in a plant, where might ponds be located?

3. Why are most ponds "facultative ponds"?

4. What is photosynthesis?

5. Where does the oxygen come from that is produced by algae in a pond?

6. What are the three types of factors that may influence pond performance?

---

20. *Pathogenic* (path-o-JEN-ick) *Organisms.*   Bacteria, viruses, protozoa, or internal parasites that can cause disease (such as giardiasis, cryptosporidiosis, typhoid fever, cholera, or infectious hepatitis) in a host (such as a person). There are many types of organisms that do not cause disease and are not called pathogenic. Many beneficial bacteria are found in wastewater treatment processes actively cleaning up organic wastes.

21. *Median.*   The middle measurement or value. When several measurements are ranked by magnitude (largest to smallest), half of the measurements will be larger and half will be smaller.

CHAPTER 9.   WASTEWATER STABILIZATION PONDS

(Lesson 2 of 3 Lessons)

9.5   STARTING THE POND

One of the most critical periods of a pond's life is the time that it is first placed in operation. If at all possible, at least one foot (0.3 m) of water should be in the pond before wastes are introduced. The water should be added to the pond in advance to prevent odor development from waste solids exposed to the atmosphere. Thus, a source of water should be available when starting a pond.

A good practice is to start ponds during the warmer months of the year because a shallow starting depth allows the contents of the pond to cool too rapidly if nights are cold. Generally, the warmer the pond contents, the more efficient the treatment processes.

ALGAL BLOOMS[22] normally will appear from seven to twelve days after wastes are introduced into a pond, but it generally takes at least sixty days to establish a thriving biological community. A definite green color is evidence that a flourishing algae population has been established in a pond. After this length of time has elapsed, bacterial decomposition of bottom solids will usually become established. This is generally indicated by bubbles coming to the surface near the pond inlet where most of the sludge deposits occur. Although the bottom is anaerobic, travel of the gas through the aerobic surface layers generally prevents odor release.

Wastes should be discharged to the pond intermittently during the first few weeks with constant monitoring of the pH. The pH in the pond should be kept above 7.5 if possible. Initially, the pH of the bottom sludge will be below 7 due to the digestion of the sludge by acid-producing bacteria. If the pH starts to drop, discharge to the pond should be diverted to another pond or diluted with makeup water (water from another source) if another pond is not available until the pH recovers. A high pH is essential to encourage a balanced anaerobic fermentation (bacterial decomposition) of the bottom sludge. This high pH also is a good indicator of high algal activity since removal of the carbon dioxide from the water in algal metabolism tends to keep the pH high. A continuing low pH indicates acid production, which will cause odors. Soda ash (sodium carbonate) may be added to the influent to a pond to increase the pH.

QUESTIONS

Please write your answers to the following questions and compare them with those on page 45.

9.5A   Why should at least one foot of fresh water cover the pond bottom before wastes are introduced?

9.5B   Why should ponds be started during the warmer months of the year if at all possible?

9.5C   What does a definite green color in a pond indicate?

9.5D   When bubbles are observed coming to the pond surface near the inlet, what is happening in the pond?

9.6   DAILY OPERATION AND MAINTENANCE

Because ponds are relatively simple to operate, they are probably neglected more than any other type of wastewater treatment process. Many of the complaints that arise regarding ponds are the result of neglect or poor housekeeping. The following is a list of day-to-day operational and maintenance duties that will help ensure peak treatment efficiency and present your plant to its neighbors as a well-run waste treatment facility. If problems develop in a pond, refer to this section and Section 9.10, "Review of Plans and Specifications," as troubleshooting guides.

9.60   Scum Control

Scum accumulation is a common characteristic of ponds and is usually greatest in the spring when the water warms and vigorous biological activity resumes. Ordinarily, wind action will break up scum accumulations and cause them to settle; however, in the absence of wind or in sheltered areas, other means must be used. If scum is not broken up, it will dry on top and become crusted.

---

22. *Algal* (AL-gull) *Bloom.*   Sudden, massive growths of microscopic and macroscopic plant life, such as green or blue-green algae, which can, under the proper conditions, develop in lakes, reservoirs, and ponds.

Not only is the scum more difficult to break up then, but a species of blue-green algae is likely to become established on the scum, which can produce disagreeable odors. If scum is allowed to accumulate, it can reach proportions where it cuts off a significant amount of sunlight from the pond. When this happens, the production of oxygen by algae is reduced and odor problems can result. Rafts of scum cause a very unsightly appearance in ponds and can quite likely become a source of botulism that will have a devastating effect on any waterfowl and shore birds that may be attracted to the pond.

Scum is broken up most easily if it is attended to promptly. Many methods of breaking up scum have been used, including agitation with garden rakes from the shore, spraying jets of water from pumps or tank trucks, and using outboard motors on boats in large ponds. Outboard motors should be of the air-cooled type to avoid plugging the cooling system with algae and scum.

9.61   Odor Control

Eventually, odors probably will come from a wastewater treatment pond no matter what kind of process is used. Most odors are caused by overloading (see Section 9.109 to determine pond loading) or poor housekeeping practices and can be remedied by taking corrective measures. If a pond is overloaded, stop loading and divert influent to other ponds, if available, until the odor problem stops. Then, gradually start loading the pond again. Once a pond develops odor problems, it is more apt to cause trouble than other ponds.

There are times such as when unexpected shutdowns of mechanical equipment occur that pond processes may be upset and cause odors. For these unexpected occurrences, it is strongly advised that a careful plan for emergency odor control be available. Odors usually occur during the spring warm-up in colder climates because biological activity was reduced during cold weather. When the water warms, microorganisms in a pond become active and use up all of the available dissolved oxygen, creating anaerobic conditions in which odors may be produced.

There are several suggested ways to reduce odors in ponds, including recirculation from aerobic units, using floating aerators, and heavy chlorination. Recirculation from an aerobic pond to the inlet of an anaerobic pond (one part recycle flow to six parts influent flow) will reduce or eliminate odors. If a recirculation program is used, (1) the recirculation rate must not raise the flow rate through the pond to a level that will cause a pond overflow, and (2) the recirculation water should be drawn from the surface of the source pond to ensure the highest possible dissolved oxygen levels in the recirculated water. Usually, floating aeration and chlorination equipment are too expensive to have standing idle waiting for an odor problem to develop. Instead, small gasoline-powered pumps have been used to recirculate pond contents and help ponds recover. These pumps also have been used to spray pond water on the surface of the pond

to add additional oxygen to the top layer of a pond to assist recovery. Odor-masking chemicals also have been promoted for this purpose and have some uses for concentrated sources of specific odors. However, in almost all cases, using the types of procedures mentioned previously is preferable. In any event, it is poor procedure to wait until an emergency arises to plan for odor control. Often, several days are needed to receive delivery of materials or chemicals, if they are required, so having alternate methods of control ready to go in case they are needed is recommended.

In some areas, sodium nitrate has been added to ponds as a source of oxygen for microorganisms rather than sulfate compounds. This practice prevents odors created by the oxidation of sulfate to sulfide. To be effective, sodium nitrate must be dispersed throughout the water in the pond. Once mixed in the pond, it acts very quickly because many common organisms (facultative groups) can use the oxygen in nitrate compounds in the absence of dissolved oxygen. The amount of sodium nitrate needed depends on the oxygen demand. Some operators dose their ponds successfully between 0.3 and 0.4 mg/L of sodium nitrate. Liquid sodium hypochlorite or chlorine solution is a faster-acting odor solution, but not necessarily the best chemical because it will interfere with biological stabilization of the wastes.

A different odor-control strategy will be needed if the ponds have already frozen over in the winter but are expected to develop odor problems when the ice thaws in the spring and the water in the pond turns over. In this situation, some operators spread sodium nitrate on the ice surface before the thaw so that the chemical will be available to supply oxygen when needed.

9.62   Weed and Insect Control

Weed control is an essential part of good housekeeping and is not a difficult task using manual removal methods, herbicides, and soil sterilants. Weeds around the edges of ponds are most problematic because they provide a sheltered area for mosquito breeding and scum accumulation. In most ponds, there is little need for mosquito control when edges are kept free of weed growth. Weeds also can hinder pond circulation. Aquatic weeds such as tules will grow in depths shallower than three feet (1 m), so having an operating pond level of at least this depth is necessary. Tules may emerge singly or be well scattered, but they should be removed promptly by hand because they will quickly multiply from the root system. One of the best methods for controlling undesirable vegetation is achieved by a daily practice of close inspection and immediate removal of the young plants (including roots). Effective weed control strategies include using RIPRAP[23] or concrete on the pond banks and maintaining two to three inches (5 to 8 cm) of gravel on the roads.

Suspended vegetation such as DUCKWEED[24] usually will not flourish if the pond is exposed to a clean sweep of the wind.

---

23. *Riprap.*   Broken stones, boulders, or other materials placed compactly or irregularly on levees or dikes for the protection of earth surfaces against the erosive action of waves.

24. *Duckweed.*   A small, green, cloverleaf-shaped floating plant, about one-quarter inch (6 mm) across, which appears as a grainy layer on the surface of a pond.

Dike vegetation control is aided by regular mowing and use of a cover grass that will crowd out undesirable growth. Because emergent weed growth will occur only when sunlight is able to reach the pond bottom, the single best preventive measure against emergent growth is to maintain a water depth of at least three feet (1 m). Due to greater water clarity, the amount of sunlight reaching the bottom will be greater in secondary or final ponds than in primary ponds. Because shallow water promotes growth, there will likely always be a battle to keep emergent weeds from becoming well established around pond banks.

Whenever emergent or suspended weeds are being pulled from the pond, protective gear such as waterproof gloves, boots, and goggles should be worn to reduce the chance of infection from pathogens that may be present in the water. Pulled weeds should be buried to prevent odor and insect problems. Although most stabilization ponds are no deeper than five feet (1.7 m), there is still sufficient depth to drown a person, especially if that person gets caught in a sticky clay liner. Using the buddy system and approved flotation devices will greatly increase the safety level when performing any pond maintenance, especially when using a boat or mowing the dike. Control measures for emergent weeds, suspended vegetation, and dike vegetation include the following:

Emergent Weeds

1. Keep the water level above three feet (1 m).

2. Pull out new (first-year) growth by hand.

3. Drown the weeds by raising the water level.

4. Lower the water level, cut the weeds or burn them with a gas burner, and raise the water level (cut and drown).

5. Use riprap (large broken stones or concrete) along the bank. If riprap is used and growth continues, herbicides are the only alternative.

6. Install a pond liner.

7. Use herbicides as a last resort.

Suspended Vegetation

1. Keep the pond exposed to a clean sweep of the wind.

2. Stock the pond with a few ducks to eat light growth of duckweed.

3. Skim small ponds with rakes or boards. This may have to be repeated.

4. Mechanically harvest excessive growth.

5. Apply herbicides as a last resort.

Dike Vegetation

1. Mow regularly during the growing season. Dike slopes may be cut using sickle bars or weed-eater equipment. When using heavy equipment, mowers designed especially for cutting slopes are preferable. Any tractor used on the dike should have a low center of gravity. The tractor must be provided with an approved roll-over protective structure and seat belts if within certain use/weight classifications. Check with your local safety regulatory agency for applicable rules.

2. Seed or reseed slopes with desirable grasses that will form a thick and somewhat impenetrable mat.

3. Use herbicides as a last resort.

Herbicides are a last resort in vegetation control for ponds, not only because of the obvious hazards facing the operator, but also because of dangers presented to the biological growth in the pond and receiving stream (if discharging). Care must be taken to follow mixing, application, and storage directions exactly. Safety for the operator and for any other persons who might come in contact with the herbicide is of utmost importance. Proper protective gear and warnings (verbal and written) are vital. Also, desirable vegetation could be killed by herbicide drift on a windy day or by other improper applications, so take those factors into consideration before applying an herbicide.

Availability, formulation, trade names, and federal clearance of state registration for certain herbicides may change. Do not use stored herbicides without checking on their current status of approval.

Some herbicides are effective only on certain plants, and many herbicides are most effective at specific stages in the life of a plant. Cattails, for instance, are usually sprayed when the shoots are 2 to 3 feet (0.7 to 1 m) tall and cattails are developing.

Perhaps the most important factor in the selection of the proper herbicide is site compatibility. In addition to the other important instructions on the product label, a comprehensive listing of approved application sites is usually provided. Check this list to verify that the product is the best choice for each specific application. Herbicides that are effective on many types of undesirable lagoon vegetation and that are approved for both pond and river (discharging lagoon) applications are often the logical choice. Some areas require an herbicide application license and complete reporting of the type and amount of material used. Verify with your regulatory agency that you will be in compliance with applicable regulations before starting an herbicide application program.

Duckweed[25] consists of tiny green plants that float on the surface of a pond. Duckweed stops sunlight penetration and hinders surface aeration, thus reducing dissolved oxygen in the pond. Rakes, water sprays, or pushing a board with a boat can be used to move the duckweed into a corner of a pond where it can be physically removed. Another approach is to install a surface discharge in the downwind corner of a pond. The prevailing winds will blow the duckweed into the corner and the duckweed will be removed by the surface discharge. The duckweed can be captured by a wire mesh bucket placed in the effluent line after the surface outlet. The bucket will need to be removed and cleaned as the duckweed accumulates.

---

25. Material in this section was adapted from an article entitled "Control of Undesirable Lagoon Vegetation," by Doug Matheny; it appeared in the Oklahoma Water Pollution Control Association's DRIP, Spring 1988.

Mosquitoes will breed in sheltered areas of standing water where there is vegetation or scum to which the egg rafts of the female mosquito can become attached. These egg rafts are fragile and will not withstand the action of disturbed water surfaces caused by wind action or normal currents. Keeping the water edge clear of vegetation and keeping any scum broken up will normally give adequate mosquito control. Shallow, isolated pools left by the receding pond level should be drained or sprayed with a larvacide. Mosquito fish (*Gambusia*) also are an effective means of controlling mosquitoes. A stocking density of 2,500 fish per acre has been reported. However, stocking levels will depend on the local situation, environmental conditions, and the desire to sustain the fish population.

Any of several tiny shrimp-like animals may infest the pond from time to time during the warmer months of the year. There are two general types of microcrustaceans: (1) those that have the appearance of crayfish without claws and (2) those that have the appearance of fleas. During periods of large blooms, these organisms will appear in large orange or red cloud-like patterns in the pond. They can prove to be a valuable part of the treatment process, especially in polishing ponds. Microcrustaceans live on algae and at times will appear in such numbers as to almost clear the pond of algae. During the more severe infestations, there will be a sharp drop in the dissolved oxygen of the pond, accompanied by a lowered pH. This is a temporary condition because the microcrustaceans will outrun the algae supply and there will be a mass die-off of the animals, followed by a rapid greening up of the pond. If the pond is operated on a batch basis, the period when the algae concentration is low may be a good time to release water due to the low suspended solids values.

Ordinarily there should be no great concern about these infestations because they soon balance themselves; however, in the case of a heavily loaded pond, a sustained low dissolved oxygen content may give rise to obnoxious odors. In that event, any of several commercial sprays can be used to control the shrimp-like animals.

Chironomid midges and other insects are often produced in wastewater ponds in sufficient numbers to be serious nuisances to nearby residential areas, farm workers, recreation sites, and industrial plants. When emerging in large numbers, they may also create traffic hazards. At present the only satisfactory control is through the use of insecticides. Control measures are time-consuming and may be difficult, particularly if there is a discharge to a receiving stream. If possible, lower the level in the ponds (one pond at a time) enough to contain a day's inflow before applying an insecticide. Holding the insecticide for at least one day will kill more insects and reduce the effect of the insecticide on receiving waters. Lowering the ponds also dries up weeds and insects. For better results, insecticides should be applied on a calm, windless day and any recirculation pumps should be stopped.

CAUTION: Before attempting to apply any insecticide or pesticide, contact your local official in charge of approving pesticide applications. This person can tell you which chemicals may be applied, the conditions of application, and safe procedures.

9.63   Levee Maintenance

Levee slope erosion caused by wave action or surface runoff from precipitation is probably the most serious maintenance problem. If allowed to continue, it will result in a narrowing of the levee crown, which will make accessibility with maintenance equipment most difficult.

If the levee slope is composed of easily erodible material, one long-range solution is to use bank-protection material such as stone riprap or broken concrete rubble. Good bank-protection materials include small pieces of broken street materials, curbs, gutters, and also bricks and other suitable materials from building demolition. Also, a semi-porous plastic sheet can be used with riprap. The sheet allows the two-way movement of air and water but prevents the movement of soils. The sheet also discourages weed growth and digging by crayfish (crawdads). Geosynthetics are an effective means of levee erosion control.

Portions of the pond levee or dike not exposed to wave action should be planted with a low-growing, spreading grass to prevent erosion by surface runoff. Native grasses may naturally seed the levees or local highway departments may be consulted about suitable grasses to control erosion. If necessary, grass may have to be mowed to prevent it from becoming too high. Do not allow large grazing animals such as cows or horses to control vegetation because they may damage the levees near the waterline and possibly complicate erosion problems, but goats are an option because they have proven effective in controlling vegetation and grasses.

Plants or grasses with long roots such as willows and alfalfa should not be allowed to grow on levees because they can damage the levees and possibly cause levee failure and costly repair. Burrowing animals such as muskrats, badgers, squirrels, and gophers also may cause levees to fail. Remove these animals from levees as soon as possible and repair their burrowed holes immediately. Local agricultural departments may be able to assist with the development of a rodent control program.

Levee tops should be crowned so that rainwater will drain over the side in a sheet flow. Otherwise, the water may flow a considerable distance along the levee crown and gather enough flow to cause erosion when it finally spills over the side and down the slope. If the levees are to be used as roadways during wet weather, they should be paved or well graveled.

If seepage or leakage from the ponds appears on the outside of levees, ask your engineer or a qualified person to investigate and solve this problem before further levee damage occurs.

9.64 Headworks and Screening Maintenance

Bar screens remove coarse material from the pond effluent. The screen should be inspected at least once or twice a day with more frequent visits during storm periods. Clean the bar screen as frequently as necessary. Screenings should be disposed of daily in a sanitary manner such as by burial to avoid odors and fly breeding. Another method of disposal is to place screenings in garbage cans and request that your local garbage service dispose of the screenings at a sanitary landfill disposal site.

Many pond installations have grit removal chambers at the headworks to protect raw wastewater lift pumps or prevent plugging of the influent lines. There are many types of grit removal equipment. Grit removed by various types of mechanical equipment or by manual means will usually contain small amounts of organic matter and should therefore be disposed of in a sanitary manner. Disposal by burial is the most common method. Disposal methods for screenings and grit must be approved by the permitting regulatory agency.

QUESTIONS

Please write your answers to the following questions and compare them with those on page 45.

9.6A Why should scum not be allowed to accumulate on the surface of a pond?

9.6B How can scum accumulations be broken up?

9.6C What are the causes of odors from a pond?

9.6D What precaution would you take to be prepared for an odor problem that might develop?

9.6E How can weeds be controlled and removed in and around ponds?

9.6F Why are weeds objectionable in and around ponds?

9.6G Why should insects be controlled?

9.6H Why should a pond be lowered before an insecticide is applied?

9.65 General Operating Hints

1. Anaerobic ponds should be covered and isolated for odor control and followed by aerobic ponds. Floating polystyrene planks can be used to cover anaerobic ponds and can be painted white for protection from the sun. These planks will help confine odors and heat and tend to make the anaerobic ponds more efficient.

2. Placing ponds in series tends to cause the first pond to become overloaded and may never allow it to recover; the overload may be carried to the next pond in series. Feeding ponds in parallel allows the distribution of the incoming load evenly between units.[26] Whether ponds are operated in series or in parallel should depend on the loading situation and *NPDES PERMIT*[27] requirements.

   When operating ponds in series, the accumulation of solids in the first pond may become a serious problem after a long period of use. Periodically, the flow should be routed around the first pond. This pond should then be drained and the solids removed and disposed of in an approved manner. The solids may be removed by dredging or pumping. A special permit from the regulatory agency may be required before burying solids removed from a pond.

3. A large amount of pond recirculation, 25 to 100 percent, can be very helpful. This allows the algae and other aerobic organisms to become thoroughly mixed with incoming raw wastewater. At the same time, good oxygen transfer can be attained by passing the incoming water over a deck and down steps or another type of aerator. However, this procedure can cause heat loss.

4. Heavy chlorination at the pond recirculation point can assist in odor control, but it will probably interfere with treatment.

5. As with any treatment process, it is necessary to measure the important water quality indicators (DO and solids) at frequent, regular intervals and to plot them so that you have some idea of the direction the process is taking in time to take corrective action when necessary. During normal pond operation, the algae produced, dissolved oxygen (DO), and solids will be highest in the summer and during daylight hours. Dissolved oxygen and solids will be lowest during the winter and at night.

6. When solids start floating to the surface of a pond during the spring or fall *OVERTURN,*[28] the pond may have to be taken out of service and cleaned. Measurement of the sludge depth on the bottom of a pond also will indicate when a pond should be cleaned. Usually, ponds are cleaned when the wet sludge is over one foot (0.3 m) deep.

26.

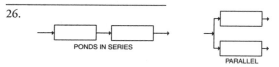

27. *NPDES Permit.* National Pollutant Discharge Elimination System permit is the regulatory agency document issued by either a federal or state agency that is designed to control all discharges of potential pollutants from point sources and stormwater runoff into US waterways. NPDES permits regulate discharges into US waterways from all point sources of pollution, including industries, municipal wastewater treatment plants, sanitary landfills, large animal feedlots, and return irrigation flows.

28. *Overturn.* The almost spontaneous mixing of all layers of water in a reservoir or lake when the water temperature becomes similar from top to bottom. This may occur in the fall/winter when the surface waters cool to the same temperature as the bottom waters and also in the spring when the surface waters warm after the ice melts. This is also called turnover.

7. Before applying insecticides or herbicides, be sure to check with appropriate authorities regarding the regulation of their use and the long-term effects of the pesticide you plan to use. Do not apply pesticides that may be toxic to organisms in the receiving waters.

8. Unfortunately, there is little an operator can do by changing pond operating procedures to effectively remove algae from a pond effluent. The best approach is to operate ponds in series and to draw off the effluent just below the surface by using a good baffling arrangement.

If the algae must be removed from the effluent of a pond, additional facilities may be designed and constructed. Techniques for eliminating algae from pond effluents are described in Section 9.11, "Eliminating Algae from Pond Effluents." Suitable effluents have been obtained under favorable field conditions. Algae removal processes include microscreening, slow sand filtration, dissolved air flotation, and algae harvesting. In final ponds that are operated in series with periodic discharges, alum may be added in doses of less than 20 mg/L to improve effluent quality before discharge. Also, algae can be removed from the effluent of polishing ponds by using agricultural pod-type pressure filters and straw.

9.66   Abnormal Operation

Abnormal operation occurs when ponds are overloaded because the BOD loads or flows are too high. Excessive BOD levels can occur when influent loads exceed design capacity due to population increases, industrial growth, or illegal industrial discharges. Under these conditions, new facilities must be constructed or the BOD loading must be reduced at the source. Repeated wide fluctuations in BOD loads over short time periods will also interfere with pond performance and can create a nearly constant state of abnormal operation.

Another type of overloading can occur when too much flow is diverted to one pond. This can happen when an operator accidentally feeds one pond more than the other or when a pipe opening is blocked by rags, solids, or grit due to low pipe velocities, and thus too much flow is diverted to another pond. When this happens and the overloaded pond starts producing odors, take the pond out of service and divert flows to the other ponds until the overloaded pond recovers. Hopefully, the ponds in service will not become overloaded. Also, be sure to remove the rags, solids, or grit that caused the overloading and inspect the other pipes to prevent this problem from happening again in the other ponds. Another source of plugging or blocking of pipes carrying water from one pond to another can be naturally occurring algae. Pipes can be cleaned on a scheduled basis using a high-velocity cleaner or hydro-jet used to clean sewers.

Usually, ponds do not become overloaded during storms and periods of high runoff because there is not a significant increase in the BOD loading on the ponds.

Large amounts of brown or black scum on the surface of a pond is an indication that the pond is overloaded. Scum on the surface of a pond often leads to odor problems. The best way to control scum is to take corrective action as soon as possible (see Section 9.60, "Scum Control").

During winter conditions, the pond can become covered with ice and snow. Sunlight is no longer available to the algae and oxygen cannot enter the water from the atmosphere. Without dissolved oxygen available for aerobic decomposition, anaerobic decomposition of the solids occurs. Anaerobic decomposition takes place slowly because of the low temperatures. By keeping the pond surface at a high level, a longer detention time will be obtained and heat losses will be minimized. During the period of ice cover, odorous gases formed by anaerobic decomposition accumulate under the ice and are dissolved into the wastewater being treated.

Some odors may be observed in the spring just after the ice cover breaks up because the pond is still in an anaerobic state and some of the dissolved gases are being released. Melting of ice in the spring provides dilution water with a high oxygen content; thus the ponds usually become facultative in a few days after breakup of the ice if they are not organically (BOD) overloaded.

9.67   Batch Operation

Some ponds do not discharge continuously. These ponds may discharge only once (fall) or twice (fall and spring) a year. Discharges should be made only when necessary and, if possible, during the nonrecreational season when flows are high in the receiving waters.

If your pond is allowed to discharge intermittently (controlled discharge), you must work closely with your pollution control agency and be sure that you are in compliance with the National Pollutant Discharge Elimination System (NPDES) permit. Before and during the discharge, samples should be collected from the pond being emptied and from the receiving waters both upstream and downstream from the point of discharge. Samples are usually analyzed for DO, BOD, pH, total suspended solids, and coliform-group bacteria.

Ponds should not be emptied too quickly. If a pond is emptied too quickly, the wet side slope along the shoreline may slide into the pond. Usually, ponds are emptied in two weeks or less, depending on how much water is to be discharged. Normally, 1.0 to 1.5 feet (0.3 to 0.45 m) of water is left in the bottom of the pond.

9.68   Controlled Discharge

A common operational modification to aerated and facultative lagoons is the controlled-discharge pond mode in which pond discharge is prohibited during the winter months in cold climates or during the peak algal growth periods in all climates. In this approach, each cell in the system is isolated and then discharged sequentially. Sufficient storage capacity is provided in the lagoon system to allow wastewater storage during winter months, peak algal growth periods, or receiving stream low-flow periods. In the Great Lakes states of Michigan, Minnesota, and Wisconsin, as well as in the province of Ontario, Canada, many pond systems are designed to discharge in the spring or fall when water quality effects are minimized. As a secondary benefit, operational costs are lower than for a continuous-discharge lagoon because of reduction in laboratory monitoring requirements and the need for less operator control.

A similar modification is called a hydrograph controlled-release lagoon (HCRL) in which water is retained until flow volume and conditions in the receiving stream are adequate for discharge, thus eliminating the need for costly additional treatment. An HCRL system has three principal components: a stream-flow monitoring system, an effluent discharge system, and a storage cell. The stream-flow monitoring system measures the flow rate in the stream and transmits the data to the effluent discharge system. The effluent discharge system consists of a controller and a discharge structure. The controller operates a discharge device, such as a motor-driven sluice gate, through which wastewater is discharged from the storage cell; however, these tasks can be manually performed.

Key considerations for the operation of HCRLs include the following:

- Having an effluent release model keyed to flow in the receiving stream (as measured by depth)

- Having outlets capable of drawing from different depths to ensure the best quality effluent

- Using water-balance equations to size storage cells

9.69   Shutting Down a Pond

Ponds may be shut down for short periods of time without any problems developing. For example, if flow to a pond must be stopped to repair a pipe or a valve, no precautionary procedures are necessary. If a pond is full and received no flows for a long period of time, start up the pond with caution and gradually increase the load. If the full load is applied immediately, the pond may become overloaded because the microorganism population in the pond is low and insufficient to treat the load.

Stop all flow to a pond before emptying it to remove bottom deposits, repair inlet or outlet structures, or repair levees. Drain it by use of discharge valves or pump water from one pond to other ponds. Feed the other ponds remaining in service equally to prevent them from becoming overloaded. Frequently, there is a time lag between the overloading of a pond and the development of problems. Therefore, watch the other ponds and lab results closely for any signs of potential problems developing (odors, low pH, low DO, drop in alkalinity, loss of green color).

9.610   Operating Strategy

To prevent ponds from developing odors or discharging an effluent in violation of your NPDES permit requirements, develop a plan to keep the ponds operating as intended.

A. Maintain constant water elevations in the ponds.

If the NPDES permit allows the discharge of a pond into receiving waters, keep a constant water level to help maintain constant loadings. When the water surface elevation starts to drop, look for the following possible causes:

1. Discharge valve open too far or a *STOP LOG*[29] is missing

2. Levees leaking due to animal burrows, cracks, soil settlement, or erosion

3. Inlet lines plugged or restricted and causing wastewater to back up into the collection system

When the water surface starts to rise, look for the following:

1. Closed discharge valve or plugged lines

2. Sources of infiltration

NOTE: Under some conditions, an operator may not want to maintain constant water levels in the ponds. An operator may allow the water surface to fluctuate under the following circumstances:

1. To control shoreline aquatic vegetation

2. To control mosquito breeding and rodent burrowing

3. To handle fluctuating inflows

4. To regulate discharge (continuous, intermittent, or seasonal)

B. Distribute inflow equally to ponds.

All ponds designed to receive the inflow should receive the same hydraulic and organic (BOD) loadings.

C. Keep pond levees or dikes in good condition.

Proper maintenance of pond levees can be a time-saving activity. Regularly inspect levees for leaks and erosion and correct any problems before they become serious. If erosion is a problem at the waterline, install riprap. Do not allow weeds to grow along the waterline and keep weeds on the levee mowed. If insect larvae are observed on the pond surface, spray with an appropriate insecticide before problems develop.

D. Observe and test pond condition.

Daily visual observations can reveal if a pond is treating the wastewater properly. The pond should be a deep green color indicating a healthy algae population. Scum and floating weeds should not be present, but if they are, remove them to allow sunlight to reach the algae in the pond.

Once or twice a week, tests should be conducted to determine the pond's dissolved oxygen level, pH, and temperature. Sampling should be conducted at the same time every day to ensure comparable results. Effluent dissolved oxygen should be measured at this time also. Other effluent tests should be conducted at least weekly and include BOD, suspended solids, dissolved solids, coliform-group bacteria, and chlorine residual. If ponds are operated on a batch or controlled-discharge basis, these effluent tests will have to be conducted only during periods of discharge.

During warm summer months, algae populations tend to be high and may cause high suspended solids concentrations

---

29. *Stop Log.*   A log or board in an outlet box or device used to control the water level in ponds and also the flow from one pond to another pond or system.

in the effluent. An advantage of ponds in arid regions is that this is also a period of high evaporation rates. Under these conditions, effluent flows may drop to almost zero or may stop. In the fall and winter months when the weather is cool and sunlight is reduced, the algae population in ponds is also reduced, thus resulting in reduced suspended solids. This situation could allow ponds to meet effluent requirements during this period.

If test results reveal that certain water quality indicators (such as DO, BOD, pH, or suspended solids) are tending to move in the wrong direction, identify the cause and take corrective action. Remember that ponds are a biological process and that changes resulting from corrective action may not occur until a week or so after the changes are made.

E.   Create a pattern of baffles.

Baffle walls placed properly in aerated ponds can ensure maximum efficiency by directing the flow along a path that will cause all of the water to pass through the area of influence in each aerator. This pattern of baffles eliminates short-circuiting through the pond.

F.   Perform troubleshooting as needed.

The Troubleshooting Guide below provides you with step-by-step procedures to follow when problems develop in a pond.

TROUBLESHOOTING GUIDE FOR PONDS[a]

Indicator	Probable Cause	Check or Monitor	Solution
1. Poor quality effluent	a. Mixing/agitation equipment failure.	a. Monitor surface aerators, rotors, or aeration equipment, and DO of pond water.	a. 1. Restart out of service mixing or aeration units. 2. Increase operating time. 3. Increase recycle flow from effluent to influent. 4. Provide additional mixing (small boat and outboard motor, one hour for every two hours of daylight, or at least three times/day)
	b. Organic overload.	b. Monitor influent laboratory data: BOD, SS, DO, pH, temperature.	b. 1. Increase or start recirculation of effluent to influent. 2. Mix pond contents hourly by surface aerators or outboard motor on small boat at least three times/day. 3. Increase run cycle on surface aeration equipment to maintain at least 1.0 mg/L DO. 4. Add chemicals to help reduce pond load by prechlorination for BOD reduction, add sodium nitrate or hydrogen peroxide for oxygen input.
	c. Excessive turbidity from scum mats.	c. Floating mats of scum or sludge on surface and corners of pond.	c. 1. Break up scum mats using water sprays, poles or rakes, mixing equipment, or outboard motors. 2. Check pond influent for excess grease or scum. 3. If seasonal temperatures change, rising sludge from pond may have to be broken up daily by mixing methods in item 1c.1.
	d. Blockage of light by excessive plant growth (tules, reeds, or grasses) near dikes.	d. Visual inspection for weed growth in and near ponds.	d. 1. Remove plant growth. 2. Schedule regular herbicide applications to levees and dikes as a last resort.
	e. Low temperature.	e. Monitor air and pond water temperatures.	e. 1. Freezing weather, raise pond levels to increase water depth. 2. Reduce recirculation rates. 3. If possible, operate ponds in series.
	f. Toxic material in influent.	f. Color change, low DO, low pH for no apparent reason.	f. 1. Sample and identify toxic material. If the evidence of the toxic material can be verified in time, isolate the toxic material in one pond until it can be identified and then detoxified or neutralized. 2. Increase recirculation from effluent to influent. 3. Increase surface aeration or pond mixing times. 4. Implement new or enforce existing sewer-use ordinances. 5. See 1b. above.

TROUBLESHOOTING GUIDE FOR PONDS[a] *(continued)*

Indicator	Probable Cause	Check or Monitor	Solution
1. Poor quality effluent *(continued)*	g. Loss of pond volume caused by sludge accumulation.	g. Sludge depth.	g. Remove sludge.
	h. Process chemicals from illegal drug laboratories dumped into rural area manholes.	h. Surveillance of rural area manholes.	h. Install locking devices on manhole to prevent access by unauthorized persons. Isolate "shocked" ponds and gradually recycle polished effluents.
2. Low dissolved oxygen in ponds.	a. Low algal growth.	a. Visual inspection, pond not green, low DO during daylight hours, none from sunup until noon. Odor.	a. 1. Recirculate last pond effluent to inlet of first pond. 2. Same as 1b.
	b. Excessive scum accumulation.	b. Pond surface conditions.	b. 1. Break up and sink resuspended scum. 2. Skim off grease balls and scum. Dispose of in landfill.
3. Odors.	a. Anaerobic conditions or spring and fall turnovers.	a. Monitor pond loading, BOD, SS, pH, DO, and temperature.	a. 1. Limit organic load by diverting influent flows to several ponds. 2. Isolate pond and remove excess sludge solids. 3. Same as 1b.
	b. Hydrogen sulfide in pond influent.	b. Monitor total dissolved H_2S.	b. 1. Check collection system. 2. Pretreat with chlorine or pre-aeration. 3. Aerate plant influent in small pond at least 30 minutes before applying to other ponds.
4. Inability to maintain sufficient liquid.	a. Leakage.	a. 1. Seepage around dikes. 2. Sampling wells located on the outside perimeter of the pond(s).	a. 1. Apply bentonite clay to the pond water to seal the leak. 2. If sampling well analysis indicates contamination, the lagoon/pond site may require lining.
	b. Excessive evaporation or percolation.	b. Detention time in pond is probably too long.	b. Divert land drainage or stream flow into pond.
5. Insect breeding.	a. Layers of scum and excessive plant growth in sheltered portions of pond.	a. Visual inspection. Mosquitoes.	a. 1. Weed and scum removal. 2. Apply approved insecticides or larvicides.
	b. Shallow pools of standing water outside pond.	b. Visual inspection.	b. Cut vegetation outside of pond and fill in nearby potholes that collect standing water.
6. Levee erosion.	a. Windy conditions.	a. Visual inspection.	a. 1. Riprap levee. 2. Construct a wind barrier around pond.
	b. Excessive surface aerator operating time.	b. Aerator operating time.	b. Reduce aerator operating time if DO levels allow.
7. Excessive weeds and tule growth.	a. Pond too shallow.	a. Visual inspection for weeds in the area.	a. Deepen all pond areas to at least 3 feet.
	b. Inadequate maintenance program to control vegetation.	b. Maintenance program.	b. 1. Correct program deficiency. 2. Install pond lining. 3. Initiate an herbicide program as a last resort.
8. Animals burrowing into the dikes.	Burrowing animals (gophers, squirrels, crayfish).	Visual inspection.	1. Alter pond level several times in rapid succession. 2. Remove animals as soon as possible. 3. Provide riprap with semiporous sheet on levee slopes.
9. Groundwater contamination.	Leakage through bottom or sides of pond.	Seepage around pond dikes.	Apply bentonite clay to pond water to seal leak.

a. Adapted from *Performance Evaluation and Troubleshooting at Municipal Wastewater Treatment Facilities,* Office of Water Program Operations, US EPA, Washington, DC.

QUESTIONS

Please write your answers to the following questions and compare them with those on page 45.

9.6I   Why are the contents of ponds recirculated?

9.6J   When do ponds that are operated on a batch basis discharge?

9.6K   What factors would you consider when developing an operating strategy for ponds?

9.7   AERATION SYSTEMS AND EQUIPMENT

9.70   Surface Aerators (Table 9.2 and Figures 9.4 and 9.5)

Surface aerators have been used in two types of applications:

1. To provide additional air for ponds during the night, during cold weather, or for overloaded ponds.

2. To provide a mechanical aeration device for ponds operated as aerated lagoons. Aerated lagoons operate like activated sludge aeration tanks without returning any settled activated sludge.

In both cases, the aerators are operated by time clocks with established ON/OFF cycles. Laboratory tests on the dissolved oxygen in a pond indicate the time period for ON and OFF

TABLE 9.2   PURPOSE OF AERATOR PARTS

Part	Purpose
PLATFORM AERATOR COMPONENTS	
Aerator	Introduces oxygen into pond.
Electric Motor	Provides energy to drive aerator.
Drive Reduction Gear Box	Converts torque from motor to drive impeller.
Draft Tube	Conveys bottom contents of pond to surface for aeration.
Discharge Guide	Regulates spray patterns for oxygen transfer to water.
Jacking Screw	Adjusts aerator impeller level in water to regulate oxygen transfer and motor loading amps.
ADDITIONAL FLOATING AERATOR COMPONENTS	
Pontoons or Floats	Provide platform for motor.
Guy Wires	Maintain aerator positions in ponds. Anchor to pond levee.
Power Cable	Conveys power to motor.
Impeller (aerator)	Pumps water into air to be aerated.

Fig. 9.4   Surface aerator
(Permission of EIMCO)

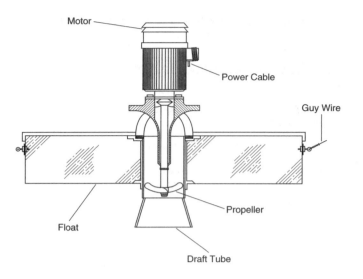

Motor

Power Cable

Guy Wire

Propeller

Float

Draft Tube

Fig. 9.5   Floating surface aerator
(Permission of Aqua-Jet)

1. Acid cleaning. Some aeration tubing systems require cleaning on a weekly basis. If excessive hardness in the wastewater leads to bicarbonate deposits, the deposits can be removed by the use of an acid wash. This can be done using an industrial solution of hydrogen chloride, commonly known as muriatic acid. This liquid acid solution is a strong acid and requires careful handling.

2. High-pressure air purging. Some aerated lagoon facilities are capable of applying 90 psi (620 kPa) of compressed air to the system just after the blowers and the air manifold. The air manifold valves and blowers should be off when performing this type of maintenance. This is not the pressure level that will reach the lagoon tubing because of the friction losses in the tubing system. High-pressure cleaning will greatly deform the slits in the tubing and break off deposits farther back in the tubing. High-pressure air may be applied for three to five minutes once a year.

3. Slit deformation. The aeration tubing slits may be allowed to deform by shutting off the air once a day for about 20 minutes. The weight of the water above the tubing will cause the slits to deform inward, thus breaking up the deposits. However, this procedure may allow some suspended material or sediment to enter the air line. This material can plug the air line when the air is turned on again.

4. Removal and cleaning of tubing. A very thorough but time-consuming job is the drawdown of the lagoon and the removal of the tubing for cleaning. The tubing can be flexed by hand or with special equipment to break up deposits. The inner surfaces of the tubing may be blown out with high-pressure air. You may be able to avoid draining the lagoon to clean the tubing by using a boat, retrieving one end of a line of tubing, and manually flexing each section of tubing.

5. Blowouts. Blowouts of aeration tubing and boils caused by slit enlargements must be repaired. The damaged area must be cut out and a special coupling inserted and clamped in place. The lagoon water level may have to be lowered so the tubing line can be hooked and drawn to the surface for repair.

Acknowledgment

Material in this section was provided by Jon Jewett.

9.8   SAMPLING AND ANALYSIS

9.80   Importance

Probably the most important sampling that can be accomplished easily by any operator is routine pH and dissolved oxygen analysis. A good practice is to take pH, temperature, and dissolved oxygen tests several times a week, and occasionally during the night. These values should be recorded because they provide a valuable measure of pond performance, indicate the status (health) of the pond, and help operators determine whether corrective action is needed. This information is also useful when preparing monitoring reports. Sampling should be conducted at the same time of day to ensure comparable results. However, the time of day should be varied occasionally for tests

cycles to maintain aerobic conditions in the surface layers of the pond. Adjustments in the ON/OFF cycles are necessary when changes occur in the quantity and quality of the influent and seasonal weather conditions. Some experienced operators have correlated their lab test results to pond appearance and regulate the ON/OFF cycles using the following rule: "If the pond has foam on the surface, reduce the operating time of the aerator; and if there is no evidence of foam on the pond surface, increase the operating time of the aerator." If there is a trace of foam on the surface, the operating time is satisfactory.

Surface aerators may be either stationary or floating. Maintenance of surface aerators should be conducted in accordance with manufacturer's recommendations. Always turn off, tag, and lock out the electric current when repairing surface aerators. Special precautions may be necessary to handle problems with icing and winter maintenance. Overhead guy wires have been used to prevent aerators from turning over when iced up. The operator should be aware that if a power failure occurs, even for a short time, the pond surface will freeze rapidly and possibly damage mechanical aerators.

9.71   Aeration Tubing

Another method of aerating ponds is by using air compressors connected to plastic tubes placed across the pond bottom. Holes are drilled in the plastic tubes, which serve as diffusers to disperse air in the pond. With this method of aeration, diffusers can become plugged and create maintenance problems. If an aeration system becomes plugged due to deposits of carbonate, try reducing the aeration.

Aeration tubing requires an effective maintenance program. The type of maintenance required depends on the type of aeration system and problems encountered. Five types of maintenance are discussed:

using *GRAB SAMPLES*[30] so that the operator becomes familiar with the pond's characteristics at various times of the day. Usually, the pH and dissolved oxygen levels will be lowest just at sunrise. Both will get progressively higher as the day goes on, reaching their highest levels in late afternoon.

Be very careful to avoid getting any atmospheric oxygen into the sample taken to measure dissolved oxygen. This is most necessary when samples are taken in the early morning or when the dissolved oxygen in the pond is low from overloading. If possible, measure the dissolved oxygen with an electric meter and probe, being careful not to allow the membrane on the end of the probe to be exposed to the atmosphere during actual DO measurement of the water sample.

Ponds often have clearly developed individuality, each being a biological community that is unique. Two adjacent ponds receiving the same influent in the same amount often have a different pH and a different dissolved oxygen content at any given time. One pond may generate considerable scum while its neighbor does not have any scum. For this reason, each pond should be tested routinely for pH and dissolved oxygen. Such testing may indicate an unequal loading because of the internal clogging of influent or distribution lines that might not be apparent from visual inspection. Tests also may indicate differences or problems that are being created by a buildup of solids or solids recycling.

When an operator becomes familiar with operating a pond, the results of some of the chemical tests can be related to visual observations. A deep green sparkling color generally indicates a high pH and a satisfactory dissolved oxygen content. A dull green color or lack of color generally indicates a declining pH and a lowered dissolved oxygen content. A gray color indicates that the pond is being overloaded or that it is not working properly.

9.81 Frequency and Location of Lab Samples

The frequency of testing and expected ranges of test results vary considerably from pond to pond, but you should establish the ranges within which your pond functions properly. Test results will also vary during the hours of the day. Table 9.3 summarizes the typical tests, locations, and frequencies of sampling. (See Chapter 14, "Laboratory Procedures," for detailed testing procedures.)

Samples should always be collected from the same point or location. Raw wastewater samples for pond influent tests may be collected either at the wet well of the influent pump station or at the inlet control structure. Samples of pond effluent should be collected from the outlet control structure or from a well-mixed point in the outfall channel. Pond samples may be taken from the four corners of the pond. The samples should be collected from a point eight feet (2.5 m) out from the water's edge and one foot (0.3 m) below the water surface. When collecting a sample, do not to stir up material from the pond bottom. Also, do not collect pond samples during or immediately after high winds or storms because solids will be stirred up by such activity.

TABLE 9.3 FREQUENCY AND LOCATION OF LAB SAMPLES[a]

Test	Frequency[b]	Location	Common Range
pH[c]	Weekly	Pond	7.5+
Dissolved Oxygen (DO)[c]	Weekly	Pond Effluent	4–12 mg/L 4–12 mg/L
Temperature	Weekly	Pond	
BOD[d]	Weekly	Influent Effluent	100–300 mg/L 20–50 mg/L
Coliform-Group Bacteria	Weekly	Effluent	MPN > 24,000/100 mL (unchlorinated)
Chlorine Residual	Daily	Effluent	0.5–2.0 mg/L
Suspended Solids[e]	Weekly	Influent Effluent	100–350 mg/L 40–80 mg/L
Dissolved Solids	Weekly	Influent	250–800 mg/L

a. Consistent sampling times are required for pond operational control and regulatory monitoring reports to establish logical interpretations of lab results and pond performance. Suggested sampling times are 1100 or 1300.

b. Tests may be less frequent for ponds with long detention times (greater than 100 days).

c. pH values above 9.0 and DO levels over 15 mg/L are not uncommon.

d. Contact your regulatory agency to determine whether effluent samples should be filtered to remove algae before testing. If the samples must be filtered, the agency will recommend the proper procedures.

e. Effluent suspended solids consist of algae, microorganisms, and other suspended matter.

BOD should be measured on a weekly basis. Samples should be taken during the day at low flow, medium flow, and high flow. The average of these three tests will give a reasonable indication of the organic load of the wastewater being treated. If it is suspected that the BOD varies sharply during the day or from day to day, or if unusual circumstances exist, the sampling frequency should be increased to obtain a clear definition of the variations. If the pond DO level is supersaturated (higher than maximum theoretical DO levels), the sample must be aerated to remove the excess oxygen before the BOD test is performed. A typical data sheet for a plant consisting mainly of ponds is provided in the Appendix at the end of this chapter. Figure 9.6 shows the influent and effluent BOD and suspended solids values from an actual pond.

A grab sample is a single sample. Grab samples are used to measure temperature, pH, dissolved oxygen, and chlorine residual. These tests must be performed immediately after the

---

30. *Grab Sample.* A single sample of water collected at a particular time and place that represents the composition of the water only at that time and place.

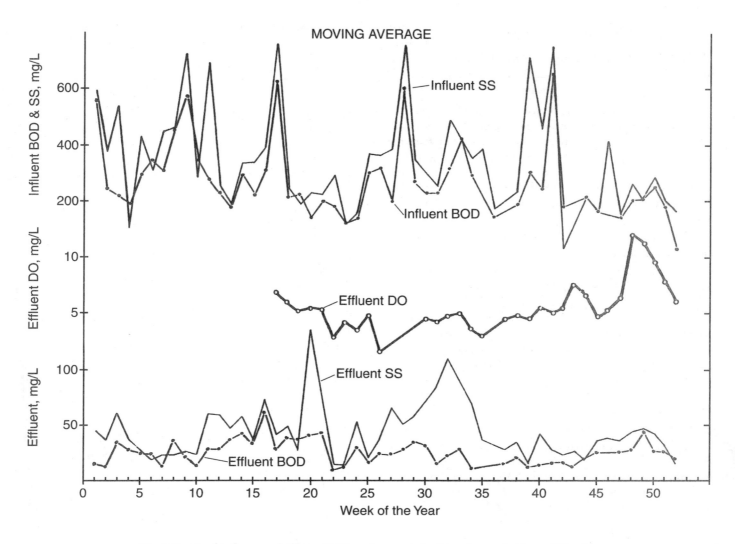

Fig. 9.6   Pond influent and effluent BOD and suspended solids values and effluent DO values

sample is collected in order to obtain accurate results. *COM-POSITE SAMPLES*[31] of pond influent or effluent are collected by gathering individual samples at regular intervals over a selected period of time. The individual samples are then mixed together in proportion to the flow at the time of sampling. Pond samples may be composited by mixing equal portions from the four corners of the pond. Composite samples should be placed in a refrigerator or ice chest as soon as possible after they are collected. Biochemical oxygen demand (BOD) and suspended solids (SS) are measured using composite samples.

Tests of pH, DO, and temperature are important indicators of the condition of the pond, whereas BOD, coliform, and SS tests measure the efficiency of the pond in treating wastes. BOD is also used to calculate the loading on the pond.

In order to estimate the organic loading on the pond, the operator must have some knowledge of the BOD of the waste and the approximate average daily flow. Influent BOD and solids will vary with time of day, day of week, and season, but a pond is a good equalizer if not overloaded.

Tests for pond alkalinity can provide helpful information to an operator. For useful results, the alkalinity test should be performed every day. After you have determined normal alkalinity levels for your pond, a sudden change in alkalinity of 10 to 20 mg/L or more may indicate that a problem is developing. A change in alkalinity may be a warning that the pH of the pond could change in a day or two if corrective action is not taken. If a trend in alkalinity changes continues in one direction for two or three days, the cause of this change should be identified.

31. *Composite (Proportional) Sample.*   A composite sample is a collection of individual samples obtained at regular intervals, usually every one or two hours during a 24-hour time span. Each individual sample is combined with the others in proportion to the rate of flow when the sample was collected. Equal volume individual samples also may be collected at intervals after a specific volume of flow passes the sampling point or after equal time intervals and still be referred to as a composite sample. The resulting mixture (composite sample) forms a representative sample and is analyzed to determine the average conditions during the sampling period.

9.82 Expected Treatment Efficiencies[32]

Table 9.4 is provided as a guide to expected ranges of removals by typical ponds.

TABLE 9.4 EXPECTED RANGES OF REMOVAL BY PONDS

Test	Detention Time	Expected Removal
BOD		50–90%
BOD (facultative pond)	50–60 days	70–80%[33]
Coliform Bacteria (facultative pond)	50–60 days	90–95%
Suspended Solids	After 3 days	90%
Dissolved Organic Solids	After 10 days	80%

The calculation of pond BOD removal efficiency is figured in terms of the percentage of BOD removed.

EXAMPLE: The influent BOD to a series of ponds is 300 mg/L, and the effluent BOD is 60 mg/L. What is the pond efficiency in removing BOD?

Known	Unknown
BOD In, mg/L = 300 mg/L	BOD Removal, %
BOD Out, mg/L = 60 mg/L	

Calculate the pond BOD removal efficiency.

$$BOD\ Removal, \% = \frac{(In - Out)}{In} \times 100\%$$

$$= \frac{(300\ mg/L - 60\ mg/L)}{300\ mg/L} \times 100\%$$

$$= \frac{240\ mg/L}{300\ mg/L} \times 100\%$$

$$= 0.80 \times 100\%$$

$$= 80\%$$

9.83 Response to Poor Pond Performance

For information about responding to poor pond performance, see Section 9.6, "Daily Operation and Maintenance," especially Section 9.65, "General Operating Hints."

QUESTIONS

Please write your answers to the following questions and compare them with those on pages 45 and 46.

9.7A   Surface aerators have been used in what two types of applications in ponds?

9.8A   Why should the pH, temperature, and dissolved oxygen be measured in a pond?

9.8B   If the color of a pond is dull green, gray, or colorless, what is happening in the pond?

9.8C   The influent BOD to a series of ponds is 200 mg/L. If the BOD in the effluent of the last pond is 40 mg/L, what is the BOD removal efficiency?

9.9 SAFETY

Even though a pond has little mechanical equipment, there still are hazards. Catwalks located over ponds should have guardrails and nonskid walking surfaces. Headworks and any enclosed *APPURTENANCES*[34] should be well ventilated to prevent dangerous gas accumulations.

WARNING

An operator should *always* be accompanied by a helper when performing any task that is dangerous because pond locations are usually quite isolated. Immediate aid might be needed to prevent serious injury or loss of life.

Be very careful when removing debris from channels and ponds. Do not attempt to lift too much at one time. Make certain you have secure footing so you will not slip and fall. Never stand or lean over too far to one side in a boat, for you could fall into the pond and possibly tip over the boat. Always wear a life jacket when in a boat.

Electrical wires and equipment are always a source of potential danger. Exercise caution when cutting weeds or removing vegetation such as trees next to electrical wires. Electrical wires in damp areas can be especially dangerous. Be careful when spraying weeds around electrical wires and equipment because the spray could act as a conductor. Always turn off, tag, and lock out electric currents when repairing surface aerators and other equipment operated by electricity.

Before applying pesticides or herbicides around a pond, verify that they are approved by the appropriate officials for your specific use. Read and follow the directions exactly. Exercise safety precautions when using these products, including following the directions for using the proper procedures for mixing or preparing the solution, applying the solution, disposing of any excess solution and the containers, and cleaning up after the task is completed. The use of protective clothing, gloves, and respiratory protection equipment may be required depending on the product, product strength, application method, and wind conditions.

---

32. Waste Removal, % = $\frac{(In - Out)}{In} \times 100\%$

33. Expected removal approximately 80 percent of the time with poorer removals during the remainder of the time.

34. *Appurtenance* (uh-PURR-ten-nans).   Machinery, appliances, structures, and other parts of the main structure necessary to allow it to operate as intended, but not considered part of the main structure.

Not only can these products kill the target pest or weed, but careless application can harm nearby grasses, plants, trees, animals, and even people (including you). Also, failing to follow directions may result in harm to the algae and microorganisms in the pond. Many areas require specific applicator training/certification. For more information, check with your local regulatory agency.

Extreme care must be exercised when working with wastewater to avoid infections and diseases. *Clostridium* is a group of obligate bacterium that are normal inhabitants of the intestinal tract of almost all animals. In the presence of free oxygen, these bacteria produce highly resistant intercellular spores. Under anaerobic conditions where nutrients and moisture are present, as in puncture wounds or severely smashed tissue, the spores can "germinate" allowing the organisms to resume their growth and reproduction, producing toxins that cause severe diseases.

Clostridia tetani release tetanus toxin that attacks the central nervous system, eventually interferes with breathing, and potentially causes death if not treated. Clostridial gas gangrene is caused by several species of *Clostridium,* most commonly *C. perfringens.* This condition causes the tissue in the area of the wound to rot and potentially causes death if not treated. Fortunately, there is an inoculation that can protect against tetanus, and gangrene infections can be treated with antibiotics.

Adequate precautions should be observed by applying first-aid procedures to all cuts and scrapes and always washing before eating or smoking. In all cases of injury involving puncture wounds or severe tissue damage, rapid medical attention must be given to prevent severe damage or death.

Fences should be installed around ponds to keep unauthorized persons and animals out of the pond area. They should be located in such a manner that they will not interfere with the mechanical or hand maintenance of levee slopes or with general plant operation and maintenance.

QUESTIONS

Please write your answers to the following questions and compare them with those on page 46.

9.9A What safety devices should be provided on catwalks located over ponds?

9.9B Why should an operator be accompanied by a helper when performing any dangerous task?

END OF LESSON 2 OF 3 LESSONS

on

WASTEWATER STABILIZATION PONDS

Please answer the discussion and review questions next.

DISCUSSION AND REVIEW QUESTIONS
Chapter 9. WASTEWATER STABILIZATION PONDS
(Lesson 2 of 3 Lessons)

Please write your answers to the following questions to determine how well you understand the material in the lesson. The question numbering continues from Lesson 1.

7. Why should water be introduced into a new pond before any wastewater?

8. Why is good housekeeping an important factor in operating a properly functioning pond?

9. Why are chlorine compounds or chlorine solution not the best chemicals for odor control in a pond?

10. What precautions should be taken when applying an insecticide?

11. What tests are indicators of the condition of a pond?

12. Estimate the BOD removal efficiency of a series of ponds if the influent BOD is 250 mg/L and the effluent BOD is 50 mg/L.

13. Why should fences be installed around ponds?

CHAPTER 9. WASTEWATER STABILIZATION PONDS

(Lesson 3 of 3 Lessons)

9.10 REVIEW OF PLANS AND SPECIFICATIONS

A careful review of the plans and specifications of a proposed pond can provide the operator an opportunity to suggest design improvements and changes before construction begins. This will allow the operator and the ponds to do their jobs better. Guidelines for reviewing plans and specifications are provided in this section. If you are having trouble with an existing pond or ponds, the items discussed in this section may help you locate possible sources of the problem.

9.100 Location

The general considerations for the location of other types of wastewater treatment plants also apply to the location of ponds. Isolation should be as great as can be economically provided. Ponds should be isolated to prevent associated nuisances such as odors and insects from disturbing residential, commercial, and recreational neighbors, as well as to prevent possible traffic hazards caused by insects. Attention to the direction of prevailing winds with due regard for present and projected downwind residential, commercial, and recreational development also is important.

Winds can have both favorable and unfavorable impacts on ponds. Winds are desirable in terms of blowing surface scum and weeds not rooted in the pond bottom to one side of the pond where they can be removed. Also, winds can be helpful by mixing the DO, algae, and incoming wastes contained in a pond. An undesirable effect of winds is the creation of waves that can erode the pond levee. Both of these factors should be considered when selecting the location of the ponds and the arrangement or length of the ponds. If high winds are to be expected in the area where ponds will be constructed, the ponds should be arranged so the winds will blow across the short width of the pond rather than its length in order to reduce levee erosion caused by waves.

9.101 Chemistry of Waste

Before the design of any pond is undertaken, determine whether the waste to be treated contains any toxic constituents that can interfere with the growth of algae or bacteria. Certain wastes such as dairy products and wine products are difficult to treat because of their low pH. Any processing waste must be carefully investigated before one can be certain that it can be successfully treated by ponds. Some process wastes contain powerful fungicides and disinfectants that may have a great inhibitive effect on the biological activity in a pond. Other wastes may have nutrient deficiencies that could inhibit the growth of desirable types of algae. Some industrial wastes are difficult to treat due to high levels of starch, sugars, fats, oils, grease, or cellulose.

In addition, some natural water supplies have a high sulfur content or other chemicals that limit the possibility of desired sludge decomposition.

9.102 Headworks and Screening

A headworks with a bar screen is recommended for removing rags, bones, and other large objects that might lodge in pipes or control structures.

A trash shredder is an optional device that may not be needed. Any material that gets past an adequate bar screen will, in all probability, not harm the influent pump. Any fecal matter that passes the screen will be pulverized when going through the pump.

Rotating screens can be very effective in removing grit, rags, plastic, and rubber goods from influent flow. Any material that can be kept out of the ponds will help enhance the ability of the pond to treat the wastewater and increase the useful life of the facility.

9.103 Flow Measuring Devices

An influent measuring device should be installed to give a direct reading on the daily volume of wastes that are introduced into the ponds. This information is used to determine pond detention time and also, along with a BOD measurement of the influent, is required to estimate the organic loading on the pond. Comparison of influent and effluent flow rates is necessary for estimating percolation and evaporation losses. A 24-hour flow rate recording instrument will show flow patterns that can be useful in showing a possible need for process adjustments.

A measuring device provides basic data for prediction of future plant expansion needs or for detecting unauthorized or abnormal flows. Reliable, well-kept records on flow volume help justify budgets and greatly assist an engineer's design of a plant expansion or new installation.

9.104 Inlet and Outlet Structures

Inlet structures should be simple, foolproof, and constructed of standard manufactured materials and devices so that replacement parts are readily available. Avoid using telescoping friction-fit tubes (see Figure 9.7) in pond inlet and outlet structures for regulating spill or discharge height because a biological growth may become attached and prevent the tubes from telescoping if they are not cleaned regularly. Occasional dosages of hypochlorite solution can effectively discourage growths. Also,

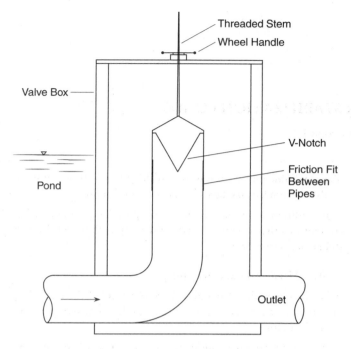

*Fig. 9.7   Telescoping friction-fit tubes for regulating discharge.
These tubes must be exercised regularly to prevent
them from becoming stuck.*

The inlet structure must provide the ability to direct the flow to any or all of the available ponds. This flexibility will make it possible to isolate ponds for maintenance or process control.

Figure 9.1 (page 12) shows four inlets to one pond. With this type of installation, usually only one inlet valve is open at a time. If all the valves were open, velocities in the pipes might become too low and solids would settle out in the pipes. To overcome low velocities, close all but one valve or recycle pond effluent back to the inlet. When all inlet valves are open, the load is more evenly distributed throughout the pond, but sufficient recycle flow is required to maintain the desired velocities in the inlet lines.

A submerged inlet will minimize the occurrence of floating material and will help conserve the heat of the pond by introducing the warmer wastewater into the depths of the pond. Warm wastewater introduced at the bottom of a cold water mass will channel to the surface and spread unless it is promptly and vigorously mixed with cold water. Warm wastewater spilled onto the surface of the pond will spread out in a thin layer on the surface and not contribute to the warmth of the lower regions of the pond where heat is needed for bacterial decomposition. Inlet and outlet structures should be so located in relation to each other as to minimize possible short-circuiting.

Outlet structures (Figures 9.8 and 9.9) should consist of a baffled and submerged pipe inlet to prevent scum and other floating surface material from leaving the pond. The actual level of the pond and rate of outflow can be controlled by the use of flash boards in the outlet structure. A rowboat may be used when access to the outlet baffle (shown in Figure 9.8) is required.

Avoid valves that have stems extending into the stream flow. Stringy material and rags will collect, form an obstruction, and may render the valve inoperative.

the formation of ice can prevent adjustment of the tubes. If freezing is a problem, place a polyethylene floating ring sprayed with urea around the friction tube of the telescopic valve to prevent freeze-ups. This device will act as a floating baffle to keep scum and floating debris from clogging up the tube, leaving the pond, and entering the effluent.

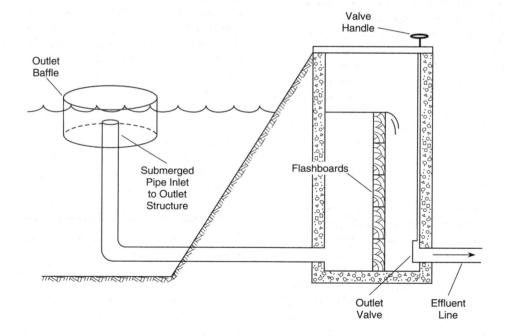

Fig. 9.8   Pond outlet structure

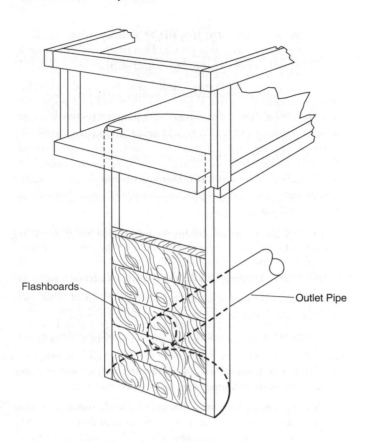

Fig. 9.9   Pond outlet control structure

Flashboards

Outlet Pipe

Avoid free overfalls (Figure 9.10) at the outlet to minimize the release of odors, foaming, and gas entrapment, which may hamper pipe flows. Free overfalls should be converted to submerged outfalls if they are causing nuisances and other problems.

If a pond has a surface outlet, floating material can be kept out of the effluent by building a simple baffle around the outlet. The baffle can be constructed of wood or other suitable material and should be securely supported or anchored.

9.105   Levees

The selection of the steepness of the levee slope must depend on several variables. A steep slope erodes more quickly from wave action unless the levee material is of a rocky nature or else protected by riprap. Trees can be planted to provide an eventual windbreak. However, a steep slope minimizes waterline weed growth. Equipment operation and the performance of routine maintenance are more difficult on steep slopes. A gentle slope will erode the least from wave wash. Also, it is easier to operate equipment and to perform routine maintenance on a gentle slope. However, waterline weed growth will have a much greater opportunity to flourish.

Ensure that provisions are made to adequately compact or seal the levee banks to prevent leaking. Pipes passing through levees should be as close to horizontal as possible to reduce the possibility of leaks. Also cut-off walls should be installed when pipes pass through levees to prevent leakage around the outside of the pipes. Proper compaction and sealing is necessary around

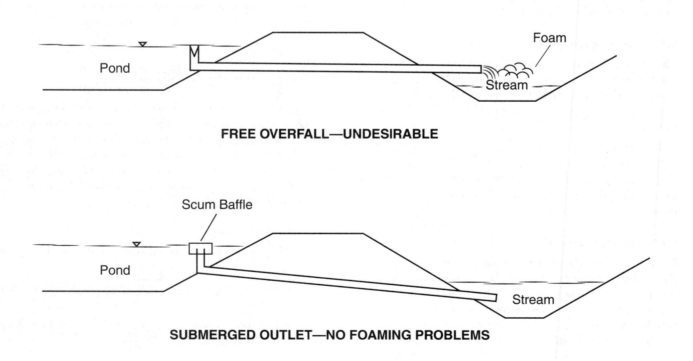

FREE OVERFALL—UNDESIRABLE

SUBMERGED OUTLET—NO FOAMING PROBLEMS

Fig. 9.10   Free overfall and submerged outlet

pipes, *SPLASH PADS,*[35] cleanouts, valves, inlet and outlet control structures, and also around recirculation, transfer, and drain lines. Once a leak develops, stopping it can be very difficult.

The top of the levee should be at least ten feet (3.1 m) wide to allow for maintenance vehicles. Pave or gravel the top of the levee surface if it will be used as a roadway during wet weather. Provisions should be made for a rounded or sloping top to allow for drainage.

9.106 Pond Depths

An observed phenomenon of lightly loaded and shallow secondary or tertiary ponds is that they are apt to become infested with filamentous algae and mosses. These limit the penetration of sunlight into the pond, hamper circulation of the pond's contents, and clog inlet and outlet structures. When the loading is increased, this condition improves because these algae and mosses require relatively clean water (low nutrients) for their environment.

Pond depths of four feet (1.2 m) or more allow a greater conservation of heat from the incoming wastes. This encourages biological activity as the ratio between pond volume and pond area is more favorable. In facultative ponds, depths over four feet (1.2 m) provide physical storage for dissolved oxygen accumulated during the day. The stored dissolved oxygen carries over through the night when no oxygen is released by the algae, unless floating algae and poor circulation keep all the oxygen near the surface. This physical storage of DO is very important during the colder months when nights are long.

A pond operating depth of at least three feet (1 m) is recommended to prevent tule and cattail growth. Weeds that emerge along the shoreline can be effectively controlled by spraying with any of several products available.

9.107 Fencing and Signs

The pond area must be surrounded by a fence capable of keeping livestock out and discouraging trespassing. A gate wide enough to allow mowing equipment and other maintenance vehicles to enter the pond facility should be provided. All access gates should have locks.

Signs should be posted along the fence around the ponds to indicate the nature of the facility and to forbid trespassing. The signs should not be more than 300 feet (90 m) apart, but the spacing should also comply with local penal codes governing trespass.

9.108 Surface Aerators

Provisions must be made for easy access to fixed aerators. Also allow sufficient space around them for maintenance and repair. Alternate anchor points should be installed in order to move floating aerators. Be sure the electrical cables are long enough to permit easy movement of the aerator and large enough to handle anticipated loads.

QUESTIONS

Please write your answers to the following questions and compare them with those on page 46.

9.10A Why are some wastes not easily treated by ponds?

9.10B Why should the influent to a pond be metered?

9.10C Why should the inlet to a pond be submerged?

9.10D Why should the outlet be submerged?

9.10E Why should free overfalls be avoided?

9.10F How could problems created by a surface outlet be reduced or corrected?

9.10G What is the minimum recommended pond operating depth?

9.109 Pond Loading (English and Metric Calculations)

The waste loading on a pond is generally spoken of in relation to its area, and may be stated in several different ways:

- "Organic (BOD) loading," which is pounds per day per acre (lbs/day/acre)

- "Hydraulic loading" or "overflow rate," which is inches (or feet) of depth added per day

- "Population loading," which is persons (or population) served per acre

Detention time is related directly to pond hydraulic loading, which is actually the rate of inflow of wastewater. Rate of inflow may be expressed as million gallons per day (MGD), or as the number of acre-inches per day or acre-feet per day. (One acre-foot is the equivalent amount or volume of water that covers one acre to a depth of one foot, or 43,560 cubic feet.) You must know the pond volume in order to determine detention time; this is most easily computed on an acre-foot basis.

Detention Time

$$\text{Detention Time, days} = \frac{\text{Pond Volume, ac-ft}}{\text{Flow Rate, ac-ft/day}}$$

This equation does not take into consideration water that may be lost through evaporation or percolation. Detention time may vary from 30 to 120 days, depending on the treatment requirements to be met. Controlled-discharge ponds may have minimum detention times of 180 days. Ponds whose discharges are disposed of by land application may have 210-day minimum detention times.

Population Loading

Loading calculated on a population-served basis is expressed simply as

$$\text{Population Loading, persons/ac} = \frac{(\text{Pop Served, persons})(43{,}560 \text{ sq ft/ac})}{\text{Pond Surface Area, sq ft}}$$

The population loading may vary from 50 to 500 persons per acre, depending on many local factors.

---

35. *Splash Pad.* A structure made of concrete or other durable material to protect bare soil from erosion by splashing or falling water.

Hydraulic Loading

Hydraulic loading refers to the flows to a treatment plant or treatment process. Detention times and surface loadings are directly influenced by flows. The hydraulic loading on a pond may vary from half an inch to several inches per day, depending on the organic load of the influent.

Hydraulic loading for ponds is calculated according to what information is known and what units are desired in the end result. In the example calculations below, we will find hydraulic loading in inches per day based on knowing the detention time.

$$\text{Hydraulic Loading, inches/day} = \frac{\text{Depth of Pond, inches}}{\text{Detention Time, days}}$$

Organic (BOD) Loading

The organic (BOD) loading is expressed as

$$\begin{array}{l}\text{Organic (BOD)} \\ \text{Loading,} \\ \text{lbs/day/ac}\end{array} = \frac{(\text{Flow, MGD})(\text{BOD, mg/L})(8.34 \text{ lbs/gal})}{\text{Pond Area, ac}}$$

Typical organic (BOD) loadings may range from 10 to 50 pounds per day per acre. An operator can use recirculation to help a pond that has an organic overload.

EXAMPLE CALCULATIONS

Basic conversion factors needed to perform the example calculations include the following:

43,560 sq ft = 1 acre	43,560 sq ft/ac
7.48 gal = 1 cu ft	7.48 gal/cu ft
12 in = 1 ft	12 in/ft
8.34 lbs = 1 gal	8.34 lbs/gal

EXAMPLE 1

To calculate the different loadings on a pond, the information listed in the known column must be available.

Known

Average Depth of Pond, ft = 4 feet

Width of Pond

   Bottom Width, ft   = 412 feet

   Pond Surface Width, ft = 428 feet

   Average Width, ft   = 420 feet

$$\begin{array}{l}\text{Average} \\ \text{Width, ft}\end{array} = \frac{\text{Bottom Width, ft + Pond Surface Width, ft}}{2}$$

$$= \frac{412 \text{ ft} + 428 \text{ ft}}{2}$$

$$= 420 \text{ ft}$$

Length of Pond

   Bottom Length, ft   = 667 feet

   Pond Surface Length, ft = 683 feet

   Average Length, ft   = 675 feet

$$\begin{array}{l}\text{Average} \\ \text{Length, ft}\end{array} = \frac{\text{Bottom Length, ft + Pond Surface Length, ft}}{2}$$

$$= \frac{667 \text{ ft} + 683 \text{ ft}}{2}$$

$$= 675 \text{ ft}$$

$$\begin{array}{l}\text{Side Slope (1 ft vertical to} \\ \text{2 ft horizontal}\end{array} = 1\!:\!2$$

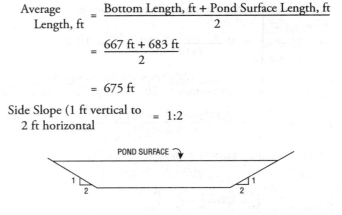

Flow, GPD	= 200,000 GPD
Flow, MGD	= 0.2 MGD
BOD, mg/L	= 200 mg/L
Population Served	= 2,000 persons

Unknown

1. Detention Time, days
2. Population Loading, persons/acre
3. Hydraulic Loading, inches/day
4. Organic (BOD) Loading, lbs/day/acre

1. Calculate the detention time in days.

   a. Estimate the average pond area in acres.

$$\text{Avg Pond Area, ac} = \frac{(\text{Avg Width, ft})(\text{Avg Length, ft})}{43,560 \text{ sq ft/ac}}$$

$$= \frac{(420 \text{ ft})(675 \text{ ft})}{43,560 \text{ sq ft/ac}}$$

$$= 6.51 \text{ ac}$$

   b. Determine the pond volume in acre-feet.

$$\text{Pond Vol, ac-ft} = (\text{Avg Pond Area, ac})(\text{Avg Depth, ft})$$

$$= (6.51 \text{ ac})(4 \text{ ft})$$

$$= 26.04 \text{ ac-ft (say 26 ac-ft)}$$

   c. Convert flow from gallons per day to acre-feet per day.

$$\text{Flow, ac-ft/day} = \frac{\text{Flow, GPD}}{(7.48 \text{ gal/cu ft})(43,560 \text{ sq ft/ac})}$$

$$= \frac{200,000 \text{ GPD}}{(7.48 \text{ gal/cu ft})(43,560 \text{ sq ft/ac})}$$

$$= 0.61 \text{ ac-ft/day}$$

   d. Calculate the detention time in days.

$$\text{Detention Time, days} = \frac{\text{Pond Volume, ac-ft}}{\text{Flow Rate, ac-ft/day}}$$

$$= \frac{26 \text{ ac-ft}}{0.61 \text{ ac-ft/day}}$$

$$= 42.6 \text{ days}$$

2. Calculate the population loading in persons per acre.

$$\text{Population Loading, persons/ac} = \frac{(\text{Pop Served, persons})(43{,}560 \text{ sq ft/ac})}{\text{Pond Surface Area, sq ft}}$$

$$= \frac{(2{,}000 \text{ persons})(43{,}560 \text{ sq ft/ac})}{(\text{Pond Surface Width, ft})(\text{Pond Surface Length, ft})}$$

$$= \frac{(2{,}000 \text{ persons})(43{,}560 \text{ sq ft/ac})}{(428 \text{ ft})(683 \text{ ft})}$$

$$= \frac{(2{,}000 \text{ persons})(43{,}560 \text{ sq ft/ac})}{292{,}324 \text{ sq ft}}$$

$$= 298 \text{ persons/ac}$$

NOTE: If there is a significant industrial waste flow mixed in with the domestic waste, an adjustment must be made to take the industrial waste into consideration. This is usually done by analyzing the industrial waste and converting it to a *POPULATION EQUIVALENT.*[36]

3. Calculate the hydraulic loading in inches per day.

$$\text{Hydraulic Loading, in/day} = \frac{(\text{Depth of Pond, ft})(12 \text{ in/ft})}{\text{Detention Time, days}}$$

$$= \frac{(4 \text{ ft})(12 \text{ in/ft})}{42.6 \text{ days}}$$

$$= 1.13 \text{ in/day}$$

4. Calculate the organic (BOD) loading in pounds per day per acre.

$$\text{Organic (BOD) Loading, lbs/day/ac} = \frac{(\text{Flow, MGD})(\text{BOD, mg/L})(8.34 \text{ lbs/gal})}{\text{Avg Pond Area, ac}}$$

$$= \frac{(0.2 \text{ MGD})(200 \text{ mg/L})(8.34 \text{ lbs/gal})}{6.51 \text{ ac}}$$

$$= 51 \text{ lbs BOD/day/ac}$$

EXAMPLE 2

Suppose that a small wastewater treatment plant must be completely shut down for major repairs that will require several months of work. Enough vacant land is near the plant to enable 16 acres of temporary ponds to be constructed as raw wastewater stabilization ponds. Determine if this is feasible. The influent flow is 1 million gallons per day and the influent BOD is 150 milligrams per liter. Average operating depth is 3.5 feet and the average pond area is 16 acres. Assume that at least a 60-day detention period (average time the wastewater must take to flow through the pond for disinfection) is desired for bacterial die-off. The organic (BOD) loading should not exceed 50 pounds per day per acre.

Known

Flow, MGD	= 1 MGD
Flow, GPD	= 1,000,000 GPD
BOD, mg/L	= 150 mg/L
Avg Depth, ft	= 3.5 ft
Avg Pond Area, ac	= 16 ac
Minimum Detention Time, days	= 60 days
Maximum Organic (BOD) Loading, lbs/day/ac	= 50 lbs/day/ac

Unknown

1. Detention Time, days
2. Organic (BOD) Loading, lbs/day
3. Organic (BOD) Loading, lbs/day/ac
4. Feasibility of Temporary Ponds

1. Calculate what the detention time would be in the pond.

   a. Determine the pond volume in acre-feet.

   $$\text{Pond Volume, ac-ft} = \text{Avg Pond Area, ac} \times \text{Avg Depth, ft}$$

   $$= 16 \text{ ac} \times 3.5 \text{ ft}$$

   $$= 56 \text{ ac-ft}$$

   b. Convert flow from gallons per day to acre-feet per day.

   $$\text{Flow, ac-ft/day} = \frac{\text{Flow, GPD}}{(7.48 \text{ gal/cu ft})(43{,}560 \text{ sq ft/ac})}$$

   $$= \frac{1{,}000{,}000 \text{ GPD}}{(7.48 \text{ gal/cu ft})(43{,}560 \text{ sq ft/ac})}$$

   $$= 3.07 \text{ ac-ft/day}$$

   c. Calculate the detention time in days.

   $$\text{Detention Time, days} = \frac{\text{Pond Volume, ac-ft}}{\text{Flow, ac-ft/day}}$$

   $$= \frac{56 \text{ ac-ft}}{3.07 \text{ ac-ft per day}}$$

   $$= 18.2 \text{ days}$$

Thus, the detention time would not be sufficient to satisfy requirements.

---

36. *Population Equivalent.* A means of expressing the strength of organic material in wastewater. In a domestic wastewater system, microorganisms use up about 0.2 pound (90 grams) of oxygen per day for each person using the system (as measured by the standard BOD test). May also be expressed as flow (100 gallons (378 liters)/day/person) or suspended solids (0.2 lb (90 grams) SS/day/person).

$$\text{Population Equivalent, persons} = \frac{\text{Flow, MGD} \times \text{BOD, mg/L} \times 8.34 \text{ lbs/gal}}{0.2 \text{ lb BOD/day/person}}$$

or

$$\text{Population Equivalent, persons} = \frac{\text{Flow, cu m/day} \times \text{BOD, mg/L} \times 10^6 \text{ L/cu m}}{90{,}000 \text{ mg BOD/day/person}}$$

2. Calculate the organic (BOD) loading in pounds per day.

Organic (BOD)
Loading,
lbs/day $= (\text{Flow, MGD})(\text{BOD, mg/L})(8.34 \text{ lbs/gal})$

$= (1 \text{ MGD})(150 \text{ mg/L})(8.34 \text{ lbs/gal})$

$= 1{,}251 \text{ lbs BOD per day}$

3. Calculate the organic (BOD) loading in pounds per day per acre.

Organic (BOD) Loading,
lbs/day/ac $= \dfrac{\text{Loading, lbs BOD/day}}{\text{Avg Pond Area, ac}}$

$= \dfrac{1{,}251 \text{ lbs BOD per day}}{16 \text{ ac}}$

$= 78.2 \text{ lbs BOD/day/ac}$

4. The temporary ponds would not be feasible because the organic (BOD) loading would exceed the desired maximum of 50 pounds per day per acre.

QUESTION

Please write your answers to the following questions and compare them with those on pages 46 and 47.

9.10H   A pond receives a flow of 2.0 MGD from 20,000 people. The influent BOD is 180 mg/L, the pond surface area is 24 acres, and the average operating depth is four feet. Determine the detention time in days, the organic (BOD) loading in pounds per day per acre, the population loading in persons per acre, and the hydraulic loading in inches per day.

EXAMPLE METRIC CALCULATIONS

EXAMPLE 1

To calculate the different loadings on a pond, the information listed in the known column must be available.

Known

Average Depth of Pond, m = 2.0 meters

Width of Pond

   Bottom Width, m   = 125 meters

   Pond Surface Width, m = 133 meters

   Average Width, m   = 129 meters

   Average
   Width, m $= \dfrac{\text{Bottom Width, m} + \text{Pond Surface Width, m}}{2}$

   $= \dfrac{125 \text{ m} + 133 \text{ m}}{2}$

   $= 129 \text{ m}$

Length of Pond

   Bottom Length, m   = 200 meters

   Pond Surface Length, m = 208 meters

   Average Length, m   = 204 meters

   Average
   Length, m $= \dfrac{\text{Bottom Length, m} + \text{Pond Surface Length, m}}{2}$

   $= \dfrac{200 \text{ m} + 208 \text{ m}}{2}$

   $= 204 \text{ m}$

Side Slope (1 m vertical to
2 m horizontal $= 1{:}2$

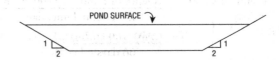

Flow, cu m/day   = 800 cu m/day

BOD, mg/L   = 200 mg/L

Population Served   = 2,000 persons

Unknown

1. Detention Time, days
2. Population Loading, persons/sq m
3. Hydraulic Loading, cm/day
4. Organic (BOD) Loading, gm/day/sq m

1. Calculate the detention time in days.

   a. Estimate the average pond area in square meters.

   Avg Pond Area, sq m $= (\text{Avg Width, m})(\text{Avg Length, m})$

   $= (129 \text{ m})(204 \text{ m})$

   $= 26{,}316 \text{ sq m}$

   b. Determine the pond volume in cubic meters.

   Pond Volume, cu m $= (\text{Avg Area, sq m})(\text{Avg Depth, m})$

   $= (26{,}316 \text{ sq m})(2 \text{ m})$

   $= 52{,}632 \text{ cu m}$

c. Calculate the detention time in days.

$$\text{Detention Time, days} = \frac{\text{Pond Volume, cu m}}{\text{Flow Rate, cu m/day}}$$

$$= \frac{52,632 \text{ cu m}}{800 \text{ cu m/day}}$$

$$= 66 \text{ days}$$

2. Calculate the population loading in persons per square meter.

$$\begin{array}{l}\text{Population}\\ \text{Loading,}\\ \text{persons/sq m}\end{array} = \frac{\text{Population Served, persons}}{\text{Pond Surface Area, sq m}}$$

$$= \frac{2,000 \text{ persons}}{(\text{Pond Surface Width, m})(\text{Pond Surface Length, m})}$$

$$= \frac{2,000 \text{ persons}}{(133 \text{ m})(208 \text{ m})}$$

$$= \frac{2,000 \text{ persons}}{27,664 \text{ sq m}}$$

$$= 0.072 \text{ persons/sq m}$$

NOTE: If there is a significant industrial waste flow mixed in with the domestic waste, an adjustment must be made to take the industrial waste into consideration. This is usually done by analyzing the industrial waste and converting it to a population equivalent.

3. Calculate the hydraulic loading in centimeters per day.

$$\begin{array}{l}\text{Hydraulic Loading,}\\ \text{cm/day}\end{array} = \frac{(\text{Depth of Pond, m})(100 \text{ cm/m})}{\text{Detention Time, days}}$$

$$= \frac{(2 \text{ m})(100 \text{ cm/m})}{66 \text{ days}}$$

$$= 3.03 \text{ cm/day}$$

4. Calculate the organic (BOD) loading in grams per day per square meter.

$$\begin{array}{l}\text{Organic (BOD)}\\ \text{Loading,}\\ \text{gm/day/sq m}\end{array} = \frac{\left(\begin{array}{l}\text{Flow,}\\ \text{cu m/day}\end{array}\right)\left(\begin{array}{l}\text{BOD,}\\ \text{mg/L}\end{array}\right)\left(\frac{1,000 \text{ L}}{1 \text{ cu m}}\right)\left(\frac{1 \text{ gm}}{1,000 \text{ mg}}\right)}{\text{Avg Pond Area, sq m}}$$

$$= \frac{800 \text{ cu m/day} \times 200 \text{ mg/L} \times \frac{1,000 \text{ L}}{1 \text{ cu m}} \times \frac{1 \text{ gm}}{1,000 \text{ mg}}}{26,316 \text{ sq m}}$$

$$= 6.1 \text{ gm BOD/day/sq m}$$

EXAMPLE 2

Suppose that a small wastewater treatment plant must be completely shut down for major repairs that will require several months of work. Enough vacant land is near the plant to enable 65,000 square meters of temporary ponds to be constructed as raw wastewater stabilization ponds. Determine if this is feasible. The influent flow is 4,000 cubic meters per day and the influent BOD is 150 milligrams per liter. Average operating depth is 1.2 meters and the average pond area is 65,000 square meters. Assume that at least a 60-day detention period (average time the wastewater must take to flow through the pond for disinfection)

is desired for bacterial die-off. The organic (BOD) loading should not exceed 5 grams per day per square meter.

Known

Flow, cu m/day	= 4,000 cu m/day
BOD, mg/L	= 150 mg/L
Avg Depth, m	= 1.2 m
Avg Pond Area, sq m	= 65,000 sq m
Minimum Detention Time, days	= 60 days
Maximum Organic (BOD) Loading, gm/day/sq m	= 5 gm/day/sq m

Unknown

1. Detention Time, days
2. Organic (BOD) Loading, gm/day
3. Organic (BOD) Loading, gm/day/sq m
4. Feasibility of Temporary Ponds

1. Calculate what the detention time would be in the pond.

a. Determine the pond volume in cubic meters.

$$\begin{array}{l}\text{Pond Volume,}\\ \text{cu m}\end{array} = \text{Avg Pond Area, sq m} \times \text{Avg Depth, m}$$

$$= 65,000 \text{ sq m} \times 1.2 \text{ m}$$

$$= 78,000 \text{ cu m}$$

b. Calculate the detention time in days.

$$\text{Detention Time, days} = \frac{\text{Pond Volume, cu m}}{\text{Flow, cu m/day}}$$

$$= \frac{78,000 \text{ cu m}}{4,000 \text{ cu m/day}}$$

$$= 19.5 \text{ days}$$

Thus, the detention time would not be sufficient to satisfy requirements.

2. Calculate the organic (BOD) loading in grams per day.

$$\begin{array}{l}\text{Organic (BOD)}\\ \text{Loading,}\\ \text{gm/day}\end{array} = \left(\begin{array}{l}\text{Flow,}\\ \text{cu m/day}\end{array}\right)\left(\begin{array}{l}\text{BOD,}\\ \text{mg/L}\end{array}\right)\left(\frac{1,000 \text{ L}}{1 \text{ cu m}}\right)\left(\frac{1 \text{ gm}}{1,000 \text{ mg}}\right)$$

$$= (4,000 \text{ cu m/day})(150 \text{ mg/L})\left(\frac{1,000 \text{ L}}{1 \text{ cu m}}\right)\left(\frac{1 \text{ gm}}{1,000 \text{ mg}}\right)$$

$$= 600,000 \text{ gm BOD/day}$$

3. Calculate the organic (BOD) loading in grams per day per square meter.

$$\begin{array}{l}\text{Organic (BOD) Loading,}\\ \text{gm/day/sq m}\end{array} = \frac{\text{Loading, gm BOD/day}}{\text{Avg Pond Area, sq m}}$$

$$= \frac{600,000 \text{ gm BOD/day}}{65,000 \text{ sq m}}$$

$$= 9.2 \text{ gm BOD/day/sq m}$$

4. The temporary ponds would not be feasible because the organic (BOD) loading would exceed the desired maximum of 5 grams per day per square meter.

9.11   ELIMINATING ALGAE FROM POND EFFLUENTS

9.110   Treatment Methods

Algae are usually present in the effluent from ponds with continuous discharges. Algae can create undesirable impacts on the receiving waters in terms of a loss of aesthetic values, increased turbidity, suspended solids, and biochemical oxygen demand. It also can cause the development of nuisance conditions. In most areas, the algae in the effluent increase the suspended solids concentration to the point that the NPDES effluent limitation on total suspended solids is exceeded, usually resulting in a permit violation.

Algal growths in polishing ponds can be greatly reduced by using microcrustaceans. When a population is established, a balance develops between the crustaceans and the algal growth, resulting in a clear, algae-free water.

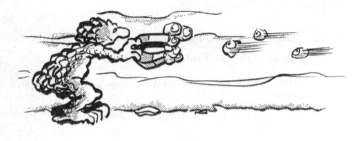

Researchers have attempted to develop cost-effective treatment processes for removing algae from pond effluents. These efforts have included the use of centrifuges, chemical coagulation, filtration, microstraining, magnetic separation, and ultrafiltration. Although some of these processes have been very effective, none have achieved a high level of acceptance due to the increased treatment costs associated with them. Three methods commonly used to eliminate algae from pond effluents are duckweed systems, intermittent sand filters, and land treatment.

9.111   Duckweed Systems

The Lemna Duckweed System is a patented process that uses aquatic duckweed plants for wastewater treatment. This system is used effectively as a polishing pond after a conventional wastewater treatment pond. The duckweed covering the polishing pond's surface prevents the penetration of sunlight and causes the algae to die and settle out of the wastewater being treated. Plastic grids (approximately 10 ft by 10 ft square) are placed on the surface of the pond to prevent the wind from blowing all the duckweed to one side of the pond. The population of duckweed within each grid reproduces and must be harvested on a regular basis for the system to be effective.

Duckweed plants are capable of removing phosphorus and nitrogen from the water. With sufficient detention time (greater than 30 days) and intensive harvesting, significant nutrient removal may be possible. Duckweed needs a water temperature of 50°F (10°C) or greater to be effective. If the water temperature drops below 50°F, duckweed will recover when the water temperature increases to 50°F.

The following are key considerations for duckweed systems:

- Allow for a 30-day retention time, shallow pond depth, and frequent harvesting (every 1 to 3 days during peak season) if nutrient removal is desired

- Have a plan in place for disposing of the harvested plants (that is, processing on site, transportation, and ultimate disposal location)

- Perform post-aeration of the effluent to meet dissolved oxygen (DO) requirements when needed

- Have the capability to draw effluent from several levels and thereby avoid high algal concentrations near the water surface during discharge

9.112   Intermittent Sand Filters

The intermittent sand filter is an outdoor, gravity-actuated, slow-rate filtration system that capitalizes on the availability of land area. It is a biological and physical wastewater treatment mechanism consisting of an underdrained bed of granular material, usually sand. The filter surface is flooded intermittently with lagoon effluent at intervals that permit the surface to drain between applications. It is recommended that the flow be directed to one filter for 24 hours. That filter is then allowed to drain and dry for one to two days while the flow is directed to an adjacent filter. It is preferable to have three filter beds where good operation and treatment can be accomplished over a three-day cycle. The system can remove suspended solids (algae) and BOD as well as convert ammonia to nitrate-nitrogen.

The preferred filter design for lagoon upgrading is as follows:

- Three cells in series, all 30 to 36 inches (76 to 91 cm) deep

- Progressively finer sands (for example, effective sizes of 0.72 mm, 0.40 mm, and 0.17 mm)

- Effluent quality of 10 mg/L of BOD_5 and SS

- Complete nitrification expected, except under extremely cold conditions

This configuration provides the most effective approach for lagoon upgrading (in terms of achieving the most favorable effluent quality attainable).

For additional information about the operation and maintenance of sand filters, see *Small Wastewater System Operation and Maintenance,* Volume I, Chapter 5, "Wastewater Treatment and Effluent Discharge Methods."

A 28-minute instructional video on intermittent sand filters is available from Orenco Systems, Inc. The video provides installation and operational guidelines as well as routine and preventive maintenance procedures. The video is available for viewing on the Orenco website at http://www.orenco.com/videos/sales/ISF_Video.html.

9.113   Land Treatment

Another method of improving effluent quality from waste stabilization ponds is to apply the pond effluent to an overland-flow treatment system followed by a constructed wetland wastewater treatment system.

Overland-flow systems consist of grass planted on slowly permeable soil and slopes typically ranging from 2 to 8 percent. The slope lengths range from 100 to 200 feet (30 to 60 m). The application rate is expressed in inches per day and depends on treatment requirements and local conditions. Application periods usually range from 6 to 12 hours per day. BOD removals are in the 75 to 90 percent range, and suspended solids are in the 50 to 90 percent range. For additional information on overland-flow treatment systems, see *Advanced Waste Treatment*, Chapter 8, "Wastewater Reclamation and Reuse," in this series of operator training manuals.

If higher levels of effluent quality are desired, effluent from an overland-flow system can be treated by constructed wetlands. Wetlands systems often consist of two cells that are either the free water surface (FWS) type or the horizontal subsurface flow (HSSF) type. The vegetation in the wetland system may consist mainly of cattails. If the cattails die off in cold winter months, the organic debris is partially burnt and the remaining unburned material is removed to prevent plugging of the drains. For additional information on constructed wetlands, see Chapter 13, "Alternative Wastewater Treatment, Discharge, and Reuse Methods."

9.114 Straw

There are many different kinds or species of algae. Field studies have shown that barley straw can be effective in controlling blue-green algae. Blue-green algae can be controlled by the active agent produced during the decomposition of barley straw. Wheat straw works, though not as well as barley straw. Straw should be applied to the final wastewater stabilization pond twice a year, once in the autumn and once again in the spring before algal growth starts. Recommended applications are about 77 pounds of straw per million gallons of water in the pond (5 grams of straw per cubic meter). More straw is needed if the water in the pond changes quickly. Water quality indicators should be monitored to ensure desired performance of the straw.

QUESTIONS

Please write your answers to the following questions and compare them with those on page 47.

9.11A What are the undesirable effects of algae on the receiving waters?

9.11B How does the Lemna Duckweed System eliminate algae from pond effluents?

9.11C What types of land treatment systems are used to improve effluent quality from waste stabilization ponds?

9.12 ARITHMETIC ASSIGNMENT

Turn to the Arithmetic Appendix at the back of this manual. Work the example problems in Section A.20, "Wastewater Stabilization Ponds," and check the arithmetic using a calculator. You should be able to get the same answers.

9.13 ACKNOWLEDGMENT

Liberal use has been made of the many papers presented by Professor W. J. Oswald of the University of California at Berkeley on the subject of the treatment of wastes by ponding.

9.14 ADDITIONAL READING

1. *"MOP 11,"* Chapter 23, "Natural Biological Processes."* *Operation of Municipal Wastewater Treatment Plants* (MOP 11). Obtain from Water Environment Federation (WEF), Publications Order Department, 601 Wythe Street, Alexandria, VA 22314-1994. Order No. WPM711. Price to members, $210.00; nonmembers, $248.00; plus shipping and handling.

2. *"New York Manual,"* page 5-43, "Stabilization Ponds."* *Manual of Instruction for Wastewater Treatment Plant Operators* (two-volume set), published by Health Education Services. No longer in print.

3. *"Texas Manual,"* Chapter 16, "Stabilization Ponds."* *Manual of Wastewater Treatment,* published by Texas Water Utilities Association. No longer in print.

4. *Stabilization Pond Operation and Maintenance.* Obtain from Minnesota Pollution Control Agency, Fiscal Services, 6th Floor, 520 Lafayette Road North, St. Paul, MN 55155. Price, $15.00.

5. *Operations Manual—Stabilization Ponds,* US Environmental Protection Agency. EPA No. 430-9-77-012. Obtain from National Technical Information Service (NTIS), 5301 Shawnee Road, Alexandria, VA 22312. Order No. PB279443. Price, $48.00, plus $6.00 shipping and handling.

*Depends on edition.

9.15 METRIC CALCULATIONS

Refer to Section 9.109, "Pond Loading (English and Metric Calculations)."

END OF LESSON 3 OF 3 LESSONS

on

WASTEWATER STABILIZATION PONDS

Please answer the discussion and review questions next.

DISCUSSION AND REVIEW QUESTIONS

Chapter 9.   WASTEWATER STABILIZATION PONDS

(Lesson 3 of 3 Lessons)

Please write your answers to the following questions to determine how well you understand the material in the lesson. The question numbering continues from Lesson 2.

14. Why is it desirable for a pond to be isolated from neighbors?

15. How can scum be prevented from leaving a pond?

16. How can erosion of levee slopes be controlled?

Use the following information to answer questions 17–20: A pond receives an inflow of 0.01 MGD from 100 people. The pond is 150 feet long, 150 feet wide, and 4 feet deep.

Influent BOD is 200 mg/L. Determine the following loading criteria:

17. Determine the detention time in days.

18. Estimate the population loading in persons per acre.

19. Calculate hydraulic loading in inches per day.

20. Calculate the organic (BOD) loading in pounds per day per acre.

21. What methods are commonly used to eliminate algae from pond effluents?

SUGGESTED ANSWERS

Chapter 9.   WASTEWATER STABILIZATION PONDS

ANSWERS TO QUESTIONS IN LESSON 1

Answers to questions on page 13.

9.1A   Wastewater stabilization ponds are advantageous for small installations where land is not costly and the location is isolated from residential, commercial, and recreational areas.

9.1B   The following are the limitations of ponds:

1. May produce odors
2. Require a large amount of land
3. Treat wastes inconsistently depending on climatic conditions
4. May contaminate groundwaters if not properly lined
5. May have high levels of suspended solids in the effluent
6. May be slow to recover from shock loads
7. Have limited process control capabilities
8. May require supplemental aeration equipment to achieve odor-free treatment

Answers to questions on page 15.

9.2A   Aerobic ponds have dissolved oxygen (DO) distributed throughout the pond; anaerobic ponds do not contain any DO. Most ponds are facultative and have aerobic conditions (have DO) on the surface and anaerobic conditions (no DO) on the bottom.

9.2B   The use of a pond will vary depending on the detention period. Ponds with a detention time of less than three days will act like sedimentation tanks. In ponds with a detention period from three to twenty days, the organic material in the influent will be converted to algae, and high concentrations of algae will be found in the effluent. Ponds with longer detention periods provide time for the reduction of algae and a better effluent due to a reduction in nutrients in the water that sustain algae.

Answers to questions on page 17.

9.3A   Organic matter in a pond is converted by aerobic bacteria to carbon dioxide and ammonia, which serve as nutrients for algae. The organic matter in anaerobic bottom sections is first converted by a group of organisms called "acid producers" to carbon dioxide, nitrogen, and organic acids. Next, a group called the "methane fermenters" breaks down the organic acids and other products of the first group to form methane gas. Another end product of organic reduction is water.

9.3B   Algae produce oxygen from the oxygen in the carbon dioxide molecule (CO_2) through photosynthesis.

9.3C   Algae occur naturally in a pond without seeding. They are found in soil, water, and air and multiply greatly under favorable conditions.

Answers to questions on page 18.

9.4A   Biological factors influencing the treatment efficiency of a pond include the type of bacteria, type and quantity of algae, activity of organisms, nutrient deficiencies, and toxic concentrations.

9.4B   Bioflocculation is the clumping together of fine, dispersed organic particles by the action of certain bacteria and algae into settleable organic solids.

9.4C   A pond is not functioning properly when it creates a visual or odor nuisance, or leaves a high BOD, solids, grease, or coliform-group bacteria concentration in the effluent unless it was designed to be anaerobic in the first stages and aerobic in later ponds for final treatment.

ANSWERS TO QUESTIONS IN LESSON 2

Answers to questions on page 19.

9.5A   At least one foot of water should cover the pond bottom before wastes are introduced to prevent decomposing solids from being exposed and causing odor problems.

9.5B   Ponds should be started during warmer months because higher temperatures are associated with efficient treatment processes.

9.5C   A definite green color in a pond indicates a flourishing algae population and is a good sign.

9.5D   When bubbles are observed coming to the pond surface near the inlet, this indicates that the solids that settled to the bottom are being decomposed anaerobically by bacterial action.

Answers to questions on page 23.

9.6A   Scum should not be allowed to accumulate on the surface of a pond because it is unsightly, may prevent sunlight from reaching the algae, and an odor-producing species of algae may develop on the scum. Also, scum can become a source of botulism.

9.6B   Scum accumulations may be broken up with rakes, jets of water, or by use of outboard motors.

9.6C   Most odors are caused in ponds by overloading or poor housekeeping.

9.6D   To prepare for an odor problem, a careful plan for emergency odor control must be developed. For example, an odor control chemical should be available before an odor problem develops. Sodium nitrate or a floating aerator will help control odors and improve treatment of the wastewater.

9.6E   Weeds may be controlled by removing them manually or by using herbicides and soil sterilants.

9.6F   Weeds are objectionable in and around ponds because they provide a shelter for the breeding of mosquitoes, promote scum accumulation, and also hinder pond circulation.

9.6G   Insects should be controlled because they may, in sufficient numbers, be a serious nuisance to nearby residential areas, farm workers, recreation sites, industrial plants, and drivers on highways.

9.6H   A pond should be lowered before the application of an insecticide to improve the destruction of insects and reduce the effect of the insecticide on the receiving waters by holding the wastewater at least one day. Lowering of the pond also will dry up weeds and insects.

Answers to questions on page 28.

9.6I   The contents of ponds are recirculated to allow algae and other aerobic organisms to become thoroughly mixed with incoming raw wastewater.

9.6J   Ponds that are operated on a batch basis may discharge only once (fall) or twice (fall and spring) a year.

9.6K   To develop an operating strategy for ponds, consider the following factors:

1. Maintain constant water elevations in the ponds.
2. Distribute inflow equally to ponds.
3. Keep pond levees or dikes in good condition.
4. Observe and test pond condition on a routine schedule and time for dissolved oxygen, pH, and temperature.
5. Place baffles properly.
6. Perform troubleshooting as needed.

Answers to questions on page 32.

9.7A   Surface aerators have been used in two types of applications: (1) to provide additional air for ponds during the night, during cold weather, and for overloaded ponds; and (2) to provide a mechanical aeration device for ponds operated as aerated lagoons.

9.8A   pH, temperature, and dissolved oxygen should be measured to provide a record of pond performance, to indicate the status (health) of the pond, and to determine whether corrective action is or may be necessary. DO may be lower at sunrise and will get progressively higher as the day goes on.

9.8B   Generally, when a pond turns dull green or colorless, the pH is declining and the DO content is low. A gray color indicates that the pond is overloaded or not working properly.

9.8C   The influent BOD to a series of ponds is 200 mg/L. If the BOD in the effluent of the last pond is 40 mg/L, what is the BOD removal efficiency?

Known	Unknown
BOD In, mg/L = 200 mg/L	BOD Removal, %
BOD Out, mg/L = 40 mg/L	

Calculate the pond BOD removal efficiency.

$$\text{BOD Removal, \%} = \frac{(\text{In} - \text{Out})}{\text{In}} \times 100\%$$

$$= \frac{(200 \text{ mg/L} - 40 \text{ mg/L})}{200 \text{ mg/L}} \times 100\%$$

$$= \frac{160 \text{ mg/L}}{200 \text{ mg/L}} \times 100\%$$

$$= 0.80 \times 100\%$$

$$= 80\%$$

Answers to questions on page 33.

9.9A   Catwalks located over ponds should have guardrails and nonskid walking surfaces.

9.9B   An operator should be accompanied by a helper when performing any dangerous task because immediate aid might be needed to prevent serious injury or loss of life.

ANSWERS TO QUESTIONS IN LESSON 3

Answers to questions on page 37.

9.10A   Some wastes are not easily treated by ponds because they contain substances with interfering concentrations that hinder algal or bacterial growth.

9.10B   The influent to a pond should be metered to justify budgets, indicate unexpected fluctuations in flows that may cause upsets, and provide data for future expansion when necessary. Influent flows also are used to determine pond detention time and to determine pond loadings.

9.10C   The inlet to a pond should be submerged to distribute the heat of the influent as much as possible and to minimize the occurrence of floating material.

9.10D   The outlet of a pond should be submerged to prevent the discharge of floating material.

9.10E   Free overfalls should be avoided to minimize odors, foaming, and gas entrapment, which may hamper the flow of water in pipes. They are generally controlled with pipes at the outfall.

9.10F   The discharge of floating material over a surface outlet may be corrected by constructing a baffle around the outlet.

9.10G   The minimum recommended pond operating depth is three feet (one meter). At shallower depths, aquatic weeds become a nuisance and pond performance is apt to be irregular.

Answers to question on page 40.

9.10H   A pond receives a flow of 2.0 MGD from 20,000 people. The influent BOD is 180 mg/L, the pond surface area is 24 acres, and the average operating depth is four feet. Determine the detention time in days, the organic (BOD) loading in pounds per day per acre, the population loading in persons per acre, and the hydraulic loading in inches per day.

Known

Flow, MGD	= 2.0 MGD
Flow, GPD	= 2,000,000 GPD
Population Served, persons	= 20,000 persons
BOD, mg/L	= 180 mg/L
Pond Surface Area, ac	= 24 ac
Avg Depth, ft	= 4 ft
One Acre	= 43,560 sq ft

Unknown

1. Detention Time, days
2. Organic (BOD) Loading, lbs/day/ac
3. Population Loading, persons/ac
4. Hydraulic Loading, in/day

1. Calculate the detention time in days.

   a. Determine the pond volume in acre-feet.

   $$\text{Pond Volume, ac-ft} = (\text{Avg Area, ac})(\text{Avg Depth, ft})$$

   $$= (24 \text{ ac})(4 \text{ ft})$$

   $$= 96 \text{ ac-ft}$$

   b. Convert flow from gallons per day to acre-feet per day.

   $$\text{Flow Rate, ac-ft/day} = \frac{\text{Flow, GPD}}{(7.48 \text{ gal/cu ft})(43,560 \text{ sq ft/ac})}$$

   $$= \frac{2,000,000 \text{ GPD}}{(7.48 \text{ gal/cu ft})(43,560 \text{ sq ft/ac})}$$

   $$= 6.1 \text{ ac-ft/day}$$

c. Calculate the detention time in days.

$$\text{Detention Time, days} = \frac{\text{Pond Volume, ac-ft}}{\text{Flow Rate, ac-ft/day}}$$

$$= \frac{96 \text{ ac-ft}}{6.1 \text{ ac-ft/day}}$$

$$= 15.7 \text{ days}$$

2. Calculate the organic (BOD) loading in pounds per day per acre.

$$\text{Organic (BOD) Loading, lbs/day/ac} = \frac{(\text{Flow, MGD})(\text{BOD, mg/L})(8.34 \text{ lbs/gal})}{\text{Avg Pond Area, ac}}$$

$$= \frac{(2.0 \text{ MGD})(180 \text{ mg/L})(8.34 \text{ lbs/gal})}{24 \text{ ac}}$$

$$= 125 \text{ lbs BOD/day/ac}$$

3. Estimate the population loading in persons per acre.

$$\text{Population Loading, persons/ac} = \frac{\text{Population Served, persons}}{\text{Pond Surface Area, ac}}$$

$$= \frac{20,000 \text{ persons}}{24 \text{ ac}}$$

$$= 833 \text{ persons/ac}$$

4. Calculate the hydraulic loading in inches per day.

$$\text{Hydraulic Loading, in/day} = \frac{(\text{Depth of Pond, ft})(12 \text{ in/ft})}{\text{Detention Time, days}}$$

$$= \frac{(4 \text{ ft})(12 \text{ in/ft})}{15.7 \text{ days}}$$

$$= \frac{48}{15.7}$$

$$= 3.06 \text{ in/day}$$

Answers to questions on page 43.

9.11A Algae can create undesirable effects on the receiving waters in terms of a loss of aesthetic values, increased turbidity, suspended solids, and biochemical oxygen demand; algae can also cause the development of nuisance conditions.

9.11B The Lemna Duckweed System eliminates algae from pond effluents by covering the pond's surface with duckweed, thus preventing the penetration of sunlight and causing algae to die and settle out of the wastewater being treated.

9.11C Land treatment systems used to improve effluent quality from stabilization ponds consist of applying the effluent to an overland-flow treatment system followed by a constructed wetland wastewater treatment system.

APPENDIX

Monthly Data Sheet

CLEANWATER, U.S.A.
WATER POLLUTION CONTROL PLANT

MONTHLY RECORD _____ 20__

OPERATOR: _____

DATE	DAY	WEATHER	TIME OF VISIT	FLOW (MGD) EFFLUENT	TEMP °F NO.1 POND	TEMP °F NO.2 POND	pH NO.1 POND	pH NO.2 POND	D.O. NO.1 POND	D.O. NO.2 POND	B.O.D. INFLUENT	B.O.D. EFFLUENT	CL2 FEED RATE LB./DAY	CL2 RESIDUAL MG./L.	E-COLI M.P.N./100	MIXING HRS NO.1 POND	MIXING HRS NO.2 POND	REMARKS
2	W	OVER CAST	9:30 A.M.	0.025	68	70	7.6	7.9	0.8	1.4	183	32	35	3.2	62	-0-	-0-	EFFL. SS = 85 mg/l
2	W	CLEAR	2:00 P.M.	0.038	76	79	7.7	8.4	1.3	8.7			35	1.4				
3	T																	
4	F																	
5	S																	
6	S																	
7	M	CLEAR	10:00 A.M.	0.026	70	74	7.6	8.1	1.2	7.6			35					
8	T	CLEAR	10:30 A.M.	0.027	69	73	7.6	8.1	1.1	7.8	158	29			2400	-0-	-0-	
9	W																	EFFL. SS = 53 mg/l
10	T	CLEAR HOT	3:15 P.M.	0.039	77	80	7.6	8.3	1.5	11.2			35	0.8				
11	F																	
12	S																	
13	S																	
14	M	CLEAR	9:30 A.M.	0.023	70	75	7.8	8.5	1.9	7.6	162	33						
15	T													40	4.2				Increased Cl₂ feed rate to 40#/day
16	W	CLEAR	1:30 P.M.	0.031	75	79	7.8	8.4	1.7	10.6			40	4.2	700	-0-	-0-	EFFL. SS = 37 mg/l
17	T											40	2.5	23			
18	F	CLEAR	3:30 P.M.	0.037	79	85	7.8	8.6	2.1	14.5			40					
19	S													40	2.8				
20	S																	
21	M																	Declared holiday — No plant check
22	W	CLEAR	9:00 A.M.	0.020	71	76	7.6	8.4	1.0	7.8	178	30	40	5.6	6	-0-	-0-	
23	W	CLEAR	10:30 A.M.	0.028	72	78	7.7	8.4	1.0	8.1			40	5.0	6			EFFL. SS = 48 mg/l
24	T																	
25	F	CLEAR	2:15 P.M.	0.035	80	87	7.8	8.5	1.3	11.4			40	3.8				
26	S																	
27	S																	
28	M	CLEAR	10:00 A.M.	0.026	73	79	7.7	8.5	1.1	8.6			40	4.7				Cl₂ feed rate back to 35#/day
29	T													35	1.1				
30	W	CLEAR	4:15 P.M.	0.038	81	88	7.8	8.5	1.8	15.3	168	35	35	1.1	6	-0-	-0-	EFFL. SS = 67 mg/l
31																		
MAX.				0.039	81	88	7.8	8.6	2.1	15.3	183	35	40	5.6	2400	-0-	-0-	
MIN.				0.020	68	70	7.6	7.9	0.8	7.6	162	29	35	0.8	6	-0-	-0-	
AVG.		CLEAR		0.030	73	78	7.7	8.3	1.3	9.2	169	31	37.2	2.7	62	-0-	-0-	

Ponds operated in series all month.
Mixers not operated during month.

SUMMARY DATA

% REMOVAL B.O.D.	81.6 %
LBS. B.O.D./ACRE/DAY	52.8
DETENTION TIME — DAYS	86.7

COST DATA

MAN DAYS 4.5 (2.5 mo.) PAYROLL	245.52
POWER PURCHASED	69.60
OTHER UTILITIES (GAS, H2O)	NONE
GASOLINE, OIL, GREASE	NONE
CHEMICALS & SUPPLIES	138.60
MAINTENANCE	NONE
VEHICLE COSTS	57.40
OTHER	NONE
TOTAL	$ 511.12
OPERATING COST/CAPITA/MO.	$ 1.68
COST/1000 GAL. TREATED	$ 0.55

FLOW METER:
LAST 232602
1st 231672
TOTAL 0.930 MG

ELECTRIC METER:
LAST 5978
1st 5817
MULT. 40 x 161 = 6440 KWH

CHAPTER 10

ACTIVATED SLUDGE

Package Plants and Oxidation Ditches

by

John Brady

TABLE OF CONTENTS
Chapter 10.   ACTIVATED SLUDGE
Package Plants and Oxidation Ditches

OBJECTIVES

Chapter 10.   ACTIVATED SLUDGE

Package Plants and Oxidation Ditches

1. Explain the principles of the activated sludge process and the factors that influence and control the process.

2. Inspect a new activated sludge facility for proper installation.

3. Place a new activated sludge process into service.

4. Schedule and conduct operation and maintenance duties.

5. Collect samples, interpret lab results, and make appropriate adjustments in treatment processes.

6. Recognize factors that indicate an activated sludge process is not performing properly, identify the source of the problem, and take corrective action.

7. Perform your job duties in a safe manner.

8. Determine aerator loadings and understand the application of different loading guidelines.

9. Maintain records for an activated sludge plant.

10. Review plans and specifications for an activated sludge plant.

WORDS
Chapter 10.   ACTIVATED SLUDGE
Package Plants and Oxidation Ditches

ABSORPTION (ab-SORP-shun)   ABSORPTION

The taking in or soaking up of one substance into the body of another by molecular or chemical action (as tree roots absorb dissolved nutrients in the soil).

ACTIVATED SLUDGE   ACTIVATED SLUDGE

Sludge particles produced in raw or settled wastewater (primary effluent) by the growth of organisms (including zoogleal bacteria) in aeration tanks in the presence of dissolved oxygen. The term activated comes from the fact that the particles are teeming with bacteria, fungi, and protozoa. Activated sludge is different from primary sludge in that the sludge particles contain many living organisms that can feed on the incoming wastewater.

ACTIVATED SLUDGE PROCESS   ACTIVATED SLUDGE PROCESS

A biological wastewater treatment process that speeds up the decomposition of wastes in the wastewater being treated. Activated sludge is added to wastewater and the mixture (mixed liquor) is aerated and agitated. After some time in the aeration tank, the activated sludge is allowed to settle out by sedimentation and is disposed of (wasted) or reused (returned to the aeration tank) as needed. The remaining wastewater then undergoes more treatment.

ADSORPTION (add-SORP-shun)   ADSORPTION

The gathering of a gas, liquid, or dissolved substance on the surface or interface zone of another material.

AERATION (air-A-shun) LIQUOR   AERATION LIQUOR

Mixed liquor. The contents of the aeration tank, including living organisms and material carried into the tank by either untreated wastewater or primary effluent.

AERATION (air-A-shun) TANK   AERATION TANK

The tank where raw or settled wastewater is mixed with return sludge and aerated. The same as aeration bay, aerator, or reactor.

AEROBES   AEROBES

Bacteria that must have dissolved oxygen (DO) to survive. Aerobes are aerobic bacteria.

AEROBIC BACTERIA (air-O-bick back-TEER-e-uh)   AEROBIC BACTERIA

Bacteria that will live and reproduce only in an environment containing oxygen that is available for their respiration (breathing), namely atmospheric oxygen or oxygen dissolved in water. Oxygen combined chemically, such as in water molecules (H_2O), cannot be used for respiration by aerobic bacteria.

AEROBIC (air-O-bick) DIGESTION   AEROBIC DIGESTION

The breakdown of wastes by microorganisms in the presence of dissolved oxygen. This digestion process may be used to treat only waste activated sludge, or trickling filter sludge and primary (raw) sludge, or waste sludge from activated sludge treatment plants designed without primary settling. The sludge to be treated is placed in a large aerated tank where aerobic microorganisms decompose the organic matter in the sludge. This is an extension of the activated sludge process.

AGGLOMERATION (uh-glom-er-A-shun)   AGGLOMERATION

The growing or coming together of small scattered particles into larger flocs or particles, which settle rapidly. Also see FLOC.

AIR LIFT PUMP

AIR LIFT PUMP

A special type of pump consisting of a vertical riser pipe submerged in the wastewater or sludge to be pumped. Compressed air is injected into a tail piece at the bottom of the pipe. Fine air bubbles mix with the wastewater or sludge to form a mixture lighter than the surrounding water, which causes the mixture to rise in the discharge pipe to the outlet.

ANAEROBES

ANAEROBES

Bacteria that do not need dissolved oxygen (DO) to survive.

ANAEROBIC (AN-air-O-bick)

ANAEROBIC

A condition in which atmospheric or dissolved oxygen (DO) is *NOT* present in the aquatic (water) environment.

ANODE (AN-ode)

ANODE

The positive pole or electrode of an electrolytic system, such as a battery. The anode attracts negatively charged particles or ions (anions).

ANOXIC (an-OX-ick)

ANOXIC

A condition in which the aquatic (water) environment does not contain dissolved oxygen (DO), which is called an oxygen deficient condition. Generally refers to an environment in which chemically bound oxygen, such as in nitrate, is present. The term is similar to ANAEROBIC.

ASHING

ASHING

Formation of an activated sludge floc in a clarifier effluent that is well oxidized and floats on the water surface (has the appearance of gray ash).

BOD (pronounce as separate letters)

BOD

Biochemical Oxygen Demand. The rate at which organisms use the oxygen in water or wastewater while stabilizing decomposable organic matter under aerobic conditions. In decomposition, organic matter serves as food for the bacteria and energy results from its oxidation. BOD measurements are used as a surrogate measure of the organic strength of wastes in water.

BOD$_5$

BOD$_5$

BOD$_5$ refers to the five-day biochemical oxygen demand. The total amount of oxygen used by microorganisms decomposing organic matter increases each day until the ultimate BOD is reached, usually in 50 to 70 days. BOD usually refers to the five-day BOD or BOD$_5$.

BACTERIAL (back-TEER-e-ul) CULTURE

BACTERIAL CULTURE

In the case of activated sludge, the bacterial culture refers to the group of bacteria classified as AEROBES and FACULTATIVE BACTERIA, which covers a wide range of organisms. Most treatment processes in the United States grow facultative bacteria that use the carbonaceous (carbon compounds) BOD. Facultative bacteria can live when oxygen resources are low. When nitrification is required, the nitrifying organisms are OBLIGATE AEROBES (require oxygen) and must have at least 0.5 mg/L of dissolved oxygen throughout the whole system to function properly.

BATCH PROCESS

BATCH PROCESS

A treatment process in which a tank or reactor is filled, the wastewater (or other solution) is treated or a chemical solution is prepared, and the tank is emptied. The tank may then be filled and the process repeated. Batch processes are also used to cleanse, stabilize, or condition chemical solutions for use in industrial manufacturing and treatment processes.

BIOASSAY (BUY-o-AS-say)

BIOASSAY

(1) A way of showing or measuring the effect of biological treatment on a particular substance or waste.

(2) A method of determining the relative toxicity of a test sample of industrial wastes or other wastes by using live test organisms, such as fish.

BIOMASS (BUY-o-mass)

BIOMASS

A mass or clump of organic material consisting of living organisms feeding on the wastes in wastewater, dead organisms, and other debris. Also see ZOOGLEAL MASS.

BULKING BULKING

Clouds of billowing sludge that occur throughout secondary clarifiers and sludge thickeners when the sludge does not settle properly. In the activated sludge process, bulking is usually caused by filamentous bacteria or bound water.

BURPING BURPING

A term used to describe what happens when billowing solids are swept by the water up and out over the effluent weirs in the discharged effluent. Billowing solids result when the settling tank sludge blanket becomes too deep (occupies too much volume in the bottom of the tank).

COD (pronounce as separate letters) COD

Chemical Oxygen Demand. A measure of the oxygen-consuming capacity of organic matter present in wastewater. COD is expressed as the amount of oxygen consumed from a chemical oxidant in mg/L during a specific test. Results are not necessarily related to the biochemical oxygen demand (BOD) because the chemical oxidant may react with substances that bacteria do not stabilize.

CATHODIC (kath-ODD-ick) PROTECTION CATHODIC PROTECTION

An electrical system for prevention of rust, corrosion, and pitting of metal surfaces that are in contact with water, wastewater, or soil. A low-voltage current is made to flow through a liquid (water) or a soil in contact with the metal in such a manner that the external electromotive force renders the metal structure cathodic. This concentrates corrosion on auxiliary anodic parts, which are deliberately allowed to corrode instead of letting the structure corrode.

CENTRIFUGE CENTRIFUGE

A mechanical device that uses centrifugal or rotational forces to separate solids from liquids.

COAGULATION (ko-agg-yoo-LAY-shun) COAGULATION

The clumping together of very fine particles into larger particles (floc) caused by the use of chemicals (coagulants). The chemicals neutralize the electrical charges of the fine particles, allowing them to come closer and form larger clumps. This clumping together makes it easier to separate the solids from the water by settling, skimming, draining, or filtering.

COMMINUTOR (kom-mih-NEW-ter) COMMINUTOR

A device used to reduce the size of the solid materials in wastewater by shredding (comminution). The shredding action is like many scissors cutting to shreds all the large solids in the wastewater.

CONTACT STABILIZATION CONTACT STABILIZATION

Contact stabilization is a modification of the conventional activated sludge process. In contact stabilization, two aeration tanks are used. One tank is for separate reaeration of the return sludge for at least four hours before it is permitted to flow into the other aeration tank to be mixed with the primary effluent requiring treatment. The process may also occur in one long tank.

DENITRIFICATION (dee-NYE-truh-fuh-KAY-shun) DENITRIFICATION

(1) The anoxic biological reduction of nitrate nitrogen to nitrogen gas.

(2) The removal of some nitrogen from a system.

(3) An anoxic process that occurs when nitrite or nitrate ions are reduced to nitrogen gas and nitrogen bubbles are formed as a result of this process. The bubbles attach to the biological floc and float the floc to the surface of the secondary clarifiers. This condition is often the cause of rising sludge observed in secondary clarifiers or gravity thickeners. Also see NITRIFICATION.

DIFFUSED-AIR AERATION DIFFUSED-AIR AERATION

A diffused-air activated sludge plant takes air, compresses it, and then discharges the air below the water surface of the aerator through some type of air diffusion device.

DIFFUSER DIFFUSER

A device (porous plate, tube, bag) used to break the air stream from the blower system into fine bubbles in an aeration tank or reactor.

DISSOLVED OXYGEN DISSOLVED OXYGEN

Molecular oxygen dissolved in water or wastewater, usually abbreviated DO.

ENDOGENOUS (en-DODGE-en-us) RESPIRATION

ENDOGENOUS RESPIRATION

A situation in which living organisms oxidize some of their own cellular mass instead of new organic matter they adsorb or absorb from their environment.

F/M RATIO

F/M RATIO

See FOOD/MICROORGANISM RATIO.

FACULTATIVE (FACK-ul-tay-tive) BACTERIA

FACULTATIVE BACTERIA

Facultative bacteria can use either dissolved oxygen or oxygen obtained from food materials such as sulfate or nitrate ions. In other words, facultative bacteria can live under aerobic, anoxic, or anaerobic conditions.

FILAMENTOUS (fill-uh-MEN-tuss) ORGANISMS

FILAMENTOUS ORGANISMS

Organisms that grow in a thread or filamentous form. Common types are *Thiothrix* and *Actinomycetes*. A common cause of sludge bulking in the activated sludge process.

FLOC

FLOC

Clumps of bacteria and particles or coagulants and impurities that have come together and formed a cluster. Found in aeration tanks, secondary clarifiers, and chemical precipitation processes.

FOOD/MICROORGANISM (F/M) RATIO

FOOD/MICROORGANISM (F/M) RATIO

Food to microorganism ratio. A measure of food provided to bacteria in an aeration tank.

$$\frac{\text{Food}}{\text{Microorganisms}} = \frac{\text{BOD, lbs/day}}{\text{MLVSS, lbs}}$$

$$= \frac{\text{Flow, MGD} \times \text{BOD, mg/L} \times 8.34 \text{ lbs/gal}}{\text{Volume, MG} \times \text{MLVSS, mg/L} \times 8.34 \text{ lbs/gal}}$$

or by calculator math system

$$= \text{Flow, MGD} \times \text{BOD, mg/L} \div \text{Volume, MG} \div \text{MLVSS, mg/L}$$

or metric

$$= \frac{\text{BOD, kg/day}}{\text{MLVSS, kg}}$$

$$= \frac{\text{Flow, ML/day} \times \text{BOD, mg/L} \times 1 \text{ kg/M mg}}{\text{Volume, ML} \times \text{MLVSS, mg/L} \times 1 \text{ kg/M mg}}$$

HEADER

HEADER

A large pipe to which the ends of a series of smaller pipes are connected. Also called a manifold.

LINEAL (LIN-e-ul)

LINEAL

The length in one direction of a line. For example, a board 12 feet (meters) long has 12 lineal feet (meters) in its length.

MANIFOLD

MANIFOLD

A large pipe to which the ends of a series of smaller pipes are connected. Also called a header.

MECHANICAL AERATION

MECHANICAL AERATION

The use of machinery to mix air and water so that oxygen can be absorbed into the water. Some examples are: paddle wheels, mixers, or rotating brushes to agitate the surface of an aeration tank; pumps to create fountains; and pumps to discharge water down a series of steps forming falls or cascades.

MICROORGANISMS (MY-crow-OR-gan-is-ums)

MICROORGANISMS

Very small organisms that can be seen only through a microscope. Some microorganisms use the wastes in wastewater for food and thus remove or alter much of the undesirable matter.

MIXED LIQUOR

MIXED LIQUOR

When the activated sludge in an aeration tank is mixed with primary effluent or the raw wastewater and return sludge, this mixture is then referred to as mixed liquor as long as it is in the aeration tank. Mixed liquor also may refer to the contents of mixed aerobic or anaerobic digesters.

MIXED LIQUOR SUSPENDED SOLIDS (MLSS)

MIXED LIQUOR SUSPENDED SOLIDS (MLSS)

The amount (mg/L) of suspended solids in the mixed liquor of an aeration tank.

MIXED LIQUOR VOLATILE SUSPENDED SOLIDS (MLVSS)

MIXED LIQUOR VOLATILE SUSPENDED SOLIDS (MLVSS)

The amount (mg/L) of organic or volatile suspended solids in the mixed liquor of an aeration tank. This volatile portion is used as a measure or indication of the microorganisms present.

NPDES PERMIT

NPDES PERMIT

National Pollutant Discharge Elimination System permit is the regulatory agency document issued by either a federal or state agency that is designed to control all discharges of potential pollutants from point sources and stormwater runoff into US waterways. NPDES permits regulate discharges into US waterways from all point sources of pollution, including industries, municipal wastewater treatment plants, sanitary landfills, large animal feedlots, and return irrigation flows.

NITRIFICATION (NYE-truh-fuh-KAY-shun)

NITRIFICATION

An aerobic process in which bacteria change the ammonia and organic nitrogen in wastewater into oxidized nitrogen (usually nitrate). The second-stage BOD is sometimes referred to as the nitrogenous BOD (first-stage BOD is called the carbonaceous BOD). Also see DENITRIFICATION.

OBLIGATE AEROBES

OBLIGATE AEROBES

Bacteria that must have atmospheric or dissolved molecular oxygen to live and reproduce.

OXIDATION

OXIDATION

Oxidation is the addition of oxygen, removal of hydrogen, or the removal of electrons from an element or compound; in the environment and in wastewater treatment processes, organic matter is oxidized to more stable substances. The opposite of REDUCTION.

POLYELECTROLYTE (POLY-ee-LECK-tro-lite)

POLYELECTROLYTE

A high-molecular-weight (relatively heavy) substance, having points of positive or negative electrical charges, that is formed by either natural or synthetic (manmade) processes. Natural polyelectrolytes may be of biological origin or obtained from starch products or cellulose derivatives. Synthetic polyelectrolytes consist of simple substances that have been made into complex, high-molecular-weight substances. Used with other chemical coagulants to aid in binding small suspended particles to larger chemical flocs for their removal from water. Often called a polymer.

POLYMER (POLY-mer)

POLYMER

A long-chain molecule formed by the union of many monomers (molecules of lower molecular weight). Polymers are used with other chemical coagulants to aid in binding small suspended particles to larger chemical flocs for their removal from water. Also see POLYELECTROLYTE.

PROTOZOA (pro-toe-ZOE-ah)

PROTOZOA

A group of motile, microscopic organisms (usually single-celled and aerobic) that sometimes cluster into colonies and generally consume bacteria as an energy source.

REAGENT (re-A-gent)

REAGENT

A pure, chemical substance that is used to make new products or is used in chemical tests to measure, detect, or examine other substances.

REDUCTION (re-DUCK-shun)
REDUCTION

Reduction is the addition of hydrogen, removal of oxygen, or the addition of electrons to an element or compound. Under anaerobic conditions (no dissolved oxygen present), sulfur compounds are reduced to odor-producing hydrogen sulfide (H_2S) and other compounds. In the treatment of metal finishing wastewaters, hexavalent chromium (Cr^{6+}) is reduced to the trivalent form (Cr^{3+}). The opposite of OXIDATION.

RISING SLUDGE
RISING SLUDGE

Rising sludge occurs in the secondary clarifiers of activated sludge plants when the sludge settles to the bottom of the clarifier, is compacted, and then starts to rise to the surface, usually as a result of denitrification, or anaerobic biological activity that produces carbon dioxide or methane.

SECCHI (SECK-key) DISK
SECCHI DISK

A flat, white disk lowered into the water by a rope until it is just barely visible. At this point, the depth of the disk from the water surface is the recorded Secchi disk transparency.

SEPTIC (SEP-tick)
SEPTIC

A condition produced by anaerobic bacteria. If severe, the sludge produces hydrogen sulfide, turns black, gives off foul odors, contains little or no dissolved oxygen, and the wastewater has a high oxygen demand.

SEQUENCING BATCH REACTOR (SBR)
SEQUENCING BATCH REACTOR (SBR)

A type of activated sludge system that is specifically designed and automated to mix/aerate untreated wastewater and allow solids flocculation/separation to occur as a batch treatment process.

SHOCK LOAD (ACTIVATED SLUDGE)
SHOCK LOAD

The arrival at a plant of a waste that is toxic to organisms in sufficient quantity or strength to cause operating problems. Possible problems include odors and bulking sludge, which will result in a high loss of solids from the secondary clarifiers into the plant effluent and a biological process upset that may require several days to a week to recover. Organic or hydraulic overloads also can cause a shock load.

SHORT-CIRCUITING
SHORT-CIRCUITING

A condition that occurs in tanks or basins when some of the flowing water entering a tank or basin flows along a nearly direct pathway from the inlet to the outlet. This is usually undesirable because it may result in shorter contact, reaction, or settling times in comparison with the theoretical (calculated) or presumed detention times.

SLUDGE AGE
SLUDGE AGE

A measure of the length of time a particle of suspended solids has been retained in the activated sludge process.

$$\text{Sludge Age, days} = \frac{\text{Suspended Solids Under Aeration, lbs or kg}}{\text{Suspended Solids Added, lbs/day or kg/day}}$$

SOLIDS CONCENTRATION
SOLIDS CONCENTRATION

The solids in the aeration tank that carry microorganisms that feed on wastewater. Expressed as milligrams per liter of mixed liquor volatile suspended solids (MLVSS, mg/L).

STABILIZED WASTE
STABILIZED WASTE

A waste that has been treated or decomposed to the extent that, if discharged or released, its rate and state of decomposition would be such that the waste would not cause a nuisance or odors in the receiving water.

STAFF GAUGE
STAFF GAUGE

A ruler or graduated scale used to measure the depth or elevation of water in a channel, tank, or stream.

STANDARDIZE
STANDARDIZE

To compare with a standard.

(1)   In wet chemistry, to find out the exact strength of a solution by comparing it with a standard of known strength. This information is used to adjust the strength by adding more water or more of the substance dissolved.

(2)   To set up an instrument or device to read a standard. This allows you to adjust the instrument so that it reads accurately, or enables you to apply a correction factor to the readings.

STEP-FEED AERATION
STEP-FEED AERATION

Step-feed aeration is a modification of the conventional activated sludge process. In step-feed aeration, primary effluent enters the aeration tank at several points along the length of the tank, rather than at the beginning or head of the tank and flowing through the entire tank in a plug flow mode.

SUPERNATANT (soo-per-NAY-tent)
SUPERNATANT

The relatively clear water layer between the sludge on the bottom and the scum on the surface of an anaerobic digester or septic tank (interceptor).

(1)   From an anaerobic digester, this water is usually returned to the influent wet well or to the primary clarifier.

(2)   From a septic tank, this water is discharged by gravity or by a pump to a leaching system or a wastewater collection system.

Also called clear zone.

TOC (pronounce as separate letters)
TOC

Total Organic Carbon. TOC measures the amount of organic carbon in water.

TURBIDITY (ter-BID-it-tee)
TURBIDITY

The cloudy appearance of water caused by the presence of suspended and colloidal matter. In the waterworks field, a turbidity measurement is used to indicate the clarity of water. Technically, turbidity is an optical property of the water based on the amount of light reflected by suspended particles. Turbidity cannot be directly equated to suspended solids because white particles reflect more light than dark-colored particles and many small particles will reflect more light than an equivalent large particle.

TURBIDITY (ter-BID-it-tee) METER
TURBIDITY METER

An instrument for measuring and comparing the turbidity of liquids by passing light through them and determining how much light is reflected by the particles in the liquid. The normal measuring range is 0 to 100 and is expressed as nephelometric turbidity units (NTUs). Also called a turbidimeter.

WET CHEMISTRY
WET CHEMISTRY

Laboratory procedures used to analyze a sample of water using liquid chemical solutions (wet) instead of, or in addition to, laboratory instruments.

ZOOGLEAL (ZOE-uh-glee-ul) MASS
ZOOGLEAL MASS

Jelly-like masses of bacteria found in both the trickling filter and activated sludge processes. These masses may be formed for or function as the protection against predators and for storage of food supplies. Also see BIOMASS.

(ZOE-uh-glee-ul)

CHAPTER 10.   ACTIVATED SLUDGE

Package Plants and Oxidation Ditches

(Lesson 1 of 3 Lessons)

10.0   THE ACTIVATED SLUDGE PROCESS

10.00   Definitions

"Activated sludge" (Figure 10.1) consists of sludge particles produced in raw or settled wastewater (primary effluent) by the growth of organisms (including zoogleal bacteria) in aeration tanks in the presence of dissolved oxygen. The term "activated" comes from the fact that the particles are teeming with bacteria, fungi, and protozoa.

The "activated sludge process" (Figure 10.2) is a biological wastewater treatment process that uses suspended growth *MICROORGANISMS*[1] to speed up the decomposition of wastes. When activated sludge is added to wastewater, the microorganisms feed and grow on waste in the wastewater. As the organisms grow and reproduce, more and more waste is removed, leaving the wastewater partially cleaned. To function efficiently, the activated sludge process needs a mass of organisms (*SOLIDS CONCENTRATION*[2]) with a steady balance of food (*FOOD/MICROORGANISM RATIO*[3]) and oxygen.

10.01   Wastewater Treatment by Activated Sludge

The activated sludge wastewater treatment process can be used as a standard wastewater treatment system to remove 90 to 98 percent of the organic carbon (carbonaceous portion), as measured by the biochemical oxygen demand test. The process also can be used to pretreat industrial wastes containing very high *BOD*[4] concentrations or be used to remove a portion of the waste load before being discharged to a collection system or other treatment system. The activated sludge process can also be used to remove specific nutrients not removed in standard wastewater treatment such as phosphorus and nitrogen, which

---

1. *Microorganisms* (MY-crow-OR-gan-is-ums).   Very small organisms that can be seen only through a microscope. Some microorganisms use the wastes in wastewater for food and thus remove or alter much of the undesirable matter.
2. *Solids Concentration.*   The solids in the aeration tank that carry microorganisms that feed on wastewater. Expressed as milligrams per liter of mixed liquor volatile suspended solids (MLVSS, mg/L).
3. *Food/Microorganism (F/M) Ratio.*   Food to microorganism ratio. A measure of food provided to bacteria in an aeration tank.

$$\frac{\text{Food}}{\text{Microorganisms}} = \frac{\text{BOD, lbs/day}}{\text{MLVSS, lbs}}$$

$$= \frac{\text{Flow, MGD} \times \text{BOD, mg/L} \times 8.34 \text{ lbs/gal}}{\text{Volume, MG} \times \text{MLVSS, mg/L} \times 8.34 \text{ lbs/gal}}$$

or by calculator math system

$$= \text{Flow, MGD} \times \text{BOD, mg/L} \div \text{Volume, MG} \div \text{MLVSS, mg/L}$$

or metric

$$= \frac{\text{BOD, kg/day}}{\text{MLVSS, kg}}$$

$$= \frac{\text{Flow, ML/day} \times \text{BOD, mg/L} \times 1 \text{ kg/M mg}}{\text{Volume, ML} \times \text{MLVSS, mg/L} \times 1 \text{ kg/M mg}}$$

4. *BOD* (pronounce as separate letters).   Biochemical Oxygen Demand. The rate at which organisms use the oxygen in water or wastewater while stabilizing decomposable organic matter under aerobic conditions. In decomposition, organic matter serves as food for the bacteria and energy results from its oxidation. BOD measurements are used as a surrogate measure of the organic strength of wastes in water.

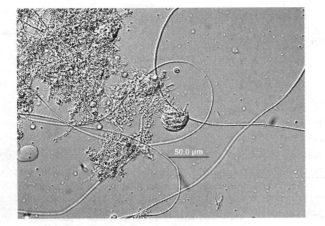

Zoogleal floc with 021N filaments; center: a "free-swimming ciliate"

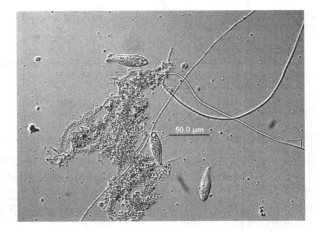

Zoogleal floc with 021N filaments and 3 "free-swimming" ciliates

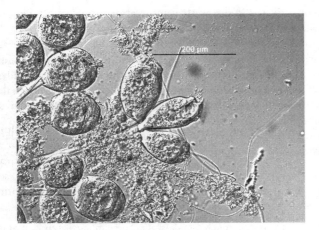

Colonial sessile ciliate

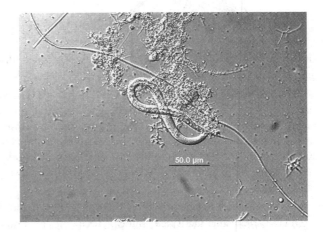

Small nematode, zoogleal floc, and filamentous bacteria

Fig. 10.1   Microorganisms in activated sludge

For information on the use of microorganisms to operate the activated sludge process, see *Advanced Waste Treatment,* Chapter 2, "Activated Sludge." Also see *Operation of Wastewater Treatment Plants,* Volume II, Section 11.10, "Microbiology for Activated Sludge," by Paul V. Bohlier.

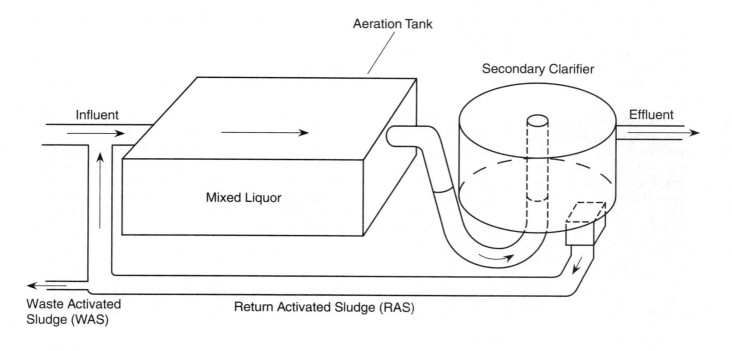

Fig. 10.2   Activated sludge process

can create undesirable effects on receiving water when discharged from a wastewater treatment plant.

The components required for an activated sludge wastewater treatment plant depend on the quantity of wastewater to be treated, the makeup of the wastewater, and whether the wastewater has received any pretreatment before entering the activated sludge system. The common feature of all activated sludge systems is that there must be a tank or basin where the following functions can occur:

- Aeration/mixing: The purpose is to mix the suspended growth organisms (activated sludge) with the incoming wastewater so they can obtain food, and to supply dissolved oxygen to the *AEROBIC*[5] and *FACULTATIVE*[6] organisms (bacteria). This process takes place in what is called an aeration tank or basin, reactor, oxidation cell, or oxidation ditch.

- Settling: After spending a prescribed amount of time in the aeration/mixing mode, the activated sludge is provided a quiescent (quiet) period. During this period, the organisms tend to clump together and form large particles (called *FLOC*[7]) that will normally settle out of the wastewater and form a blanket layer of solids on the bottom of the tank or basin. The clear liquid remaining above the activated sludge blanket is the supernatant (effluent). The settling process takes place in a secondary clarifier, settling tank, or settler.

- Decanting: In this stage, the settled activated sludge culture is separated from the clarified wastewater. The decanting stage may vary slightly depending on the layout of the treatment system. If the entire process of aeration/mixing, settling, and decanting occurs in a single tank, the decanting stage occurs in the following manner.

After the quiescent period in which solids flocculation/separation has occurred, the supernatant layer of treated wastewater (effluent) is removed from the tank, to a level just above the settled activated sludge blanket. Once the treated effluent has been removed, the tank is again filled with untreated wastewater and the process of mixing/aeration and settling is repeated. This type of activated sludge system is referred to as a "batch reactor" or, if the system is specifically designed and automated to perform the above functions, it is called a "sequencing batch reactor" (SBR).

---

5. *Aerobic Bacteria* (air-O-bick back-TEER-e-uh).   Bacteria that will live and reproduce only in an environment containing oxygen that is available for their respiration (breathing), namely atmospheric oxygen or oxygen dissolved in water. Oxygen combined chemically, such as in water molecules (H_2O), cannot be used for respiration by aerobic bacteria.

6. *Facultative* (FACK-ul-tay-tive) *Bacteria*.   Facultative bacteria can use either dissolved oxygen or oxygen obtained from food materials such as sulfate or nitrate ions. In other words, facultative bacteria can live under aerobic, anoxic, or anaerobic conditions.

7. *Floc*.   Clumps of bacteria and particles or coagulants and impurities that have come together and formed a cluster. Found in aeration tanks, secondary clarifiers, and chemical precipitation processes.

SBRs are a special modification of the activated sludge process that is used by some small systems. For details on how to operate and maintain sequencing batch reactors, see the training manual on *Operation of Wastewater Treatment Plants,* Volume II, Chapter 11, "Activated Sludge," Section 11.9, "Sequencing Batch Reactors (SBRs)."

In conventional activated sludge plants, aeration/mixing and settling occur in separate tanks, with the wastewater continuously flowing through them. In continuous-flow systems, the treated wastewater (effluent) leaves the settling tank from the surface or supernatant zone located above the activated sludge blanket in the tank. At the same time, a portion of the activated sludge blanket is removed from the bottom of the settling tank and returned to the aeration or reactor tank where it will be mixed with incoming wastewater to be treated and supplied with dissolved oxygen by the aeration system.

Figures 10.3 and 10.4, respectively, show a flow diagram and a sample plant layout of a typical activated sludge plant for a community with a conventional gravity wastewater collection system. Figures 10.5 and 10.6, respectively, show the typical flow diagram and plant layout for an activated sludge plant designed for a small community that uses septic tanks. In this example, the septic tank effluent is collected and routed to a wastewater treatment facility by means of a STEP (Septic Tank Effluent Pump) or SDGS (Small-Diameter Gravity Sewer) system.

A small wastewater treatment facility employing the activated sludge process would likely use one of the following three activated sludge systems:

1. Standard-rate activated sludge

2. Package plant or field-fabricated plant

3. Oxidation ditch

10.02 Process Description

The purpose of secondary treatment in the form of an activated sludge process (Figures 10.3 and 10.4) is *OXIDATION* [8] and the removal of soluble or finely divided suspended materials that were not removed by previous treatment. Aerobic organisms do this in a few hours as wastewater flows through an aeration tank. The aerobic organisms *STABILIZE* [9] soluble or finely divided suspended solids by converting them into nutrients, which become available to other bacteria and protozoans in the sludge. The bacteria and protozoans are removed by sedimentation in a clarifier. Some ongoing organism growth and reproduction releases a portion of the waste in the form of carbon dioxide, water, and sulfate and nitrate compounds. The remaining solids are changed to a form that can be settled and removed as sludge during sedimentation.

After the aeration period, the wastewater is routed to a secondary settling tank for a liquid-organism (water-solids) separation. Settled organisms in the final clarifier are in a deteriorating condition due to a lack of oxygen and food and should be returned to the aeration tank as quickly as possible. The remaining clarifier effluent is usually chlorinated, dechlorinated, and discharged from the plant.

Conversion of dissolved and suspended material to settleable solids is the main objective of high-rate activated sludge processes, while low-rate processes focus on oxidation. In the activated sludge process, the biochemical oxidation carried out by living organisms is the focus. The same organisms also are effective in converting substances to settleable solids if the plant is operated properly.

When wastewater enters the aeration tanks, it is mixed with the activated sludge to form a mixture of sludge, carrier water, and influent solids. These solids come mainly from the discharges from homes, factories, and businesses. The activated sludge that is added contains many different types of helpful living organisms that were grown during previous contact with wastewater. These organisms are the workers in the treatment process. They use the incoming wastes for food and as a source of energy for their life processes and for the reproduction of more organisms. The activated sludge also forms a lacy network, or floc mass, that entraps materials not used as food.

Some organisms (workers) will require a long time period to use the available food in the wastewater at a given waste concentration; however, the organisms will compete with each other in the use of available food (waste) and thereby shorten the time factor and increase the portion of waste stabilized. The ratio of food to organisms is a primary control in the activated sludge process.

---

8. *Oxidation.*   Oxidation is the addition of oxygen, removal of hydrogen, or the removal of electrons from an element or compound; in the environment and in wastewater treatment processes, organic matter is oxidized to more stable substances. The opposite of REDUCTION.

9. *Stabilized Waste.*   A waste that has been treated or decomposed to the extent that, if discharged or released, its rate and state of decomposition would be such that the waste would not cause a nuisance or odors in the receiving water.

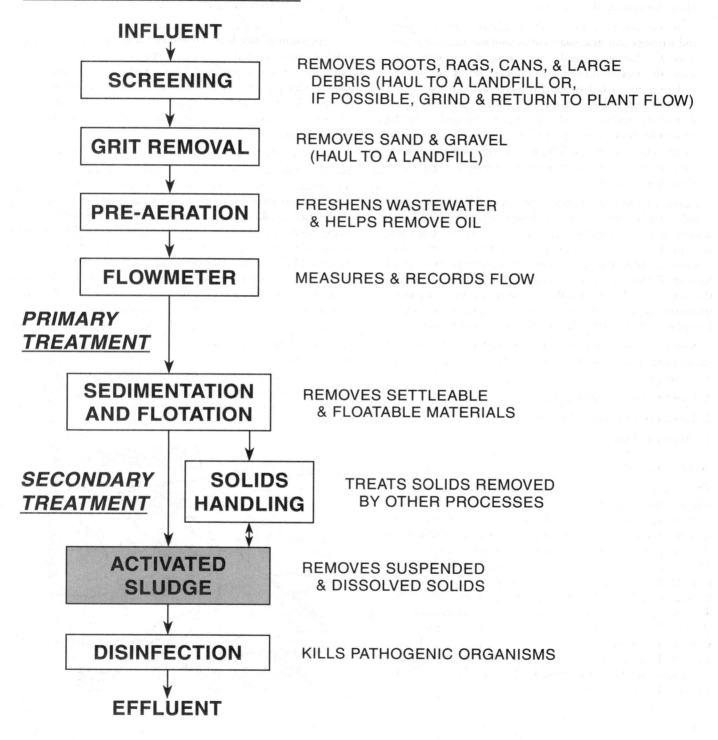

TREATMENT PROCESS   **FUNCTION**

PRELIMINARY TREATMENT

INFLUENT

| SCREENING | REMOVES ROOTS, RAGS, CANS, & LARGE DEBRIS (HAUL TO A LANDFILL OR, IF POSSIBLE, GRIND & RETURN TO PLANT FLOW) |

| GRIT REMOVAL | REMOVES SAND & GRAVEL (HAUL TO A LANDFILL) |

| PRE-AERATION | FRESHENS WASTEWATER & HELPS REMOVE OIL |

| FLOWMETER | MEASURES & RECORDS FLOW |

PRIMARY TREATMENT

| SEDIMENTATION AND FLOTATION | REMOVES SETTLEABLE & FLOATABLE MATERIALS |

SECONDARY TREATMENT

| SOLIDS HANDLING | TREATS SOLIDS REMOVED BY OTHER PROCESSES |

| ACTIVATED SLUDGE | REMOVES SUSPENDED & DISSOLVED SOLIDS |

| DISINFECTION | KILLS PATHOGENIC ORGANISMS |

EFFLUENT

Fig. 10.3   Flow diagram of a typical plant

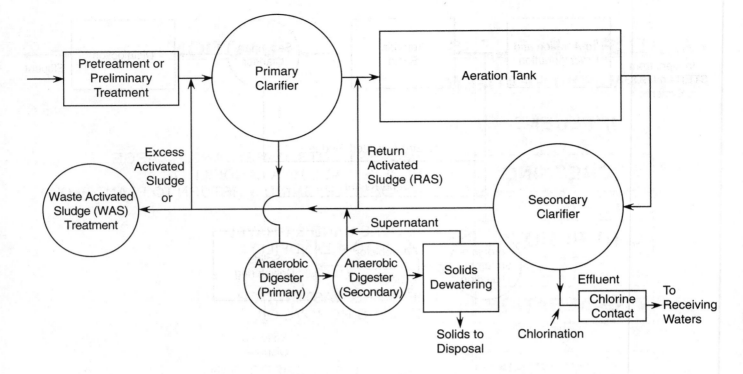

Fig. 10.4   Plan layout of a typical activated sludge plant

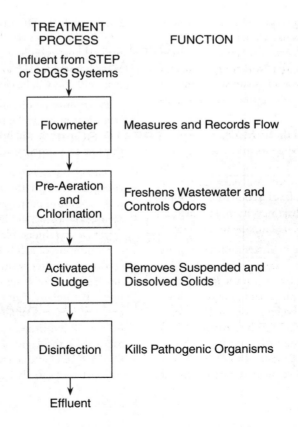

Fig. 10.5   Flow diagram of a typical activated sludge plant
treating wastewater from STEP or SDGS system

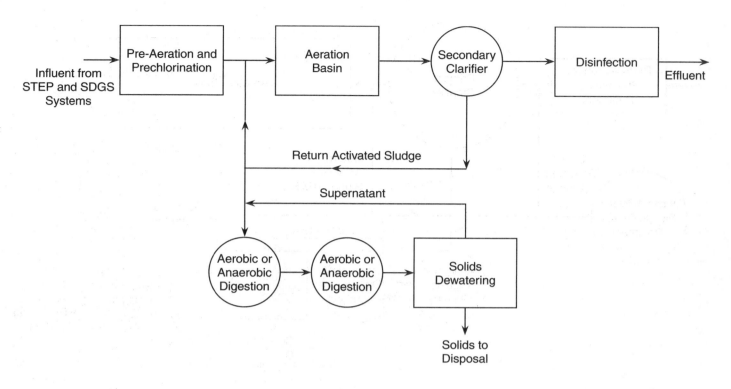

Fig. 10.6   Plant layout of a typical activated sludge plant for STEP and SDGS systems

Organism populations tend to increase with waste (food) load and time spent in the aeration tank. Under favorable conditions, the operator will remove the excess organisms in order to maintain the required number of active worker organisms for effective waste treatment. The removal of organisms from the treatment process (sludge wasting) is an important control technique.

Air is added to the aeration tank in the activated sludge process to provide oxygen to sustain the living organisms as they oxidize wastes to obtain energy for growth. Insufficient oxygen will slow down the activity of aerobic organisms, make facultative organisms work less efficiently, and favor production of foul-smelling intermediate products of decomposition and incomplete reactions. Adding air also encourages mixing in the aerator.

Air (oxygen) requirements increase in an aeration tank when the strength (BOD) of the incoming wastewater increases. This increase is needed because more food (waste) in the influent encourages more organism activity and more oxidation (reproduction and respiration). An excess of oxygen is required for complete waste stabilization. Therefore, the dissolved oxygen (DO) content in the aeration tank is an essential control test. Some minimum level of oxygen must be maintained to favor the desired type of organism activity to achieve the necessary treatment efficiency. If the DO in the aeration tank is too low, *FILAMENTOUS ORGANISMS*[10] will thrive and the sludge floc will not settle in the secondary clarifier. Therefore, the proper DO level must be maintained so that solids will settle properly and the plant effluent will be clear.

Flows must be distributed evenly among two or more similar treatment units. If the plant is equipped with a splitter box or a series of boxes, it will be necessary to check periodically to verify that the flow is being split as intended.

The operator needs to determine the activated sludge solids concentrations in the aerator and the secondary clarifier for process control purposes. Solids are in a deteriorating condition as long as they remain in the secondary clarifier. Controlling the depth of the sludge blanket in the secondary clarifier and the concentration of solids in the aerator is very important for successful wastewater treatment. *CENTRIFUGE*[11] tests will give a quick estimate of solids concentrations and their locations in the units. Before any changes are made in the mode of operation, precise solids tests should be conducted periodically for comparison with centrifuge solids tests. Settleability tests on aerator solids show the degree and volume of solids settling that may be obtained in a secondary clarifier; however, visual plant checks show what is actually happening.

10. *Filamentous* (fill-uh-MEN-tuss) *Organisms.*   Organisms that grow in a thread or filamentous form. Common types are Thiothrix and Actinomycetes. A common cause of sludge bulking in the activated sludge process.
11. *Centrifuge.*   A mechanical device that uses centrifugal or rotational forces to separate solids from liquids.

Primary clarifiers are designed to remove material that settles to the bottom of the tank or floats to the water's surface. Activated sludge helps this process along by collecting and *AGGLOMERATING*[12] the tiny particles in the primary effluent or raw wastewater so that they settle better. If, for some reason, the organisms fail to make this change in the soluble solids, then the secondary clarifier effluent quality will not be satisfactory. Some activated sludge plants do not have a primary clarifier. For the activated sludge process to work properly, the operator must maintain the proper solids (floc mass) concentration in the aerator for the waste (food) inflow. This is done by adjusting the waste sludge pumping rate and regulating the oxygen supply to maintain a satisfactory level of dissolved solids in the process. When these critical factors in the aeration tank are under proper control, the organisms in the tank convert soluble solids and agglomerate the fine particles into a floc mass.

A floc mass is made up of millions of organisms (10^{12} to 10^{18}/100 mL in a good activated sludge), including bacteria, fungi, yeast, protozoa, and worms. When a floc mass is returned to the aerator from the final clarifier, the organisms grow as a result of taking food from the inflowing wastewater. The irregular surface of the activated sludge floc mass promotes the transfer of wastewater pollutants into the solids by means of mechanical entrapment, absorption, adsorption, or adhesion. Many substances not used as food also are transferred to the floc mass, thus improving the quality of the plant effluent.

Material taken into the floc mass is partially oxidized by the organisms to form cell mass and oxidation products. Ash or inorganic material (silt and sand) taken in by the floc mass increase the density of the mass. Mixing the contents of the aerator causes the floc masses to bump into each other and form larger clumps. Eventually, these masses become heavy enough to settle to the bottom of the secondary clarifier where they can be removed easily. The resulting sludge now contains most of the organisms and waste material that had been mixed in the wastewater.

The next step in the activated sludge process is the removal of sludge from the secondary clarifier. Some of the material is converted and released to the atmosphere in the form of stripped gases (carbon dioxide or volatile gases not converted and released in the aeration tank), leaving water and sludge solids. A certain amount of the solids (return activated sludge) will be returned to the aerator to treat incoming wastewater. The operator must pump these solids to the aerator. The rest of the activated sludge must be wasted or removed and disposed of so that it does not continue in the plant flow. After the sludge solids have been removed from the final clarifier, the treated wastewater (clarifier effluent) moves to advanced waste treatment processes or the disinfection process.

The successful operation of an activated sludge plant requires the operator to be aware of the many factors influencing the process and to check them repeatedly. To keep the organisms working in the activated sludge, the operator must provide a suitable environment. Factors that create an unsuitable environment and prevent organisms from working in the activated sludge process of an aeration tank include the following:

1. High concentrations of acids, bases, and other toxic substances that may kill the working organisms

2. Uneven wastewater flows that may cause overfeeding, starvation, and other problems that upset the activated sludge process

3. Failure to supply enough oxygen that may result in decreased organism activity

4. Failure to maintain the appropriate food to microorganism (F/M) ratio, which is a measure of food provided to bacteria in an aeration tank

While successful operation of an activated sludge plant involves an understanding of many factors, actual control of the process as outlined in this section is relatively simple. Control consists of maintaining the proper solids (floc mass or organisms) concentration in the aerator for the waste (food) inflow by adjusting the waste sludge pumping rate and regulating the oxygen (air) supply to maintain a satisfactory level of dissolved oxygen in the process.

QUESTIONS

Please write your answers to the following questions and compare them with those on page 111.

10.0A   What is the activated sludge process?

10.0B   What is a stabilized waste?

10.0C   Why is air added to the aeration tank in the activated sludge process?

10.0D   What happens to the air requirement in an aeration tank when the strength (BOD) of the incoming wastewater increases?

10.0E   What factors could cause an unsuitable environment for the activated sludge process in an aeration tank?

10.1   REQUIREMENTS FOR CONTROL

Effective control of the activated sludge process depends on the operator's ability to interpret and adjust several interrelated factors, including the following:

1. Effluent quality requirements

2. Flow, concentration, and characteristics of the wastewater received

3. Amount of activated sludge (containing the working organisms) to be maintained in the process relative to inflow

4. Amount of oxygen required to stabilize wastewater oxygen demands and to maintain a satisfactory level of dissolved oxygen to meet organism requirements

---

12. *Agglomeration* (uh-glom-er-A-shun).   The growing or coming together of small scattered particles into larger flocs or particles, which settle rapidly. Also see FLOC.

5. Equal division of plant flow and waste load between duplicate treatment units (two or more clarifiers or aeration tanks)

6. Transfer of the pollutional material (food) from the wastewater to the floc mass (solids or active worker organisms) and separation of the solids from the treated wastewater

7. Effective control and disposal of in-plant residues (solids, scums, and supernatants) to accomplish ultimate disposal in a nonpolluting manner

8. Provisions for maintaining a suitable environment for the work force of living organisms treating the wastes to keep them healthy and active

Effluent quality requirements may be stated by regulatory agencies in terms of percentage removal of wastes or allowable quantities of wastes that may be discharged. These quantities are based on flow and concentrations of significant items such as solids, oxygen demand (BOD_5[13]), coliform bacteria, nitrogen, and oil, as specified by the regulatory agencies in your *NPDES PERMIT*.[14]

The effluent quality requirements in your NPDES permit usually determine what kind of activated sludge operation you can use and how tightly you must control the process. For example, if an effluent containing 50 mg/L of suspended solids and BOD_5 is satisfactory, a high-rate activated sludge process will probably meet your needs. If the limit is 10 mg/L, the high-rate process alone would not be suitable. If a high degree of treatment is required, very close process control and additional treatment after the activated sludge process may be needed. Today, secondary treatment plants are expected to remove 85 percent of the BOD and provide an effluent with a 30-day average BOD of less than 30 mg/L.

The treatment plant operator has little control over the makeup or amount of influent entering the treatment plant. However, harmful industrial waste discharges may be regulated by municipal ordinances to prevent industries from dumping substances or wastes into the collection system if they could seriously damage treatment facilities or create safety hazards. Even with these ordinances in place to protect the collection system, you may need to develop some type of inspection program (see *Pretreatment Facility Inspection* in this series of training manuals). It also may be necessary for you to plan for pretreatment or controlled discharge in order to protect the activated sludge process and the treatment plant.

QUESTIONS

Please write your answers to the following questions and compare them with those on page 111.

10.1A   What two different ways can effluent quality requirements be stated by regulatory agencies?

10.1B   How might harmful industrial waste discharges be regulated to protect an activated sludge process?

END OF LESSON 1 OF 3 LESSONS

on

ACTIVATED SLUDGE

Please answer the discussion and review questions next.

DISCUSSION AND REVIEW QUESTIONS

Chapter 10.   ACTIVATED SLUDGE

Package Plants and Oxidation Ditches

(Lesson 1 of 3 Lessons)

At the end of each lesson in this chapter, you will find discussion and review questions. Please write your answers to these questions to determine how well you understand the material in the lesson.

1. Define activated sludge.

2. Sketch the activated sludge process.

3. Define facultative bacteria.

4. Why should activated sludge in the final clarifier be returned to the aeration tank as soon as possible?

5. Some activated sludge plants do not have a primary clarifier. True or False?

6. How can the operator control the activated sludge process?

---

13. *BOD₅*.   BOD₅ refers to the five-day biochemical oxygen demand. The total amount of oxygen used by microorganisms decomposing organic matter increases each day until the ultimate BOD is reached, usually in 50 to 70 days. BOD usually refers to the five-day BOD or BOD₅.

14. *NPDES Permit*.   National Pollutant Discharge Elimination System permit is the regulatory agency document issued by either a federal or state agency that is designed to control all discharges of potential pollutants from point sources and stormwater runoff into US waterways. NPDES permits regulate discharges into US waterways from all point sources of pollution, including industries, municipal wastewater treatment plants, sanitary landfills, large animal feedlots, and return irrigation flows.

CHAPTER 10.   ACTIVATED SLUDGE

Package Plants and Oxidation Ditches

(Lesson 2 of 3 Lessons)

10.2   ACTIVATED SLUDGE PACKAGE PLANTS

10.20   Purpose of Activated Sludge Package Plants

The activated sludge treatment process is widely used in small systems or communities. It provides high-quality biological treatment when operated properly in the extended aeration mode.

This type of plant comes in many sizes and arrangements, but basically there are just two to four compartments made from one large tank. Environmental equipment suppliers fabricate these plants from steel with all of the necessary parts. Table 10.1 lists the parts of an activated sludge plant and briefly describes their purpose. Figure 10.7 shows a two-compartment plant and Figure 10.8 shows a three-compartment plant.

The complete unit is delivered to the community's treatment site by truck. The plant may be set above or on the ground on a prepared pad. Most installations are placed in excavations with the top of the tank 6 to 12 inches (15 to 30 cm) above normal ground level to prevent surface water inflow and to provide better access for operational personnel to perform routine cleaning and maintenance. Before backfilling the excavation around the tank, check the following items because they are important to the plant's operation and life span.

1. The tank must be absolutely level in both directions; one side or end lower than the other will hinder proper operation of the aeration equipment or cause *SHORT-CIRCUITING*[15] through the tank's compartments, thus creating process problems.

TABLE 10.1   PURPOSE OF ACTIVATED SLUDGE PLANT PARTS

Part	Purpose
Bar Rack	Catches rags and large debris and prevents them from entering the aeration tank. Also called bar screen.
Aeration Tank	Provides detention time during which activated sludge microorganisms treat wastewater.
Blower	Supplies air to the plant. Air is used for mixing, oxygenation, and pumping by air lift action.
Diffusers	Transfer oxygen from air to water for respiration by microorganisms. Also keep the contents of the aeration tank mixed and moving.
Settling Tank	Allows activated sludge to be separated from the water being treated. Clear effluent leaves the plant and settled activated sludge is either returned to the aeration tank or wasted to the aerobic digester to maintain the desired F/M ratio.
Air Lifts	Pump air to move settled activated sludge from one compartment to another.
Weir	Controls flow from one compartment to another.
Air Filter	Filters inlet air to blowers. Protects blowers from internal damage or excessive wear from dirt, and prevents diffusers from clogging internally.
Time Clock	Controls operation of blower. In normal operation, the blower runs several times per hour; it is rarely run continuously unless an operational problem requires it.
Froth Control	Some plants are equipped with effluent pumps to supply water pressure to spray heads along the perimeter (outer edges) of the aeration tank to control foam or froth.

---

15. *Short-Circuiting.*   A condition that occurs in tanks or basins when some of the flowing water entering a tank or basin flows along a nearly direct pathway from the inlet to the outlet. This is usually undesirable because it may result in shorter contact, reaction, or settling times in comparison with the theoretical (calculated) or presumed detention times.

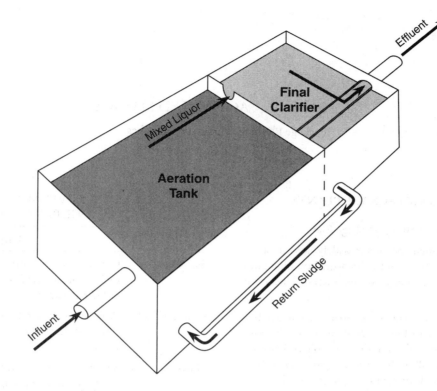

Fig. 10.7   Package plant (two compartments)

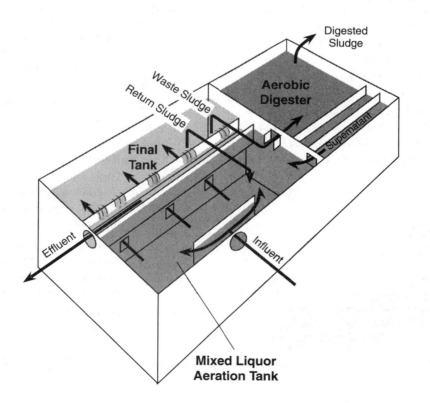

Fig. 10.8   Package plant (three compartments)

2. The tank's exterior corrosion protection material should be checked for any damage that may have occurred during transport or placement. Scratches from cables or equipment used during placement of the tank should be re-coated with the proper protective material.

3. Most installations include *CATHODIC PROTECTION*.[16] Check to be sure the *ANODES*[17] are properly placed and that the electrical leads are well bonded to the tank and anodes.

4. Buried tanks must be placed on a foundation of sand bedding material.

5. The space between the sides of the tank and the walls of the excavation should be filled using Class A material.

After backfilling, the soil around the plant should be sloped away from the tank to provide good drainage and to allow for easy access during wet weather. After the plant is placed in its proper location, remove all debris, piping supports, and equipment tie downs that were installed in the compartments by the manufacturer for shipping purposes. To prevent flotation of the plant from high groundwater, it is good practice to fill the package plant with water as soon as possible after the space between the excavation walls and the plant's walls have been backfilled with soil and compacted. If the plant ever is floated, it must be re-excavated and leveled again to function properly.

The main parts of an activated sludge package plant are the two to four compartments that provide the following process functions:

• Aeration tank or reactor—This is usually the largest compartment of the plant. Its purpose is to provide oxygen and mixing energy to the activated sludge and incoming wastewater.

• Secondary clarifier or settler—This compartment is next to the aeration tank and receives flow (mixed liquor) from the aeration tank. The settling compartment is normally one-third the size of the aeration tank and has a hopper on the bottom. The tank provides the quiescent period to the mixed liquor to separate the activated sludge from wastewater by settling. The wastewater effluent flows over a weir and normally leaves the plant to either be further treated or disinfected and discharged. The settled activated sludge is returned to the aeration tank.

The clarifier compartment also provides a reserve supply of activated sludge solids that may be returned to the aeration tank as seed culture in cases of plant upsets caused by hydraulic overload, toxic discharge, or an unexpected high-organic load (shock load). The availability of this reserve sludge reduces the time required to establish a new culture from days to a few hours.

All activated sludge package plants at least have the two compartments described above; some plants are provided with a third compartment, the aerobic digester.

• Aerobic digester—This compartment is about the same size as the clarifier compartment. The aerobic digester serves three functions: solids concentration, solids stabilization, and microorganism storage, all of which are very beneficial to the process and the operator. During the time period that excess activated sludge solids are being held in the clarifier compartment, the concentration of solids increases because the clarified supernatant is frequently drawn off (decanted). In addition, microorganisms in the tank continue to feed on and digest solid material in the wastewater, thereby reducing some of the solids through *ENDOGENOUS RESPIRATION*.[18]

• Some plants also have a fourth compartment. Depending on the specific needs of the particular treatment plant, the compartment may be located at the beginning (inlet) or end (effluent) of the plant. When the tank is located at the plant inlet, it could serve as a rock or debris trap, a primary clarifier, or a pre-aeration or other pretreatment unit. If the tank is located at the effluent plant, it may be used as a holding chamber to chemically dose post-treatment systems, as a post-aeration chamber before discharge, or as a disinfection chamber for the treated effluent.

10.200   Use of Activated Sludge Package Plants

The type of plant described in the previous section was a prefabricated activated sludge treatment unit. Other systems are sometimes constructed using separate pre-built tanks for the various functions; the individual tanks in these systems are connected with pipelines or channels to convey wastewater from one unit process to another. In other instances, small plants are constructed on site to handle slightly larger wastewater flows or provide a system for a specific activated sludge process. Some examples of the plant configurations available are shown in Figure 10.9.

Most activated sludge package plants are designed to operate in the extended aeration mode. Conventional activated sludge plants are normally operated with a *SLUDGE AGE*[19] ranging

---

16. *Cathodic* (kath-ODD-ick) *Protection.*   An electrical system for prevention of rust, corrosion, and pitting of metal surfaces that are in contact with water, wastewater, or soil. A low-voltage current is made to flow through a liquid (water) or a soil in contact with the metal in such a manner that the external electromotive force renders the metal structure cathodic. This concentrates corrosion on auxiliary anodic parts, which are deliberately allowed to corrode instead of letting the structure corrode.

17. *Anode* (AN-ode).   The positive pole or electrode of an electrolytic system, such as a battery. The anode attracts negatively charged particles or ions (anions).

18. *Endogenous* (en-DODGE-en-us) *Respiration.*   A situation in which living organisms oxidize some of their own cellular mass instead of new organic matter they adsorb or absorb from their environment.

19. *Sludge Age.*   A measure of the length of time a particle of suspended solids has been retained in the activated sludge process.

$$\text{Sludge Age, days} = \frac{\text{Suspended Solids Under Aeration, lbs or kg}}{\text{Suspended Solids Added, lbs/day or kg/day}}$$

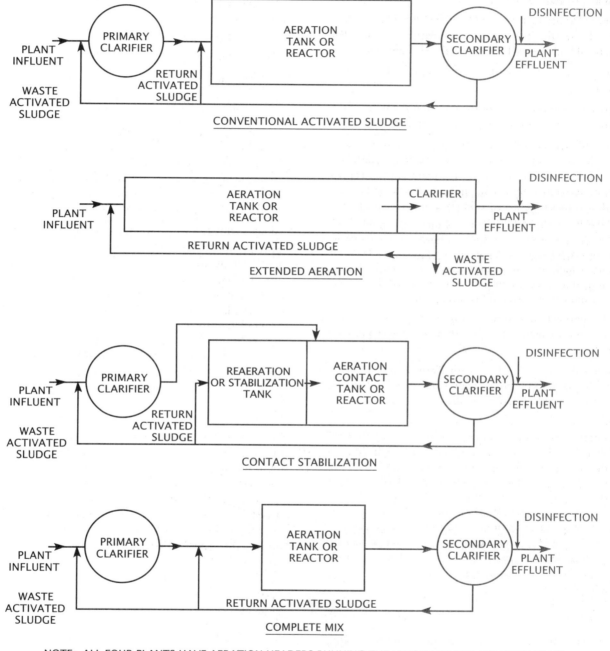

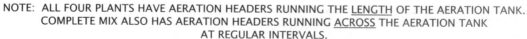

NOTE: ALL FOUR PLANTS HAVE AERATION HEADERS RUNNING THE <u>LENGTH</u> OF THE AERATION TANK.
COMPLETE MIX ALSO HAS AERATION HEADERS RUNNING <u>ACROSS</u> THE AERATION TANK
AT REGULAR INTERVALS.

Fig. 10.9   Types of package aeration plants

from three to seven days and a *MIXED LIQUOR SUSPENDED SOLIDS (MLSS)*[20] concentration (a measure of the microorganisms in the biological culture) of 2,000 to 3,000 mg/L. Extended aeration package plants operate with a sludge age ranging from ten to twenty days. At this sludge age, mixed liquor suspended solids (MLSS) are maintained at a concentration of 2,000 to 6,000 mg/L. The longer sludge age and higher MLSS concentration makes this process mode very reliable and less likely to be upset by shock loads or operational errors. The operation of these plants is similar to the operation of any other activated sludge plant. A high-quality effluent requires attention, understanding, and good plant operation.

You may be assigned to operate a small activated sludge plant. You may not have laboratory facilities and may not be supplied with equipment for simple tests such as dissolved oxygen (DO), pH, or settleable solids. These tests should be run occasionally, at least three to four times a year. If necessary, the tests can be run by an outside laboratory or a local high school or college instructor may be able to assist you in conducting periodic analyses. If you do not have access to a laboratory, this manual will provide you with basic procedures for performing some simple tests that will help you effectively operate your plant. The tests described in this manual will not provide precise information; however, the results will be a useful guide for operating your plant if the tests are performed the same way each time and run at least once a week, and the results are recorded for comparison with previous test results. For details on how to perform the settleability and suspended solids tests, see Chapter 14, "Laboratory Procedures."

10.201 Package Plant Treatment Process Modes (Figure 10.9)

The most common modes of operation for activated sludge treatment processes are extended aeration, contact stabilization, and complete mix. All three of these process modes are essentially modifications of the conventional activated sludge process. The name of each mode basically describes the structural arrangement of the aeration tank(s) as well as the various arrangements of process streams that are used to provide process flexibility. Realistically, almost all package activated sludge plants are of the extended aeration type. Their common characteristics include long solids retention times, high mixed liquor suspended solids, and low food/microorganism ratios.

EXTENDED AERATION

Extended aeration is similar to conventional activated sludge except that the organisms are retained in the aeration tank longer and do not get as much food. The organisms get less food because there are more of them to feed. Mixed liquor suspended solids concentrations range from 2,000 to 6,000 mg/L. In addition to consuming the incoming food, the organisms in the aeration tank also consume any stored food in dead organisms. The new products are carbon dioxide, water, and a biologically inert residue. Extended aeration does not produce as much waste sludge as other processes because much of the material is oxidized to soluble compounds or gases. However, wasting still is necessary to maintain proper control of the process.

CONTACT STABILIZATION

Contact stabilization is similar to conventional activated sludge except that the capture of the waste material and the digestion of that material by the organisms are accomplished in two different aeration tanks. In both the conventional and the contact stabilization modes, the activated sludge flows to the clarifier to be separated from the wastewater. The organisms can *ADSORB*[21] the waste material on their cell walls in only fifteen to thirty minutes, but it takes several hours to *ABSORB*[22] the material through the cell walls. In conventional activated sludge, adsorption and absorption both take place in one tank; therefore, the wastewater has to remain in the tank for a longer time. In the contact stabilization mode, however, the settled organisms are moved to another separate aeration tank (called a stabilization or re-aeration tank) where they digest their food and then are returned hungry to the original aeration tank (contact tank) ready to eat more food. The mixed liquor suspended solids (MLSS) concentration of the contact tank should be maintained around 1,500 to 2,000 mg/L. If the MLSS concentration gets too high, the sludge that the microorganisms form is disposed of or "wasted" as in conventional activated sludge. In most package plants, contact stabilization is not used and the adsorption/oxidation is achieved in one tank.

COMPLETE MIX

In an ideal complete-mix activated sludge plant, the contents of the aeration tank are completely mixed; that is, the MLSS are uniformly distributed throughout the entire aeration tank. To ensure that this is achieved, special arrangements are often used to uniformly distribute the influent and withdraw the effluent from the aeration tank. Attention to the tank shape and to intensive mixing is important. There are some means that the operator may use to evaluate the degree to which a particular process operates in the complete-mix mode. First and foremost, the contents of the tank should be as uniform as possible throughout the tank. The operator can determine this by measuring the DO and suspended solids in the tank. If the tank is thoroughly mixed, these measurements should be nearly uniform. If the DO or suspended solids measurements are far from uniform, incomplete mixing is indicated and the operator should work with an engineer or the package plant manufacturer to determine the cause. The settleability of the complete-mix sludges is generally well within the accepted range of normal operation. The MLSS in the aeration tank ranges from 2,000 to 5,000 mg/L.

---

20. *Mixed Liquor Suspended Solids (MLSS).* The amount (mg/L) of suspended solids in the mixed liquor of an aeration tank.
21. *Adsorption* (add-SORP-shun). The gathering of a gas, liquid, or dissolved substance on the surface or interface zone of another material.
22. *Absorption* (ab-SORP-shun). The taking in or soaking up of one substance into the body of another by molecular or chemical action (as tree roots absorb dissolved nutrients in the soil).

Most package plants have one influent line and one influent port. The aeration compartment has one aeration *HEADER*[23] (Figure 10.10) and the tank contents are completely mixed in a very short time. A limitation of the complete-mix mode of operation is that the process may be more susceptible to short-circuiting.

10.202 Aeration Methods

Two methods are commonly used to supply oxygen from the air to the bacteria: mechanical aeration and diffused aeration.

Mechanical aeration devices agitate the water's surface in the aerator to cause spray and waves by paddle wheels (Figure 10.11), mixers, rotating brushes (Figure 10.12), or some other method of splashing water into the air or vice versa where oxygen can be absorbed. Mechanical aerators in the tank tend to be lower in installation and maintenance costs. Usually, they are more versatile in terms of mixing, production of surface area of bubbles, and oxygen transfer per unit of applied power.

Diffused-air systems are more common to manufactured package activated sludge plants. A device called a diffuser (Figure 10.13) is used to break up the air stream from the blower system into fine bubbles in the mixed liquor. The smaller the bubble, the greater the oxygen transfer due to the greater surface area of rising air bubbles surrounded by water. Unfortunately, fine bubbles tend to regroup into larger bubbles while rising unless they are broken up by suitable mixing energy and turbulence, which is caused by the volume and pressure of the air being supplied from the compressor to the diffusers.

There are a number of devices (Figure 10.14) that can be used to diffuse (break up) an air stream into small bubbles to obtain a high efficiency of oxygen transfer from the air bubbles into the water. To obtain this highest possible oxygen transfer efficiency, the diffuser material must be very fine, such as in a Carborundum plate diffuser or a tightly wound diffuser. Unfortunately, these types of fine diffusers require an air supply that is very, very clean because dust particles will clog the pores of the diffuser internally. This will create back pressure on the air system and reduce oxygen transfer to the water. These diffusers also

plug very quickly if, for any reason, the air supply to the diffuser is disrupted. When this occurs, activated sludge liquor may flow back into the diffusers. For these reasons, diffusers that generate larger air bubbles and are less efficient in oxygen transfer (3 to 5 percent of oxygen in air) are often used because they do not require frequent cleaning or other maintenance.

Most diffused-air package plants are equipped with a diffuser known as a sparger, disk, or comb (Figure 10.15). These diffusers are placed on air pipes that are connected to the main air manifold from the air blowers. These pipes may be equipped with knuckle joints so that an array of diffusers may be mechanically lifted from the aeration tank for maintenance work such as cleaning or replacement.

Another option is to install a pipe union on the air main near the top of the aeration tank. Another pipe union is used to attach a one-inch-diameter pipe on which two to four diffusers have been installed. This arrangement (Figure 10.10) makes it possible to disconnect the pipe to raise the bottom portion of the pipe and diffusers out of the aeration tank for maintenance.

All diffusers for mixing and oxygen transfer in an aeration tank are located approximately 18 inches (46 cm) or less above the bottom of the tank to facilitate rolling the contents to provide the mixing action needed. All the diffusers in the aeration tank must be positioned at the same elevation to ensure an even distribution of air leaving the diffusers. If the diffusers are used for other purposes in the tank, they must have different air pressures supplying them.

QUESTIONS

Please write your answers to the following questions and compare them with those on page 111.

10.2A How many compartments are there in a typical activated sludge package plant? What is the purpose of each compartment?

10.2B Why should package plants have cathodic protection?

10.2C What are some common characteristics of all package aeration plants?

10.21 Pre-Start Check-Out

While the tank is empty, check the following:

1. Is the tank level from one end to the other?

2. If the tank is constructed of metal, is cathodic protection required for corrosion control?

3. What is the condition of paint on exterior and interior?

4. Have all rocks and debris been removed from both compartments?

5. If equipped with *COMMINUTOR*[24] or grinder, is there adequate lubrication?

---

23. *Header.* A large pipe to which the ends of a series of smaller pipes are connected. Also called a manifold.
24. *Comminutor* (kom-mih-NEW-ter). A device used to reduce the size of the solid materials in wastewater by shredding (comminution). The shredding action is like many scissors cutting to shreds all the large solids in the wastewater.

Air headers (package plant)

Air header (grit chamber)

1. Distribution system connector fitting
2. Header valve—regulating and isolation
3. Union
4. Riser pipe
5. Horizontal air header
6. Diffuser

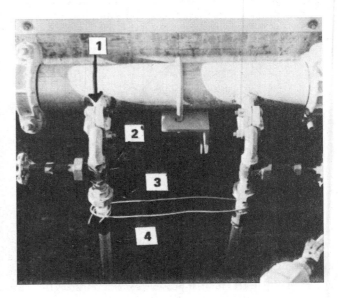

Air header (channel aeration)

Fig. 10.10   Fixed headers

Fig. 10.11   Mechanical aeration device
(Permission of INFILCO INC.)

Brush rotor

Brush rotor

Outboard bearing assembly

Brush rotor drive motor and
gear reducer assembly

Fig. 10.12   Brush rotor used in oxidation ditches

NOTE: Diffuser does not appear to be level. Most of the air is leaving from the higher left side, which has a lower head of water over it. Diffuser should be level to ensure even distribution of air leaving the diffuser.

Fig. 10.13   Air diffuser
(Courtesy Paul Hallbach, National Training Center, Water Quality Office/EPA)

1. Threaded stub for header mounting
2. Air outlet orifices (four)

1. Threaded stub for header mounting
2. Air outlet orifices (four)
3. Air deflector crown

1. Air inlet and header mounting surfaces
2. Air outlet orifices (eight)
3. Plunge tubes

Fig. 10.14   Coarse bubble diffusers

Air headers

Air distribution piping and air headers

Swing header

1. Distribution system connector fittings
2. Header valve—regulating and isolation
3. Double pivot upper swing joint
4. Upper riser pipe
5. Lower riser pipe
6. Pivot elbow
7. Leveling tee
8. Horizontal air header
9. Air blowoff leg
10. Hoist

Fig. 10.15   Swing headers

6. Check aeration device:

   a. Lubrication of motors

   b. Direction of rotation

   c. Mechanical aeration—proper agitator depth

   d. Compressor, if diffused air

     (1) Air filter and oil bath

     (2) Air header and valves

     (3) Air lift tubes and valves

     (4) Diffusers installed

     (5) Swing header lifts easily and free

     (6) Fixed headers or air drop legs secure, unions tight, and isolation valves operate freely

7. Record and file the following data:

   a. Plant model and serial number

   b. Two copies of plant operation and maintenance manual

   c. Nameplate data from equipment

     (1) Comminutor

     (2) Comminutor motor

     (3) Aeration motor

     (4) Compressor or agitator

     (5) Amperage at design load

     (6) Oils and greases specified for each unit

     (7) Sizes and types of blower inlet air filters

     (8) If blower belt driven, size of belts; if multiple sheaves (two or more belts driving blower from motor), belts must be ordered in matched sets. If one belt breaks, all belts must be replaced at the same time.

8. Check influent gate or valve for proper operation.

10.22 Starting the Plant

If the plant previously has not been filled with fresh water, the aeration blower should be started and air introduced to the diffusers prior to the introduction of wastewater. Some types of diffusers could become plugged if wastewater enters the tank before air is coming out of the diffusers. In a package plant, the aeration blower is intended to operate intermittently. The installed diffusers are ones not likely to be affected by wastewater flowing back into the diffusers when the blower is idle. If the plant is the diffused-air type with air lifts for return sludge, the air line valve to the air lifts will have to be closed until the settling compartment is filled. Otherwise, all the air will attempt to go to the empty compartment and none will go to the diffusers. Once the settling compartment is filled from the overflow from the aeration tank, the air lift valves may be opened. They will have to be adjusted to return a constant stream of water and solids to the aeration tank. This adjustment usually requires two to three turns open on the air valve to each air lift. If the plant is equipped with an aerobic digestion compartment, the valves on the air headers in that compartment should be closed. There will be no excess sludge to digest for some time so there is no need to aerate that compartment. If the plant is started in cold weather or if mosquito breeding is a problem, the air valves to the diffusers should be opened slightly to obtain a slight roll of the water to prevent ice formation or to reduce insect breeding.

You may have several options when starting a new plant. It is going to require a week or more to establish an activated sludge culture. The quicker a culture is established, the sooner the plant will produce an acceptable effluent. The fastest way to establish an activated sludge culture would be to acquire seed sludge from a nearby activated sludge plant. A septic tank pumper tank truck can transport 1,000 or 1,500 gallons of concentrated return sludge from another wastewater treatment plant and dump it in your aeration compartment (with the aeration system running). Aerate the new sludge for an hour, then admit raw wastewater slowly into the aeration tank. Once the clarifier compartment is filled, open the air valves on the air lift return lines or start the return sludge pumps. During start-up mode, it is good practice to return as much settled activated sludge as possible to the aeration tank, even when it has a very low concentration of MLSS. The reason is to get the organisms back into an aerobic environment with a food source as soon as possible. However, care must be taken to be sure that the combined flow of RAS (return activated sludge) and raw influent does not exceed the design capacity (shorten the detention time) of the clarifier. If the total flow is too high, the sludge will not have time to settle out and will flow out of the plant.

During the first few days of the start-up period, the aeration blower should be operated on a continuous basis to adequately supply oxygen and mixing. After several days of operation and when you can observe at least 20 percent by volume of MLSS in a jar test, then you can program the time clock to gradually reduce blower operation. The blower should never be off for longer than one hour. In the aeration compartment during normal operation the blower should operate for 10 to 15 minutes, then turn off for 5 to 10 minutes, and then turn back on for 10 to 15 minutes to repeat the cycle. Blower operating time is such that the aeration tank dissolved oxygen (DO) level during blower operation is 2 to 3 mg/L, and the mixed liquor DO does not drop below 1.0 mg/L during the OFF cycle of the blower. If the DO drops below this level, the blower must be operated longer or more frequently.

If you do not have access to seed activated sludge for start-up then you must raise your own activated sludge culture. If there were existing treatment facilities before the new plant was installed, that is, if the new plant is replacing a leach field or some other treatment method, then continue to use that facility. Start the new plant by admitting a small amount of raw wastewater (filling the aeration and clarifier tanks only). Run the aeration equipment and return sludge system on a continuous recycle basis. The next day, add additional raw wastewater (20 percent of the aeration tank volume) at two-hour intervals. After the third day, the plant should be capable of receiving the full flow. The effluent will not be of the highest quality until sufficient MLSS have been established in the aeration tank.

In cases where there are no prior wastewater treatment facilities, then the flow is admitted to the new plant and the highest return sludge rates possible are maintained. Once sufficient MLSS are established, return flows are reduced to start concentrating solids in the clarifier to obtain a higher concentration of solids in the return sludge.

The last two methods described previously usually will result in development of a white foam on the aeration tank during start-up. An accelerated growth of filamentous organisms on the tank walls and weirs also will appear. The effluent will have a milky appearance. All of these problems will occur while the F/M ratio is too high during the start-up period and while the MLSS concentration is low, meaning there is more food in the tank than the organisms can digest.

Foam can be controlled by the froth sprays (if the tank is so equipped) or by using a water hose with a sprinkler. The slime growths can be brushed or hosed off the walls and weirs. Once an activated sludge culture is established, the effluent will become clear and lose the milky or cloudy appearance and the foam and filamentous growths should disappear.

See Chapter 14, "Laboratory Procedures," for procedures to determine solids, DO, and pH.

10.23 Operation of Aeration Equipment

In conventional activated sludge plants equipped with diffused air, aeration equipment is operated continuously to reduce clogging of the diffusers. In mechanically aerated plants, the aeration equipment may or may not be operated on a continuous basis. This depends on the process mode and the waste being treated. For economic efficiency, small activated sludge package plants with blowers and air diffusers quite frequently are operated intermittently; again, blower operating time will depend on the process mode and the waste being treated.

Intermittent blower operation extends equipment life and is more economical than continuous operation in areas having high electrical rates. The prime requirement when determining the length and cycle of aeration equipment operation is that DO levels be maintained in the aeration tank as described previously. All valves to the aeration tank diffusers should be kept wide open. Only air valves on the air lifts or diffusers in the aerobic digestion tank are throttled for control purposes.

The blower discharge air manifold is equipped with a pressure relief valve that generally is set at two psig (pounds per square inch gauge pressure) above the blower manifold discharge pressure. The purpose of the pressure relief valve is to prevent excessive loading on the blower. If the air valves are throttled and restricting air flow, the air vents through the pressure relief valve, which is not efficient. The pressure relief valve may open on blower start-up if there is wastewater in the diffusers or drop pipes. It may take a minute or so for the water to be expelled from the pipes and diffusers, but the pressure relief valve should not continue to vent after start-up. If it does continue to vent, check all diffuser control valves for open positions and check for blocked or plugged diffusers. If both of these items are OK, check the pressure relief valve setting with a pressure gauge and adjust the springs or weights to let the valve open at two psig above normal manifold discharge pressure.

Blowers are controlled by a time clock. In units where two blowers are used to supply air, the clock alternates the RUN and OFF operation of the units. Some plants are equipped with two time clocks. This is especially needed if the plant is subject to intermittent loading, such as wastewater flow from a school. One clock is programmed for the school week when students are in attendance and the other clock is programmed for the weekends when little or no flow is discharged to the plant. The second clock provides a backup for the first clock in case of failure. Installing two clocks also eliminates the need for an operator to visit the plant on weekends to manually program a single clock for weekend or weekday operation. Clocks can also control the operation of other devices such as pumps, chlorinators, lighting, and other process units.

You can judge how well the aeration equipment is working by the appearance of the water in the settling compartment and the effluent that goes over the weir. The aeration device is operated by adjusting the air rates until the solids settle properly. If the water in the aeration tank is murky or cloudy and has a rotten egg odor (H_2S), then not enough air is being supplied. To resolve this problem, increase the air supply (aeration rate) slightly each day until the water is clear in the settling compartment. If the water is clear in the settling compartment, the aeration rate is probably sufficient. If you have a DO probe or lab equipment to measure the DO, work to maintain a DO level of around 2 mg/L throughout the aeration tank. Measure the DO at different locations in the aeration tank as well as from top to bottom.

QUESTIONS

Please write your answers to the following questions and compare them with those on page 111.

10.2D Why should the valve to the air lift pumps be closed until the settling compartment is filled?

10.2E What would you do if the water in the aeration compartment was murky or cloudy and the aeration compartment had a rotten egg odor (H_2S)?

10.24 Wasting Sludge

Many older package plants were not equipped with facilities for wasting sludge. The reason was that the extended aeration system was considered capable of stabilizing the aeration tank suspended solids to a level that was acceptable for release to the receiving waters. Today, a higher degree of treatment and effluent quality is required. Therefore, package plants should have provisions for routinely wasting sludge (removing solids) to achieve optimum plant performance and to protect the receiving waters. For best effluent quality, waste about five percent of the solids each week during summer operation to prevent excessive solids *BURPING*.[25]

---

25. *Burping.* A term used to describe what happens when billowing solids are swept by the water up and out over the effluent weirs in the discharged effluent. Billowing solids result when the settling tank sludge blanket becomes too deep (occupies too much volume in the bottom of the tank).

To waste the excess activated sludge, the approximate amount of MLSS in the aeration tank should be determined. If laboratory equipment is not available, a simple test using a quart jar can be performed. Fill the jar with mixed liquor as it flows from the aeration tank into the settling tank. Let the jar sit without mixing for 30 minutes. If at least 50 percent of the jar is sludge after 30 minutes, then a portion of the sludge should be wasted to the aerobic digester. Mixed liquor solids in the aeration tank should never be less than 20 percent in the jar test to have a good, healthy activated sludge culture.

If the plant is not equipped with an aerobic digester, sludge wasting should be postponed until the jar test indicates a well-compacted sludge volume of at least 70 percent. Once that amount of MLSS is accumulated, sludge wasting should occur. Arrange for a septic tank pump truck to haul off the wasted activated sludge.

Continuously aerate the aeration tank for four to six hours before the arrival of the septic tank pumper. During this aeration period, turn off the return sludge air lifts or pumps so that sludge will become more concentrated in the settling tank. The septic tank pumper should pump the waste activated sludge out of the settling hoppers and dispose of it at a larger treatment plant, in a sanitary landfill, or at some other approved site.

In plants equipped with an aerobic digester, waste activated sludge can be returned to the digester. However, the mixed liquor solids in the aeration tank should not be reduced by more than 10 percent per day to prevent a system overload (shock loading) in the digester.

Another method of sludge wasting is to turn off the return pumps or air lifts for one hour and continue to let the rest of the plant function. After one hour of not returning sludge, use a portable pump to pump about five percent of the waste solids from the settling compartment to a sand or soil drying bed. The amount of solids pumped is determined by measuring the depth of the sludge blanket and then reducing it five percent. Record the pumping time and the amount of weekly waste solids for this time period if results are satisfactory.

If your package activated sludge plant does not have an aerobic digester, applying waste activated sludge to drying beds may cause odor problems. If odors from waste sludge drying beds are a problem, consider the following solutions:

1. Waste the excess activated sludge into an aerated holding tank. This tank can be pumped out and the sludge disposed of in a manner approved by the regulatory agency. If aerated long enough, the sludge could be applied to drying beds.

2. Have the excess or waste activated sludge removed by a septic tank pumper truck and disposed of in a manner approved by the regulatory agency.

3. Arrange for disposal of the excess activated sludge at a larger treatment plant nearby.

Additional solutions for controlling odors from drying beds include annually checking the bottom of the hoppers for rocks, sticks, and grit deposits by probing or dewatering the tank. Also check the tail pieces of the air lifts to be sure that they are clear of rags and rubber goods and are in proper working condition.

After several months of operation, you may want to revise the frequency and amount of wasting by examining the following:

1. The amount of carryover of solids in the effluent

2. The depth to which the solids settle in the aeration compartment when the aeration device is off (should be greater than one-third of the distance from top to bottom)

3. The appearance of floc and foam in the aeration compartment (color, settleability, foam makeup) and excess solids on the water surface in the tank

4. The results of laboratory testing (refer to Section 10.26, "Laboratory Testing for Small Activated Sludge Plants")

White, fluffy foam indicates low solids content in the aerator while a brown, leathery foam indicates high solids concentrations. If you notice high effluent solids levels at the same time each day, the solids loading may be too great for the final clarifier. Excessive solids indicate that the mixed liquor suspended solids concentration is too high for the flows and more solids should be wasted. Sometimes, a loss of solids over the weirs is caused by a combination of high solids and high flows. Turning off foam sprays and hoses can increase the settling time and may allow the sludge to better settle.

10.25   Operation of Activated Sludge Package Plants

These small activated sludge package plants are intended to function in the following ranges:

1. Plant Flow	0.005 to 0.1 MGD (19 to 380 cu m/day)
2. BOD Loading	10 to 30 lbs/day per 1,000 cubic feet (200 to 500 kg/day per 1,000 cu m) of aeration tank
3. F/M Ratio	0.05 to 0.3 lb BOD/day/lb MLVSS
4. MLSS Concentration	2,000 to 6,000 mg/L
5. Sludge Age	20 to 35 days
6. Dissolved Oxygen in Aeration Tank	1.0 to 3.0 mg/L
7. BOD Removal	85 to 90%
8. Suspended Solids Removal	90 to 95%

10.250   Normal Operation

As part of normal operation, package activated sludge plants should be visually checked every day. Each visit should include the following:

1. Check the appearance of aeration tank and final clarification compartment.

2. Check the aeration unit for proper operation and lubrication.

3. Check the return sludge line for proper operation. If the air lift is not flowing properly, briefly close the outlet valve to force the air to go down and out the tail piece. Usually, this will blow it out and clear any obstructions. Re-open the discharge valve and adjust to desired return sludge flow.

CAUTION: Back flushing the air lift will disturb the settled sludge blanket in the clarifier hopper when the tail piece is cleared and this may cause solids to be discharged from the clarifier into the plant effluent. Back flushing should be very brief. If repeated clogging of the air lift occurs, the settling tank should be dewatered and the debris removed from the hopper.

4. Check the comminuting device for lubrication and operation. Clean the bar screen by removing captured rags and debris, and hose down the screen and inlet to aeration tank.

5. Hose down the aeration tank and final clarification compartment. Remove any sludge accumulations from the walls and hoppers of the clarifier tank with a squeegee. Sludge deposits left for long periods of time without oxygen will become *SEPTIC*[26] and produce odors caused by anaerobic decomposition.These septic deposits in either the aeration tank or the clarifier may be harmful to the activated sludge process.

6. Brush weirs when necessary to remove algae and captured materials.

7. Skim off grease and other floating material such as plastic and rubber goods. Dispose of this material, along with captured bar screen debris, at an approved site.

8. Perform necessary maintenance and operational adjustments to other process units (chlorination/disinfection equipment), if used. Check the plant discharge for proper appearance, grease, or material of wastewater origin that is not desirable.

10.251 Abnormal Operation

Remember that changing or abnormal conditions can upset the microorganisms in the aeration tank. As the temperature changes from season to season, the activity of the organisms speeds up or slows down. Also, the flows and waste (food as measured by BOD and suspended solids) in the plant influent change seasonally. All of these factors require the operator to gradually adjust aeration rates, return sludge rates, and wasting rates. Abnormal conditions may consist of high flows or solids concentrations as a result of storms or weekend loads. These problems require the operator and the plant to be prepared and to do the best possible job with the available facilities.

Toxic wastes such as pesticides, detergents, solvents, or high or low pH levels can upset or kill the microorganisms in the aeration tank. The plant effluent usually does not deteriorate until after the toxic substance has passed through the plant. To correct problems caused by toxic substances, locate the source and prevent future discharges.

If the microorganisms in the aeration tank have been killed by toxic wastes, build up the microorganisms as if you were working with a new plant. If the plant is equipped with an aerobic digester, use a portion of the sludge solids in the digester compartment. Turn off the air to the aerobic digester diffusers, let the sludge solids settle out, decant the liquid above the sludge blanket, and pump the settled sludge back into the aeration tank. This sludge will introduce a large number of organisms required to treat the wastewater and will permit the plant to return rapidly to normal operation. This is one of the largest benefits of an activated sludge plant having aerobic digestion. Operate the blower continuously for several hours to supply dissolved oxygen to the aeration tank. Once the activated sludge culture is well established, which may require several days, intermittent blower operation can be resumed. Any time solids are lost for any reason, this method should permit rapid recovery of the treatment process.

An example of a common toxic substance added by operators is chlorine for odor control (prechlorination). Chlorine is a toxicant and should not be allowed to enter the activated sludge process in uncontrolled amounts because it is not selective with respect to the types of organisms it will kill or inactivate. Although chlorine is used in larger activated sludge plants for the control of *BULKING*[27] sludge, its use in package plants is not recommended. It may kill the organisms that should be retained as workers. Chlorine is effective in disinfecting the plant effluent after treatment by the activated sludge process.

10.252 Troubleshooting

When problems develop in the activated sludge process, identify the cause of the problems and select the best possible solutions. Table 10.2 lists a number of problems that may be encountered when operating an activated sludge package plant and suggests the probable cause of each problem, what should be checked or monitored, and possible solutions to the problem. Remember that the activated sludge process is a biological process and may require from three to seven days or longer to show any response to the proper corrective action. Allow seven or more days for the process to stabilize after making a change in the treatment process. The following are some of the problems that may develop.

1. Solids in the effluent

   a. If the effluent appears turbid (muddy or cloudy), the return activated sludge pumping rate is out of balance. Try increasing the return sludge rate. Also, consider the possible presence of something toxic to the microorganisms or the possibility of a hydraulic overload washing out some of the solids.

   b. If the activated sludge is not settling in the clarifier (sludge bulking), several possible factors could be causing this problem. Look for too low a solids level in the system; low dissolved oxygen concentrations in the aeration tank; strong, stale septic influent; high grease levels in the influent; or alkaline wastes from a laundry.

---

26. *Septic* (SEP-tick).   A condition produced by anaerobic bacteria. If severe, the sludge produces hydrogen sulfide, turns black, gives off foul odors, contains little or no dissolved oxygen, and the wastewater has a high oxygen demand.

27. *Bulking.*   Clouds of billowing sludge that occur throughout secondary clarifiers and sludge thickeners when the sludge does not settle properly. In the activated sludge process, bulking is usually caused by filamentous bacteria or bound water.

TABLE 10.2   ACTIVATED SLUDGE PACKAGE PLANT TROUBLESHOOTING GUIDE

Indicator	Probable Cause	Check or Monitor	Solution
Rocks, mud, debris	Broken collection main	Check collection system	Repair broken main.
	Open manhole	Check collection system	Replace manhole cover—add riser ring to cone to raise cover above existing grade to prevent future drainage.
		Check aeration tank for rocks, mud, debris	Remove debris from aeration tank.
Rags (mops), large pieces of plastic in aeration tank	Bar screen	Check bar screen for rags and debris	Clean bar screen more frequently.
		Check screen spacing	Reduce bar openings to capture more material.
	Comminutor	Comminutor operation	Comminutor operating in correct direction—replace broken combs on comminutor screen.
		Comminutor cutters	Cutters dull—replace.
Aeration tank—no mixing, low DO	Blower off, will not run	Blower operation	Check power supply to time clock and blower motors. Reset main power breaker.
		Time clock	Program time clock to correct run times.
		Blower motor	Check motor for operation. Replace if burned out.
	Blower motor runs	Blower drive belts broken	Replace broken belt. If more than one belt on drive, replace all belts.
		Blower drive belts loose or slipping	Adjust belt tension between motor and blowers.
		Blower drive belts burned/worn.	Check sheave alignment. Realign motor and blower. Replace belt.
		Direct drive or gear reducer	Check coupling. Check gear reducer output—if low, replace.
	Blower operation normal	Blower inlet air filter dirty	Clean filter or replace filter cartridge.
	Air manifold pressure lower than normal	Blower discharge check valve	Check valve not opening. Remove obstruction or replace valve.
		Blower is hot	Check blower lubricating oil level. Adjust oil to proper operating level, as needed. Check blower enclosure ventilation screens for obstructions that restrict cooling air.
		Blower worn out	Replace blower.
Large air boil	Broken pipe/diffuser	Check drop pipe and diffusers	Replace pipe, diffuser valves, or missing diffuser.
Air valve malfunction	Air manifold pressure higher than normal	Air valves	Repair air valves.
	Air release valve venting	Air valves	Open all air diffuser valves in aeration tank to full open.
		Diffusers plugged	Remove drop pipes and diffusers. Flush drop pipes with clean water. Clean or replace diffusers. Remove rags or plastic wrapped around diffusers. Check comminutor or bar screen.
High DO	Blower operating time	Check time clock	Program time clock for shorter operating time.
No return sludge flow	Air valve closed to air lift	Operate valve to air lift	Open air control valve so return pipe runs at least ¼ full.
	Tail pipe of air lift plugged	No return with air control valve full open	Tap on air line to tail pipe. Back flush by closing several air valves for 30 seconds on aeration tank diffuser drop pipes.
	Debris in settling tank hopper	Sound hopper (measure sludge depth in hopper)	Dewater settling tank. Remove debris from tank and hopper. Clean air lift pipes.

TABLE 10.2   ACTIVATED SLUDGE PACKAGE PLANT TROUBLESHOOTING GUIDE (continued)

Indicator	Probable Cause	Check or Monitor	Solution
Return has low solids, almost clear liquid	Low solids content	Check MLSS concentration	Build up solids. If aerobic digester, transfer solids to aeration tank.
	Solids amount normal—sludge blanket bridged	Sound settling hoppers if sludge blanket bridged or coned	Squeegee or probe settling hopper, breaking bridge or channel so solids can be removed.
Corrosion	Electrolysis	Sacrificial anodes	Check anodes. If absorbed, replace with new anodes.
Rusting of metal pipes, tuberculations	Exposed pipes	Protective coating	Repaint metal surfaces, particularly those under water in aeration tanks.
Low DO in aeration tank	Blower not operating long enough	DO level in aeration tank	Check DO hourly.
		Blower tank clock setting	Increase blower run times.
Septic sludge	Insufficient aeration	Blower operation	Increase blower run time.
	Return sludge rate too low	Sound sludge blanket for depth and solids condition	Increase return sludge rate.
	Shock load/toxic waste	Sample influent to plant—solids, pH, temperature	Stop sludge wasting.
		Heavy organic load—septic tank pumped into collection system	Increase aeration time.
White foam on aeration tank	Low MLSS	Run jar test	If necessary, add solids from aerobic digester.
Dark, leathery froth	High MLSS	Run jar test, if more than 70% settled sludge	Waste sludge to aerobic digester.
Odors	Low DO, shock load	Run DO test, check pH	Increase aeration.
Scum on clarifier	Skimmer operation	Check surface skimmers	Check air flow to surface skimmer, clean inlets. Hand skim tank. Dispose of screenings.
Clarifier effluent—ashing; sludge clumps, dark brown ¼–1″ in size, rise to surface	Too much solids in system	Run jar test—over 70% settled sludge	Waste sludge to normal level concentration.
Pin floc—pin head size of floc, turbid effluent	Sludge too old	Run jar test—50% settles in 5 minutes, rapid settling of granular particles	Waste some of the return sludge.
Rising sludge or clumping—large patches of sludge rise to clarifier surface	Sludge is denitrifying	Run jar test—settled sludge rises to surface in two hours after settling	Increase return sludge rate. Decrease blower operating time slightly.
Bulking sludge does not settle	Low DO, pH, septicity, grease, industrial waste, F/M ratio out of proportion	Run test on influent, mixed liquor	Increase aeration. Adjust pH to near 7.0. Reduce return sludge rate slightly.
Straggler floc—fluffy cream floc lost over weir in effluent	Sludge age too low or too high	Run jar test—clear effluent.	Increase MLSS.
		Run jar test—MLSS dark brown, scummy foam.	Decrease MLSS by wasting sludge.

c. If the solids level is too high in the sludge compartment of the secondary clarifier, solids will appear in the effluent. Try increasing the return sludge pumping rate. Remember, excessive return flows can cause a hydraulic overload on the clarifier.

d. If odors are present and the aeration tank mixed liquor appears black as compared with the usual brown color, try increasing the aeration rates and look for septic dead spots.

e. If light-colored floating sludge solids are observed on the clarifier surface, try reducing the aeration rates. Work to maintain the dissolved oxygen at around 2 mg/L throughout the entire aeration tank.

f. Long sludge retention times in the clarifier will result in anoxic/anaerobic conditions, which release gases that will be trapped in the sludge and cause it to float.

2. Odors

a. If the effluent is turbid and the aeration tank mixed liquor appears black as compared with the usual brown color, low aeration rates may be the cause. Try increasing aeration rates and look for septic dead spots.

b. If clumps of black solids appear on the clarifier surface, low return sludge rates may be the cause. Try increasing the return sludge rate. Also, be sure the sludge return line is not plugged and that there are no septic dead spots around the edges or elsewhere in the clarifier.

c. Poor methods of wasting and disposing of waste activated sludge could be the source of the odors.

d. Poor housekeeping procedures could result in odors. Do not allow solids to accumulate or debris removed from the wastewater to sit around the plant in open containers.

3. Foaming/Frothing

Foaming is usually caused by too low a solids level, while frothing is caused by too long a solids retention time.

a. If too much activated sludge was wasted, reduce wasting rate.

b. If over-aeration caused excessive foaming, reduce aeration rates.

c. If the plant is recovering from overload or septic conditions, allow time for recovery.

d. If frothing is occurring and building up on the clarifier tank, skim off the floating froth and dispose of it with grease and other debris at a landfill. Do not return the floating froth from the clarifier back to the aeration tank. This seeds the culture and prolongs the frothing problem.

Foaming can be temporarily controlled by water sprays or commercially available defoaming agents until the cause is corrected by reducing or stopping wasting and building up solids levels in the aeration tank.

To learn more about the operation of an activated sludge process under both normal and abnormal conditions, refer to another training manual in this series, *Operation of Wastewater Treatment Plants,* Volume II, Chapter 11, "Activated Sludge." There you will also find a troubleshooting guide for activated sludge plants.

10.253 Shutdown

Shutdown procedures depend on whether the plant is being shut down because of operational problems, for maintenance and repairs, or for the off-season (such as in a resort area). Activated sludge microorganisms die quickly from lack of oxygen if the aeration system is out of service for even a short time. Whenever the tank must be drained, first determine the groundwater level because a high groundwater level can float a tank and cause considerable damage to structures and pipes. Diffusers should be cleaned before the tank is returned to service. If the package plant is shut down during the off-season, mothball the equipment to prevent damage from weather and moisture. Exact procedures will depend on location and climate.

10.254 Operational Strategy

This section provides a brief summary of the basic concepts in the operation and control of the activated sludge process as it relates to both package plants and oxidation ditches. The following list outlines items that you, the operator, must consider in the day-to-day operation of your treatment plant.

Keep a log book with notes about the conditions, activities, and changes at the plant. Any equipment changes or adjustments should be recorded, such as adjustments of operating times (time clock settings at various durations during the year) and return sludge rates (air valve position, pipe volume). All such information will be a valuable resource you can refer to in the future, especially if problems develop and corrective actions have been taken. Flow conditions, temperature, rain, snow, and test results should all be recorded. The more information you keep and the more accurate and complete the information is, the easier it is to solve similar operation or maintenance problems in the future. Maintenance records are also required and should include lubrication, oil changes, filter changes, belt replacement, equipment repair, and painting. By reviewing your recorded data and answering the following questions, you develop an operating strategy to improve plant performance.

1. Do the influent flow characteristics vary significantly during the year?

2. Is your activated sludge process operation adequate to provide suitable treatment for these variations?

3. Is adequate pretreatment and collection system monitoring being practiced to avoid downstream mechanical or process failure?

4. Are routine solids tests performed (centrifuge, settleability, depth of blanket, visual observations) with results plotted on graphs to assist you in determining if a change in the process mode or operation is necessary?

5. Is suitable aeration time and mixing being provided to allow adequate oxidation, conversion, and floc formation of the solids?

6. Is adequate sludge wasting being practiced to properly maintain a favorable food to microorganism balance throughout the system?

7. If an increase or decrease in organisms occurs, is the oxygen level adjusted accordingly to maintain proper solids settling and production of a clear final effluent?

8. Does the return sludge flow rate produce a high concentration of solids with the minimum amount of water being returned to the aerator?

9. Do you visit your plant on a regular basis to observe process conditions, check equipment for proper operation, lubricate and maintain equipment, and clean process tanks and related equipment?

10. When a problem develops in your activated sludge process, do you refer to Section 10.252, "Troubleshooting"?

11. Do you avoid injuries by removing hazards and following safe procedures?

12. Before leaving your plant for the day, do you make a final and detailed check of the equipment for proper operation? Do you ensure that flow rates are set properly, that flow gates are set for possible storm or high-flow conditions, that timer-controlled equipment and equipment alarms are set correctly, that equipment is stored properly, and that buildings and gates are securely locked?

10.26 Laboratory Testing for Small Activated Sludge Plants

Testing of plant wastestreams is very important. The dissolved oxygen level may be determined with a DO analyzer. This is an instrument that gives an immediate reading of the dissolved oxygen level when a probe is inserted into the flow stream or into a grab sample. DO analyzers require periodic standardization. The probes may become coated with grease or oil, membranes may be damaged, or a substance toxic to the probe will give an erroneous reading. Most DO analyzers are delicate instruments that must be handled with care.

DO may also be determined by *WET CHEMISTRY.*[28] Certain *REAGENTS*[29] must be mixed and *STANDARDIZED*[30] to perform accurate testing. The wet chemistry procedures are not difficult but care must be exercised in making the reagents and performing the test. When wet chemistry (Winkler Method) is used in mixed liquor, an inhibitor must be added in the procedure to stop the organisms from using oxygen during the test; otherwise, inaccurate readings will be obtained. Dissolved oxygen kits that have prepackaged reagents are available for running the test.

pH is a useful test but is not really a necessity unless an alkaline or acid waste is being treated. The pH can be determined by simple color comparator units or electronic instruments.

Chlorine concentrations may be determined by color comparator kits or wet chemistry methods.

Suspended solids determinations provide information as to plant loadings coming in and solids being discharged in the effluent. The suspended solids test is also used to determine mixed liquor solids concentrations as well as return sludge concentrations. Accurate suspended solids information is determined by filtering a precise volume of well-mixed sample, drying the filtered residue, and very accurately weighing the residue.

The centrifuge test is a method used for approximating the suspended solids in the mixed liquor or return sludge (not on the influent or effluent). Some centrifuges are operated by an electric motor that spins the sample tubes; other units are turned by a hand-operated crank. A well-mixed sample is taken and two centrifuge tubes are filled to the upper fill mark on each tube. The tubes are placed in a centrifuge and spun at a given speed for a set time. A hand-operated centrifuge is usually cranked at 1,800 RPM for one minute. After the centrifuge stops spinning, the tubes are removed and the solids portion is

---

28. *Wet Chemistry.* Laboratory procedures used to analyze a sample of water using liquid chemical solutions (wet) instead of, or in addition to, laboratory instruments.

29. *Reagent* (re-A-gent). A pure, chemical substance that is used to make new products or is used in chemical tests to measure, detect, or examine other substances.

30. *Standardize.* To compare with a standard. (1) In wet chemistry, to find out the exact strength of a solution by comparing it with a standard of known strength. This information is used to adjust the strength by adding more water or more of the substance dissolved. (2) To set up an instrument or device to read a standard. This allows you to adjust the instrument so that it reads accurately, or enables you to apply a correction factor to the readings.

read on the tube. The amount of the tube that the solids occupy in mL or in percent can be converted to a comparison parts per million (ppm) or milligrams per liter (mg/L) that would approximate a Gooch crucible test by using a graph like the one shown in Figure 10.16.

Another way to measure suspended solids is the jar test method described in Chapter 14, "Laboratory Procedures." Note, however, that the jar test method is only an approximation. This method can be used for plant control of mixed liquor or return sludge, but it cannot supply accurate enough information for NPDES reports or other reports requiring accurate information.

BOD (biochemical oxygen demand) is a wet chemistry *BIOASSAY*[31] test. It requires the accurate measurement of dissolved oxygen at the beginning and end of the test period. A DO probe or wet chemistry is used for this determination. The BOD test must also be accurately performed and incubated at a precise temperature (20°C) for a prescribed time period, usually five days. Even under ideal conditions, the BOD test is an approximation.

Depending on the treatment plant and the degree of treatment provided, other tests may be required, such as settleable solids, grease, nitrogen, total Kjeldahl nitrogen (TKN), and phosphorus. After disinfection, the effluent is tested for coliform organisms. The *SECCHI DISK*[32] test is used to measure the clarity of the plant effluent.

Testing for solids condition may be accomplished by the settling test. Using a quart jar, take a sample from the aeration compartment after the aeration device has been operating for 10 to 15 minutes and fill the jar to the top. Let the jar stand and watch the floc form and settle to the bottom of the jar. At the end of 30 minutes, the jar should be approximately half full of the settled solids, or slightly less, and have a chocolate-brown color with clear water above it. The solids should have a curled or granular appearance. If the solids do not settle to the bottom half of the jar and the water above them is cloudy or murky in appearance, a longer aeration period, more air, or solids wasting is needed. Also check the mixed liquor suspended solids (MLSS) concentration for the proper level. If the solids settle to less than one quarter of the jar's depth and the water above the solids is murky or cloudy, no wasting of solids should be done, and the solids level in the aerator should be allowed to increase.

An indication of good operation is when the solids (1) settle to the bottom of the jar, leaving a clear liquor on top; (2) stay down for one hour; and (3) come up after two hours. Solids should never be allowed to remain in the settling compartment longer than two hours. Solids rising in one hour is an indication, usually, of too much air or too many solids. Make a slight adjustment to reduce the air to the aeration compartment or increase the return sludge rate.

Another possible cause of solids rising in one hour could be that there are not enough solids under aeration. When this happens, the sludge will rise because of the high respiration (breathing) rate of the overtaxed organisms in the settling compartment. Under these circumstances, increase or build up the mixed liquor suspended solids. To identify the cause of problems and select the proper solution, you must keep good records and observe what is happening in your plant, as explained in the previous section.

The final clarifier should be equipped with a scum baffle. A properly operated plant will produce some light, oxidized floc that will float to the surface of the settling compartment. A scum baffle will prevent this flow from leaving the compartment in the plant effluent. The better the treatment, the more likely scum froth will develop, unless the unit is septic.

The frequency of performing each test will depend on how the plant is operated, how it performs, and the regulatory requirements for monitoring the discharge.

QUESTIONS

Please write your answers to the following questions and compare them with those on page 111.

10.2F   If it becomes necessary to waste sludge, how much and when should it be wasted?

10.2G   How frequently should a package plant be visually checked by an operator?

10.2H   What should you do if you take a sample in a jar from the aeration compartment and after 30 minutes
a. solids do not settle to the bottom half of the jar?
b. solids settle to the bottom and then float to the top?

10.27 Safety

Operators of wastewater treatment plants work in hazardous conditions and have had one of the worst safety records in the US. If you are the operator of a small package plant, you must be extremely careful. Frequently, you may be the only person at the plant and there will be no one nearby to help you if you are hurt or seriously injured. Therefore, practice safety, avoid hazardous conditions, and use safe procedures. Prepare and rehearse a safety plan for all hazards that exist at your plant.

There are many hazardous conditions an operator needs to be aware of, including deadly toxic gases (hydrogen sulfide), electric shock, drowning, or suffocation in an oxygen-deficient atmosphere. Slippery surfaces, falls, and attempting to lift heavy objects also can cause serious injury. Cuts and bruises can lead to infection, and pathogens (disease-causing bacteria) in wastewater can make you sick. Chlorine and dangerous lab chemicals can blind you or seriously burn your skin. Whenever you must enter a confined space, additional personnel must be present.

---

31. *Bioassay* (BUY-o-AS-say).   (1) A way of showing or measuring the effect of biological treatment on a particular substance or waste. (2) A method of determining the relative toxicity of a test sample of industrial wastes or other wastes by using live test organisms, such as fish.

32. *Secchi* (SECK-key) *Disk*.   A flat, white disk lowered into the water by a rope until it is just barely visible. At this point, the depth of the disk from the water surface is the recorded Secchi disk transparency.

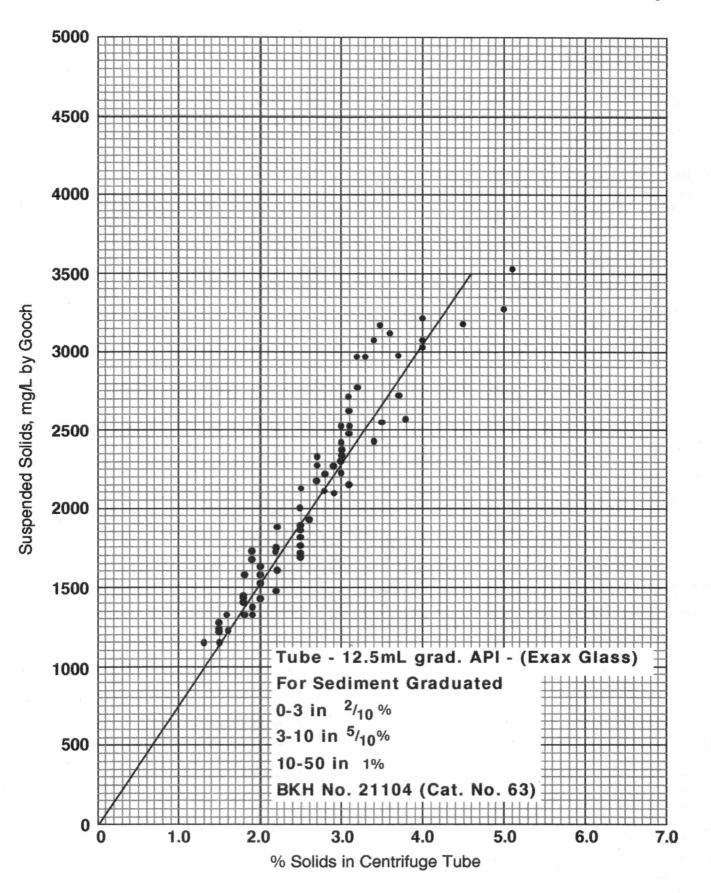

Fig. 10.16   Estimation of suspended solids using centrifuge

All work in confined spaces must be done in full compliance with your safety regulatory agency's requirements. Do electrical troubleshooting *ONLY* if you are qualified and authorized to do so. If you must work alone doing routine operation and maintenance, phone your office at regular intervals and have someone check on you if you fail to report. You can avoid injuries if you avoid hazards and follow safe procedures.

10.28   Maintenance

Follow manufacturer instructions when performing equipment maintenance in package plants. Items requiring maintenance attention include the following:

1. Plant cleanliness. Wash down tank walls, weirs, and channels to reduce the collection of odor-causing materials

2. Aeration equipment

    a. Air blowers and air diffusion units

    b. Mechanical aerators

3. Air-lift pumps

4. Scum skimmer

5. Sludge scrapers

6. Froth spray system

7. Weirs, gates, and valves

8. Raw wastewater pumps

10.29   Additional Reading

Package Treatment Plants—Operations Manual, US Environmental Protection Agency. EPA No. 430-9-77-005. Obtain from National Technical Information Service (NTIS), 5301 Shawnee Road, Alexandria, VA 22312. Order No. PB279444. Price, $60.00, plus $6.00 shipping and handling.

NOTE: This is an outstanding publication and if you are the operator of a package activated sludge plant, you should obtain this Operations Manual.

QUESTIONS

Please write your answers to the following questions and compare them with those on page 111.

10.2I   How can operators avoid being injured?

10.2J   What safety precautions should be taken if you must work alone?

END OF LESSON 2 OF 3 LESSONS

on

ACTIVATED SLUDGE

Please answer the discussion and review questions next.

DISCUSSION AND REVIEW QUESTIONS

Chapter 10.   ACTIVATED SLUDGE

Package Plants and Oxidation Ditches

(Lesson 2 of 3 Lessons)

Please write your answers to the following questions to determine how well you understand the material in the lesson. The question numbering continues from Lesson 1.

7. How would you operate the aeration device in a package plant?

8. How would you waste sludge in a package plant?

9. What items can cause problems for the operator of a package activated sludge plant?

10. What is a temporary way to control foaming in the aeration compartment of a package plant?

CHAPTER 10.   ACTIVATED SLUDGE

Package Plants and Oxidation Ditches

(Lesson 3 of 3 Lessons)

10.3   OXIDATION DITCHES

10.30   Use of Oxidation Ditches

10.300   Flow Path for Oxidation Ditches

The oxidation ditch discussed and shown in Table 10.3 and Figure 10.17 is a modified form of the activated sludge process and is usually operated in the extended aeration mode.

TABLE 10.3   PURPOSE OF OXIDATION DITCH PARTS

Part	Purpose
Oxidation Ditch	Provides detention time where activated sludge microorganisms treat wastewater.
Rotor	Causes surface aeration, which transfers oxygen from air to water for respiration by microorganisms. Keeps contents of the ditch mixed and moving.
Level Control Weir	Regulates how deep rotors sit in the flow of wastewater. This affects the amount of oxygen dissolved in or transferred to the water being treated. Overflow goes to final settling tank.
Final Settling Tank	Allows activated sludge to be separated from water being treated. Clear effluent leaves plant and settled activated sludge is either returned to oxidation ditch or wasted.
Return Sludge Pump	Returns settled activated sludge from final settling tank to oxidation ditch or to excess sludge handling facilities.
Excess Sludge Handling Facilities (not shown)	Treat waste activated sludge for ultimate disposal.

The main parts of the oxidation ditch are the aeration basin, which generally consists of two channels placed side by side and connected at the ends to produce one continuous loop of the wastewater flow, a brush rotor assembly (see Figure 10.12, page 80), settling tank, return sludge pump, and excess sludge handling facilities.

Oxidation ditches usually do not have a primary settling tank or grit removal system. Inorganic solids such as sand and grit are captured in the oxidation ditch and removed during sludge wasting or cleaning operations. The raw wastewater passes directly through a bar screen to the ditch. The bar screen is necessary for the protection of the mechanical equipment such as the rotor and pumps. Comminutors or barminutors may be installed after or instead of a bar screen. The oxidation ditch forms the aeration basin in which the raw wastewater is mixed with previously formed active organisms. The rotor is the aeration device that transfers the necessary oxygen into the liquid for microbial life and keeps the contents of the ditch mixed and moving. The velocity of the liquid in the ditch must be maintained to prevent settling of solids, normally at 1.0 to 1.5 feet per second (0.3 to 0.45 m/sec). The ends of the ditch are well rounded to prevent eddying and dead areas, and the outside edges of the curves have erosion protection.

The mixed liquor flows from the ditch to a clarifier for separation. The clarified water passes over the effluent weir and is chlorinated. Plant effluent is discharged to either a receiving stream, percolation ditches, or a subsurface disposal or leaching system. In approved systems, the treated water may be used for irrigation of hay crops, golf courses, or roadside landscaping. If irrigation is practiced, the water is usually delivered to a holding pond and then pumped to the irrigation site as needed. In some areas, the regulatory agency may require that all runoff irrigation water be captured and recycled. The settled sludge is removed from the bottom of the clarifier by a pump and is returned to the ditch or wasted. Scum that floats to the surface of the clarifier is removed and either returned to the oxidation ditch for further treatment or disposed of by burial.

Because the oxidation ditch is operated as a closed system, the amount of volatile suspended solids will gradually increase. It will periodically become necessary to remove some sludge from the process. Wasting of sludge lowers the MLSS concentration in the ditch and keeps the microorganisms more active. Control of sludge concentration by wasting of excess sludge is one of the reasons for the high reductions made possible by this process. Excess sludge may be dried directly on sludge drying beds or stored in a holding tank or in sludge lagoons for later disposal to larger treatment plants or approved sanitary landfills.

The basic process design results in simple, easy operation. A high MLSS concentration, usually between 2,000 to 6,000 mg/L (some plants carry 6,000 to 8,000 mg/L MLSS), is carried in the ditch and the plant may be capable of handling shock and peak loads without upsetting plant operation. Unlike other types of activated sludge plants, there is no foam problem after solids

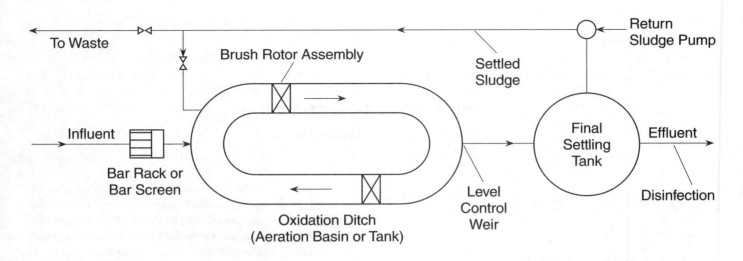

Fig. 10.17   Oxidation ditch plant
(Source: "Oxidation Ditch" prepared by William L. Berk for the
New England Regional Wastewater Institute, South Portland, Maine 04106. August 1970.)

build up in an oxidation ditch. Cold weather operation has less effect on plant efficiency than other processes because of the large number of microorganisms in the ditch. However, in some hot climates, excessive foam can occur when the MLSS is high. Also, the solids will have poor settling characteristics with solids being lost over the weirs. Lowering the MLSS usually corrects these problems.

Operating plants in the United States are achieving BOD removals that average 90 percent and go as high as 98 percent.

10.301   Description of Oxidation Ditches

Operational guidelines for an oxidation ditch plant are as follows:

1. Plant Flow — 0.2 to 20.0 MGD (750 to 75,000 cu m/day)

2. BOD Loading — 10 to 50 lbs/day per 1,000 cubic feet of ditch (200 to 800 kg/day per 1,000 cu m)

3. F/M Ratio[33] — 0.03 to 0.1 lb BOD/day/lb MLVSS

4. MLSS Concentration — 2,000 to 6,000 mg/L

5. Sludge Age — 20 to 35 days

6. Ditch Detention Time — 3 to 24 hours

7. Minimum Velocity — 1.0 fps (0.3 mps)

8. DO Levels — 0.5 to 3.0 mg/L

9. Liquid Level — 3.0 to 7.0 feet (1 to 2 m)

See Section 10.35, "Operational Guidelines," for procedures to calculate these items.

---

33. *Food/Microorganism (F/M) Ratio.*   Food to microorganism ratio. A measure of food provided to bacteria in an aeration tank.

$$\frac{Food}{Microorganisms} = \frac{BOD,\ lbs/day}{MLVSS,\ lbs}$$

$$= \frac{Flow,\ MGD \times BOD,\ mg/L \times 8.34\ lbs/gal}{Volume,\ MG \times MLVSS,\ mg/L \times 8.34\ lbs/gal}$$

or metric

$$= \frac{BOD,\ kg/day}{MLVSS,\ kg}$$

$$= \frac{Flow,\ ML/day \times BOD,\ mg/L \times 1\ kg/M\ mg}{Volume,\ ML \times MLVSS,\ mg/L \times 1\ kg/M\ mg}$$

10.31 Safety

Lost time, injury, and even death can all result when an operator does not apply the rules of safety to all activities involved in operating and maintaining a plant. Practicing safety is not just knowing what to do; it is a mindset. Operators must not only practice safety, but also must know what to do if an accident occurs. Refer to Volume I, Chapter 3, "Safety," for more information.

The following are some safety precautions that operators should observe at all times when working in a treatment plant:

1. Wear safety shoes with steel toes, shanks, and soles that retard slipping. Cork-inserted composition soles provide the best traction for all-around use.

2. Scrub off and wash away slippery algal growths whenever they appear.

3. Keep all areas clear of spilled oil or grease. Use soap and water, not gasoline or solvents, for cleaning.

4. Wear gloves when working with equipment or while in direct contact with wastewater.

5. Do not leave tools, equipment, or materials where they could create a safety hazard.

6. Provide adequate lighting for night work and in areas with limited lighting.

7. Wear spiked shoes in icy winter conditions, and sand icy areas if the ice cannot be thawed away with wash water.

8. Remove only sections of handrails, deck plates, or grating necessary for the immediate job. Removed sections should be properly stored out of the way and properly secured against falling into tanks. The area should be barricaded to prevent unauthorized personnel entry and possible injury.

9. Do not walk on top of the oxidation ditch sidewalls; you may slip and fall into the ditch.

10. Prepare an emergency response plan for each hazard in the plant. Outline each response action to be taken. Rehearse the response plan.

QUESTIONS

Please write your answers to the following questions and compare them with those on page 112.

10.3A   List the major components of an oxidation ditch treatment process.

10.3B   Why are the ends of oxidation ditches well rounded?

10.3C   Why should operators not walk on top of oxidation ditch sidewalls?

10.32 Start-Up

Prior to start-up, prepare a punch list for use in recording any deficiencies that may be revealed. The punch list is a good way to identify problems when everyone is observing the deficiencies. Make no mechanical changes or repairs until you have signed off on the project. Any work may invalidate warranties on the equipment.

There are two primary objectives of start-up. One is to make certain that all mechanical equipment is operating properly. The second is to develop a proper microbial floc (activated sludge) in the oxidation ditch. Floc development is essential for the plant to succeed in reducing the quantities of polluting materials in the raw wastewater.

The start-up procedures described here should be used along with the manufacturer's start-up procedures for all components of the plant. The plant operator, contractor, engineer, equipment manufacturer's representative, and regulatory agency's representative should all be present at the start-up to respond to problems that may be identified.

During start-up of the plant, some construction may still be in progress. Special care should be taken to ensure that all safety procedures are followed at all times.

10.320 Pre-Start Inspection

The following items should be part of the pre-start inspection of an oxidation ditch:

Check all equipment to verify that it is properly installed and secured in place.

Check the inlet structure for debris and clean all debris from the structure prior to start-up.

Clean all debris from the oxidation ditch structure prior to start-up. Check the walkways to ensure there is no debris that can later fall into the channel. Also, inspect the influent and effluent lines to be sure that they are free of debris.

If you are preparing to start a rotor that is a new installation, one that has not been in operation for some time, or one that has just been overhauled, a thorough inspection check should be made prior to starting the rotor to prevent damage to the rotor and injury to personnel.

ROTOR PRE-START INSPECTION CHECKLIST

1. ON/OFF switch is OFF and the main breaker is locked out and tagged

2. Motor is secured to gear reducer

3. Gear reducer assembly is secured to the mounting platform

4. Rotor cylinder shaft is secured to reducer coupling bolts

5. Rotor blades and teeth are secured to the cylinder

6. Driver, rotor, bearings, and stand(s) are properly aligned

7. Rotor turns with a reasonable pull on the cylinder

8. Rotor anti-rotation screws are properly adjusted

9. Proper oil type and quantity is in the reducer

10. All bearings are greased with the proper lubricant

11. All lube oil line fittings are tight

12. Gear reducer housing air vent is open

13. All bolts are tightened

14. All tools and foreign material are clear of the rotor assembly and inlet piping

15. Safety guards over moving parts are properly installed and secure.

Following completion of the pre-start checks, the rotor assembly is ready to start. Be sure all personnel are clear of the rotor assembly. Turn the main power breaker ON. The ON/OFF switch should now be positioned to ON and the rotor will start. Activate the various switches in sequence, starting at the main breaker and working to the local ON/OFF switch. The main breaker should never be used for an ON/OFF switch.

If the rotor assembly is part of a new oxidation ditch installation, the following additional steps should be taken:

1. Check the rotation of the rotor and run the rotor for at least one hour (see "bump start" below).

2. Start the rotor(s) with the rotor(s) out of the water.

3. Check and record the motor amperage and voltage on each phase.

4. Recheck the rotor support bearing(s) and drive alignment. Realign, if necessary, according to manufacturer's installation drawings.

5. Tighten all nuts, bolts, and set screws to prepare the rotor for regular operation. Check and record the mechanical items noted under "Records" in Section 10.330, "Normal Operation."

NOTE: When starting a new or recently overhauled rotor assembly, a "bump start" is recommended BEFORE a routine full start and run is attempted.

A "bump start" allows operation of the rotor assembly for 2 to 3 seconds and is accomplished by briefly positioning the ON/OFF switch to ON and immediately returning it to the OFF position.

This short run-time will allow you to determine if the rotor assembly operates freely and properly, observe whether it operates in the proper rotation, and helps avoid extensive damage to the rotor assembly if it is not properly installed.

Make sure the adjustable weir operates freely and does not bind. Set the weir at the proper elevation in accordance with the O&M manual or the manufacturer's instructions.

The clarifier structure and piping should be inspected and cleared of all debris. All control gates and valves should be checked for smooth operation and proper seating.

The return sludge and waste sludge systems should be examined for leakage and all valves should be operated one complete cycle and set for normal operation. Pumps should be manually operated with liquid in them to check proper operation. One pump should be operated and checked for vibration, excessive noise, or overheating, and the amperage reading should be recorded. The same procedure should then be repeated for the second pump with the first pump shut off. If your plant has waste sludge treatment facilities (holding tanks, lagoons, or drying beds), operate the necessary valves to allow pumping to this part of the plant. Operate each pump manually and note the discharge. Shut off both pumps and return the valves to the normal flow position to return the settled sludge back to the oxidation ditch.

10.321   Plant Start-Up

Start to fill the ditch with water or wastewater. If possible, add a water tank or two full of healthy seed activated sludge from a nearby plant. Divert all wastewater to be treated into the ditch. If the ditch was initially filled with wastewater and it took one or two days to fill the ditch during hot weather, odor problems could develop. Start the rotors when the water level reaches the bottom of the rotor blades.

During plant start-up, a dark gray color may be seen in the developing MLSS, which usually indicates a lack of bacterial buildup in the mixed liquor. If this condition continues for more than several days, check your return sludge system to see that it is operating properly.

Do not start discharging to the clarifier when the water reaches the rotors. Allow the wastewater to continue to fill the ditch and to be treated. When the water level approaches the maximum submergence of the rotors and the peak motor load, start allowing some of the water to be discharged to the clarifier. During start-up, an unstable clarifier effluent may result due to inadequate biological treatment. As this effluent is generally the discharged product (as final effluent), chlorination should be used to reduce health hazards on the receiving waters during this time. State regulatory agencies should be contacted to ensure that the receiving waters will not be harmed as a result of heavily chlorinating the plant effluent. Effluent dechlorination using sulfur dioxide may be required.

During the period of start-up, wastewater testing procedures should be initiated as soon as possible. The actual flow rates should be recorded along with the incoming BOD and COD[34] levels.

Building up of the MLSS concentration is the most important activity to be performed during the start-up process of an oxidation ditch. At least three and possibly up to fifteen days are required to build up the MLSS concentration. In the event that the actual MLSS concentration cannot be determined on a daily basis, you should at least record daily the results of the 30-minute sludge settleability tests. (See Section 10.26, "Laboratory Testing for Small Activated Sludge Plants," and Chapter 14, "Laboratory Procedures," for details on how to conduct

---

34. *COD* (pronounce as separate letters).   Chemical Oxygen Demand. A measure of the oxygen-consuming capacity of organic matter present in wastewater. COD is expressed as the amount of oxygen consumed from a chemical oxidant in mg/L during a specific test. Results are not necessarily related to the biochemical oxygen demand (BOD) because the chemical oxidant may react with substances that bacteria do not stabilize.

sludge settleability tests.) During this period, maintain the highest possible return activated sludge rate.

Dissolved oxygen (DO) should be measured at a sampling location approximately 15 feet (4.5 m) upstream from the rotor. Until the desired MLSS concentration is reached and the 30-minute sludge settleability volume reaches 20 percent, a minimum DO concentration of 2.0 mg/L should be kept in the ditch. Following this period, a lesser DO concentration may be desired, but never less than 0.5 mg/L at a point 15 feet (4.5 m) upstream of the rotors.

Following start-up, when the plant has stabilized, the solids should settle rapidly in the clarifier, leaving a clear, odorless, and stable effluent. The solids should look like particles, golden to rich dark brown in color, with sharply defined edges.

You should not expect immediate results from the start-up procedures. Plant start-up takes time, sometimes over a month if "seed" activated sludge from another treatment plant is not available. Also, some conditions may occur during start-up that would, under normal conditions, indicate a poorly operating process, such as light foaming in the ditch or a cloudy supernatant in the settleable solids tests. These conditions should only be temporary if the information in this section is applied properly.

QUESTIONS

Please write your answers to the following questions and compare them with those on page 112.

10.3D   What are the two primary objectives of start-up?

10.3E   What items or structures should be inspected during the pre-start inspection?

10.3F   During plant start-up, when should the rotors be started?

10.3G   When an oxidation ditch is operating properly, how should the activated sludge solids appear?

10.33   Operation

10.330   Normal Operation

The process controls and normal operation of an oxidation ditch are similar to the activated sludge process. To obtain maximum performance efficiency, the following control methods must be maintained:

1. Proper food supply (measured as BOD or COD and no toxicants) for the microorganisms

2. Proper DO levels in the oxidation ditch

3. Proper ditch environment (no toxicants, and sufficient microorganisms to treat the wastes)

4. Proper ditch detention time to treat the wastes by control of the adjustable weir

5. Proper water/solids separation in the clarifier

Each of these control methods are discussed in detail in this section.

PROPER FOOD SUPPLY FOR THE MICROORGANISMS

Influent flows and waste characteristics are subject to limited control by the operator. Municipal ordinances may prohibit discharge to the collection system of materials damaging to treatment structures or human safety. Control over wastes dumped into the collection system requires a pretreatment facility inspection program to ensure compliance. Alternate means of disposal, pretreatment, or controlled discharge of significantly damaging wastes may be required in order to permit dilution to an acceptable level by the time the waste arrives at the treatment plant. For waste discharge control methods, refer to Chapter 4, "Preventing and Minimizing Wastes at the Source," in *Industrial Waste Treatment,* Volume I, or the *Pretreatment Facility Inspection* manual in this series of operator training manuals.

PROPER DO LEVELS

Proper operation of the process depends on the rotor assembly supplying the right amount of dissolved oxygen to the waste flow in the ditch. For the best operation, a DO concentration of 0.5 to 2.0 mg/L should be maintained just upstream (15 feet or 4.5 m) of the rotors. Over-oxygenation wastes power and excessive DO levels can cause a pinpoint floc to form that does not settle and is lost over the weir in the settling tank.

The DO in the oxidation ditch is regulated by raising or lowering the ditch outlet level control weir. This weir controls the level of water in the ditch, which, in turn, regulates the degree of submergence of the rotors. The level or elevation of the rotors is fixed, but the deeper the rotors sit in the water, the greater the transfer of DO from the air to the water.

PROPER ENVIRONMENT

The oxidation ditch process with its long-term aeration basin is designed to carry MLSS concentrations of 2,000 to 6,000 mg/L. This provides a large organism mass in the system.

The performance of the ditch and ditch environment can be evaluated by conducting a few simple tests and by general observations. The color and characteristics of the floc in the ditch as well as the clarity of the effluent should be observed and recorded daily. Typical tests are settleable solids, DO upstream of the rotor, pH, and residual chlorine in the plant effluent. These test procedures are outlined in Chapter 14, "Laboratory Procedures." Laboratory tests such as BOD, COD, suspended solids, volatile solids, total solids, and microscopic examinations should be performed periodically by the plant operator or an outside laboratory. These test procedures are outlined in Chapter 16, "Laboratory Procedures and Chemistry," in *Operation and Maintenance of Wastewater Treatment Plants,* Volume II. The results will aid you in determining the actual operating efficiency and performance of the process.

Oxidation ditch solids are controlled by regulating the return sludge rate and waste sludge rate. Remember that solids continue to deteriorate as long as they remain in the clarifier. Therefore, adjust the return sludge rate to return the microorganisms in a healthy condition from the final settling tank to the oxidation ditch. If dark solids appear in the settling tank, either the return sludge rate should be increased because the solids are remaining in the clarifier too long, or the DO levels are too low in the oxidation ditch.

Adjusting the waste sludge rate regulates the solids concentration (number of microorganisms) in the oxidation ditch. The appearance of the oxidation ditch surface can be a helpful indication of whether the sludge wasting rate should be increased or decreased. If the surface of the ditch has a white, crisp foam, reduce the sludge wasting rate. Some plants operate successfully with an MLSS concentration of 6,000 to 8,000 mg/L and very little wasting of solids. If the surface has a thick, dark foam, increase the wasting rate. Waste activated sludge may be removed from the ditch by pumping to a sludge holding tank, to sludge drying beds, to sludge lagoons, or to a tank truck. Ultimate disposal may be to larger treatment plants.

Because an oxidation ditch is a biological treatment process, several days may be required before the process responds to operation changes. Make your changes slowly, be patient, and observe and record the results. Allow seven or more days for the process to stabilize after making a change. For additional information on the regulation of the process, see Section 10.26, "Laboratory Testing for Small Activated Sludge Plants," which discusses the use of the settling test for operational control. For more details on controlling the activated sludge process, see *Operation and Maintenance of Wastewater Treatment Plants,* Volume II, Chapter 11, "Activated Sludge."

PROPER TREATMENT TIME AND FLOW VELOCITIES

Treatment time is directly related to the flow of wastewater and is controlled by an adjustable weir. Velocities in the ditch should be maintained at 1.0 to 1.5 feet per second (0.3 to 0.45 m/sec) to prevent the deposition of floc. With this in mind, the ditch contents should travel the complete circuit of the ditch, or from rotor to rotor, every 3 to 6 minutes. If the rotors are operated by time clocks (30 minutes off and 30 minutes on, for example), the velocities in the ditch must be sufficient to resuspend any settled material. Settled sludge that is not resuspended forms sludge deposits that can become anaerobic and toxic to the activated sludge organisms.

PROPER WATER/SOLIDS SEPARATION

MLSS that have entered and settled in the secondary clarifier are continuously removed from the clarifier and returned to the oxidation ditch or to the waste sludge handling facility. Usually, all sludge formed by the process and settled in the clarifier is returned to the ditch, except when wasting sludge. Scum that is captured on the surface of the clarifier also is removed from the clarifier and either returned to the oxidation ditch for further treatment or disposed of by burial. All disposal methods for waste sludge, screenings, foam, or other debris must be approved by the regulatory agency.

OBSERVATIONS

You can control and adjust some aspects of your oxidation ditch plant operation with the help of some general observations. General observations of the plant are important to help you determine whether or not your oxidation ditch is operating as intended. These observations include color of the mixed liquor in the ditch, odor at the plant site, and clarity of the ditch and sedimentation tank surfaces.

- Color. You should note the color of the mixed liquor in the ditch on a daily basis. A properly operating oxidation ditch plant mixed liquor should have a medium to rich dark brown color. If the MLSS, following proper start-up, change color from a dark brown to a light brown and the MLSS appear to be thinner than before, the sludge waste rate may be too high, which may cause the plant to lose efficiency in removing waste materials. By decreasing sludge waste rates before the color lightens too much, you can ensure that the plant effluent quality will not deteriorate due to low MLSS concentrations.

  If the MLSS become black, the ditch is not receiving enough oxygen and has gone "anaerobic." The oxygen output of the rotors must be increased to eliminate the black color and return the process to normal aerobic operation. This is done by increasing the submergence level of the rotor.

- Odor. When the oxidation ditch plant is operating properly, there will be little or no odor. Odor, if detected, should have an earthy smell. If an odor other than this is present, you should determine the cause. Odor similar to rotten eggs indicates that the ditch may be going anaerobic, requiring more oxygen or a higher ditch velocity to prevent the deposit of solids. The color of the MLSS could be black if the ditch is turning anaerobic.

  Odor may also be a sign of poor housekeeping. Grease and solids buildup on the edge of the ditch or settling tank will go anaerobic and cause odors. With an oxidation ditch, odors are much more often caused by poor housekeeping than poor operation. Regularly verify that any odors are not being generated in the collection system.

- Clarity. In a properly operating oxidation ditch a layer of clear water or supernatant is usually visible a few feet upstream from the rotor. The depth of this relatively clear water may vary from almost nothing to as much as two or more inches (5 cm) above the mixed liquor. The clarity will depend on the ditch velocity and the settling characteristics of the activated sludge solids. Flocculated particles may be seen in the clarified water.

  Two other good indications of a properly operating oxidation ditch are the clarity of the settling tank water surface and the oxidation ditch surface free of foam buildup. Foam buildup in the ditch (normally not enough to be a nuisance) is usually caused by an insufficient MLSS concentration. Most frequently, foam buildup is only seen during plant start-up and will gradually disappear.

  The best indication of plant performance is the clarity of the effluent from the secondary clarifier that is discharged over the weirs. A very clear effluent shows that the plant is achieving excellent pollutant removals. A cloudy effluent often indicates a problem with the plant operation.

- Foam. Heavy foam can build up if the MLSS concentration is too high and the biological activity is operating at a high level of respiration.

RECORDS

Accurate records are essential and invaluable in evaluating rotor efficiency and in establishing normal operating conditions. The items listed below should be checked daily, or as scheduled, and a record made of these checks.

ITEM	CONDITION
Mechanical	
1. Motor	a. High or uneven amperage
	b. High temperature
	c. Unusual noise
	d. Operating hours
2. Gear reducer	a. Unusual bearing or gear noise
	b. Proper oil level (use sight glass or dipstick to check)
3. Gear reducer housing	Air vent pipe "open"
4. Outboard bearing	Unusual bearing noise
5. Rotor	a. Unusual noise or vibration
	b. Remove any debris caught in rotor blades such as rags, weeds, or plastic goods.
Operation	
1. DO concentration	Use portable DO probe or perform Winkler Method lab test. DO should be taken approximately 15 feet (4.5 m) upstream of the rotor(s). Maintain 0.5 to 2.0 mg/L at this point.
2. Ditch velocity	1.0 to 1.5 fps (0.3 to 0.45 mps)
3. Ditch water level	Read STAFF GAUGE[35] in ditch. Convert this reading to brush aerator submergence.
4. Ditch surface condition	Neither white foam nor a heavy brown scum
5. Mixed liquor suspended solids concentration (MLSS)	2,000 to 6,000 mg/L

FINAL PLANT SURVEY

Before leaving for the day, make one final inspection around the plant. Addressing the following questions may help you leave the plant in proper operating condition:

1. Are any pieces of equipment operating poorly (hot bearings, loose belts)? Will they need to be checked before the next scheduled day of operator attendance?

2. Are return sludge rates set to the correct level?

3. Are flowmeters clean and operating properly?

4. Are inlet gates set properly in case of high flows?

5. Has the rotor level of submergence been set properly?

6. If some equipment is time-clock controlled, are the time clocks set?

7. If remote alarms are used to warn operators about equipment failures, are these set properly?

8. Is equipment stored and locked to prevent vandalism or theft?

9. Are outside lights on or set to come on automatically?

PERFORMANCE

Figure 10.18 shows the performance record of an actual oxidation ditch.

QUESTIONS

Please write your answers to the following questions and compare them with those on page 112.

10.3H   How is the DO in the oxidation ditch regulated?

10.3I   What should be the velocity in the ditch?

10.3J   What observations should be made daily to help indicate the performance of the oxidation ditch?

10.331   Abnormal Operation

While checking the mechanical items noted under "Records" in Section 10.330, "Normal Operation," you may occasionally find some abnormal conditions. Serious damage to the rotor assembly may result if the abnormal conditions are not corrected as soon as possible.

Rotor assembly operation is essential for the efficient operation of the oxidation ditch. Loss of rotor-generated mixing and air entrainment to the mixed liquor in the ditch for an extended period of time will turn your oxidation ditch activated sludge process into an upset, bulking activated sludge process.

Table 10.4 lists some abnormal brush rotor conditions, possible causes, and suggested operator responses to the conditions that will aid you in the safe and efficient operation of the rotor.

Environmental factors that affect the wastewater treatment process include temperature and precipitation. The wastewater temperature affects the activity of the microorganisms. During cold weather, this reduced activity might lower the efficiency of the treatment system. Besides biological effects of temperature,

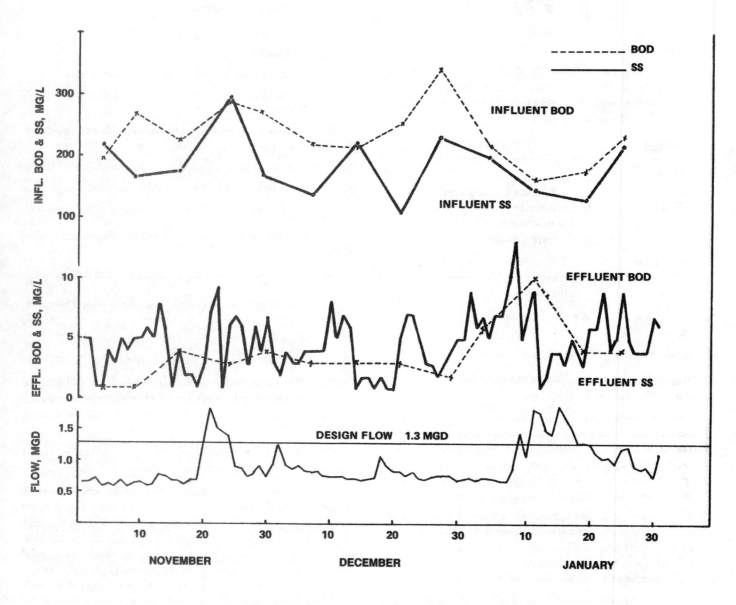

Fig. 10.18   Performance record of an actual oxidation ditch

TABLE 10.4 ABNORMAL BRUSH ROTOR OPERATION

Item	Abnormal Condition	Possible Cause	Operator Response
Motor	Motor off	Ambient temperature in switchboard panel or room too high.	Check for overloading. Reduce temperature with fans.
		Tripped breakers.	Reset breakers.
		Degree of rotor submergence results in excessive amperage draw.	Adjust rotor submergence.
		Tripped breakers.	Reset breakers.
		Motor shorted or burned out.	Check for overloading. Repair motor.
	Motor hot to the touch	Dry bearings.	Lubricate bearings.
		Excessive grease.	Remove all grease and then grease motor properly.
		Worn bearings.	Replace bearings.
		Excessive amperage draw (motor load).	Adjust rotor submergence.
Gear Reducer	Grinding, chipping, whirring, or whining noise	Low oil level.	Drain and refill. Check for leakage and make repairs.
Bearings	Grinding or bumping noise	Insufficient bearing grease.	Grease bearing(s) routinely.
		Worn bearing(s)	Replace bearing(s).

the flocculation and sedimentation of the mixed liquor solids are not as effective at lower temperatures.

Ice buildup will hinder or stop altogether the proper operation of mechanical parts such as rotors and sludge scraper mechanisms. Chunks of ice may develop that float in the ditch and eventually enter the area of the rotor assembly. Unless adequate safeguards are provided, serious damage to the rotor will result. Below are some of the safeguards that must be considered:

1. The oxidation ditch in cold weather areas should be operated for the maximum detention time practical in order to conserve as much heat in the wastewater as possible. This action will inhibit ice chunk formation. Set the effluent control weir at the highest possible level.

2. The splashing and spraying action produced by the rotor will allow ice to form on the rotor assembly. Serious consideration must be given to covering the whole rotor assembly with a structure made of wood, fiberglass, or other suitable material that will bridge over the ditch. Installations of this type generally are not heated.

NOTE: If the rotor is shut off for maintenance, ice will form on it. Prior to restarting the rotor, hose it off with water to thaw the ice. Otherwise, the ice could cause vibrations that can damage the rotor.

CAUTION: Ice conditions in winter require spiked shoes and sanding icy areas if the ice cannot be thawed with wash water.

In some treatment plants, very heavy rains or snow melts cause flows to the treatment plant to exceed the design flow by three to four times. This is generally accompanied by a weaker wastewater in terms of BOD and COD due to the dilution effect of the stormwater. These hydraulic overloads may exceed the capacity of the clarifier to settle sludge solids properly. When this occurs, extremely high BOD, COD, and suspended solids concentrations will be discharged to the receiving stream in the final effluent. Possible process upsets can occur if corrective action is not taken in this situation. Coagulant chemicals such as alum, ferric chloride, and *POLYMERS* [36] can be added to the final settling tank to assist in settling solids during abnormal conditions. When chemicals such as alum are used, the volume

---

36. *Polymer* (POLY-mer). A long-chain molecule formed by the union of many monomers (molecules of lower molecular weight). Polymers are used with other chemical coagulants to aid in binding small suspended particles to larger chemical flocs for their removal from water. Also see POLYELECTROLYTE.

of return sludge is increased and the pH of the sludge may be reduced.

Another method of preventing hydraulic overloads from causing this discharge of high BOD, COD, and suspended solids concentrations is to shut down one or more of the rotor assemblies in the ditch. This will allow the ditch to act as a large settling tank, keeping the MLSS from flowing into the clarifier where they could be washed out. When the plant flow decreases to more normal levels, the rotor(s) can then be restarted to resume normal operation.

Other than these corrective actions, there is not much the operator can do to offset the changes in treatment efficiency caused by temperature changes and high flows during abnormal conditions without plant modifications. However, ice buildup can be controlled by frequent observation and removal during cold periods. For persistent cold weather problems, constructing a lightweight building over the ditch and clarifier may be more economical in the long run than fighting ice repeatedly. The normal wastewater temperature will generally supply sufficient heat inside a building to prevent ice from forming. However, the moist atmosphere over the water surface may lead to fogging, excess condensation, and even ice buildup on uninsulated surfaces inside the building.

Usually, it is necessary to vary the amount of MLSS in the ditch as seasons change. If complete *NITRIFICATION* [37] is desired, more microorganisms are needed to treat the same amount of wastes during cold weather than during warm weather because the colder the water, the less active the microorganisms.

10.332 Shutdown

Shutdown of the oxidation ditch may be necessary for emergency repairs. Schedule any shutdowns so that the activated sludge microorganisms will be without air for the shortest possible time period. Problems such as odors and a loss of the microorganism culture can start within two hours after the rotors have been shut off. The microorganism culture may start to deteriorate within 10 minutes, but can recover quickly if the downtime does not exceed four hours. If possible, keep one rotor operating at all times.

To prevent injury to personnel, take the following steps to shut down the rotor assembly whenever any maintenance function is performed.

1. Turn the local ON/OFF switch to OFF.

2. Turn any control panel or motor control center ON/OFF switches to the OFF position.

3. Turn the main power breaker OFF.

4. Lock out and tag the main power breaker in the OFF position.

Maintenance work may now be performed.

If the oxidation ditch treats seasonal loads and is shut down during the off-season, protect equipment from the weather and moisture.

10.333 Troubleshooting

Refer to Section 10.252, "Troubleshooting." The problems of package aeration plants and the solutions are very similar to those of oxidation ditches.

If floatables appear in the final settling tank, examine the baffle around the level control weir. Because oxidation ditches do not have primary clarifiers, plastic goods and other floatables can be a problem if the baffle is not properly adjusted.

10.334 Operational Strategy

Refer to Section 10.254, "Operational Strategy." to learn more about the items considered in the operational strategy for package aeration plants, which are the same as those considered for oxidation ditches.

10.34 Maintenance

10.340 Housekeeping

A general daily cleanup at the plant is important. This helps your plant perform better and gives you a more pleasant place to work. Daily cleanup related to plant maintenance usually includes removing and disposing of debris that may have accumulated on the bar screen, removing grease and scum from the surface of the clarifier, and washing or brushing down the ditch and clarifier weirs and walls.

10.341 Equipment Maintenance

Regularly scheduled equipment maintenance must be performed. To make this task easier, read and familiarize yourself with the equipment manufacturers' instruction manuals. Check each piece of equipment daily to see that it is functioning properly. You may have very few mechanical devices in your oxidation ditch plant, but they are all important. The rotors and pumps should be inspected to see that they are operating properly. If pumps are clogged, the obstructions should be removed. Listen for unusual noises and identify their sources. Check for loose bolts and tighten as needed. Look for evidence of excess heat such as burnt paint. Be aware of unusual odors that may indicate overheated devices or switches. Uncovering a mechanical problem in its early stages could prevent a costly repair or replacement at a later date.

Lubrication should also be performed on a fixed operating schedule and properly recorded. Make sure that the proper lubricants are used. Avoid overlubrication, which is wasteful and reduces the effectiveness of lubricant seals and may cause overheating of bearings or gears. Follow the lubrication and maintenance instructions furnished with each piece of equipment. If

---

37. *Nitrification* (NYE-truh-fuh-KAY-shun).   An aerobic process in which bacteria change the ammonia and organic nitrogen in wastewater into oxidized nitrogen (usually nitrate). The second-stage BOD is sometimes referred to as the nitrogenous BOD (first-stage BOD is called the carbonaceous BOD). Also see DENITRIFICATION.

you are unable to find the instructions, write to the manufacturer for a new set.

Equipment should be painted periodically. In addition to beautifying your plant, it gives a good protective coating on all iron or metal surfaces and will prolong the life of the metal.

Never paint over equipment identification tags! This important information needs to be visible at all times.

You should know who is the manufacturer and local supplier of every piece of equipment in your plant. Knowing how to quickly contact the manufacturer may save important time and money when equipment breakdowns occur. Your equipment maintenance should include the following:

1. Motors

   Motors should be greased after about 2,000 hours of operation or as often as conditions and/or the manufacturer recommends. *The motor must be stopped when greasing begins.*

   Remove the filler and drain plugs, free the grease holes of any hardened grease, and then add new grease through the filler hole until it starts to come out of the drain hole. Start the motor and let it run for about 15 minutes to expel any excess grease. Stop the motor again and reinstall the filler and drain plugs.

   Rotor assembly motors are generally exposed to a high degree of moisture. For this reason, the motor should be checked at least yearly by an electrician to be sure that all parts are in good working condition.

   See *Operation of Wastewater Treatment Plants,* Volume II, Chapter 15, "Maintenance," for more information.

2. Gear Reducer

   Generally, all new oil-lubricated equipment has a break-in period of about 400 hours. After this time, the oil should be drained from the gear reducer, the unit flushed, and new oil added. This procedure removes fine metal particles that have worn off the internal components as a result of the initially close tolerances as the equipment was broken in. If large quantities of fine metal particles are found after the break-in period, the manufacturer should be consulted. A small magnet can be used to check the sediment in the old oil for iron particles, which will be attracted to the magnet.

   Oil in the gear reducers should be checked to see if water has entered the unit. If water is in the oil, it will cause the oil to have a cloudy tan or cocoa brown color.

3. Bearings

   a. Gear Reducer Bearings

      These bearings are generally greased twice per week while the rotor assembly is in operation to ensure proper distribution of the lubricant.

      Extension tubes or pipes are usually attached to the bearing cap. A grease fitting is then installed on the other end of the tube or pipe. This allows for safety when greasing the bearings while the rotor is operating by extending the grease fitting away from the hazardous area of the operating rotor. This grease fitting modification may be used in other equipment applications where lubrication of running equipment would present a safety hazard.

   b. Rotor Intermediate and Outboard Bearings

      These bearings are generally lubricated daily while the rotor assembly is in operation.

      In most bearing lubrication applications, overlubrication is wasteful and can reduce or destroy the effectiveness of lubricant seals and may cause overheating of bearings. Rotor intermediate and outboard bearings are generally the exception. As a rule, these bearings cannot be overlubricated.

      Some rotor bearings are protected by neoprene seals that retain lubricant and keep out moisture. The rotor shaft may have "slingers" (rings on a shaft that throw oil from the shaft before it gets to the oil seal), that protect the bearing, and a detachable shield may cover the bearing.

4. Lubrication

   Lubricating equipment in locations that have extreme weather variations is critical. Oil and grease must be changed, using the proper type and grade for the expected weather conditions, as determined by the equipment manufacturer.

   Using the proper type and grade of oil is essential. If the oil is too thin or too thick, it will interfere with the proper functioning of bearings and gears.

   Bearing grease changes may be accomplished by flushing the grease housing with a 30- to 90-weight oil. The oil is added in the same manner as the grease.

   Lubricant suppliers can provide you with a list of oils and greases that are equal to the product recommended by the equipment manufacturer.

QUESTIONS

Please write your answers to the following questions and compare them with those on page 112.

10.3K   What problems can be caused by ice during cold weather?

10.3L   Why are more microorganisms (higher MLSS) needed in the ditch during cold weather than during warm weather?

10.3M   Why is a general cleanup each day at the plant important?

10.35   Operational Guidelines

10.350   English System

Section 10.301, "Description of Oxidation Ditches," lists operational guidelines for an oxidation ditch plant. This section outlines the procedures to follow to calculate the guidelines for your oxidation ditch. The first step is to draw a sketch of your oxidation ditch to obtain the important dimensions and to record the appropriate flow and wastewater characteristics.

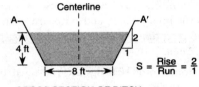

CROSS SECTION OF DITCH

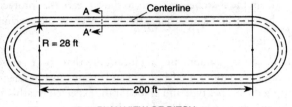

PLAN VIEW OF DITCH

Known

Oxidation Ditch Dimensions

Length, ft	= 200 ft
Bottom, ft	= 8 ft
Depth, ft	= 4 ft
Radius, ft	= 28 ft
Slope	= 2

Flow and Waste Characteristics

Flow, MGD	= 0.2 MGD
BOD, mg/L	= 200 mg/L
Influent SS, mg/L	= 200 mg/L
MLSS, mg/L	= 4,000 mg/L
Volatile Matter in MLSS	= 70% or 0.70

Unknown

1. Ditch Volume, cu ft and MG
2. BOD Loading, lbs/day and lbs/day/1,000 cu ft
3. F/M, lbs BOD/day/lb MLVSS
4. Sludge Age, days
5. Detention Time, hours

1. Calculate the oxidation ditch volume in cubic feet and million gallons.

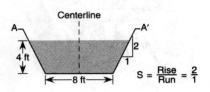

a. Find the average width of the water in the ditch in feet.

$$\text{Width, ft} = \text{Bottom, ft} + \frac{\text{Depth, ft}}{\text{Slope}}$$

$$= 8\text{ ft} + \frac{4\text{ ft}}{2}$$

$$= 8\text{ ft} + 2\text{ ft}$$

$$= 10\text{ ft}$$

b. Find the cross-sectional area of the water in the ditch in square feet.

$$\text{Area, sq ft} = \text{Width, ft} \times \text{Depth, ft}$$

$$= 10\text{ ft} \times 4\text{ ft}$$

$$= 40\text{ sq ft}$$

c. Find the length of the two circular ends of the ditch in feet.

$$\text{Ends Length, ft} = 2 \times \pi \times \text{Radius, ft}$$

$$= 2 \times 3.14 \times 28\text{ ft}$$

$$= 176\text{ ft}$$

d. Determine the total centerline length of the ditch in feet.

$$\text{Total Length, ft} = \text{Ends Length, ft} + (\text{Length, ft} \times 2)$$

$$= 176\text{ ft} + (200\text{ ft} \times 2)$$

$$= 576\text{ ft}$$

e. Calculate the oxidation ditch volume in cubic feet and million gallons.

$$\text{Vol, cu ft} = \text{Total Length, ft} \times \text{Area, sq ft}$$

$$= 576\text{ ft} \times 40\text{ sq ft}$$

$$= 23{,}040\text{ cu ft}$$

or

$$= 23.04 \bullet (1{,}000\text{ cu ft}) \text{ (read as 23.04 one thousand cubic feet)}$$

$$\text{Vol, MG} = \text{Vol, cu ft} \times 7.48\text{ gal/cu ft}$$

$$= 23{,}040\text{ cu ft} \times 7.48\text{ gal/cu ft}$$

$$= 172{,}339\text{ gal}$$

$$= 0.172\text{ MG}$$

2. Calculate the BOD loading in pounds per day and pounds per day per 1,000 cubic feet of ditch volume.

$$\text{BOD Loading, lbs/day} = (\text{Flow, MGD})(\text{BOD, mg/L})(8.34\text{ lbs/gal})$$

$$= (0.2\text{ MGD})(200\text{ mg/L})(8.34\text{ lbs/gal})$$

$$= 334\text{ lbs BOD/day}$$

$$\text{BOD Loading, lbs/day/} 1{,}000\text{ cu ft} = \frac{\text{BOD Loading, lbs/day}}{\text{Ditch Volume, 1,000 cu ft}}$$

$$= \frac{334\text{ lbs BOD/day}}{23.04 \bullet 1{,}000\text{ cu ft}}$$

$$= 14\text{ lbs BOD/day/1,000 cu ft}$$

3. Determine the food/microorganism ratio in pounds BOD per day per pound MLVSS.

a. Microorganisms are measured as pounds of mixed liquor volatile suspended solids (MLVSS) under aeration in the oxidation ditch. VM means volatile matter. Determine MLVSS in pounds.

$$\text{MLVSS, lbs} = (\text{Vol, MG})(\text{MLSS, mg/L})(\text{VM})(8.34 \text{ lbs/gal})$$

$$= (0.172 \text{ MG})(4{,}000 \text{ mg/L})(0.70)(8.34 \text{ lbs/gal})$$

$$= 4{,}017 \text{ lbs MLVSS}$$

b. Determine the food/microorganism ratio in pounds BOD per day per pound MLVSS.

$$\text{F/M, lbs BOD/day/lb MLVSS} = \frac{\text{BOD, lbs/day}}{\text{MLVSS, lbs}}$$

$$= \frac{334 \text{ lbs BOD/day}}{4{,}017 \text{ lbs MLVSS}}$$

$$= 0.08 \text{ lb BOD/day/lb MLVSS}$$

4. Determine the sludge age in days.

a. Calculate the pounds of solids under aeration.

$$\text{Aeration Solids, lbs} = (\text{Ditch Vol, MG})(\text{MLSS, mg/L})(8.34 \text{ lbs/gal})$$

$$= (0.172 \text{ MG})(4{,}000 \text{ mg/L})(8.34 \text{ lbs/gal})$$

$$= 5{,}738 \text{ lbs}$$

b. Calculate the solids fed to the ditch in pounds per day.

$$\text{Solids Added, lbs/day} = (\text{Flow, MGD})(\text{Infl SS, mg/L})(8.34 \text{ lbs/gal})$$

$$= (0.2 \text{ MGD})(200 \text{ mg/L})(8.34 \text{ lbs/gal})$$

$$= 334 \text{ lbs/day}$$

c. Calculate the sludge age in days.

$$\text{Sludge Age, days} = \frac{\text{Aeration Solids, lbs}}{\text{Solids Added, lbs/day}}$$

$$= \frac{5{,}738 \text{ lbs}}{334 \text{ lbs/day}}$$

$$= 17 \text{ days}$$

5. Calculate the ditch detention time in hours.

$$\text{Detention Time, hours} = \frac{(\text{Ditch Volume, MG})(24 \text{ hr/day})}{\text{Flow, MGD}}$$

$$= \frac{(0.172 \text{ MG})(24 \text{ hr/day})}{0.2 \text{ MGD}}$$

$$= 20.6 \text{ hours}$$

10.351 *Metric System*

This section outlines the procedures for you to follow to calculate the guidelines for your oxidation ditch using the metric system. The first step is to draw a sketch of your oxidation ditch to obtain the important dimensions and to record the appropriate flow and wastewater characteristics.

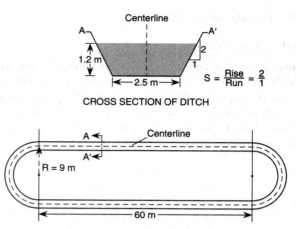

CROSS SECTION OF DITCH

PLAN VIEW OF DITCH

Known

Oxidation Ditch Dimensions

Length, m	= 60 m
Bottom, m	= 2.5 m
Depth, m	= 1.2 m
Radius, m	= 9 m
Slope	= 2

Flow and Waste Characteristics

Flow, cu m/day	= 750 cu m/day
BOD, mg/L	= 200 mg/L
Influent SS, mg/L	= 200 mg/L
MLSS, mg/L	= 4,000 mg/L
Volatile Matter in MLSS	= 70% or 0.70

Unknown

1. Ditch Volume, cu m
2. BOD Loading, kg/day and kg/day/1,000 cu m
3. F/M, kg BOD/day/kg MLVSS
4. Sludge Age, days
5. Detention Time, hours

1. Calculate the oxidation ditch volume in cubic meters.

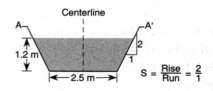

a. Find the average width of the water in the ditch in meters.

$$\text{Width, m} = \text{Bottom, m} + \frac{\text{Depth, m}}{\text{Slope}}$$

$$= 2.5 \text{ m} + \frac{1.2 \text{ m}}{2}$$

$$= 2.5 \text{ m} + 0.6 \text{ m}$$

$$= 3.1 \text{ m}$$

b. Find the cross-sectional area of the water in the ditch in square meters.

$$\text{Area, sq m} = \text{Width, m} \times \text{Depth, m}$$
$$= 3.1 \text{ m} \times 1.2 \text{ m}$$
$$= 3.72 \text{ sq m}$$

c. Find the length of the two circular ends of the ditch in meters.

$$\text{Ends Length, m} = 2 \times \pi \times \text{Radius, m}$$
$$= 2 \times 3.14 \times 9 \text{ m}$$
$$= 56.5 \text{ m}$$

d. Determine the total centerline length of the ditch in meters.

$$\text{Total Length, m} = \text{Ends L, m} + (\text{Length, m} \times 2)$$
$$= 56.5 \text{ m} + (60 \text{ m} \times 2)$$
$$= 176.5 \text{ m}$$

e. Calculate the oxidation ditch volume in cubic meters.

$$\text{Vol, cu m} = \text{Total Length, m} \times \text{Area, sq m}$$
$$= 176.5 \text{ m} \times 3.72 \text{ sq m}$$
$$= 657 \text{ cu m}$$
$$= 0.657 \bullet (1{,}000 \text{ cu m}) \text{ (read as 0.657 one thousand cubic meters)}$$

2. Calculate the BOD loading in kilograms per day and kilograms per day per 1,000 cubic meters of ditch volume.

$$\text{BOD Loading, kg/day} = (\text{Flow, cu m/day})(\text{BOD, mg/L})\left(\frac{1 \text{ kg}}{1{,}000{,}000 \text{ mg}}\right)\left(\frac{1{,}000 \text{ L}}{1 \text{ cu m}}\right)$$
$$= (750 \text{ cu m/day})(200 \text{ mg/L})\left(\frac{1 \text{ kg}}{1{,}000{,}000 \text{ mg}}\right)\left(\frac{1{,}000 \text{ L}}{1 \text{ cu m}}\right)$$
$$= 150 \text{ kg BOD/day}$$

$$\text{BOD Loading, kg/day/1{,}000 cu m} = \frac{\text{BOD, kg/day}}{\text{Ditch Volume, 1{,}000 cu m}}$$
$$= \frac{150 \text{ kg BOD/day}}{0.657 \bullet 1{,}000 \text{ cu m}}$$
$$= 228 \text{ kg BOD/day/1{,}000 cu m}$$

3. Determine the food/microorganism ratio in kilograms BOD per day per kilogram MLVSS.

a. Microorganisms are measured as kilograms of mixed liquor volatile suspended solids (MLVSS) under aeration in the oxidation ditch. VM means volatile matter. Determine MLVSS in kilograms.

$$\text{MLVSS, kg} = (\text{Vol, cu m})(\text{MLSS, mg/L})(\text{VM})\left(\frac{1 \text{ kg}}{1{,}000{,}000 \text{ mg}}\right)\left(\frac{1{,}000 \text{ L}}{1 \text{ cu m}}\right)$$
$$= (657 \text{ cu m})(4{,}000 \text{ mg/L})(0.70)\left(\frac{1 \text{ kg}}{1{,}000{,}000 \text{ mg}}\right)\left(\frac{1{,}000 \text{ L}}{1 \text{ cu m}}\right)$$
$$= 1{,}840 \text{ kg MLVSS}$$

b. Determine the food/microorganism ratio in kilograms BOD per day per kilogram MLVSS.

$$\text{F/M, kg BOD/day/kg MLVSS} = \frac{\text{BOD, kg/day}}{\text{MLVSS, kg}}$$
$$= \frac{150 \text{ kg BOD/day}}{1{,}840 \text{ kg MLVSS}}$$
$$= 0.08 \text{ kg BOD/day/kg MLVSS}$$

4. Determine the sludge age in days.

a. Calculate the kilograms of solids under aeration.

$$\text{Aeration Solids, kg} = (\text{Ditch Vol, cu m})(\text{MLSS, mg/L})\left(\frac{1 \text{ kg}}{1{,}000{,}000 \text{ mg}}\right)\left(\frac{1{,}000 \text{ L}}{1 \text{ cu m}}\right)$$
$$= (657 \text{ cu m})(4{,}000 \text{ mg/L})\left(\frac{1 \text{ kg}}{1{,}000{,}000 \text{ mg}}\right)\left(\frac{1{,}000 \text{ L}}{1 \text{ cu m}}\right)$$
$$= 2{,}628 \text{ kg}$$

b. Calculate the solids fed to the ditch in kilograms per day.

$$\text{Solids Added, kg/day} = (\text{Flow, cu m/day})(\text{Infl SS, mg/L})\left(\frac{1 \text{ kg}}{1{,}000{,}000 \text{ mg}}\right)\left(\frac{1{,}000 \text{ L}}{1 \text{ cu m}}\right)$$
$$= (750 \text{ cu m/day})(200 \text{ mg/L})\left(\frac{1 \text{ kg}}{1{,}000{,}000 \text{ mg}}\right)\left(\frac{1{,}000 \text{ L}}{1 \text{ cu m}}\right)$$
$$= 150 \text{ kg/day}$$

c. Calculate the sludge age in days.

$$\text{Sludge Age, days} = \frac{\text{Aeration Solids, kg}}{\text{Solids Added, kg/day}}$$
$$= \frac{2{,}628 \text{ kg}}{150 \text{ kg/day}}$$
$$= 17.5 \text{ days}$$

5. Calculate the ditch detention time in hours.

$$\text{Detention Time, hours} = \frac{(\text{Ditch Volume, cu m})(24 \text{ hr/day})}{\text{Flow, cu m/day}}$$
$$= \frac{(657 \text{ cu m})(24 \text{ hr/day})}{750 \text{ cu m/day}}$$
$$= 21 \text{ hours}$$

QUESTION

Please write your answers to the following questions and compare them with those on page 112.

10.3N Determine the BOD loading on an oxidation ditch in (1) pounds per day and (2) kilograms per day. The inflow is 0.8 MGD and the influent BOD is 250 mg/L. 1 MGD = 3,785 cu m/day.

10.4 REVIEW OF PLANS AND SPECIFICATIONS

As an operator, you can be very helpful to design engineers in pointing out some design features that would make your job easier. This section lists some of the items that you should look for when reviewing plans and specifications for expansion of existing facilities or construction of a new package plant or oxidation ditch plant.

10.40 Package Plants

1. Is the plant designed for implementation of modifications to the activated sludge process (adequate flexibility to accommodate future treatment requirements)?

2. Are adequate standby units (equipment) designed into the system?

3. Are ladders, railings, and walkways provided to allow safe, easy access to equipment, pipes, and valves for normal operation, routine maintenance, or repair?

4. Are adequate remote and local controls provided for the mechanical equipment?

5. Are the equipment and related instrumentation designed to operate at low flows and load levels common in the early stages of plant operation?

6. Are adequate dewatering systems provided to permit rapid servicing of submerged equipment?

7. Is the chlorination facility flexible enough to allow for prechlorination, and are adequate control devices provided?

8. Is the treatment plant's total connected horsepower adequate to allow operation of all equipment in parallel?

9. Are flow equalization facilities provided to handle high flows during wet weather or industrial discharges?

10. Is adequate support equipment provided to allow easy and safe removal of aeration diffusers?

11. Is a laboratory provided with appropriate equipment for conducting at least the minimum process control tests?

12. Are adequate sludge drying beds provided with consideration given to wet weather conditions?

13. Is a sludge transfer pump provided to transfer sludge to the drying beds or solids handling facilities?

14. Are the drying beds designed for easy removal of dried sludge and are there provisions for proper disposal of the sludge that comply with regulatory agency requirements?

15. Is there a provision for waste or digested sludge liquid storage in place of or in addition to sludge drying beds?

16. Is standby or auxiliary power provided?

10.41 Oxidation Ditches

1. Influent and return activated sludge should enter an oxidation ditch just upstream of a rotor assembly to afford immediate mixing with mixed liquor in the channel.

2. Effluent should exit the oxidation ditch upstream of the rotor and far enough upstream from the injection of the influent and return activated sludge to prevent short-circuiting.

3. Water level in the aeration channel should be controlled by an adjustable weir. In calculating weir height or "set point," use maximum raw flow plus maximum recirculated flow to prevent excessive rotor immersions.

4. Walkways with railings must be provided across the aeration channel to allow access to the rotor for maintenance. The normal location is upstream of the rotor. Location should be such to prevent spray from the rotor on the walkway. Approved flotation devices should be provided at strategic locations if accidental entry into the channel is possible.

5. All basins with water depths over 6 feet (1.8 m) should use horizontal baffles, placed within 15 feet (4.5 m) downstream of the rotor, to provide proper mixing in the entire depth of the basin.

6. In a single oxidation ditch or aeration channel, the rotor drive assembly should be on the outboard side for ease of access.

7. The ditch should be constructed with some type of lining. Consideration should be given to the most economical means of lining available in the particular plant location. Options include gunite or shotcrete, poured concrete, asphalt, precast concrete, or tile liners.

8. All drive and gear assemblies should be elevated out of the water and placed in safe and easy-to-access locations for maintenance.

9. Standby or auxiliary power must be provided to operate critical equipment.

10. In order to maintain adequate treatment, floating aerators and related equipment should be provided in case of rotor failure.

11. Secondary clarifier.

    a. Surface Rate: 600 gal/day/sq ft (24 cu m/day/sq m) based upon plant design flow. Where wide variations in plant hydraulic loading are expected, care should be taken to limit the maximum instantaneous surface rate to 1,200 gal/day/sq ft (49 cu m/day/sq m).

    b. Solids Loading: Normal oxidation ditch operation and surface loading rates will prevent excessive solids loading to the final clarifier. However, if complete nitrification is required 12 months per year, the maximum instantaneous solids loading must not exceed 35 lbs/day/sq ft (150 kg/day/sq m).

    c. Detention Time: Three hours based on plant design flow. In no case, however, should the final clarifier have a side water depth of less than eight feet (2.4 m).

12. Effluent discharge.

   a. Receiving Stream: The discharge pipe should be located where flood water will not flow back into the plant if there is a power outage or pump failure. A tide gate or check valve can be installed in the effluent line to prevent backflows. Be sure the outlet in the receiving waters is submerged to reduce foam and scum problems.

   b. Percolation Ditch: The design should be based on a percolation rate of gallons per *LINEAL*[38] foot of ditch. Provisions must be made to prevent blowing sand or dirt from entering the ditch.

13. Cold climates.

   a. Provisions should be made to maximize detention times in order to conserve as much heat as possible in the wastewater.

   b. Determine if it is practical to cover the oxidation ditch. If not, consider a large and a small ditch or a single ditch that can be modified to use only half its total capacity.

   c. Be sure that all equipment requiring normal maintenance is housed and/or heated. This will extend the useful life of the equipment and facilitate service work. Changing a gear box, repairing a pipe, or installing electrical components becomes a major task at very low temperatures.

   d. The rotor assembly should be provided with a lightweight cover to prevent rotor icing.

   e. Equip the clarifier with a lightweight tarpaulin to keep heavy snow out of the clarifier and to reduce problems resulting from freezing.

   f. A subsurface discharge pipeline and percolation field should be provided for effluent discharge in winter months. A percolation ditch will freeze over in the winter.

QUESTIONS

Please write your answers to the following questions and compare them with those on page 112.

10.4A   What provisions should be made to permit rapid servicing of submerged equipment?

10.4B   What items would you check when reviewing the location of a walkway across an aeration channel that provides access to the rotor for maintenance?

10.5   ARITHMETIC ASSIGNMENT

Turn to the Arithmetic Appendix at the back of this manual and work the example problems in Section A.21, "Activated Sludge (Oxidation Ditches)"; check the arithmetic using your calculator.

10.6   METRIC CALCULATIONS

Refer to Section 10.35, "Operational Guidelines," for metric calculations.

END OF LESSON 3 OF 3 LESSONS
on
ACTIVATED SLUDGE

Please answer the discussion and review questions next.

DISCUSSION AND REVIEW QUESTIONS

Chapter 10.   ACTIVATED SLUDGE

Package Plants and Oxidation Ditches

(Lesson 3 of 3 Lessons)

Please write your answers to the following questions to determine how well you understand the material in the lesson. The question numbering continues from Lesson 2.

11. What happens to inorganic solids such as grit, sand, and silt that enter an oxidation ditch plant?

12. During plant start-up, what does a dark gray color in the MLSS indicate and what would you do if this condition persists for more than several days?

13. During plant start-up, when should water be discharged to the clarifier?

14. How can an operator determine if the sludge wasting rate should be increased or decreased?

15. What would you do if solids were in the effluent of the final settling tank during high flows caused by storms or during high influent solids levels caused by the cleaning of the collection system sewers?

---

38. *Lineal* (LIN-e-ul).   The length in one direction of a line. For example, a board 12 feet (meters) long has 12 lineal feet (meters) in its length.

SUGGESTED ANSWERS
Chapter 10.   ACTIVATED SLUDGE
Package Plants and Oxidation Ditches

ANSWERS TO QUESTIONS IN LESSON 1

Answers to questions on page 70.

10.0A The activated sludge process is a biological wastewater treatment process that uses suspended growth microorganisms to speed up the decomposition of wastes.

10.0B A stabilized waste is a waste that has been treated or decomposed to the extent that, if discharged or released, its rate and state of decomposition would be such that the waste would not cause a nuisance or odors in the receiving waters.

10.0C Air is added to the aeration tank in the activated sludge process to provide oxygen to sustain the living organisms as they oxidize wastes to obtain energy for growth. Adding air also encourages mixing in the aerator.

10.0D Air (oxygen) requirements increase in an aeration tank when the strength (BOD) of the incoming wastewater increases. This increase is needed because more food (waste) in the influent encourages more organism activity and more oxidation (reproduction and respiration).

10.0E Factors that could cause an unsuitable environment for the activated sludge process in an aeration tank include the following:

1. High concentrations of acids, bases, and other toxic substances that may kill working organisms
2. Uneven flows of wastewater that cause overfeeding or starvation and other problems that upset the activated sludge process
3. Failure to supply enough oxygen, resulting in decreased organism activity
4. Failure to maintain the F/M ratio

Answers to questions on page 71.

10.1A Effluent quality requirements may be stated by regulatory agencies in terms of percentage removal of wastes or allowable quantities of wastes that may be discharged.

10.1B The treatment plant operator has little control over the makeup or amount of influent entering the treatment plant. However, harmful industrial waste discharges may be regulated by municipal ordinances to prevent industries from dumping substances or wastes into the collection system if they could seriously damage treatment facilities or create safety hazards. Even with these ordinances in place to protect the collection system, you may need to develop some type of inspection program. It also may be necessary for you to plan for pretreatment or controlled discharge in order to protect the activated sludge process and the treatment plant.

ANSWERS TO QUESTIONS IN LESSON 2

Answers to questions on page 77.

10.2A A typical activated sludge package plant may have either two or three compartments. The purpose of each compartment is:

1. Aeration. Aerate and mix waste to be treated with activated sludge.
2. Clarification and Settling. Allow activated sludge to be separated from wastewater being treated. Clarified effluent leaves plant and settled activated sludge is returned to aeration tank to treat more wastewater.
3. Aerobic digestion (optional). Treatment of waste activated sludge.

10.2B Package plants should have cathodic protection to prevent rust, corrosion, and pitting of metal surfaces in contact with water, wastewater, or soil.

10.2C Common characteristics of package aeration plants include long solids retention times, high mixed liquor suspended solids, and low food/microorganism ratios.

Answers to questions on page 85.

10.2D The valve to the air lift pumps must be closed until the settling compartment is filled or all of the air will attempt to flow out the air lifts and no air will flow out the diffusers.

10.2E If the water in the aeration compartment is murky or cloudy and the aeration compartment has a rotten egg odor (H_2S), you should increase the aeration rate.

Answers to questions on page 92.

10.2F For best effluent quality, waste about five percent of the solids each week during warm weather operation.

10.2G A package plant should be visually checked by an operator every day.

10.2H a. If the solids do not settle in the jar, the aeration rate should be increased and the MLSS concentration checked for proper level.
b. If the solids settle and then float to the surface, the aeration rate should be reduced a little each day until the solids settle properly. Also, reduce the wasting rate to increase the MLSS content.

Answers to questions on page 94.

10.2I Operators can avoid being injured on the job by practicing safety, by avoiding hazardous conditions, and by using safe procedures.

10.2J If you must work alone, phone your office at regular intervals and have someone check on you if you fail to report.

ANSWERS TO QUESTIONS IN LESSON 3

Answers to questions on page 97.

10.3A   The major components of an oxidation ditch treatment process include the following:

1. Oxidation ditch
2. Rotor
3. Level control weir
4. Final settling tank
5. Return sludge pump
6. Excess sludge handling facilities

10.3B   The ends of oxidation ditches are well rounded to prevent eddying and dead areas.

10.3C   Operators should not walk on top of oxidation ditch sidewalls to avoid slipping and falling into the ditch.

Answers to questions on page 99.

10.3D   The two primary objectives of start-up are to make certain that all mechanical equipment is operating properly and to develop a proper microbial floc (activated sludge) in the oxidation ditch.

10.3E   The following items or structures should be inspected during the pre-start inspection:

1. Inlet structure
2. Oxidation ditch structure
3. Rotor
4. Adjustable weir
5. Clarifier structure and piping
6. Return sludge and waste sludge systems

10.3F   During plant start-up, start the rotors when the water level in the ditch reaches the bottom of the rotor blades.

10.3G   When an oxidation ditch is operating properly, the activated sludge solids should look like particles, golden to rich dark brown in color, with sharply defined edges.

Answers to questions on page 101.

10.3H   The DO in the oxidation ditch is regulated by raising or lowering the ditch outlet level control weir. This weir controls the level of water in the ditch, which, in turn, regulates the degree of submergence of the rotors. The level or elevation of the rotors is fixed, but the deeper the rotors sit in the water, the greater the transfer of DO from the air to the water.

10.3I   The ditch velocity should be maintained between 1.0 and 1.5 fps (0.3 to 0.45 mps) to prevent the deposition of floc.

10.3J   Daily observations of ditch color, odors, lack of foam on the aerator surface, and settling tank clarity help indicate the performance of the oxidation ditch.

Answers to questions on page 105.

10.3K   During cold weather, ice buildup will hinder or stop altogether the proper operation of mechanical parts such as rotors and sludge scraper mechanisms.

10.3L   If complete nitrification is desired, more microorganisms are needed to treat the same amount of wastes during cold weather than during warm weather because the colder the water, the less active the microorganisms.

10.3M   A general daily cleanup at the plant is important because it improves plant performance and provides a more pleasant place to work.

Answers to question on page 108.

10.3N   Determine the BOD loading on an oxidation ditch in (1) pounds per day and (2) kilograms per day. The inflow is 0.8 MGD and the influent BOD is 250 mg/L. 1 MGD = 3,785 cu m/day.

Known	Unknown
Flow, MGD = 0.8 MGD	1. BOD Loading, lbs/day
BOD, mg/L = 250 mg/L	
1 MGD = 3,785 cu m/day	2. BOD Loading, kg/day

1. Determine the BOD loading in pounds per day.

$$\text{BOD Loading, lbs/day} = (\text{Flow, MGD})(\text{BOD, mg/L})(8.34 \text{ lbs/gal})$$

$$= (0.8 \text{ MGD})(250 \text{ mg/L})(8.34 \text{ lbs/gal})$$

$$= 1,668 \text{ lbs/BOD/day}$$

2. Determine the BOD loading in kilograms per day.

a. Convert flow from MGD to cubic meters per day.

$$\text{Flow, cu m/day} = (\text{Flow, MGD})\left(\frac{3,785 \text{ cu m/day}}{1 \text{ MGD}}\right)$$

$$= \frac{0.8 \text{ MGD} \times 3,785 \text{ cu m/day}}{1 \text{ MGD}}$$

$$= 3,028 \text{ cu m/day}$$

b. Calculate the BOD loading in kilograms per day.

$$\text{BOD Loading, kg/day} = (\text{Flow, cu m/day})(\text{BOD, mg/L})\left(\frac{1 \text{ kg}}{1,000,000 \text{ mg}}\right)\left(\frac{1,000 \text{ L}}{1 \text{ cu m}}\right)$$

$$= (3,028 \text{ cu m/day})(250 \text{ mg/L})\left(\frac{1 \text{ kg}}{1,000,000 \text{ mg}}\right)\left(\frac{1,000 \text{ L}}{1 \text{ cu m}}\right)$$

$$= 757 \text{ kg BOD/day}$$

Answers to questions on page 110.

10.4A   Adequate dewatering systems should be provided to permit rapid servicing of submerged equipment.

10.4B   When reviewing the location of a walkway intended to provide access to the rotor for maintenance, check the following:

1. Normal location is upstream from rotor
2. Spray from the rotor will not fall on the walkway

CHAPTER 11

ROTATING BIOLOGICAL CONTACTORS (RBCs)

by

Richard Wick

Revised by

John Brady

TABLE OF CONTENTS
Chapter 11.   ROTATING BIOLOGICAL CONTACTORS (RBCs)

OBJECTIVES

Chapter 11. ROTATING BIOLOGICAL CONTACTORS (RBCs)

1. Describe a rotating biological contactor and the purpose of each major part.

2. Start up and operate a rotating biological contactor.

3. Operate a rotating biological contactor under abnormal conditions.

4. Shut down and restart a rotating biological contactor.

5. Maintain and troubleshoot a rotating biological contactor.

6. Safely perform the operator duties for a rotating biological contactor.

7. Review the plans and specifications for a rotating biological contactor.

8. Calculate the hydraulic and organic loadings on a rotating biological contactor.

WORDS
Chapter 11.   ROTATING BIOLOGICAL CONTACTORS (RBCs)

ACUTE HEALTH EFFECT ACUTE HEALTH EFFECT

An adverse effect on a human or animal body, with symptoms developing rapidly.

BIODEGRADABLE (BUY-o-dee-GRADE-able) BIODEGRADABLE

Organic matter that can be broken down by bacteria to more stable forms that will not create a nuisance or give off foul odors is considered biodegradable.

CHRONIC HEALTH EFFECT CHRONIC HEALTH EFFECT

An adverse effect on a human or animal body with symptoms that develop slowly over a long period of time or that recur frequently.

COMPOSITE (PROPORTIONAL) SAMPLE COMPOSITE (PROPORTIONAL) SAMPLE

A composite sample is a collection of individual samples obtained at regular intervals, usually every one or two hours during a 24-hour time span. Each individual sample is combined with the others in proportion to the rate of flow when the sample was collected. Equal volume individual samples also may be collected at intervals after a specific volume of flow passes the sampling point or after equal time intervals and still be referred to as a composite sample. The resulting mixture (composite sample) forms a representative sample and is analyzed to determine the average conditions during the sampling period.

DENITRIFICATION (dee-NYE-truh-fuh-KAY-shun) DENITRIFICATION

(1) The anoxic biological reduction of nitrate nitrogen to nitrogen gas.

(2) The removal of some nitrogen from a system.

(3) An anoxic process that occurs when nitrite or nitrate ions are reduced to nitrogen gas and nitrogen bubbles are formed as a result of this process. The bubbles attach to the biological floc and float the floc to the surface of the secondary clarifiers. This condition is often the cause of rising sludge observed in secondary clarifiers or gravity thickeners. Also see NITRIFICATION.

GRAB SAMPLE GRAB SAMPLE

A single sample of water collected at a particular time and place that represents the composition of the water only at that time and place.

HARMFUL PHYSICAL AGENT
  or TOXIC SUBSTANCE
HARMFUL PHYSICAL AGENT
  or TOXIC SUBSTANCE

Any chemical substance, biological agent (bacteria, virus, or fungus), or physical stress (noise, heat, cold, vibration, repetitive motion, ionizing and non-ionizing radiation, hypo- or hyperbaric pressure) that:

(1) Is regulated by any state or federal law or rule due to a hazard to health

(2) Is listed in the latest printed edition of the National Institute of Occupational Safety and Health (NIOSH) Registry of Toxic Effects of Chemical Substances (RTECS)

(3) Has yielded positive evidence of an acute or chronic health hazard in human, animal, or other biological testing conducted by, or known to, the employer

(4) Is described by a Material Safety Data Sheet (MSDS) available to the employer that indicates that the material may pose a hazard to human health

Also see ACUTE HEALTH EFFECT and CHRONIC HEALTH EFFECT.

MPN

MPN is the Most Probable Number of coliform-group organisms per unit volume of sample water. Expressed as a density or population of organisms per 100 mL of sample water.

NAMEPLATE

A durable, metal plate found on equipment that lists critical installation and operating conditions for the equipment.

NEUTRALIZATION (noo-trull-uh-ZAY-shun)

Addition of an acid or alkali (base) to a liquid to cause the pH of the liquid to move toward a neutral pH of 7.0.

NITRIFICATION (NYE-truh-fuh-KAY-shun)

An aerobic process in which bacteria change the ammonia and organic nitrogen in wastewater into oxidized nitrogen (usually nitrate). The second-stage BOD is sometimes referred to as the nitrogenous BOD (first-stage BOD is called the carbonaceous BOD). Also see DENITRIFICATION.

PYROMETER (pie-ROM-uh-ter)

An apparatus used to measure high temperatures.

SEPTIC TANK

A system sometimes used where wastewater collection systems and treatment plants are not available. The system is a settling tank in which settled sludge and floatable scum are in intimate contact with the wastewater flowing through the tank and the organic solids are decomposed by anaerobic bacterial action. Used to treat wastewater and produce an effluent that is usually discharged to subsurface leaching. Also referred to as an interceptor; however, the preferred term is septic tank.

SEPTIC TANK EFFLUENT PUMP (STEP) SYSTEM

A facility in which effluent is pumped from a septic tank into a pressurized collection system that may flow into a gravity sewer, treatment plant, or subsurface leaching system.

SLOUGHED or SLOUGHING (SLUFF-ing)

The breaking off of biological or biomass growths from the fixed film or rotating biological contactor (RBC) media. The sloughed growth becomes suspended in the effluent and is later removed in the secondary clarifier as sludge.

SMALL-DIAMETER GRAVITY SEWER (SDGS) SYSTEM

A type of collection system in which a series of septic tanks discharge effluent by gravity, pump, or siphon to a small-diameter wastewater collection main. The wastewater flows by gravity to a lift station, a manhole in a conventional gravity collection system, or directly to a wastewater treatment plant.

SOLUBLE BOD

Soluble BOD is the BOD of water that has been filtered in the standard suspended solids test. The soluble BOD is a measure of food for microorganisms that is dissolved in the water being treated.

SUPERNATANT (soo-per-NAY-tent)

The relatively clear water layer between the sludge on the bottom and the scum on the surface of an anaerobic digester or septic tank (interceptor).

(1)   From an anaerobic digester, this water is usually returned to the influent wet well or to the primary clarifier.

(2)   From a septic tank, this water is discharged by gravity or by a pump to a leaching system or a wastewater collection system.

Also called clear zone.

MPN

NAMEPLATE

NEUTRALIZATION

NITRIFICATION

PYROMETER

SEPTIC TANK

SEPTIC TANK EFFLUENT PUMP (STEP) SYSTEM

SLOUGHED or SLOUGHING

SMALL-DIAMETER GRAVITY SEWER (SDGS) SYSTEM

SOLUBLE BOD

SUPERNATANT

TOXIC TOXIC

A substance that is poisonous to a living organism. Toxic substances may be classified in terms of their physiological action, such as irritants, asphyxiants, systemic poisons, and anesthetics and narcotics. Irritants are corrosive substances that attack the mucous membrane surfaces of the body. Asphyxiants interfere with breathing. Systemic poisons are hazardous substances that injure or destroy internal organs of the body. Anesthetics and narcotics are hazardous substances that depress the central nervous system and lead to unconsciousness.

TOXIC SUBSTANCE TOXIC SUBSTANCE

See HARMFUL PHYSICAL AGENT and TOXIC.

TRICKLING FILTER TRICKLING FILTER

A treatment process in which wastewater trickling over media enables the formation of slimes or biomass, which contain organisms that feed upon and remove wastes from the water being treated.

CHAPTER 11. ROTATING BIOLOGICAL CONTACTORS (RBCs)

11.0 DESCRIPTION OF RBCs

RBCs are a secondary biological treatment process for domestic and *BIODEGRADABLE*[1] industrial wastes. For rotating biological contactors to treat wastewater efficiently, the wastewater must first receive at least primary treatment (flow through a primary clarifier) where the major portion of the settled solids are removed for treatment and disposal. For wastewater transported in conventional gravity sewers or pressure sewers where grinder pumps discharge to the wastewater collection system, treatment is commonly accomplished by routing the wastewater flow through a wastewater treatment facility as shown in Figure 11.1. Figure 11.2 indicates a typical flow route through a wastewater treatment facility where wastewater flows from septic tanks (interceptors) and is discharged to small-diameter gravity sewers, or from septic tank effluent pump system sewers. The difference is that septic tanks in small systems perform similarly to primary clarifiers in larger systems. Figure 11.3 illustrates how rotating biological contactors are used in small systems with septic tanks, collection systems, treatment, disinfection, and discharge systems.

The major requirement for efficient RBC performance is that some form of sedimentation (primary treatment) has been provided to the wastewater for removal of settleable and floatable solids before reaching the rotating biological contactors (reactors) for further treatment.

Additional preliminary treatment or pretreatment may or may not be needed. The makeup of the wastewater, type of collection system used to convey the wastewater to the treatment facility, and other factors such as length of travel time of the flow through the system to the treatment plant, temperature of the wastewater, and condition of the collection system, will dictate pretreatment requirements.

Biological contactors have a rotating shaft surrounded by plastic disks called the "media." The shaft and media combined are called the "drum" (Figures 11.4 and 11.5). A biological slime grows on the media of an RBC when conditions are suitable. This process is very similar to a trickling filter where the biological slime grows on rock or other media and settled wastewater (primary clarifier effluent) is applied over the media. With rotating biological contactors, the biological slime or film grows on the surface of the plastic disk media. The slime is rotated into the settled wastewater and then into the atmosphere to provide oxygen for the organisms (Figure 11.4). The wastewater being treated either flows parallel to the rotating shaft or perpendicular to the shaft as it flows from stage to stage or tank to tank.

The rotating biological contactor process uses several plastic media drums. The plastic disk media are made of high-density plastic circular sheets usually 12 feet (3.6 m) in diameter. These sheets are bonded and assembled onto horizontal shafts up to 25 feet (7.5 m) in length. Spacing between the sheets provides the hollow (void) space for distribution of wastewater and air (Figures 11.5 and 11.6). The media provide the required surface area for the slimes containing the active biological growth. Concrete or coated steel tanks usually hold the wastewater being treated. The media rotate at about 1.5 RPM while approximately 40 percent of the media surface is immersed in the wastewater (Figure 11.6).

As the drum rotates, the media pick up a thin layer of wastewater, which flows over the biological slimes on the disks. Organisms living in the slimes use organic matter from the wastewater for food and dissolved oxygen from the air, thus removing wastes from the water being treated. As the attached slimes pass through the wastewater, some of the slimes are *SLOUGHED*[2] from the media as the media rotates downward into the wastewater being treated. The effluent with the sloughed slimes flows to the secondary clarifier where the slimes are removed from the effluent by settling. Figure 11.7 shows the location of a rotating biological contactor process in a wastewater treatment plant. The process is located in the same position as the trickling filter or activated sludge aeration basin. Usually, the process operates on a "once-through" scheme, with no recycling of effluent or sludge, which makes it a simple process to operate. Figure 11.8 shows a treatment plant without a primary clarifier because the solids and floatables are removed by septic tanks (interceptors).

---

1. *Biodegradable* (BUY-o-dee-GRADE-able). Organic matter that can be broken down by bacteria to more stable forms that will not create a nuisance or give off foul odors is considered biodegradable.
2. *Sloughed* or *Sloughing* (SLUFF-ing). The breaking off of biological or biomass growths from the fixed film or rotating biological contactor (RBC) media. The sloughed growth becomes suspended in the effluent and is later removed in the secondary clarifier as sludge.

TREATMENT PROCESS   FUNCTION

PRELIMINARY TREATMENT

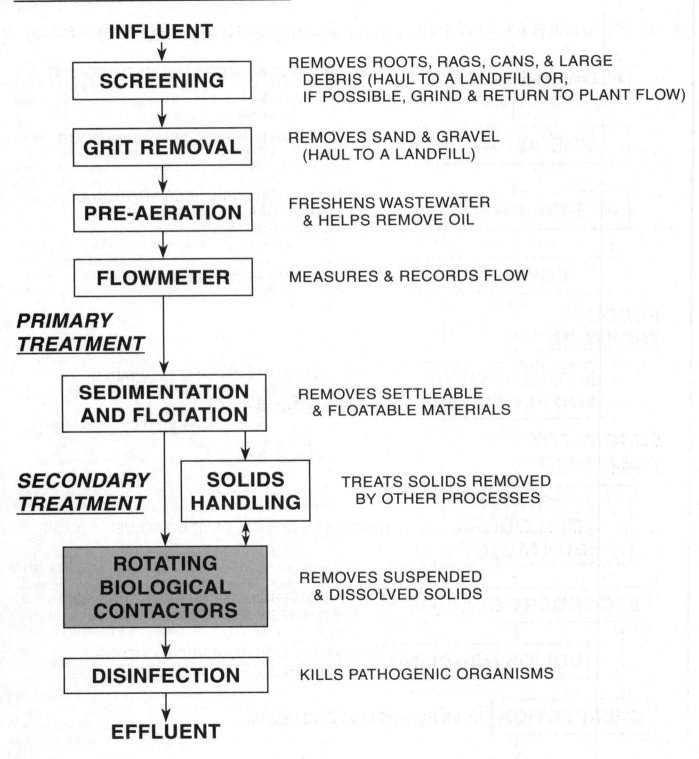

INFLUENT

SCREENING — REMOVES ROOTS, RAGS, CANS, & LARGE DEBRIS (HAUL TO A LANDFILL OR, IF POSSIBLE, GRIND & RETURN TO PLANT FLOW)

GRIT REMOVAL — REMOVES SAND & GRAVEL (HAUL TO A LANDFILL)

PRE-AERATION — FRESHENS WASTEWATER & HELPS REMOVE OIL

FLOWMETER — MEASURES & RECORDS FLOW

PRIMARY TREATMENT

SEDIMENTATION AND FLOTATION — REMOVES SETTLEABLE & FLOATABLE MATERIALS

SECONDARY TREATMENT

SOLIDS HANDLING — TREATS SOLIDS REMOVED BY OTHER PROCESSES

ROTATING BIOLOGICAL CONTACTORS — REMOVES SUSPENDED & DISSOLVED SOLIDS

DISINFECTION — KILLS PATHOGENIC ORGANISMS

EFFLUENT

Fig. 11.1   Flow diagram of treatment plant

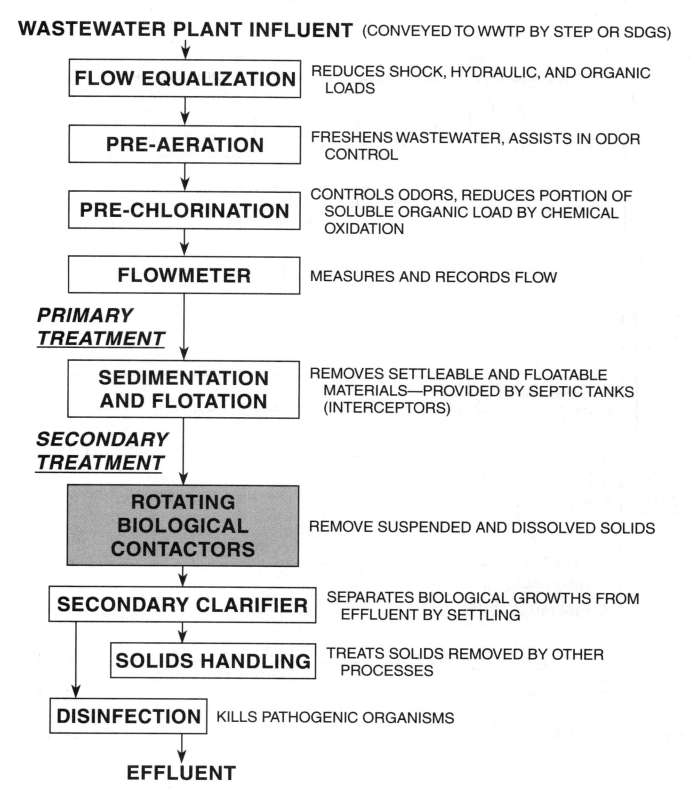

Fig. 11.2   *Flow diagram of treatment plant treating wastewater from community served by septic tanks (interceptors)*

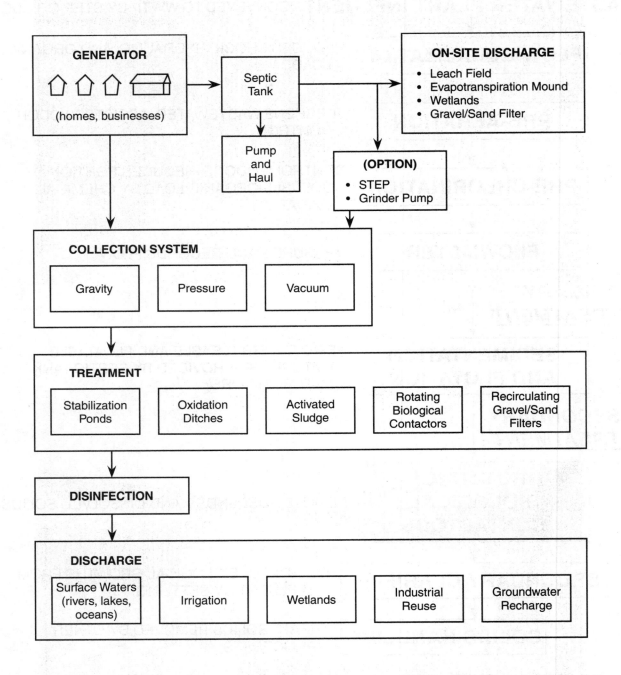

Fig. 11.3 Flow diagram of small collection, treatment, and discharge systems

Fig. 11.4   Rotating biological contactors
(Permission of Autotrol Corporation)

Fig. 11.5   Plastic disk media and biological contactor drum
(Permission of Autotrol Corporation)

Media cross section

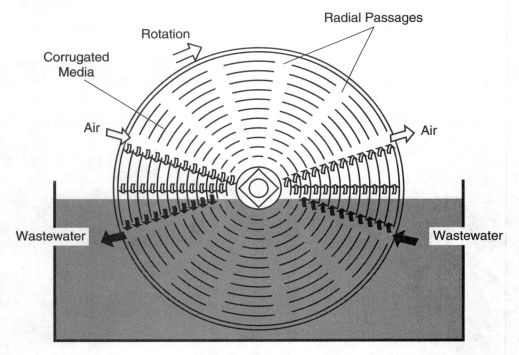

End-view sketch illustrates exchange
of air and wastewater

Fig. 11.6   Sections of the plastic disk media
(Permission of Autotrol Corporation)

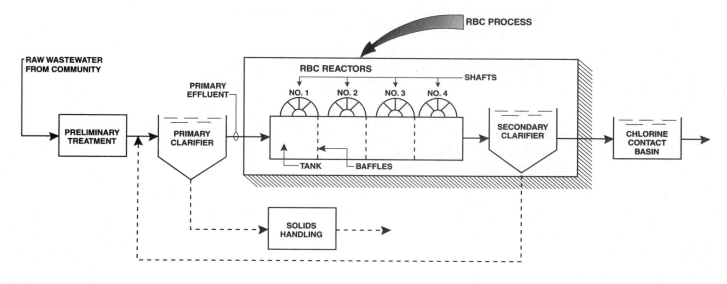

Fig. 11.7   Typical rotating biological contactor (reactor) treatment plant

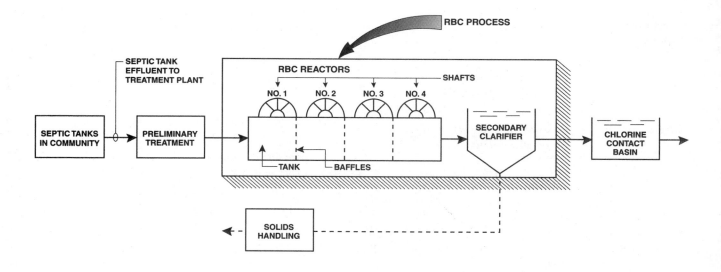

Fig. 11.8   Typical rotating biological contactor (reactor) treatment plant
treating wastewater from septic tanks (interceptors)

Table 11.1 lists the parts of an RBC and their purpose. The concrete or steel tanks are commonly shaped to conform to the general shape of the media. This shape eliminates dead spots where solids could settle out and cause odors and septic conditions.

TABLE 11.1　PURPOSE OF RBC PARTS

Part	Purpose
Concrete or Steel Tank Divided Into Bays (Sections) by Baffles (Bulkheads)	Tank holds the wastewater being treated and allows it to come into contact with the organisms on the disks.
	Bays and baffles prevent short-circuiting of wastewater.
Orifice or Weir Located in Baffle	Controls flow from one stage to the next stage or from one bay to the next bay.
Rotating Media	Provide support for organisms. Rotation provides food (from wastewater being treated) and air for organisms.
Cover Over Contactor	Protects organisms from severe weather fluctuations, especially freezing. Also contains odors.
Drive Assembly	Rotates the media.
Influent Lines With Valves	Influent lines transport wastewater to be treated to the RBC bays.
	Influent valves regulate influent to contactor and also isolate contactor for maintenance.
Effluent Lines With Valves	Effluent lines convey treated wastewater from the RBC to the secondary clarifier.
	Effluent valves regulate effluent from the RBC and isolate RBC for maintenance.
Underdrains	Allow for the removal of solids that may settle out in tank.

The rotating biological contactor process is usually divided into multiple stages (up to four) with three possible layouts (Figure 11.9). Each stage is separated by a removable baffle, concrete wall, or cross-tank bulkhead. Wastewater flow can be either parallel or perpendicular to the shaft. Each bulkhead or baffle has an underwater orifice or hole to permit flow from one stage to the next. Each section of media between bulkheads acts as a separate stage of treatment.

The RBC process is divided into multiple stages to increase the effectiveness of a given amount of media surface area by separating competing biomass populations. Organisms on the first-stage media are exposed to high levels of BOD and reduce the BOD at a high rate. As the BOD levels decrease from stage to stage, the rate at which the organisms can remove BOD decreases and *NITRIFICATION*[3] starts.

Treatment plants requiring four or more shafts of media usually are arranged so that each shaft serves as an individual stage of treatment. The shafts are arranged so that the flow is perpendicular to the shafts (Figure 11.9, Layout No. 3). The RBC media should move against the direction of flow to provide better treatment of the wastewater. Plants with fewer than four shafts are usually arranged with the flow parallel to the shaft (Figure 11.9, Layout No. 1).

Rotating biological contactors are covered for several reasons relating to climatic conditions:

1. Protecting biological slime growths (organisms) from severe changes in weather, especially freezing

2. Preventing heavy rains from washing off some of the slime growth

3. Stopping exposure of media to direct sunlight to prevent growth of algae

4. Avoiding exposure of media to sunlight, which may cause it to become brittle

5. Providing protection for operators from sun, rain, snow, or wind while maintaining equipment

Fiberglass covers in the shape of the media are easily removed for maintenance. In some areas, the rotating biological contactors are covered by a building. In other areas only a roof is placed over the media for protection against sunlight. The type of cover used depends on climatic conditions.

---

3. *Nitrification* (NYE-truh-fuh-KAY-shun). An aerobic process in which bacteria change the ammonia and organic nitrogen in wastewater into oxidized nitrogen (usually nitrate). The second-stage BOD is sometimes referred to as the nitrogenous BOD (first-stage BOD is called the carbonaceous BOD). Also see DENITRIFICATION.

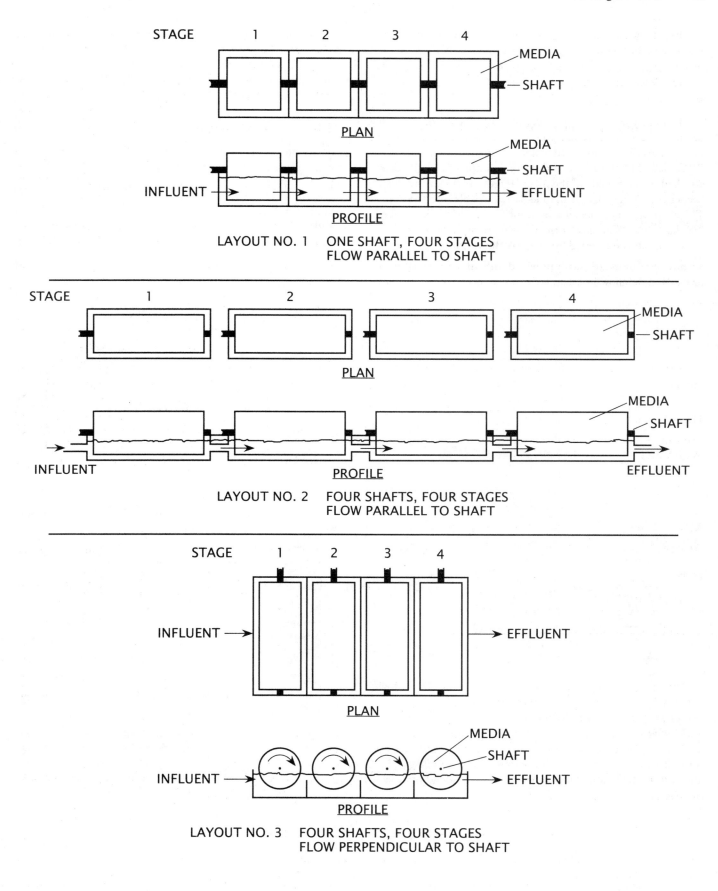

Fig. 11.9   Rotating biological contactor layouts

Three types of drive assemblies are used to rotate the shafts supporting the media:

1. Motor with chain drive (Figure 11.10)
2. Motor with direct shaft drive
3. Air drive (Figure 11.11)

The first type of drive assembly consists of a motor and belt or chain drive. The second type of drive system consists of a motor, belt drive, and speed reducer mounted directly to the RBC shaft. The third type of drive unit consists of plastic cups attached to the outside of the media (Figure 11.11). A small air header below the edge of the media releases air into the cups. The air in the cups creates a buoyant force, which then makes the shaft turn. With any of the three types of drive assemblies, the mainshaft is supported by two main bearings.

Individual units are usually provided with influent and effluent line valving to allow isolation for maintenance reasons. Usually, the units are not shut down during low-flow conditions because power consumption is minimal and, as the flows in the RBC decrease, the percent of BOD removal increases.

QUESTIONS

Please write your answers to the following questions and compare them with those on page 149.

11.0A   How does a rotating biological contactor (RBC) treat wastewater?

11.0B   Why is the RBC process divided into multiple stages?

11.0C   What is the purpose of a cover over the RBC unit?

11.0D   What are the three types of drive assemblies used to rotate the shafts supporting the media in an RBC?

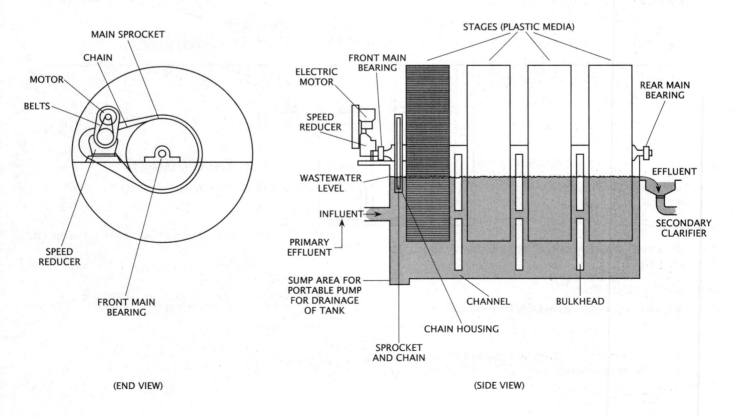

Fig. 11.10   Motor with chain drive unit

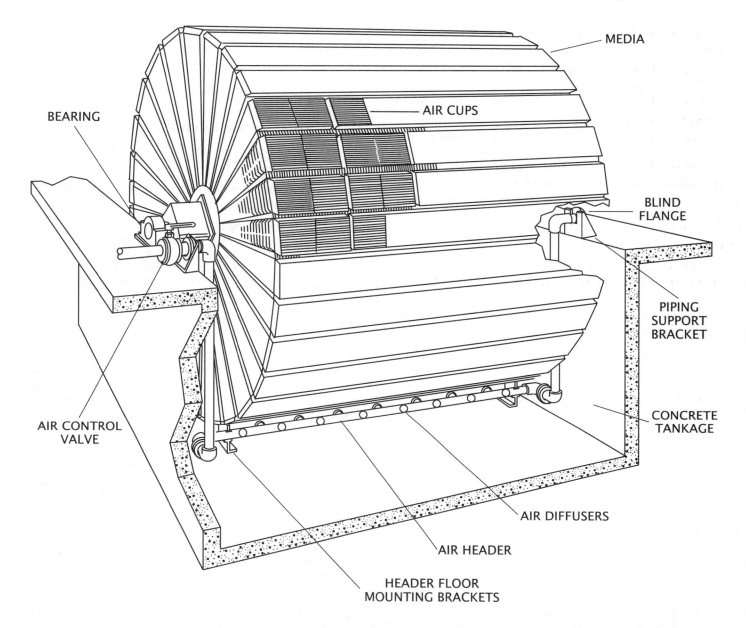

MEDIA

Media with polyethylene air cups attached to the outer perimeter capture air as it is released from the air header.

AIR CUPS

Air cups, attached to the media, capture air, which in turn rotates the media.

AIR CONTROL VALVE

Butterfly control valve controls inlet air supply to each unit.

AIR HEADER

Lightweight headers that carry the air through the system run the length of the media assembly and are easily removable for cleaning.

HEADER FLOOR MOUNTING BRACKETS

Brackets secure header to the floor of the tank.

AIR DIFFUSER

Coarse-bubble air diffusers distribute air from the header into the air cups and media.

PIPING SUPPORT BRACKET

Bracket on each end of the header holds unit in place.

Fig. 11.11   Air drive unit
(Permission of Autotrol Corporation)

11.1   PROCESS OPERATION

Plants have been designed to treat flows ranging from 18,000 GPD to 50 MGD (70,000 liters/day to 200 million liters/day), however, the majority of plants treat flows of less than 5 MGD (20 MLD). Typical operating and performance characteristics are as follows:

Characteristic	Range
Hydraulic Loading[4]	
BOD Removal	1.5 to 6 GPD/sq ft[5]
	(0.06 to 0.24 cu m/day/sq m)[5]
Nitrogen Removal	1.5 to 1.8 GPD/sq ft[5]
	(0.06 to 0.07 cu m/day/sq m)[5]
Organic Loading[4]	
Soluble BOD[6,7]	2.5 to 4 lbs/day/1,000 sq ft[5]
	(15 to 25 gm/day/sq m)[5]
BOD Removal	80 to 95 percent
Effluent Total BOD	10 to 30 mg/L
Effluent Soluble BOD	5 to 15 mg/L
Effluent NH_3-N	1 to 10 mg/L
Effluent NO_3-N	2 to 7 mg/L

See Section 11.5, "Loading Calculations," for procedures showing how to calculate the hydraulic and organic loadings on rotating biological contactors. Both hydraulic and organic loads to a rotating biological contactor should be evaluated. Experience indicates that the organic load on the RBC (especially to the first stage) often controls the performance of the RBC. Performance by rotating biological contactors also is affected by hydraulic loadings and temperatures below 55°F (13°C).

The advantages of rotating biological contactors over *TRICKLING FILTERS*[8] include the elimination of the rotating distributor, the elimination of problems caused by ponding on the media, and reduction of problems with filter flies. More efficient use of the media is achieved due to the even or uniform rotation of the media into the wastewater being treated. Another advantage of RBCs is the lack of anaerobic conditions in properly designed facilities as compared with anaerobic conditions found in the bottom of trickling filters.

A limitation of the process, as compared with trickling filters, is the lack of flexibility due to the absence of provisions for recirculation; however, in most installations recirculation is not needed but may improve effluent quality when used. Rotating biological contactors are more sensitive to industrial waste shock loading impacts than trickling filters. Care must be taken to ensure that industrial organic loadings to the RBCs do not cause low dissolved oxygen conditions in the treatment system.

11.10   Pretreatment Requirements

Rotating biological contactors are usually preceded by preliminary treatment processes consisting of screening, grit removal, and primary settling. Debris, grit, and suspended solids should be removed to prevent them from settling beneath the drums and forming sludge deposits. These sludge deposits can reduce the effective tank volume, produce septic conditions, physically damage the media, and possibly stall the unit.

Some rotating biological contactor plants have aerated flow equalization tanks between the primary clarifiers and the rotating biological contactors. Flow equalization tanks may be installed to equalize or balance highly fluctuating flows and to allow for the dilution of strong wastes and neutralization of highly acidic or alkaline wastes. These equalization tanks are capable of reducing shock loads and providing pre-aeration to the RBC.

11.11   Start-Up

Prior to plant start-up, become familiar with the contents of your plant's O&M manual. If you have any questions, be sure to ask the design engineer or manufacturer's representative. Both of these persons should instruct the operator on the proper operation of the plant and the maintenance of the equipment. If possible, visit an active operating system and talk to an experienced RBC operator whose experience and knowledge could be very helpful during start-up and operation of your new plant.

11.110   *Pre-Start Checks for New Equipment*

Before starting any equipment or allowing any wastewater to enter the treatment process, record pertinent data including motor data and the sizes of drive belts, chains, and bearings. Also, check the following items before starting any equipment:

1. *TIGHTNESS OF BOLTS AND PARTS*

    Inspect the following parts for tightness in accordance with the manufacturer's recommendations.

    a. Anchor bolts

    b. Mounting studs

    c. Bearing caps and set screws (check any torque limitations)

    d. Locking collars

    e. Jacking screws

---

4.   Hydraulic and organic loadings depend on influent flow, influent soluble BOD, effluent BOD, temperature, and surface area of plastic media. Manufacturers provide charts converting flow to hydraulic and organic loadings for their media.

5.   All areas refer to media surface area.

6.   *Soluble Bod.*   Soluble BOD is the BOD of water that has been filtered in the standard suspended solids test. The soluble BOD is a measure of food for microorganisms that is dissolved in the water being treated.

7.   BOD loading for first stage or first shafts only.

8.   *Trickling Filter.*   A treatment process in which wastewater trickling over media enables the formation of slimes or biomass, which contain organisms that feed upon and remove wastes from the water being treated.

f. Roller chain (be sure chain is properly aligned)

g. Media (unbalanced media may cause slippage)

h. Belts (use matched sets on multiple-belt drives)

2. *LUBRICATION OF EQUIPMENT*

Be sure the following parts have been correctly lubricated with proper lubricants in accordance with the manufacturer's recommendations.

a. Mainshaft bearings

b. Roller chain

c. Speed reducer

3. *CLEARANCES FOR MOVING PARTS*

a. Between media and tank wall

b. Between media and baffles or cover support beams

c. Between chain casing and media

d. Between roller chain, sprockets, and chain casing

4. *ALIGNMENT OF DRIVE UNIT*

Check drive unit alignment to drum shaft.

5. *FLOW CONTROL EQUIPMENT*

a. Influent and effluent gates

b. Any weirs for proper level

c. Drain plugs and ports

6. *ELECTRICAL CONTROLS*

a. Check all breakers

b. Check START/STOP switches for proper operation

7. *SAFETY GUARDS*

a. Be sure safety guards are properly installed over chains and other moving parts.

b. Be sure safety rails and toe plates are properly installed.

11.111   *Procedure for Starting Unit*

Actual start-up procedures for a new unit should be included in the plant O&M manual provided by the manufacturer. A typical starting procedure is outlined next.

1. Switch on power, allow shaft to rotate one turn, then turn off the power, lock out, and tag the main breaker. Inspect and correct the following if necessary during this rotation:

a. Movement of chain casing.

b. Unusual noises.

c. Direction of media rotation. Where wastewater flow is parallel to the rotating media shaft, the direction of rotation is not critical. If the wastewater flow is perpendicular to the rotating media shaft, the media should be moving through the wastewater against the direction of flow (see Figure 11.9).

2. Switch on power and allow shaft to rotate for 15 minutes, then inspect the following:

a. Chain drive sprocket alignment.

b. Noises in bearings, chain drives, and drive package.

c. Motor amperage; compare with value on *NAMEPLATE*.[9]

d. Temperature of mainshaft bearing and drive package pillow block bearing (check by hand). If too hot for your hand, use a *PYROMETER*[10] or thermometer. Temperature should not exceed 200°F (93°C).

e. Tightness of shaft bearing cap bolts. Tighten to the manufacturer's recommended torque.

f. Determine the number of revolutions per minute for the drum and record the information for future reference.

3. Open the inlet valve and allow wastewater to fill the tank (all four stages if in one tank). Open the outlet valve to allow water to flow through the tank. Make inspections listed in steps 1 and 2 again while drum is rotating. Shut off power, lock out, and tag the main breaker to make any corrections.

4. Check the relationship between the clarifier inlet and the rotating biological contactor outlet for hydraulic balance. This means that you want to be sure the tank containing the biological contactor will not overflow, possibly causing equipment damage.

5. Take an initial reading of the hydraulic or electronic load cell, if so equipped. Record operating weight while unit is rotating.

6. Turn on and check the air supplied to the air drive of the unit for uniform distribution and proper blower operation.

7. See Section 11.20 for break-in maintenance instructions, which apply after eight hours of operation.

Encourage the development of biological slimes by regulating the flow rate and strength of the wastewater applied at nearly constant levels by the use of recirculation, if available. Maintaining the wastewater temperatures at 55°F (15°C) or higher during start-up will help. The best rotating speed for the media of an RBC unit is a speed that shears off excess growth. The media on the drum is usually operated at about 1.5 RPM, while approximately 40 percent of the media surface is immersed in the wastewater being treated.

Allow one to two weeks for an even growth of biological slimes (biomass) to develop on the surface of the media with normal strength wastewater. After start-up, a slimy growth (biomass) will appear. During the first week, excessive sloughing will

---

9. *Nameplate.*   A durable, metal plate found on equipment that lists critical installation and operating conditions for the equipment.

10. *Pyrometer* (pie-ROM-uh-ter).   An apparatus used to measure high temperatures.

occur naturally. This sloughing is normal and the sloughed material is soon replaced with a fairly uniform, shaggy, brown-to-gray appearing biomass with very few or no bare spots.

Follow the same start-up procedures whether a plant is starting at less than design flow or at full design flow. Start-up of an RBC during cold weather takes longer than in warm weather because the organisms in the slime growth (biomass) are not as active and require more time to grow and reproduce.

The RBC works best when operating at or near design BOD and hydraulic loadings. If the plant is currently over-designed to handle expansion of the system or population growth, it may be possible to operate a portion of units at the beginning of start-up. If all units are not needed (based on loading calculations), the operating units should be switched between running and not running RBC trains to protect them from deterioration.

11.12   Operation

Rotating biological contactor treatment plants are not difficult to operate. They produce a good effluent provided the operator properly and regularly performs the duties of inspecting and maintaining the equipment, testing the influent and effluent, observing the media, and taking corrective action when necessary.

11.120   Inspecting Equipment

This treatment process has relatively few moving parts. There is a drive train to rotate the shaft and there are bearings upon which the shaft rotates. Check the following items when inspecting equipment:

1. Touch the outer housing of the shaft bearing to determine if it is running hot. Use a pyrometer or thermometer, if the temperature is too hot for your hand. If the temperature exceeds 200°F (93°C), the bearings may need to be replaced. Also, check for proper lubrication and be sure the shaft is properly aligned. The longer the shaft, the more critical the alignment. See the manufacturer's manual for procedures to check alignment and allowable tolerances.

2. Listen for, locate, and correct any unusual noises in the motor bearings.

3. Touch the motors to determine if they are running hot. If hot, determine the cause and correct. If hot, check the motor nameplate for permissible degrees of temperature rise (typically 40°C). If temperature rise is greater than the nameplate rating, check the following items:

   a. Amperage draw on each leg

   b. Bearing lubrication and temperature

   c. Motor alignment to drive unit

4. Look around the drive train and shaft bearing for oil spills. If oil is visible, check oil levels in the speed reducers and chain drive system. Also, look for damaged or worn out gaskets or seals. Clean up leaking oil for safety and general cleanliness.

5. Inspect the chain drive for alignment, tightness, and sprocket wear.

6. Inspect belts for proper tension.

7. Be sure all guards located over moving parts and equipment are in place and properly installed.

8. Clean up any spills, messes, or debris.

Damage to RBC equipment will be minimized if required maintenance is done as soon as the need is identified.

11.121   Testing Influent and Effluent

RBC influent and effluent wastewater analysis is required to monitor overall plant and process performance. Because there are few process control functions to be performed, only a minimal analysis is required to monitor and report daily performance. To determine if the rotating biological contactors are operating properly, you should measure BOD, suspended solids, pH, and dissolved oxygen (DO). Performance is best monitored by analysis of a 24-hour *COMPOSITE SAMPLE*[11] for BOD and suspended solids on a daily basis. DO and pH should be measured using *GRAB SAMPLES*[12] at specific times. The actual frequency of tests may depend on how often you need the results for plant control and also how often your NPDES permit requires you to sample and analyze the plant effluent.

Sampling points should include the following:

1. Plant influent: BOD and suspended solids (SS)

2. Plant effluent: BOD and SS

3. RBC effluent: BOD, DO, SS, and pH

4. Secondary clarifier effluent: BOD, SS, and DO

BOD, SS, and DO tests performed on samples taken at or after each drum provide useful information on how well the system is working.

DISSOLVED OXYGEN

The DO in the wastewater being treated beneath the rotating media will vary from stage to stage. A plant designed to treat primary effluent for BOD and suspended solids removal will usually have 0.5 to 1.0 mg/L DO in the first stage. The DO level should increase to 1 to 3 mg/L at the end of the first stage. A plant designed for nitrification to convert ammonia and organic nitrogen compounds to nitrate usually will have four stages and DO levels of 4 to 8 mg/L. The difference between an

---

11. *Composite (Proportional) Sample.*   A composite sample is a collection of individual samples obtained at regular intervals, usually every one or two hours during a 24-hour time span. Each individual sample is combined with the others in proportion to the rate of flow when the sample was collected. Equal volume individual samples also may be collected at intervals after a specific volume of flow passes the sampling point or after equal time intervals and still be referred to as a composite sample. The resulting mixture (composite sample) forms a representative sample and is analyzed to determine the average conditions during the sampling period.

12. *Grab Sample.*   A single sample of water collected at a particular time and place that represents the composition of the water only at that time and place.

RBC unit designed for BOD removal and one designed for nitrification is that the design flow applied per square foot of media surface area is lower for nitrification. DO in the first stage of a nitrification unit will be more than 1 mg/L DO and often as high as 2 to 3 mg/L.

EFFLUENT VALUES

Typical BOD, suspended solids, and ammonia and nitrate effluent values for rotating biological contactors depend on NPDES permit requirements and design effluent values. As flows increase in an RBC plant, effluent values increase, lowering treatment constituent removal efficiencies because a greater flow is applied to each square foot of media while the time the wastewater is in contact with the slime growths is reduced. Also, the greater the levels of BOD, suspended solids, and nitrogen in the

influent, the greater the levels in the plant effluent. Figure 11.12 shows typical influent and effluent values for a rotating biological contactor. The influent and effluent data plotted are seven-day moving averages, which smooth out daily fluctuations and reveal trends. Procedures for calculating moving averages are explained in Chapter 18, "Analysis and Presentation of Data," in *Operation of Wastewater Treatment Plants*, Volume II.

If analysis of samples reveals a decrease in process efficiency, look for three possible causes:

1. Lowered wastewater temperatures

2. Unusual variations in flow or organic loadings

3. High or low pH values (less than 6.5 or greater than 8.5)

Once the cause of the problem has been identified, possible solutions can be considered and the problem corrected.

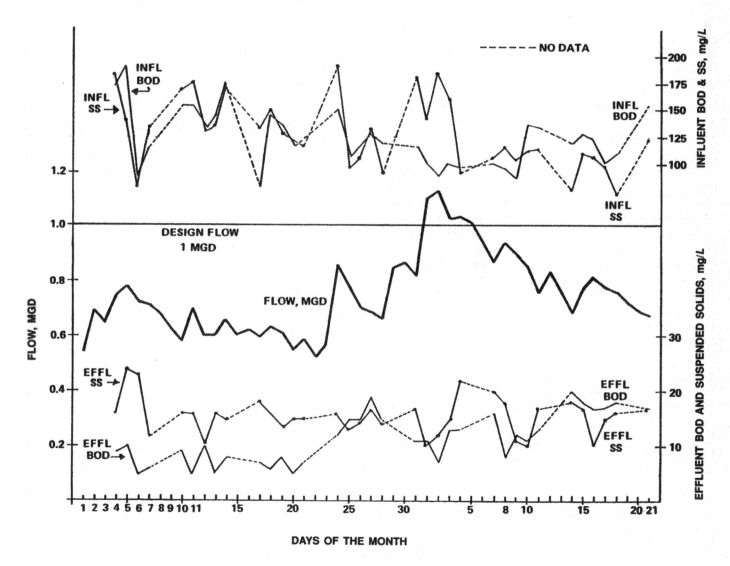

NOTE: Influent data (top lines) are for conventional wastewater before any preliminary or primary treatment. Typical wastewater flowing from septic tanks will have 40 to 100 mg/L BOD and 30 to 80 mg/L suspended solids.

Fig. 11.12 Typical BOD and suspended solids values for an RBC unit

TEMPERATURE

Wastewater temperatures below 55°F (13°C) will result in a reduction of biological activity and in a decrease in BOD or organic material removal. Not much can be done by the operator except to wait for the temperatures to increase again. Under severe conditions, provisions can be made to heat the building, the air inside the RBC unit cover, or the RBC unit influent.

Solar heat can be used effectively to maintain temperature in buildings and enclosures without drying out the biological slime growths. Ceilings should be kept low to use available heat effectively. If existing buildings have high ceilings, large-vaned fans can be mounted on the ceilings to direct heat downward.

INFLUENT VARIATIONS

When large daily influent flow or organic (BOD) variations occur, a reduction in process efficiency is likely to result. Before corrective steps are taken, the exact extent of the problem and resulting change in process efficiency must be determined. In most cases, when the influent flow or organic peak loads are less than three times the daily average values during a 24-hour period, little decrease in process efficiency will result.

In treatment plants where the influent flow or organic loads exceed design values for a sustained period, the effluent BOD and suspended solids must be measured regularly during this period to determine if corrective action is required.

During periods of severe organic overload, the bulkhead or baffle between stages one and two may be removed. This procedure provides a greater amount of media surface area for the first stage of treatment. If the plant is continuously overloaded and the effluent violates the NPDES permit requirements, additional treatment units should be installed. A possible short-term solution to an organic overload problem might be the installation of facilities to recycle effluent; however, this would cause a greater increase of any hydraulic overload. Supplemental aeration is an effective method of controlling excess biomass and improving treatment efficiency.

pH

Every body of wastewater has an optimum pH level for best treatability. Domestic wastewater pH varies between 6.5 and 8.5 and will have little effect on organic waste removal efficiency. However, if this range is exceeded at any time (due to industrial waste discharges, for example), a decrease in efficiency is likely.

To adjust the pH toward 7.0, either pre-aerate the influent or add chemicals. If the pH is too low, add sodium bicarbonate or lime.[13] If the pH is too high, add acetic acid. The amount of chemical to be added depends on the characteristics of the water and can best be determined by adding chemicals to samples in the lab and measuring the change in pH. Before chemicals are used to adjust the pH, run tests in the lab to determine how much of each chemical being considered will be needed to raise the pH to the desired level. Do not guess how much chemical may be needed. Always wear appropriate safety gear (goggles or face shield, impervious gloves, protective clothing) when handling chemicals.

When dealing with RBC nitrification, pH and alkalinity are very critical. The pH should be kept as close as possible to a value of 8.4 when nitrifying. The alkalinity level in the raw wastewater should be maintained at a level at least 7.1 times the influent ammonia concentration to allow the reaction to go to completion without adversely affecting the microorganisms. Effluent alkalinity of 50 to 80 mg/L is required to avoid pH depression in the effluent. Sodium bicarbonate can be used to increase both the alkalinity and pH.

Another cause of pH variations could be the addition of SUPERNATANT[14] from an aerobic or anaerobic digester. The supernatant should be tested for pH and suspended solids. Without testing the supernatant, you will not know what kind of load you are placing on the rest of the plant. Sometimes, it is best to return supernatant at low flows to the plant. Caution should be taken to avoid overloading the process. If the supernatant pH is too low, supernatant could be drawn off during high flows when these flows can be used for dilution and NEUTRALIZATION.[15]

11.122 Observing the Media

Rotating biological contactors use bacteria and other living organisms growing on the media to treat wastes. The appearance and odor of the bacteria and organisms can help you identify problems. Under normal circumstances, the slime growth, or biomass, should have a brown-to-gray color, no algae present, a shaggy appearance with a fairly uniform coverage, and very few or no bare spots. The biomass thickness should not be excessive. The odor should not be offensive, and certainly there should be no sulfide (rotten egg) smells. A black or white appearance in the slime growth indicates abnormal conditions and that something in the influent is disrupting normal slime growth.

BLACK APPEARANCE

If the media appear black and odors that are not normal occur, then this could be an indication of solids or BOD overloading. These conditions would probably be accompanied by low DO

---

13. Lime is relatively insoluble and yields less alkalinity control per pound added than sodium bicarbonate. Lime often settles in the bottom of the tanks and basins and eventually needs to be removed.

14. *Supernatant* (soo-per-NAY-tent).   The relatively clear water layer between the sludge on the bottom and the scum on the surface of an anaerobic digester or septic tank (interceptor). (1) From an anaerobic digester, this water is usually returned to the influent wet well or to the primary clarifier. (2) From a septic tank, this water is discharged by gravity or by a pump to a leaching system or a wastewater collection system. Also called clear zone.

15. *Neutralization* (noo-trull-uh-ZAY-shun).   Addition of an acid or alkali (base) to a liquid to cause the pH of the liquid to move toward a neutral pH of 7.0.

in the plant effluent. Compare previous influent suspended solids and BOD values with current test results to determine if there is an increase. To solve this problem, place another rotating biological contactor unit in service, if possible, operate or install supplemental air, or pre-aerate the influent to the RBC unit. Also, review the operation of the primary clarifiers and sludge digesters to be sure they are not the source of the overload.

WHITE APPEARANCE

A white appearance on the disk surface might be present during high loading conditions. This might be due to a type of bacteria that feeds on sulfur compounds. The overloading could result from industrial discharges containing sulfur compounds upon which certain sulfur-loving bacteria thrive and produce a white slime biomass. Corrective action consists of placing another RBC unit in service, operating supplemental air, or pre-aerating the influent to the unit. During periods of severe organic or sulfur overloading, remove the bulkhead or baffle between stages one and two. Prechlorination of plant influent also will control sulfur-loving bacteria.

Another cause of overloading may be sludge deposits that have been allowed to accumulate in the bottom of the bays. To remove these deposits, drain the bays, wash the sludge deposits out, and return the unit to service. Be sure the orifices in the baffles between the bays are clear.

The biomass on the surface of the media and on the ends of the drums is made up of many different types of organisms. These organisms can be observed using a microscope that has at least 400 times magnification. Organism observations should be made whenever there is a change in the appearance of the RBC biomass or a change in the effluent quality. These organism observations can be related to DO at the point where the samples are taken.

SLOUGHING

If severe sloughing or loss of biomass occurs after the start-up period and process difficulty arises, the causes may be due to the influent wastewater containing toxic substances that kill the organisms in the biomass or restrict their ability to treat wastes. Microscopic observation of the biomass will indicate whether the organisms are dead or alive. If dead, then this indicates the presence of a toxic substance. To solve this toxicity problem, steps must be taken to eliminate the toxic substance even though this may be very difficult and costly. Biological processes will never operate properly as long as they attempt to treat toxic wastes. Until the toxic substance can be located and eliminated, loading peaks should be dampened (reduced) and a diluted, uniform concentration of the toxic substance allowed to reach the media in order to minimize harm to the biological culture. While the corrections are made at the plant, dampening may be accomplished by regulating inflow to the plant. Be careful not to flood any homes or overflow any low manholes. Toxic wastes may be diluted in the plant using plant effluent. Pump any toxic wastes to emergency storage to prevent the toxic substances from being discharged to receiving waters.

Another problem that could cause loss of biomass is an unusual variation in flow and/or organic loading. In small communities, one cause may be high flow during the day and near zero flow at night. During the day, the biomass is receiving food and oxygen and starts growing; then the night flow drops to near zero—available food is reduced and nearly stops. The biomass starts sloughing off again due to lack of food. When diurnal swings (day–night variations) occur, investigate the possibility of taking a portion of the system out of service at night and then putting it back on line in the morning. If the system has been designed with this degree of flexibility, this is a good way to treat load variations.

Possible solutions to sloughing of the biomass due to excessive variations in plant flow or organic loading include throttling peak conditions and recycling from the secondary clarifier or RBC effluent during low flows. Be very careful when throttling plant inflows so that low-elevation homes are not flooded or that manholes do not overflow. Usually, RBC units do not have provisions for any recycling from the secondary clarifier. If low flows at night are creating operational problems due to lack of organic matter, a possible solution is to install a pump to recirculate water from the secondary clarifier. If recirculation is provided, maintain a hydraulic loading rate of greater than 1.0 to 1.5 GPD/sq ft (40 to 60 liters per day/sq m). A flow equalization tank can be used to provide fairly continuous or even flows.

11.123   Weighing the Biomass

The shaft and media of an RBC are designed to carry biomass, but it is possible that the design load carrying capacity of the shaft and media could be exceeded by excessive biomass growth, which may lead to shaft failure. Refer to the Appendix at the end of this chapter for a description of the equipment and procedures that can be used to determine the total operating weight on the RBC shaft.

11.124   Controlling Snails

Snails can be a problem where the RBC is expected to remove nitrogenous biochemical oxygen demand (NBOD) because snails remove slow-growing nitrifying bacteria and interfere with nitrification. Snail shells are a problem when they clog tanks, pipes, and pumps. Snails are not a problem in RBCs used to remove carbonaceous biochemical oxygen demand (CBOD) because the growth of "bugs" (microbes) removing CBOD is high and the microbial slime (treating the wastewater) consumed by snails is quickly replaced by new growth.

Chlorination is commonly used to control snails on RBCs. One approach is to take an RBC train (the RBCs treating a particular slug or flow of wastewater) off line. Add chlorine to the water to a concentration of 60 to 70 mg/L, rotate the RBC in the superchlorinated solution for two to three days, and then return the RBC to service. Chlorination also can be used to control filamentous microorganisms. The superchlorinated water can be dechlorinated before it is discharged from the facility by applying sulfur dioxide. When the RBC is returned to service, it will take a period of time for the biomass to recover and reach full treatment capacity.

Another approach to control snails is to increase the pH to 10. A pH of 10 will kill snails without harming the microbial growth on the RBCs. The pH can be increased by adding caustic soda (sodium hydroxide) or soda ash (sodium carbonate) and maintaining the RBC exposure for eight hours. Operators may have to increase the pH to 10 every one to two months to control the snails.

Also, if secondary solids are being recycled to the head of the plant, snail eggs and snails may be recycled along with the solids. Some operators reduce this transfer of snails by sending the solids from the secondary clarifier directly to anaerobic digesters.

11.125   Operational Troubleshooting

Indicators of possible rotating biological contactor process operational problems, probable causes, and solutions are summarized in Table 11.2.

The top 10 tasks operators can do to improve RBC plant performance are as follows:

- Distribute influent flows to RBCs accurately

- Adjust staging on RBCs

- Aerate RBCs

- Inspect/clean RBC tank solids deposits

- Optimize secondary sludge removals

- Control/minimize/equalize in-plant returns, for example, digester supernatant, thickener overflows, belt filter press precoat chemicals, and filter backwash; avoid BOD and ammonia spikes to RBCs

- Control equalization tank more efficiently

- Recirculate biological sludge to aerated RBCs

- Recirculate RBC effluent to primary anaerobic treatment tank (septic tank) for denitrification and alkalinity recovery

- Recirculate biological sludge to aerated equalization tank

11.13   Abnormal Operation

Abnormal operating conditions may develop under the following circumstances:

1. High or low flows

2. High or low solids loading

3. Power outages

When your plant must treat high or low flows or solids (organic) loads, abnormal conditions develop as the treatment efficiency drops. For solutions to these problems, refer to Section 11.12, "Operation," and Table 11.2. One advantage of RBC units is that high flows usually do not wash the slime growths off the media; consequently the organisms are present and treating the wastewater during and after the high flows.

A power outage requires the operator to take certain precautions to protect the equipment and the slime growths while no power is available. If the power is off for less than eight hours, nothing needs to be done. If the power outage lasts longer than eight hours, the RBC shaft needs to be turned about one-half turn every eight to ten hours. The shaft can be rotated at shorter intervals and smaller increments, if desired, but must equal the same distance in relation to time as previously noted (that is, one-half turn every eight to ten hours). Turning prevents all the slime growth from accumulating on the bottom portion of the plastic disk media. Before attempting to turn the shaft during a power outage, lock out and tag the power in case the outage ends abruptly. To turn the shaft, remove the belt guard using extreme care. Turn the shaft by using the spokes on the pulley or a strap wrench on the shaft. Do not expose your fingers or hands to the pinch point between the belt(s) and the pulley(s). Place a wedge-shaped block between the belts and belt pulley or sheave to hold the shaft and media in the desired location.

NOTE: The shaft is very delicately balanced and easy to rotate. Do not try to weld handles or brackets to the shaft to facilitate turning because this will throw the shaft off balance.

WARNING

If the shaft starts to roll back to its original position before you get the block properly inserted, do not try to stop the shaft. Let it roll back and stop. If you try to stop the shaft from rolling back, you could injure yourself and also damage the belts and the pulley or sheave.

During a power outage, gently spray water on the slime growth that is not submerged frequently enough to keep the biomass moist whenever the drum is not rotating.

If the power outage lasts longer than 12 hours, more than normal sloughing will occur from the media when the unit is placed back in service. When the sloughing becomes excessive, increase the sludge pumping rate from the secondary clarifier.

TABLE 11.2   TROUBLESHOOTING GUIDE—ROTATING BIOLOGICAL CONTACTORS[a]

Indicator	Probable Cause	Check or Monitor	Solution
1. Decreased treatment efficiency.	a. Organic overload.	a. Check peak organic loads—BOD, SS, DO, pH, temperature.	a. 1. Improve pretreatment of plant. 2. Place another RBC in service, if available. 3. Remove bulkhead between stages 1 and 2 for larger first stage. 4. Recycle effluent as a possible short-term solution. 5. Install or operate supplemental aeration.
	b. Hydraulic overload.	b. Check peak hydraulic loads—if less than twice the daily average, should not be the cause.	b. 1. Flow equalization; eliminate source of excessive flow. 2. Balance flows between reactors. 3. Store peak flows in collection system, monitor possible overflows of collection system.
	c. pH too high or too low.	c. Desired range is 6.5–8.5 for secondary treatment; 8–8.5 for nitrification.	c. 1. Eliminate source of undesirable pH or add acid or base to adjust pH. When nitrifying, maintain alkalinity at 7 times the influent NH_3 concentration. 2. Sodium bicarbonate can be used to increase both pH and alkalinity.
	d. Low wastewater temperatures.	d. Temperatures of less than 55° F (13° C) will reduce efficiency.	d. 1. Cover RBC to contain heat of wastewater. 2. Heat influent to unit or building.
2. Excessive sloughing of biomass from disks.	a. Toxic materials in influent.	a. Determine material and its source.	a. 1. Eliminate toxic material if possible—if not, use flow equalization to reduce variations in concentration so biomass can acclimate. Monitor industrial contributors for flow variations and toxic discharges. 2. Recycle effluent for dilution.
	b. Excessive pH variations.	b. pH below 5 or above 10 can cause sloughing.	a. Eliminate source of pH variations or maintain control of influent pH.
	c. Unusual variation in flow or organic loading.	c. Influent flow rate(s) and organic strength.	c. Eliminate/reduce variations by throttling peak conditions and recycling from the secondary clarifier or RBC effluent during low flows.
3. Development of white biomass over most of disk area.	a. Septic influent or high H_2S concentrations.	a. Influent odor.	a. Pre-aerate wastewater or add sodium nitrate or hydrogen peroxide, or place another RBC unit in service. Prechlorination of influent also controls sulfur-loving bacteria. 　　If STEP or SDGS systems, check septic tanks for excessive solids buildup. If excessive solids, pump solids out of septic tank.
	b. First stage is overloaded organically.	b. Organic loading on first stage.	b. 1. Improve pretreatment of plant. 2. Place another RBC in service, if available. 3. Adjust baffles between first and second stages to increase total surface area in first stage. 4. Install or operate supplemental aeration.
4. Solids accumulation in reactors.	a. Inadequate pretreatment.	a. Determine if solids are grit or organic.	a. Remove solids from reactors and provide improved grit removal or primary settling.

a. Adapted from *Performance Evaluation and Troubleshooting at Municipal Wastewater Treatment Facilities,* Office of Water Program Operations, US EPA, Washington, DC.

11.14   Shutdown and Restart

The rotating biological contactor may be stopped by turning off the power to the drive package. If the process is to be stopped for longer than eight hours, follow the precautions listed in Section 11.13, "Abnormal Operation," when a power outage occurs. Do not allow one portion of the media to be submerged in the wastewater being treated for more than eight hours. Occasionally spray the media that is not submerged to prevent the slime growth from drying out whenever the drum is not rotating.

If the shaft cannot be rotated, or if freezing temperatures exist and the flow is stopped, the tank must be drained or pumped out so that the water level remains below the RBC media to prevent equipment damage. A portable sump pump may be used to drain the tank. A sump is usually located at the end of the unit by the motor. Pump the water either to the primary clarifier or to the inlet end of an RBC unit in operation. A trough running the full length of the tank allows the solids to be pumped out. While the tank is empty, inspect it for cracks and any other damage and make necessary repairs.

Keep the slime growths moist to eliminate sloughing and to minimize the reduction in organism activity when the process starts again. A loss in process efficiency can result if the slimes are washed off the media. Do not wash the slime growth off the media because you will be washing away the organisms that treat the wastewater. The exception to this is that if the unit will be out of service for longer than one day, the slimes may be washed off the media to prevent the development of odor problems.

Restart rotation by applying power to the drive unit. Before applying power, inspect the shaft and drive unit for possible interference from such items as tools or bulkheads. If slippage occurs from an unbalanced media, inspect and adjust the alignment and tension. Check the lubrication of the shaft bearings and the drive assembly and bearings.

11.15   Decommissioning RBCs for Extended Shutdowns

There are a variety of situations in which RBCs are taken out of service for months each year. Examples are summer camps, RV parks, ski resorts, marinas, and other seasonally operated facilities. Some larger plants may have standby capacity or seasonal effluent limits so RBCs are treatment alternatives that are put into and taken out of service at various times of the year. To shut down an RBC for an extended period, use the following procedures:

1. Biofilm thickness on the RBC media should be reduced (or minimized) before extended or seasonal shutdowns to prevent accumulation of air-dried solids in the media or basin during the shutdown period. Reduction of biofilm can be accomplished in varying degrees as follows:

    - Stop wastewater flow to the RBC while the RBC continues rotating in idle wastewater. The process will eventually deplete nutrients in the wastewater and the biofilm will starve and slough off.
    - Stop wastewater flow to the RBC, shut off the RBC, drain the tank, fill the basin with clean water (or secondary effluent), and restart the RBC. The biofilm will starve and eventually slough off.

    - Use chlorine (maintain a chlorine residual of 4 mg/L) to strip the biomass or add sodium hydroxide (maintain a pH of 10 to 12) water in the basin to chemically kill the biofilm and produce sloughing. Chemical cleaning is normally not indicated for routine seasonal shutdowns.

    Using water is more effective than using wastewater during cleaning. The effectiveness of the biomass reduction procedure can be assessed visually or determined with load cell equipment (see the Appendix at the end of this chapter). In all cases, intermittent scouring with supplemental aeration will accelerate biomass removal.

    NOTE: Under no circumstances should an RBC be operated for more than a few revolutions if there is a normal wet biofilm growth on the media and the basin is not filled to normal operating level. RBC design relies on the buoyancy of the tank liquid to support a portion of the loaded shaft.

2. After biomass reduction, drain or pump out the RBC tank, flush, wash, or otherwise remove any sludge or solids from the basin. The tank should remain empty and clean for the duration of shutdown; do not allow an accumulation of incidental water (which might freeze) or debris.

3. Protect the RBC from sunlight.

4. During shutdown, the following precautions are recommended:

    - If located in a dusty environment, cover drive and RBC shaft bearings with plastic or similar material to prevent dust and dirt accumulation.
    - Check lubrication of motor, reducer, and bearings during shutdown and every six months.
    - At one-month intervals, after removing dust cover material, using the drive, rotate the RBC for five minutes. On three-month intervals, lubricate the RBC shaft bearings while rotating the RBC.
    - Maintain a rust-preventive coating on RBC shaft ends and bearings.
    - Refer to the manufacturer's O&M manual for lubrication and process information on equipment restart.

RBCs have the advantage of being convenient to take out of or put back into service. During initial start-up, it takes several weeks to reach biomass equilibrium because the new media tends to be very smooth and there is no existing biomass population. However, establishing a biomass on a previously operated RBC occurs more quickly. Accordingly, when taking an RBC out of service for a season, it is neither necessary nor advantageous to get back to squeaky clean media.

QUESTIONS

Please write your answers to the following questions and compare them with those on page 149.

11.1A   Why should debris, grit, and suspended solids be removed before the wastewater being treated reaches the RBC unit?

11.1B   List the major items for a pre-start check.

11.1C   What are the main operational duties for an operator of an RBC unit?

11.2 MAINTENANCE

Rotating biological contactors have few moving parts and require minor amounts of preventive maintenance. Chain drives, belt drives, sprockets, rotating shafts, and any other moving parts should be inspected and maintained in accordance with manufacturers' instructions or your plant's O&M manual. All exposed parts, bearing housing, shaft ends, and bolts should be painted or covered with a layer of grease to prevent rust damage. Motors, speed reducers, and all other metal parts should be painted for protection. Do not paint over nameplate data.

Maintenance also includes the repair or replacement of broken parts. A preventive maintenance program that keeps equipment properly lubricated and adjusted to help reduce wear and breakage requires less time and money than a program that waits for breakdowns to occur before taking any action. The frequency of inspection and lubrication is usually provided by manufacturer's instructions and also may be found in the plant O&M manual. The following sections indicate a typical maintenance program for a rotating biological contactor treatment process. More details can be found in a plant O&M manual. Also see Section 11.23, "Troubleshooting Guides," for specific assistance with different parts of the RBC.

11.20 Break-In Maintenance

AFTER 8 HOURS OF OPERATION

1. Recheck tightening torque of capscrews in all split-tapered bushings in the drive package.

2. Visually inspect hubs and capscrews for general condition and possibility of rubbing against an obstruction.

3. Inspect belt drive (drive package) and tighten as needed.

AFTER 24 HOURS OF OPERATION

1. Inspect all chain drives.

AFTER 40 HOURS OF OPERATION

1. Inspect all belt drives in drive packages.

AFTER 100 HOURS OF OPERATION

1. Change oil in speed reducer. Use manufacturer's recommended lubricants.

2. Clean magnetic drain plug in speed reducer. If a large amount of metal particles is found in the magnetic plug, open the gear box and look for chipped or broken gear teeth. If damaged teeth are found, contact the engineer or contractor who is responsible for repair of equipment under warranty.

3. Check tightness of all capscrews on split-tapered bushings.

4. Check tightness of set screws and bearings on drive package output sprockets.

5. Inspect belt drive of drive package.

AFTER 3 WEEKS OF OPERATION

1. Change oil in chain casing. Be sure oil level is at or above the mark on the dipstick. Use manufacturer's recommended lubricants.

11.21 Preventive Maintenance Program

Interval	Procedure
Daily	1. Check for overheated shaft and bearings. If hot, check bearing lubrication and alignment. Replace bearings if temperature exceeds 200°F (93°C) and there are problems.
Daily	2. Listen for unusual noises in shaft and bearings. Identify the cause of the noise and correct, if necessary.
4 wk.	3. Inspect all chain drives.
4 wk.	4. Inspect mainshaft bearings and drive bearings.
3 mo.	5. Grease the mainshaft bearings and drive bearings. Use manufacturer's recommended lubricants. Add grease slowly while shaft rotates. When grease begins to ooze from the housing, the bearings contain the correct amount of grease. Add six full strokes where bearings cannot be seen. Lubricant providers can supply a list of "or equal" products that meet specifications. If you use an alternate brand lubricant, have the provider give you a letter of assurance for an "or equal" product.
3 mo.	6. Check for unusual odors that may indicate overheated painted surfaces or hot wire insulation. Find the cause of any odors.
3 mo.	7. Change oil in chain casing. Use manufacturer's recommended lubricants. Be sure oil level is at or above the mark on the dipstick, or to a level that covers the lowest point on the shaft sprocket.
3 mo.	8. Inspect belt drive.
6 mo.	9. Apply a generous coating of general purpose grease to mainshaft stub ends, mainshaft bearings, and end collars.
6 mo.	10. Change oil in speed reducer. Use manufacturer's recommended lubricants. If speed reducer is a variable drive unit, operate unit through full range once per week to prevent locking of disk drive sheaves.
6 mo.	11. Clean magnetic drain plug in speed reducer. Check for excess metallic particles.
6 mo.	12. Purge the grease in the double-sealed shaft seals of the speed reducer by removing the plug located 180 degrees from the grease fitting on both the input and output seal cages. Pump grease into the seal cages and then replace the plug. Use manufacturer's recommended grease.
12 mo.	13. Grease motor bearings. Use manufacturer's recommended grease. To grease motor bearings, stop motor and remove drain plugs. Inject new grease with pressure gun until all old grease has been forced out of the bearing through the grease drain. Run motor until all excess grease has been expelled. This may require up to several hours running time for some motors. Replace drain plugs.

11.22   Housekeeping

Properly designed RBC systems have sufficient turbulence to prevent solids or sloughed slime growths from settling out on the bottom of the bays. If grease balls appear on the water surface in the bays, they should be removed with a dip net or screen device.

If the media comes apart on welded media systems, squeeze the two unbonded sections together with a pair of pliers. Take another pair of pliers and force a heated nail through the media. The heat from the nail will melt the plastic and make a plastic weld between the two sections of media.

11.23   Troubleshooting Guides

Use the information in the following troubleshooting guides (Tables 11.3–11.7) to identify causes and take corrective action.

TABLE 11.3   TROUBLESHOOTING GUIDE—ROLLER CHAIN DRIVE

Trouble	Probable Cause	Corrective Action
1. Noisy Drive	a. Moving parts rub against stationary parts	a. Tighten and align casing and chain. Remove dirt or other interfering matter.
	b. Chain does not fit sprockets	b. Replace with correct parts.
	c. Loose chain	c. Maintain a taut chain at all times.
	d. Faulty lubrication	d. Lubricate properly.
	e. Misalignment or improper assembly	e. Correct alignment and assembly of the drive.
	f. Worn parts	f. Replace worn chain or bearings. Reverse worn sprockets before replacing.
2. Rapid Wear	a. Faulty lubrication	a. Lubricate properly.
	b. Loose or misaligned parts	b. Align and tighten entire drive.
3. Chain Climbs Sprockets	a. Chain does not fit sprockets	a. Replace chain or sprockets.
	b. Worn-out chain or worn sprockets	b. Replace chain. Reverse or replace sprockets.
	c. Loose chain	c. Tighten.
4. Stiff Chain	a. Faulty lubrication	a. Lubricate properly.
	b. Rust or corrosion	b. Clean and lubricate.
	c. Misalignment or improper assembly	c. Correct alignment and assembly of the drive.
	d. Worn-out chain or worn sprockets	d. Replace chain. Reverse or replace sprockets.
5. Broken Chain or Sprockets	a. Shock or overload	a. Repair or replace broken parts. Add more RBC units or remove baffles between stages 1 and 2.
	b. Wrong size chain, or chain that does not fit sprockets	b. Replace chain. Reverse or replace sprockets.
	c. Rust or corrosion	c. Replace parts. Correct corrosive conditions.
	d. Misalignment	d. Correct alignment.
	e. Interferences	e. Make sure no solids get between chain and sprocket teeth. Loosen chain if necessary for proper clearance over sprocket teeth.

TABLE 11.4   TROUBLESHOOTING GUIDE—BELT DRIVE

Trouble	Probable Cause	Corrective Action
1. Excessive edge wear	Misalignment of non-rigid centers	Check alignment and reinforcement mounting.
2. Jacket wear on pressure-face side of belt tooth*	Bent flange	Straighten flange.
3. Excessive jacket wear between belt teeth (exposed tension members)*	Excessive overload and/or excessive belt tightness	Reduce installation tension and/or increase drive load-carrying capacity.
4. Cracks in Neoprene backing	a. Excessive installation tension b. Exposure to excessively low temp (below –30°F or –35°C)	a. Reduce installation tension. b. Eliminate high temperature and oil condition or consult factory for proper belt construction.
5. Softening of Neoprene backing	Exposure to excessive heat (+200°F or 90°C) or oil	Eliminate high temperature and oil condition or consult factory for proper belt construction.
6. Tensile or tooth shear failure*	a. Small or sub-minimum diameter pulley b. Belt too narrow	a. Increase pulley diameter. b. Increase belt width.
7. Excessive pulley tooth wear (on pressure face or OD)*	a. Excessive overload or excessive belt tightness b. Insufficient hardness of pulley material	a. Reduce installation tension and/or increase drive load-carrying capacity. b. Surface-harden pulley or use harder material.
8. Unmounting of flange	a. Incorrect flange installation b. Misalignment	a. Reinstall flange correctly. b. Correct alignment.
9. Excessive drive noise	a. Misalignment b. Excessive installation tension c. Sub-minimum pulley diameter	a. Correct alignment. b. Reduce tension. c. Increase pulley diameter.
10. Tooth shear*	a. Less than six teeth in mesh (TIM) b. Excessive load	a. Increase TIM or use next smaller pitch. b. Increase drive load-carrying capacity.
11. Apparent belt strength	Reduction of center distance or non-rigid mounting	Re-tension drive and/or reinforce mounting.
12. Cracks or premature wear at belt tooth root*	Improper pulley groove top radius	Regroove or install new pulley.
13. Tensile break	a. Excessive load b. Sub-minimum pulley diameter	a. Increase load-carrying capacity of drive. b. Increase pulley diameter.

*Pertains to a timing belt system only. Recent systems use a V-belt drive.

TABLE 11.5   TROUBLESHOOTING GUIDE—V-BELT DRIVE

Trouble	Probable Cause	Corrective Action
1. Short belt life	a. Spin burns from belt slipping on driver sheave under stalled load conditions or when starting	a. Tension belts.
	b. Gouges or extreme cover wear caused by belts rubbing on drive	b. Eliminate obstruction or realign drive to provide clearance.
	c. High ambient temperature	c. Use gripnotch belts. Provide ventilation. Shield belts.
	d. Grease or oil on belts	d. Check for leaky bearings. Clean belts and sheaves.
	e. Worn sheaves	e. Replace sheaves.
2. Belt turns over in grooves	a. Damaged cord section in belts. Frayed or gouged belts	a. Replace belts.
	b. Excessive vibration	b. Tension belts. Replace belts if damaged.
	c. Flat idler pulley misaligned	c. Realign idler.
	d. Worn sheaves	d. Replace sheaves.
	e. Sheave misalignment	e. Realign sheaves.
3. Belt squeals	a. High starting load. Belts not tensioned properly Excessive overload	a. Tension drive or redesign and replace drive.
	b. Insufficient arc of contact	b. Increase center distance or use gripnotch belts.
4. Belt breakage	a. Foreign material in drive	a. Provide drive guard.
	b. Belts damaged during installation	b. Follow manufacturer's installation instructions.
	c. Shock or extreme overload	c. Eliminate overload cause or redesign drive.
5. Belt stretch beyond take-up	a. Worn sheaves	a. Replace sheaves.
	b. Under-designed drive	b. Redesign and replace drive.
	c. Take-up slipped	c. Reposition take-up.
	d. Drive excessively tensioned	d. Properly tension drive.
	e. Damaged cord section during installation	e. Replace belts and properly install.
6. Excessive vibration	a. Damaged belt cord section	a. Replace belts.
	b. Loose belts	b. Tension drive.
	c. Belts improperly tensioned	c. Tension drive with slack of each belt on the same side of drive.
7. Belt too long (or short) at installation	a. Insufficient take-up	a. Use shorter (or longer) belts.
	b. Drive improperly set up	b. Recheck driver and driven machine setup.
	c. Wrong size belts	c. Use correct size belts.
8. Belt mismatched at installation	a. Belts matched by code number only	a. Replace belts with machine-matched belts.
	b. Old and new belts used together on same drive	b. Replace with all new belts.
	c. Different brand name belts used on same drive	c. Replace with set of machine-matched belts.
	d. Driver and driven shafts not parallel	d. Realign drive.
	e. Worn sheaves	e. Replace sheaves.
9. Belt mismatched after service	a. Belts improperly tensioned, causing more stretch of some belts than others	a. Replace belts and tension drive with slack of each belt on the same side of the drive.
	b. Old and new belts used together on same drive	b. Replace with all new belts.
	c. Different brand name belts used on same drive	c. Replace with set of machine-matched belts.
	d. Driver and driven shafts shifted from parallel	d. Realign drive.
	e. Belt cord section damaged during installation	e. Replace belts and install properly.

TABLE 11.6   TROUBLESHOOTING GUIDE—BEARINGS AND MOTORS

Trouble	Probable Cause	Corrective Action
1. Shaft bearings running hot or failing. If temperature exceeds 200°F (93°C), bearings may need to be replaced	a. Inadequate lubrication	a. Lubricate bearings per manufacturer's instructions. Check tightness (torque) and alignment of bearings.
	b. Shaft misalignment	b. Align shaft properly. The longer the shaft, the more critical the alignment.
2. Motors running hot	a. Inadequate maintenance	a. Lubricate per manufacturer's instructions. Maintain correct oil level and oil viscosity in speed reducer.
	b. Improper chain drive alignment	b. Align properly.
	c. Excessive ambient temperature	c. Control/lower ambient temperature.

TABLE 11.7   TROUBLESHOOTING GUIDE—RBC SHAFT

Trouble	Probable Cause	Corrective Action
1. Noisy bearing	a. Rollers or bearing race damaged	a. Replace bearing cartridge.
	b. No grease between bearing cartridge housing and pillow block housing	b. Lubricate outside diameter of cartridge housing.
2. Hot bearing	a. No grease in bearing	a. See item 4 below.
	b. Bearing rollers or cups and cones damaged	b. Replace bearing cartridge assembly.
3. Bearing leaks grease	Grease escapes between the seal and locking collar when lubricating the bearing or when grease expands due to an increase in ambient temperature	None—this is normal.
4. Excessive grease usage (4 oz. of grease is required to lubricate the bearing. Using more than this amount may cause trouble)	Worn or damaged seals or seal lip contact area on bearing	Replace seals. Inspect bearing rollers, races, and seal lip contact area. If worn or damaged, replace bearing cartridge assembly. Reduce amount of grease used. Note amount needed to keep equipment operating properly.
5. Evidence of water in bearing (purged grease is milky or rusty)	a. Water trapped between locking collar and seal	a. Wipe clean and relubricate.
	b. Worn or damaged seals or seal lip contact area on bearings	b. Replace seals. Inspect bearing rollers, races, and seal lip contact area. If worn or damaged, replace bearing cartridge assembly.
6. Excessive shaft weight and biomass thickness	Organic loading too high	Decrease organic loading.

11.3   SAFETY

Any equipment with moving parts or electrical components should be considered a potential safety hazard. Always shut off the power to the unit, tag the switch, and lock the main breaker in the OFF position before working on a unit.

11.30   Slow-Moving Equipment

Slow-moving equipment does not appear to be dangerous. Unfortunately, moving parts such as the sprockets, chains, belts, and sheaves can cause serious injury by tearing or crushing your hands or legs.

11.31   Wiring and Connections

Wiring and connections should be inspected regularly for potential hazards such as loose connections and bare wires. Again, always shut off, tag, and lock out the main breaker before working on a unit.

11.32   Slippery Surfaces

Caution must be taken on slippery surfaces. Falls can result in serious injuries. Any spilled oil or grease must be cleaned up immediately. If covers over the media allow sufficient space for walkways, condensed moisture on surfaces will create slippery places.

If the temperature of the air within the enclosure can be kept several degrees higher than the temperature of the wastewater, then condensation is significantly reduced. This condensation cannot be avoided completely so walk carefully at all times. The application of a nonskid material on walkways can greatly reduce the potential for slips and falls.

11.33   Infections and Diseases

Precautions must be taken to prevent infections in cuts or open wounds and illnesses from waterborne diseases. After working on a unit, always wash your hands before smoking or eating. Good personal hygiene must be practiced by all operators at all times.

11.4   REVIEW OF PLANS AND SPECIFICATIONS

When reviewing plans and specifications, be sure the following items are included in the design of rotating biological contactors.

1. Enclosure to protect biomass from freezing temperatures. Enclosures should be constructed of suitable corrosion-resistant materials and have windows or louvered structures in sides for ventilation. Forced ventilation is not necessary.

2. Heating. A heat source is helpful during winter operation to minimize the corrosion caused by condensation and to improve operator comfort. If the temperature of the air within the enclosure is kept several degrees above the temperature of the wastewater, condensation is significantly reduced. Ceilings should be kept low to effectively use available heat.

3. Recirculation. Provisions for recirculating secondary effluent and also secondary sludge will allow for flexibility and can improve performance.

4. Supplemental aeration. Better RBC media systems are designed to accommodate the benefits of air scour and biomass control afforded by supplemental aeration. Consider installation of headers and diffusers with future blower hookup if the cost of the blower is not warranted during initial installation.

5. Provisions should be made to catch and store oil that is drained from the gear boxes and other units.

QUESTIONS

Please write your answers to the following questions and compare them with those on page 149.

11.2A   How often should maintenance be performed on an RBC unit during start-up?

11.3A   List some possible safety hazards operators encounter when working around RBC units.

11.5   LOADING CALCULATIONS

11.50   Typical Loading Rates

Hydraulic and organic loadings on rotating biological contactors depend on influent soluble BOD, effluent total and soluble BOD requirements, and wastewater temperature. The loading rates given here are for typical values. Values for your plant could be different.

Characteristic	Range
Hydraulic Loading	
BOD Removal	1.5 to 6 GPD/sq ft
Nitrogen Removal	1.5 to 1.8 GPD/sq ft
Organic Loading	
Soluble BOD	2.5 to 4 lbs/day/1,000 sq ft
Total BOD	6 to 8 lbs/day/1,000 sq ft

11.51   Computing Hydraulic Loading

Hydraulic loading on a rotating biological contactor is the amount (gallons) of wastewater per day that flows past the rotating media. To calculate the hydraulic loading, we must have the following information:

1. Gallons per day treated by the rotating biological contactor

2. The surface area of the media in square feet

EXAMPLE 1

A rotating biological contactor treats a flow of 3.5 MGD. The surface area of the media is 1,000,000 square feet (provided by manufacturer). What is the hydraulic loading in GPD/sq ft?

Known		Unknown
Flow, MGD	= 3.5 MGD	Hydraulic Loading, GPD/sq ft
Surface Area, sq ft	= 1,000,000 sq ft	

Calculate the hydraulic loading in gallons per day of wastewater per square foot of media surface.

$$\text{Hydraulic Loading, GPD/sq ft} = \frac{\text{Flow, gal/day}}{\text{Surface Area, sq ft}}$$

$$= \frac{3,500,000 \text{ GPD}}{1,000,000 \text{ sq ft}}$$

$$= 3.5 \text{ GPD/sq ft}$$

11.52   Computing Organic (BOD) Loading

Organic loadings on rotating biological contactors are based on soluble BOD. Soluble BOD is the BOD of water that has been filtered in the standard suspended solids test. If soluble BOD information is not available, soluble BOD may be estimated on the basis of the total BOD and the suspended solids as follows:

$$\text{Soluble BOD, mg/L} = \text{Total BOD, mg/L} - \text{Suspended BOD, mg/L}$$

where

Suspended BOD, mg/L = $K \times$ Suspended Solids, mg/L

Therefore,

$$\text{Soluble BOD, mg/L} = \text{Total BOD, mg/L} - (K \times \text{Suspended Solids, mg/L})$$

where

K = 0.5 to 0.7 for most domestic wastewaters

NOTE: The K value of a wastewater is the ratio of suspended BOD to suspended solids; it is obtained by dividing suspended BOD by suspended solids. By calculating K values for many samples we obtain a range of K for most domestic wastewaters of from 0.5 to 0.7. K is useful in estimating the soluble BOD when only the total BOD and suspended solids are known.

EXAMPLE 2

The rotating biological contactor in Example 1 treats an industrial influent with a total BOD of 200 mg/L and suspended solids of 250 mg/L. Assume a K value of 0.5 to calculate the soluble BOD. What is the organic (BOD) loading in pounds of soluble BOD per day per 1,000 square feet of media surface?

Known		Unknown
Flow, MGD	= 3.5 MGD	Organic (BOD)
Surface Area, sq ft	= 1,000,000 sq ft	Loading, lbs/day/
Total BOD, mg/L	= 200 mg/L	1,000 sq ft
SS, mg/L	= 250 mg/L	
K	= 0.5	

1. Estimate the soluble BOD treated by the rotating biological contactor in milligrams per liter.

$$\text{Soluble BOD, mg/L} = \text{Total BOD, mg/L} - (K \times \text{Suspended Solids, mg/L})$$

$$= 200 \text{ mg/L} - (0.5 \times 250 \text{ mg/L})$$

$$= 200 \text{ mg/L} - 125 \text{ mg/L}$$

$$= 75 \text{ mg/L}$$

2. Determine the soluble BOD applied to the rotating biological contactor in pounds of soluble BOD per day.

$$\text{Soluble BOD, lbs/day} = (\text{Flow, MGD})(\text{Soluble BOD, mg/L})(8.34 \text{ lbs/gal})$$

$$= (3.5 \text{ MGD})(75 \text{ mg/L})(8.34 \text{ lbs/gal})$$

$$= 2,189 \text{ lbs soluble BOD/day}$$

3. Calculate the organic (BOD) loading in pounds of soluble BOD per day per 1,000 square feet of media surface. Because the media surface area is 1,000,000 square feet, there are 1,000 • 1,000 square feet of media surface.

$$\text{Organic (BOD) Loading, lbs/day/ 1,000 sq ft} = \frac{\text{Soluble BOD, lbs/day}}{\text{Surface Area of Media (in 1,000 sq ft)}}$$

$$= \frac{2,189 \text{ lbs Soluble BOD/day}}{1,000 \bullet 1,000 \text{ sq ft}}$$

$$= 2.2 \text{ lbs BOD/day/1,000 sq ft}$$

11.53 Typical Loading Rates (Metric)

The next three sections show typical loading rates for rotating biological contactors and how to calculate the hydraulic and organic loading using the metric system.

Characteristic	Range
Hydraulic Loading	
BOD Removal	0.06 to 0.24 cu m/day/sq m
Nitrogen Removal	0.06 to 0.07 cu m/day/sq m
Organic Loading	
Soluble BOD	15 to 25 gm/day/sq m
Total BOD	36 to 50 gm/day/sq m

11.54 Computing Hydraulic Loading (Metric)

Hydraulic loading on a rotating biological contactor is the amount (cubic meters) of wastewater per day that flows past the rotating media. To calculate the hydraulic loading, we must have the following information:

1. Cubic meters per day treated by the rotating biological contactor

2. The surface area of the media in square meters

EXAMPLE 3

A rotating biological contactor treats a flow of 15,000 cubic meters per day. The surface area of the media is 100,000 square meters (provided by the manufacturer). What is the hydraulic loading in cu m/day/sq m?

Known		Unknown
Flow, cu m/day	= 15,000 cu m/day	Hydraulic Loading,
Surface Area, sq m	= 100,000 sq m	cu m/day/sq m

Calculate the hydraulic loading in cubic meters per day of wastewater per square meter of media surface.

$$\text{Hydraulic Loading, cu m/day/sq m} = \frac{\text{Flow, cu m/day}}{\text{Surface Area, sq m}}$$

$$= \frac{15,000 \text{ cu m/day}}{100,000 \text{ sq m}}$$

$$= 0.15 \text{ cu m/day/sq m}$$

11.55   Computing Organic (BOD) Loading (Metric)

EXAMPLE 4

The rotating biological contactor in Example 3 treats an industrial influent with a soluble BOD of 75 mg/L. What is the organic (BOD) loading in grams of soluble BOD per day per square meter of media surface?

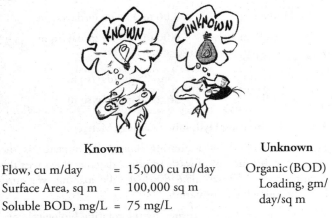

Known		Unknown
Flow, cu m/day	= 15,000 cu m/day	Organic (BOD)
Surface Area, sq m	= 100,000 sq m	Loading, gm/
Soluble BOD, mg/L	= 75 mg/L	day/sq m

1. Determine the soluble BOD applied to the rotating biological contactor in grams of soluble BOD per day.

$$\text{Soluble BOD, gm/day} = \left(\text{Flow, cu m/day}\right)\left(\text{Soluble BOD, mg/L}\right)\left(\frac{1,000 \text{ kg}}{\text{cu m}}\right)\left(\frac{1,000 \text{ gm}}{\text{kg}}\right)$$

$$= \left(\frac{15,000 \text{ cu m}}{\text{day}}\right)\left(\frac{75 \text{ mg}}{1,000,000 \text{ mg}}\right)\left(\frac{1,000 \text{ kg}}{\text{cu m}}\right)\left(\frac{1,000 \text{ gm}}{\text{kg}}\right)$$

$$= 1,125,000 \text{ gm soluble BOD/day}$$

2. Calculate the organic (BOD) loading in grams of soluble BOD per day per square meter of media surface.

$$\text{Organic (BOD) Loading, gm/day/sq m} = \frac{\text{Soluble BOD, gm/day}}{\text{Surface Area of Media, sq m}}$$

$$= \frac{1,125,000 \text{ gm Soluble BOD/day}}{100,000 \text{ sq m}}$$

$$= 11.2 \text{ gm BOD/day/sq m}$$

QUESTIONS

Please write your answers to the following questions and compare them with those on pages 149 and 150.

A rotating biological contactor treats a flow of 2.0 MGD with an influent soluble BOD of 100 mg/L. The surface area of the media is 500,000 square feet.

11.5A   What is the hydraulic loading in gallons per day per square foot of media surface?

11.5B   What is the organic (BOD) loading in pounds per day per 1,000 square feet of media surface?

11.5C   If the first-stage surface area is 250,000 square feet, what is the first-stage organic (BOD) loading in pounds day per 1,000 square feet of first-stage media surface?

11.6   ARITHMETIC ASSIGNMENT

Turn to the Arithmetic Appendix at the back of this manual and work all the problems in Section A.22, "Rotating Biological Contactors." Check the arithmetic with your calculator.

11.7   ACKNOWLEDGMENTS

Jerry Greene, Autotrol Corporation, reviewed this section and provided helpful suggestions, photographs, drawings, and procedures. His contributions are greatly appreciated.

Mark Gehring with US Filter, John R. Harrison, and Mark Lambert also reviewed this revised chapter and their comments and suggestions are sincerely appreciated.

11.8   ADDITIONAL READING

1. *MOP 11,* Chapter 21, "Rotating Biological Contactors."*

2. *New York Manual,* Chapter 5, "Secondary Treatment."*

3. *Biofilm Reactors* (MOP 35). Obtain from Water Environment Federation (WEF), Publications Order Department, PO Box 18044, Merrifield, VA 22118-0045. Order No. W100061. Price to members, $115.00; nonmembers, $130.00; plus shipping and handling.

*Depends on edition.

Please answer the discussion and review questions next.

DISCUSSION AND REVIEW QUESTIONS

Chapter 11.   ROTATING BIOLOGICAL CONTACTORS (RBCs)

Please write your answers to the following discussion and review questions to determine how well you understand the material in the chapter.

1. Describe the rotating biological contactor (RBC) process and discuss how it works.

2. List the tasks that should be included when performing a regular inspection of rotating biological contactor equipment.

3. What water quality indicators would you test for in the effluent from an RBC treatment plant?

4. How do the slime growths (biomass) on the plastic media look under (a) normal conditions and (b) abnormal conditions?

5. What factors can cause decreased treatment efficiency of an RBC unit, and how can these problems be corrected?

SUGGESTED ANSWERS

Chapter 11. ROTATING BIOLOGICAL CONTACTORS (RBCs)

Answers to questions on page 130.

11.0A A rotating biological contactor treats wastewater by allowing the slime growths to develop on the surface of plastic media. The slime growths contain organisms that remove the organic materials from the wastewater.

11.0B The RBC process is divided into multiple stages to increase the effectiveness of a given amount of media surface area. Organisms on the first-stage media are exposed to high levels of BOD and reduce the BOD at a high rate. As the BOD levels decrease from stage to stage, the rate at which the organisms can remove BOD decreases and nitrification starts.

11.0C RBC units are covered to protect organisms from severe changes in weather, especially freezing. Other reasons include preventing heavy rains from washing off slime growth and protecting operators from weather while maintaining equipment. Covers also contain odors and prevent the growth of algae on the media.

11.0D There are three types of drive assemblies used to rotate the shafts supporting the media in an RBC:

1. Motor with chain drive
2. Motor with direct shaft drive
3. Air drive

Answers to questions on page 140.

11.1A Debris, grit, and suspended solids should be removed to prevent them from settling beneath the drum and forming sludge deposits. These sludge deposits can reduce the effective tank volume, produce septic conditions, physically damage the media, and possibly stall the unit.

11.1B The following are the major items for a pre-start check:

1. Tightness of bolts and parts
2. Lubrication of equipment
3. Clearances for moving parts
4. Drive unit alignment to drum shaft
5. Flow control equipment
6. Electrical controls
7. Proper installation of safety guards

11.1C The following are the main operational duties for an operator of an RBC unit:

1. Inspecting equipment
2. Testing influent and effluent
3. Observing the media
4. Maintaining the equipment
5. Taking corrective action when needed

Answers to questions on page 146.

11.2A During start-up, maintenance should be performed on an RBC unit

1. after 8 hours of operation,
2. after 24 hours of operation,
3. after 40 hours of operation,
4. after 100 hours of operation, and
5. after 3 weeks of operation.

11.3A The following are possible safety hazards to operators working around RBC units:

1. Equipment with slow-moving parts or electrical components
2. Electrical wires and connections
3. Slippery surfaces
4. Infections in cuts and open wounds
5. Illnesses from waterborne diseases

Answers to questions on page 148.

Data for questions 11.5A, 11.5B, and 11.5C

A rotating biological contactor treats a flow of 2.0 MGD with an influent soluble BOD of 100 mg/L. The surface area of the media is 500,000 square feet.

Known

Flow MGD	= 2.0 MGD
Soluble BOD, mg/L	= 100 mg/L
Surface Area, sq ft	= 500,000 sq ft
Surface Area, sq ft (First Stage)	= 250,000 sq ft

Unknown

11.5A Hydraulic Loading, GPD/sq ft

11.5B Organic (BOD) Loading, lbs/day/1,000 sq ft

11.5C First-Stage Organic (BOD) Loading, lbs/day/1,000 sq ft

11.5A Calculate the hydraulic loading in gallons per day of wastewater per square foot of media surface.

$$\text{Hydraulic Loading, GPD/sq ft} = \frac{\text{Flow, gal/day}}{\text{Surface Area, sq ft}}$$

$$= \frac{2,000,000 \text{ gal/day}}{500,000 \text{ sq ft}}$$

$$= 4 \text{ GPD/sq ft}$$

11.5B   1. Determine the soluble BOD applied to the rotating biological contactor in pounds of soluble BOD per day.

Soluble
  BOD, = (Flow, MGD)(Soluble BOD, mg/L)(8.34 lbs/gal)
  lbs/day

        = (2.0 MGD)(100 mg/L)(8.34 lbs/gal)

        = 1,668 lbs soluble BOD/day

        2. Calculate the organic (BOD) loading in pounds per day per 1,000 square feet of media surface.

Organic (BOD)
  Loading, lbs/day/ = $\dfrac{\text{Soluble BOD, lbs/day}}{\text{Surface Area of Media (in 1,000 sq ft)}}$
  1,000 sq ft

                    = $\dfrac{1{,}668 \text{ lbs Soluble BOD/day}}{500 \bullet 1{,}000 \text{ sq ft}}$

                    = 3.3 lbs BOD/day/1,000 sq ft

11.5C   Calculate the first-stage organic (BOD) loading in pounds per day per 1,000 square feet of first-stage media surface.

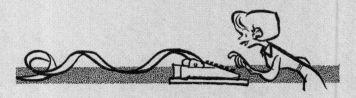

Organic (BOD)
  Loading, lbs/day/ = $\dfrac{\text{Soluble BOD, lbs/day}}{\text{First-Stage Surface Area of Media (in 1,000 sq ft)}}$
  1,000 sq ft

                    = $\dfrac{1{,}668 \text{ lbs Soluble BOD/day}}{250 \bullet 1{,}000 \text{ sq ft}}$

                    = 6.6 lbs BOD/day/1,000 sq ft

NOTE: Soluble BOD loading on the first stage should be less than 4 lbs BOD/day/1,000 sq ft.

APPENDIX

QUANTITATIVE BIOMASS ASSESSMENT AND CORRECTIVE ACTIONS

(Adapted from operations manual, courtesy of US Filter, Envirex Products)

The shaft and media of a rotating biological contactor (RBC) are designed to carry biomass. It is possible for the design load carrying capacity of the shaft and media to be exceeded by excessive biomass growth, which may lead to shaft failure. The hydraulic weighing apparatus and bearing load cell are used to determine the total weight on the shaft.

The RBC weighing apparatus consists of a hand-operated hydraulic pump, pressure gauge, flexible hose, and fittings needed to connect the hose to the quick-disconnect coupler that is attached to the bearing housing (Figure 11.13).

The operation of the RBC weighing apparatus is used in conjunction with the load cell bearing located on the expansion bearing end (idle end) of the RBC unit. A load cell kit is used on an expansion-type bearing. The kit is to be installed in the field prior to setting the bearings on the base plates.

The bearing load cell is the most accurate method of determining biomass weight. By weighing the RBC while in operation, shaft weight, media weight, biomass, and any entrapped water are measured. Data have been developed to convert a pounds per square inch (psi) reading obtained from the hydraulic weighing device to the total operating weight of the shaft.

SET-UP PROCEDURE

Weigh the RBC unit while it is rotating.

Loosen the nuts securing the bearing housing to the base plate until they can be turned by hand. Tighten each nut finger-tight, then back off one turn to limit the bearing movement.

Connect the hose to the coupler located on the bearing housing. With the relief valve open, stroke the pump a few times to purge air from the pump cylinder. Close the valve and proceed.

Apply pressure with slow, steady strokes of the hand pump. The pressure will rise steadily until the bearing is lifted off the base. With continued stroking of the pump, the pressure should remain constant until the bearing reaches the limiting nuts, after which it increases sharply. Stop pumping before the bearing reaches the nuts to prevent O-ring blow-by.

If the bearing is against the limiting nuts (nuts cannot be rotated by hand), release the pressure in small increments until the nuts can be rotated by hand.

PRESSURE READING

Observe the pressure required to keep the bearing floating between the bearing base and the limiting nuts.

Most RBCs are not perfectly balanced and a pressure swing likely will be observed during rotation. Record this pressure range.

If the load cell bearing alternately bottoms out and hits the limiting nuts while the RBC is rotating, there is a large volume of air trapped in the pressure system. In this case, the extreme pressure readings are inaccurate and only the relatively stable mean pressure (average) should be recorded.

If one side of the bearing housing touches the limiting nuts before the other side, release the pressure and re-pressurize the system. (If, after several attempts, the condition continues, record the pressure reading and note this condition.)

If the load cell bearing settles as it rotates, this is due to a loss in system pressure. The pump must be continually stroked to keep the bearing floating to record an accurate pressure reading.

Pressure readings can be converted to RBC weight from the curve(s) supplied by the manufacturer. Note the load limitation(s).

RE-SETTING BEARING

Open the relief valve slowly to lower bearing. Check the bearing to ensure that it is level and properly aligned. The bearing housing should rest evenly on the base plate. The bearing cartridge should be centered in the housing.

Disconnect the hose from the coupler. Re-torque the nuts securing the bearing housing to the base plate at 160 lbs/ft. (The threads must be clean, dry, and free from oil and grease.)

FREQUENCY FOR LOAD CELL READINGS

Load cell readings should be taken twice a month and become part of the plant's permanent records. Overloading condition will require corrective action. Contact the RBC manufacturer if the following overload corrective measures do not resolve overloading conditions.

Continual overload condition may shorten equipment life and may decrease process efficiency. In addition, sustained overloaded

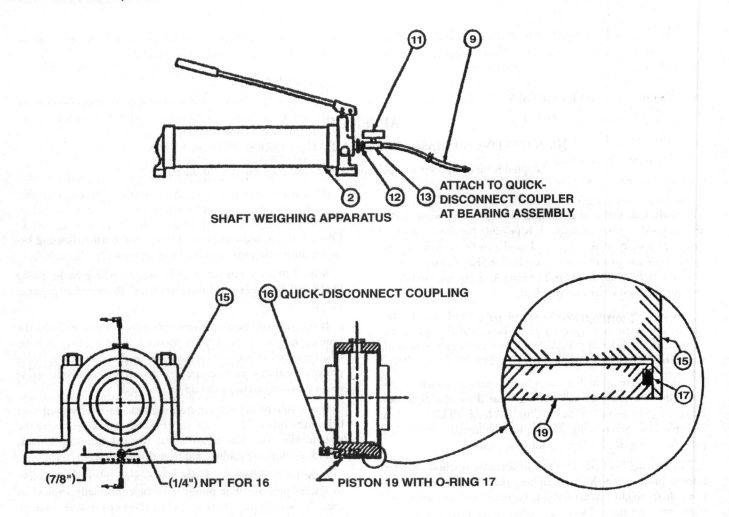

SHAFT WEIGHING APPARATUS

ATTACH TO QUICK-
DISCONNECT COUPLER
AT BEARING ASSEMBLY

QUICK-DISCONNECT COUPLING

(7/8")   (1/4") NPT FOR 16   PISTON 19 WITH O-RING 17

LEGEND

Item	Description
2	Hydraulic pump (hand-operated)
9	Flexible hose with male coupler
11	Pressure gauge
12	Hex, nipple
13	Pipe, tee
15	Pillow-block bearing
16	Quick-disconnect coupler with dust cap
17	O-ring
19	Piston

NOTE: Install O-ring, quick-disconnect coupler, and piston before setting shaft on bearing base. Do not twist O-ring when installing it on piston. Lubricate O-ring with clean oil or light grease before assembling piston into bearing base.

Fig. 11.13   RBC shaft weighing apparatus installation diagram
(Courtesy of US Filter Envirex Products)

condition will void the equipment warranty. Shaft, media, or support fatigue failures may occur. Media clogging may also result.

CAUSES OF EXCESSIVE BIOMASS

- High stage of system SBOD loadings
- Extended stage detention times
- Cold wastewater temperature
- Specific organic makeup of SBOD
- Precipitation of inorganic salts, for example, lime precipitates as $CaCO_3$
- Polymer coagulation (from upstream or side stream wastes, for example, belt press or centrifuge returns)
- De-emulsification of solvents carrying fats, oil, and grease (FOGs)

OVERLOAD CORRECTIVE MEASURES

The most common causes of overloading (excessive shaft weight) are (1) organic loading rate in excess of design and (2) improper staging of RBCs.

Total shaft weight reduction can be achieved by either changing the organic loading or increasing the shearing forces on the biomass. This can be accomplished by one of the following overload correction methods:

1. Changing the staging configuration
2. Step feeding the influent
3. Recirculation (organic load dilution)
4. Diffused aeration
5. Starvation
6. Chemical stripping

1. Changing the Staging Configuration

Increasing the media surface area in a given stage by changing the baffle location will reduce the overall organic load and subsequent biomass weight for that stage. This method is limited to installations with removable baffles. Note that increasing the area of a stage will either decrease the size or number of the stages that follow, which changes downstream loading, removal, and biomass.

2. Step Feeding

Feeding a portion of the RBC influent to stages other than the influent stages will have essentially the same effect as changing baffle configuration.

3. Recirculation

Organic load dilution can be affected by recirculation of the effluent mixed liquor or secondary clarifier effluent through the RBC(s). Although organic loading is not decreased using this method, the strength of the influent waste is decreased, which can have a positive influence on biomass thickness reduction in the initial stage(s). This method also has the effect of decreasing the size and number of the following stages, which will change the downstream loading, removal, and biomass.

4. Diffused Aeration

The addition of diffused aeration enhances the biological process while stripping biomass. It is most effective when applied intermittently at high rates, for example, 10 to 12 cfm/ft of RBC media length. This tool represents a long-term solution for controlling periods of excessive biomass growth.

5. Starvation

This method requires the RBC to be taken out of service so it may be impractical in a single- or two-shaft plant. The rate of decay is only about 10 percent that of biomass growth.

6. Chemical Stripping

Chemical stripping is most effective, but it is caustic, dangerous to use, and difficult to dispose of. The RBC must be taken out of service and treatment, particularly nitrification, will not recover for weeks after chemical stripping.

Failure to use some method of positive biofilm control may have undesirable consequences. Combinations of these methods may be tried until a proven method of overload correction is identified. In any event, this procedure should allow the operator to control biofilm to maintain a relatively thin, uniform biofilm layer.

CHAPTER 12

DISINFECTION AND CHLORINATION

by

John Brady

Much of this text is taken from *Operation of Wastewater Treatment Plants*, Volume I, Chapter 10, "Disinfection Processes," by Leonard W. Hom, revised by Tom Ikesaki, with a special section by J. L. Beals.

TABLE OF CONTENTS

Chapter 12.   DISINFECTION AND CHLORINATION

OBJECTIVES
Chapter 12.　DISINFECTION AND CHLORINATION

1. Explain the principles of wastewater disinfection with chlorine.

2. Control the chlorination process to obtain the desired effluent disinfection.

3. Handle chlorine safely.

4. Inspect new chlorination facilities for proper installation.

5. Schedule and conduct chlorination operation and maintenance duties.

6. Recognize factors that indicate the chlorination process is not performing properly, identify the source of the problem, and take corrective action.

7. Conduct your duties in a safe manner.

8. Determine chlorine dosages.

9 Explain chlorine's applications and limitations for uses other than disinfection.

10. Keep records of chlorination operation.

NOTE: This chapter does not contain information on the safe operation and maintenance of gaseous chlorinators or sulfur dioxide dechlorinators. For information on these units, see *Operation of Wastewater Treatment Plants,* Volume I, Chapter 10, "Disinfection Processes."

WORDS

Chapter 12.   DISINFECTION AND CHLORINATION

AMPEROMETRIC (am-purr-o-MET-rick)   AMPEROMETRIC

A method of measurement that records electric current flowing or generated, rather than recording voltage. Amperometric titration is a means of measuring concentrations of certain substances in water.

BACTERIA (back-TEER-e-uh)   BACTERIA

Bacteria are living organisms, microscopic in size, that usually consist of a single cell. Most bacteria use organic matter for their food and produce waste products as a result of their life processes.

BREAKOUT OF CHLORINE   BREAKOUT OF CHLORINE
  (CHLORINE BREAKAWAY)   (CHLORINE BREAKAWAY)

A point at which chlorine leaves solution as a gas because the chlorine feed rate is too high. The solution is saturated and cannot dissolve any more chlorine. The maximum strength a chlorine solution can attain is approximately 3,500 mg/L. Beyond this concentration molecular chlorine will break out of solution and cause off-gassing at the point of application.

BREAKPOINT CHLORINATION   BREAKPOINT CHLORINATION

Addition of chlorine to water or wastewater until the chlorine demand has been satisfied. At this point, further additions of chlorine will result in a free chlorine residual that is directly proportional to the amount of chlorine added beyond the breakpoint.

CHLORAMINES (KLOR-uh-means)   CHLORAMINES

Compounds formed by the reaction of hypochlorous acid (or aqueous chlorine) with ammonia.

CHLORINATION (klor-uh-NAY-shun)   CHLORINATION

The application of chlorine to water or wastewater, generally for the purpose of disinfection, but frequently for accomplishing other biological or chemical results.

CHLORINE CONTACT CHAMBER   CHLORINE CONTACT CHAMBER

A baffled basin that provides sufficient detention time of chlorine contact with wastewater for disinfection to occur. The minimum contact time is usually 30 minutes. Also commonly referred to as basin or tank.

CHLORINE DEMAND   CHLORINE DEMAND

Chlorine demand is the difference between the amount of chlorine added to water or wastewater and the amount of chlorine residual remaining after a given contact time. Chlorine demand may change with dosage, time, temperature, pH, and nature and amount of the impurities in the water.

    Chlorine Demand, mg/L = Chlorine Applied, mg/L – Chlorine Residual, mg/L

CHLORINE REQUIREMENT   CHLORINE REQUIREMENT

The amount of chlorine that is needed for a particular purpose. Some reasons for adding chlorine are reducing the MPN (Most Probable Number) of coliform bacteria, obtaining a particular chlorine residual, or oxidizing some substance in the water. In each case, a definite dosage of chlorine will be necessary. This dosage is the chlorine requirement.

CHLORINE RESIDUAL   CHLORINE RESIDUAL

The concentration of chlorine present in water after the chlorine demand has been satisfied. The concentration is expressed in terms of the total chlorine residual, which includes both the free and combined or chemically bound chlorine residuals. Also called residual chlorine.

CHLORINE RESIDUAL ANALYZER CHLORINE RESIDUAL ANALYZER

An instrument used to measure chlorine residual in water or wastewater. This instrument also can be used to control the chlorine dose rate.

CHLORORGANIC (klor-or-GAN-ick) CHLORORGANIC

Organic compounds combined with chlorine. These compounds generally originate from, or are associated with, living or dead organic materials, including algae in water.

COLIFORM (KOAL-i-form) COLIFORM

A group of bacteria found in the intestines of warm-blooded animals (including humans) and also in plants, soil, air, and water. The presence of coliform bacteria is an indication that the water is polluted and may contain pathogenic (disease-causing) organisms. Fecal coliforms are those coliforms found in the feces of various warm-blooded animals, whereas the term coliform also includes other environmental sources.

COMBINED AVAILABLE CHLORINE COMBINED AVAILABLE CHLORINE

The total chlorine, present as chloramine or other derivatives, that is present in a water and is still available for disinfection and for oxidation of organic matter. The combined chlorine compounds are more stable than free chlorine forms, but they are somewhat slower in disinfection action.

COMBINED AVAILABLE CHLORINE RESIDUAL COMBINED AVAILABLE CHLORINE RESIDUAL

The concentration of chlorine residual that is combined with ammonia, organic nitrogen, or both in water as a chloramine (or other chloro derivative) and yet is still available to oxidize organic matter and help kill bacteria.

COMBINED CHLORINE COMBINED CHLORINE

The sum of the chlorine species composed of free chlorine and ammonia, including monochloramine, dichloramine, and trichloramine (nitrogen trichloride). Dichloramine is the strongest disinfectant of these chlorine species, but it has less oxidative capacity than free chlorine.

DNA DNA

Deoxyribonucleic acid. A chemical that encodes genetic information that is transmitted between generations of cells.

DECHLORINATION (DEE-klor-uh-NAY-shun) DECHLORINATION

The removal of chlorine from the effluent of a treatment plant. Chlorine needs to be removed because chlorine is toxic to fish and other aquatic life.

DISINFECTION (dis-in-FECT-shun) DISINFECTION

The process designed to kill or inactivate most microorganisms in water or wastewater, including essentially all pathogenic (disease-causing) bacteria. There are several ways to disinfect, with chlorination being the most frequently used in water and wastewater treatment plants. Compare with STERILIZATION.

EDUCTOR (e-DUCK-ter) EDUCTOR

A hydraulic device used to create a negative pressure (suction) by forcing a liquid through a restriction, such as a Venturi. An eductor or aspirator (the hydraulic device) may be used in the laboratory in place of a vacuum pump. As an injector, it is used to produce vacuum for chlorinators. Sometimes used instead of a suction pump.

ELECTROLYSIS (ee-leck-TRAWL-uh-sis) ELECTROLYSIS

The decomposition of material by an outside electric current.

END POINT END POINT

The completion of a desired chemical reaction. Samples of water or wastewater are titrated to the end point. This means that a chemical is added, drop by drop, to a sample until a certain color change (blue to clear, for example) occurs. This is called the end point of the titration. In addition to a color change, an end point may be reached by the formation of a precipitate or the reaching of a specified pH. An end point may be detected by the use of an electronic device, such as a pH meter.

ENZYMES (EN-zimes)   ENZYMES

Organic substances (produced by living organisms) that cause or speed up chemical reactions. Organic catalysts or biochemical catalysts.

FOOT VALVE   FOOT VALVE

A special type of check valve located at the bottom end of the suction pipe on a pump. This valve opens when the pump operates to allow water to enter the suction pipe but closes when the pump shuts off to prevent water from flowing out of the suction pipe.

FREE AVAILABLE CHLORINE   FREE AVAILABLE CHLORINE

The amount of chlorine available in water. This chlorine may be in the form of dissolved gas (Cl_2), hypochlorous acid (HOCl), or hypochlorite ion (OCl^-), but does not include chlorine combined with an amine (ammonia or nitrogen) or other organic compound.

FREE AVAILABLE CHLORINE RESIDUAL   FREE AVAILABLE CHLORINE RESIDUAL

The amount of chlorine available in water at the end of a specified contact period. This chlorine may be in the form of dissolved gas (Cl_2), hypochlorous acid (HOCl), or hypochlorite ion (OCl^-), but does not include chlorine combined with an amine (ammonia or nitrogen) or other organic compound.

GAS/LIQUID   GAS/LIQUID

Gaseous/Liquid. Gaseous/liquid chlorination refers to the fact that free chlorine is delivered to small treatment plants in containers that hold liquid chlorine with a free chlorine gas above the liquid in the container. The release of chlorine gas from the liquid chlorine surface depends on the temperature of the liquid and the pressure of the chlorine gas on the liquid surface. As chlorine gas is removed from the container, the gas pressure drops and more liquid chlorine becomes chlorine gas.

HEPATITIS (HEP-uh-TIE-tis)   HEPATITIS

Hepatitis is an inflammation of the liver caused by an acute viral infection. Yellow jaundice is one symptom of hepatitis.

HYPOCHLORINATION (HI-poe-klor-uh-NAY-shun)   HYPOCHLORINATION

The application of hypochlorite compounds to water or wastewater for the purpose of disinfection.

HYPOCHLORINATORS (HI-poe-KLOR-uh-nay-tors)   HYPOCHLORINATORS

Chlorine pumps, chemical feed pumps, or devices used to dispense chlorine solutions made from hypochlorites, such as bleach (sodium hypochlorite) or calcium hypochlorite into the water being treated.

HYPOCHLORITE (HI-poe-KLOR-ite)   HYPOCHLORITE

Chemical compounds containing available chlorine; used for disinfection. They are available as liquids (bleach) or solids (powder, granules, and pellets) in barrels, drums, and cans. Salts of hypochlorous acid.

MPN   MPN

MPN is the Most Probable Number of coliform-group organisms per unit volume of sample water. Expressed as a density or population of organisms per 100 mL of sample water.

MSDS   MSDS

See MATERIAL SAFETY DATA SHEET (MSDS).

MATERIAL SAFETY DATA SHEET (MSDS)   MATERIAL SAFETY DATA SHEET (MSDS)

A document that provides pertinent information and a profile of a particular hazardous substance or mixture. An MSDS is normally developed by the manufacturer or formulator of the hazardous substance or mixture. The MSDS is required to be made available to employees and operators or inspectors whenever there is the likelihood of the hazardous substance or mixture being introduced into the workplace. Some manufacturers are preparing MSDSs for products that are not considered to be hazardous to show that the product or substance is not hazardous.

MICRON (MY-kron)   MICRON

µm, Micrometer or Micron. A unit of length. One millionth of a meter or one thousandth of a millimeter. One micron equals 0.00004 of an inch.

NITROGENOUS (nye-TRAH-jen-us)   NITROGENOUS

A term used to describe chemical compounds (usually organic) containing nitrogen in combined forms. Proteins and nitrate are nitrogenous compounds.

OXIDIZING AGENT   OXIDIZING AGENT

Any substance, such as oxygen (O_2) or chlorine (Cl_2), that will readily add (take on) electrons. When oxygen or chlorine is added to water or wastewater, organic substances are oxidized. These oxidized organic substances are more stable and less likely to give off odors or to contain disease-causing bacteria. The opposite is a REDUCING AGENT.

PARASITIC (pair-uh-SIT-tick) BACTERIA   PARASITIC BACTERIA

Parasitic bacteria are those bacteria that normally live off another living organism, known as the host.

PATHOGENIC (path-o-JEN-ick) ORGANISMS   PATHOGENIC ORGANISMS

Bacteria, viruses, protozoa, or internal parasites that can cause disease (such as giardiasis, cryptosporidiosis, typhoid fever, cholera, or infectious hepatitis) in a host (such as a person). There are many types of organisms that do not cause disease and are not called pathogenic. Many beneficial bacteria are found in wastewater treatment processes actively cleaning up organic wastes.

POSTCHLORINATION   POSTCHLORINATION

The addition of chlorine to the plant discharge or effluent, following plant treatment, for disinfection purposes.

POTABLE (POE-tuh-bull) WATER   POTABLE WATER

Water that does not contain objectionable pollution, contamination, minerals, or infective agents and is considered satisfactory for drinking.

PRECHLORINATION   PRECHLORINATION

The addition of chlorine in the collection system serving the plant or at the headworks of the plant prior to other treatment processes mainly for odor and corrosion control. Also applied to aid disinfection, to reduce plant BOD load, to aid in settling, to control foaming in Imhoff units, and to help remove oil.

RNA   RNA

Ribonucleic acid. A chemical that provides the structure for protein synthesis (building up).

REAGENT (re-A-gent)   REAGENT

A pure, chemical substance that is used to make new products or is used in chemical tests to measure, detect, or examine other substances.

REDUCING AGENT   REDUCING AGENT

Any substance, such as base metal (iron) or the sulfide ion (S^{2-}), that will readily donate (give up) electrons. The opposite is an OXIDIZING AGENT.

RESIDUAL ANALYZER, CHLORINE   RESIDUAL ANALYZER, CHLORINE
See CHLORINE RESIDUAL ANALYZER.

RESIDUAL CHLORINE   RESIDUAL CHLORINE

The concentration of chlorine present in water after the chlorine demand has been satisfied. The concentration is expressed in terms of the total chlorine residual, which includes both the free and combined or chemically bound chlorine residuals. Also called chlorine residual.

ROTAMETER (ROTE-uh-ME-ter)   ROTAMETER

A device used to measure the flow rate of gases and liquids. The gas or liquid being measured flows vertically up a tapered, calibrated tube. Inside the tube is a small ball or bullet-shaped float (it may rotate) that rises or falls depending on the flow rate. The flow rate may be read on a scale behind or on the tube by looking at the middle of the ball or at the widest part or top of the float.

SAPROPHYTES (SAP-row-fights)   SAPROPHYTES

Organisms living on dead or decaying organic matter. They help natural decomposition of organic matter in water or wastewater.

SEPTICITY (sep-TIS-uh-tee)   SEPTICITY

The condition in which organic matter decomposes to form foul-smelling products associated with the absence of free oxygen. If severe, the wastewater produces hydrogen sulfide, turns black, gives off foul odors, contains little or no dissolved oxygen, and the wastewater has a high oxygen demand.

SET POINT   SET POINT

The position at which the control or controller is set. This is the same as the desired value of the process variable. For example, a thermostat is set to maintain a desired temperature.

SPORE   SPORE

The reproductive body of certain organisms, which is capable of giving rise to a new organism either directly or indirectly. A viable (able to live and grow) body regarded as the resting stage of an organism. A spore is usually more resistant to disinfectants and heat than most organisms. Gangrene and tetanus bacteria are common spore-forming organisms.

STERILIZATION (STAIR-uh-luh-ZAY-shun)   STERILIZATION

The removal or destruction of all microorganisms, including pathogens and other bacteria, vegetative forms, and spores. Compare with DISINFECTION.

TITRATE (TIE-trate)   TITRATE

To titrate a sample, a chemical solution of known strength is added drop by drop until a certain color change, precipitate, or pH change in the sample is observed (end point). Titration is the process of adding the chemical reagent in small increments (0.1–1.0 milliliter) until completion of the reaction, as signaled by the end point.

TOTAL CHLORINE   TOTAL CHLORINE

The total concentration of chlorine in water, including the combined chlorine (such as inorganic and organic chloramines) and the free available chlorine.

TOTAL CHLORINE RESIDUAL   TOTAL CHLORINE RESIDUAL

The total amount of chlorine residual (including both free chlorine and chemically bound chlorine) present in a water sample after a given contact time.

V-NOTCH WEIR   V-NOTCH WEIR

A triangular weir with a V-shaped notch calibrated in gallons (liters) per minute readings. The weir can be placed in a pipe or open channel. As the flow passes through the V-notch, the depth of water flowing over the weir can be measured and converted to a flow in gallons (liters) per minute.

CHAPTER 12.   DISINFECTION AND CHLORINATION

(Lesson 1 of 3 Lessons)

12.0   NEED FOR DISINFECTION

12.00   Removing Microorganisms by Various Treatment Processes

Homes, hospitals, and industrial facilities all discharge liquid and solid waste materials into the wastewater system. Diseases from human discharges may be transmitted by wastewater. Typical waterborne disease-causing microorganisms include bacteria, viruses, and parasites. These microorganisms are commonly referred to as *PATHOGENIC* (disease-causing) *ORGANISMS*.[1] These microorganisms can cause the following types of illnesses and infections:

Bacteria-Caused	Internal Parasite-Caused
Anthrax	Amoebic Dysentery
Bacillary Dysentery	*Cryptosporidium*
Cholera	Giardiasis
Gangrene	Various Roundworms (nematodes)
Paratyphoid	Various Tapeworms (cestodes)
Salmonellosis	
Shigellosis	**Virus-Caused**
Tetanus	Infectious Hepatitis
Typhoid Fever	Polio

NOTE: To date there is no record of an operator becoming infected by the HIV virus or developing AIDS due to on-the-job duties.

Wastewater must be disinfected because disease-producing microorganisms are potentially present in all wastewaters. These microorganisms must be removed or killed before treated wastewater can be discharged to the receiving waters and, in some instances, subsurface ground injection. The purpose of disinfection is to destroy or inactivate pathogenic microorganisms and thus prevent the spread of waterborne diseases.

The following wastewater treatment processes usually remove some of the pathogenic microorganisms:

1. Physical removal through sedimentation and filtration

2. Natural die-away or die-off of microorganisms in an unfavorable environment

3. Destruction through chemical treatment

The effectiveness of treatment processes in killing or removing microorganisms depends on the type of treatment process and the hydraulic and organic loadings. Typical levels of microorganism destruction or removal by various processes are summarized next.

Treatment Process	Microorganism Removal, %
Screening	10–20
Grit Channel	10–25
Primary Sedimentation/Septic Tank	25–75
Oxidation Lagoons	80–95
Gravel/Sand Filters, Single Pass	90–95
Rotating Biological Contactors	90–95
Activated Sludge (SBRs)	90–98
Recirculating Gravel/Sand Filters	95–99
Disinfection	98–99
Ozone	98.5–99.5[a]
UV Radiation	99–99.9%[b]

a   Depends on microorganisms and ct values.
b   Depends on microorganisms and on UV intensity × time.

Although pathogenic microorganisms are reduced in number by the various treatment processes listed and by natural die-away or die-off in unfavorable environments, many microorganisms still remain. Disinfection is practiced to ensure that all pathogenic microorganisms are destroyed in the effluents that are surface discharged from community wastewater treatment plants.

In the case of a rural home using a septic tank with on-site subsurface effluent discharge, a specified distance must be maintained between the wastewater effluent discharge area and a

---

1. *Pathogenic* (path-o-JEN-ick) *Organisms.*   Bacteria, viruses, protozoa, or internal parasites that can cause disease (such as giardiasis, cryptosporidiosis, typhoid fever, cholera, or infectious hepatitis) in a host (such as a person). There are many types of organisms that do not cause disease and are not called pathogenic. Many beneficial bacteria are found in wastewater treatment processes actively cleaning up organic wastes.

water source such as a domestic water well, water supply line, stream, or lake. These distances are established by the local health or environmental control agency and are based on the type of soil and distance to the water table at the site. The purpose is to prevent the spread of waterborne diseases and to provide a zone (area) for filtering action as the wastewater passes through the soil.

12.01   Disinfection Using Chlorine

Two terms you should understand are "disinfection" and "sterilization." *DISINFECTION* is the destruction or inactivation of pathogenic microorganisms, while *STERILIZATION* is the destruction of all microorganisms.

The main objective of disinfection is to prevent the spread of waterborne diseases by protecting:

1. Public water supplies

2. Receiving waters used for recreational purposes

3. Shellfish growing areas

Disinfection is effective because pathogenic microorganisms are more sensitive to destruction by chemicals than nonpathogens. Many nonpathogenic microorganisms are classified as *SAPROPHYTES*[2] and are essential to wastewater treatment processes. Since chlorine gas or hypochlorite solutions are the most widely used chemical for disinfection, this chapter will deal primarily with the principles and practices of chlorine disinfection (Figures 12.1 and 12.2). Chlorination for disinfection purposes results in killing essentially all of the pathogens in the plant effluent. No attempt is made to sterilize wastewater because it is unnecessary and impractical and may be detrimental to other treatment processes that are dependent on the activity of nonpathogenic saprophytes. Many other sensitive organisms in contact with chlorine are destroyed too. Chlorine is a nonselective killer. Organisms are affected by chlorine on the basis of their sensitivity and growth rate and also on the concentration of chlorine and exposure time. This is why dechlorination is practiced to protect fish and other aquatic organisms from being harmed by chlorinated effluents that are discharged to receiving waters. For information on effluent dechlorination, see *Operation of Wastewater Treatment Plants,* Volume I, Chapter 10, "Disinfection Processes."

One of the main uses of chlorine in wastewater treatment is for disinfection. Chlorine is relatively cheap to manufacture and easy to obtain in comparison with other chemicals. Even at relatively low dosages, chlorine is extremely effective.

Today, people are living in closer contact with wastewater than ever before. Wastewater effluent may be used for irrigating lawns, parks, cemeteries, freeway plantings, golf courses, college campuses, and other public areas. Recreational lakes used for boating, swimming, fishing, and other water sports are frequently made up partially and, in a few cases, solely of treated effluents. As public contact has increased and diluting waters have decreased or declined in quality, more consideration must be given to disinfection practices.

QUESTIONS

Please write your answers to the following questions and compare them with those on page 204.

12.0A   How are pathogenic bacteria destroyed or removed from wastewater?

12.0B   What is disinfection? What is its main objective?

12.0C   Why is wastewater not sterilized?

12.0D   Why is chlorine used for disinfection?

12.02   How Is Disinfection Achieved?

12.020   Filtration

There are a number of permeable membranes that can be used as filters to remove a large portion of pathogens. These membranes must be very tight and are required to remove particles sized in the *MICRON*[3] range. When the openings in the membrane are kept small they are subject to fouling by oils, greases, and other particulate matter found in wastewater effluents. For wastewater treatment systems that have any significant flows or treat flows that are of poor quality, this filtration method is not practical and is seldom used.

The most widely practiced filtration method is the percolation of wastewater effluents through soil. Specific soils make an excellent filtering media that retains small micro-size particles, yet permits the treated wastewater to pass through. For small-flow systems with the proper soil types and sufficient land area, percolation through soil is an economical and efficient disinfection and disposal system.

In small wastewater treatment systems consisting of individual septic tanks that use leach field discharge of effluent, the removal of pathogens can be achieved by filtration. In treated wastewater effluent from these systems, the pathogenic organisms have been greatly reduced by the septic tank due to settling of the solids and decomposition of the organic matter through biological oxidation. The septic tank environment is not an ideal environment for the pathogens due to adverse temperatures and the presence of other organisms that prey upon the pathogens during the detention time of the wastewater in the septic tank. Any pathogens that survive and remain in the septic tank effluent discharged to a subsurface disposal system, such as a leach field, are further subjected to a harsh environment. As the wastewater moves through the soil, the pathogens are filtered out, trapped in the soil, and either die or become food for some other plant or organism. If the soil barrier that the wastewater flows

2. *Saprophytes* (SAP-row-fights).   Organisms living on dead or decaying organic matter. They help natural decomposition of organic matter in water or wastewater.

3. *Micron* (MY-kron).   μm, Micrometer or Micron. A unit of length. One millionth of a meter or one thousandth of a millimeter. One micron equals 0.00004 of an inch.

TREATMENT PROCESS   FUNCTION

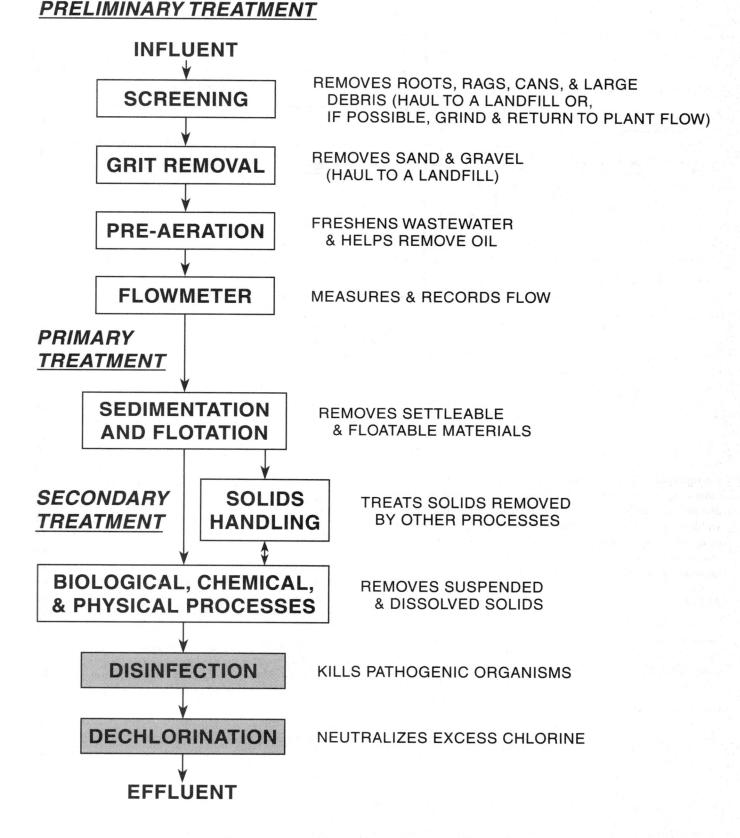

PRELIMINARY TREATMENT

INFLUENT

SCREENING — REMOVES ROOTS, RAGS, CANS, & LARGE DEBRIS (HAUL TO A LANDFILL OR, IF POSSIBLE, GRIND & RETURN TO PLANT FLOW)

GRIT REMOVAL — REMOVES SAND & GRAVEL (HAUL TO A LANDFILL)

PRE-AERATION — FRESHENS WASTEWATER & HELPS REMOVE OIL

FLOWMETER — MEASURES & RECORDS FLOW

PRIMARY TREATMENT

SEDIMENTATION AND FLOTATION — REMOVES SETTLEABLE & FLOATABLE MATERIALS

SECONDARY TREATMENT

SOLIDS HANDLING — TREATS SOLIDS REMOVED BY OTHER PROCESSES

BIOLOGICAL, CHEMICAL, & PHYSICAL PROCESSES — REMOVES SUSPENDED & DISSOLVED SOLIDS

DISINFECTION — KILLS PATHOGENIC ORGANISMS

DECHLORINATION — NEUTRALIZES EXCESS CHLORINE

EFFLUENT

Fig. 12.1   Typical flow diagram of wastewater treatment plant

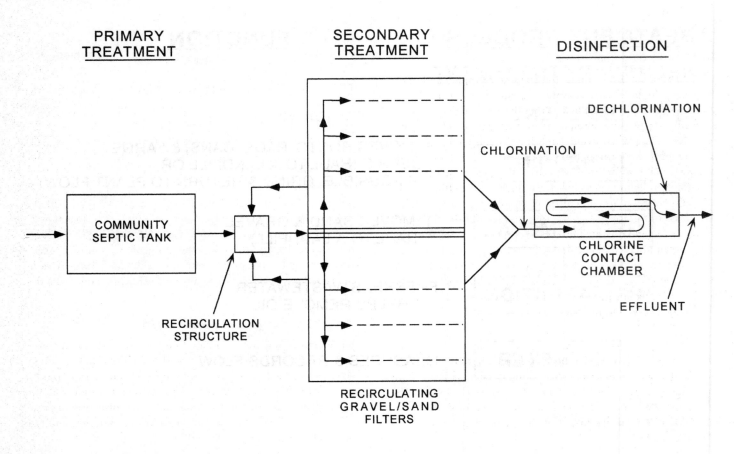

Fig. 12.2   Typical flow diagram of a small wastewater treatment facility

through retains the water long enough before it enters into another water source, the pathogen numbers are so greatly reduced that the remaining few are unlikely to create any public health problem. Not only are the pathogens reduced in number but many other constituents, such as minerals or small quantities of contaminants such as some heavy metals, are also removed from the wastewater as it filters through the soil.

12.021   Chemicals

A number of chemicals can be used to disinfect wastewater effluents with chlorine being the most commonly used. Chlorine is available as free chlorine (*GAS/LIQUID*[4] in the same container) or as a *HYPOCHLORITE*,[5] which may be obtained as a liquid or a dry product in granules, pellets, and tablets. Iodine, bromine, potassium permanganate, and hydrogen peroxide have been

tried for wastewater disinfection, but in most instances given up in favor of chlorine. Chlorination is the most economical process for disinfection of *POTABLE WATER*[6] and wastewater.

Large wastewater treatment plants commonly use free chlorine (gas/liquid) for disinfection and application in other in-plant processes. Smaller plants typically use hypochlorite systems (liquid solution containing chlorine) for safety reasons.

Wastewater treatment facilities that are described in this manual typically use a hypochlorite disinfection system of tablet chlorinators or a liquid solution pumping system. If the plant is slightly larger, discharging effluent in excess of 100,000 gallons per day, and has a well-trained operator, cylinder-mounted gas chlorinators may be used. Only the hypochlorite disinfection systems will be discussed in detail in this manual.

---

4. *Gas/Liquid.*   Gaseous/Liquid. Gaseous/liquid chlorination refers to the fact that free chlorine is delivered to small treatment plants in containers that hold liquid chlorine with a free chlorine gas above the liquid in the container. The release of chlorine gas from the liquid chlorine surface depends on the temperature of the liquid and the pressure of the chlorine gas on the liquid surface. As chlorine gas is removed from the container, the gas pressure drops and more liquid chlorine becomes chlorine gas.

5. *Hypochlorite* (HI-poe-KLOR-ite).   Chemical compounds containing available chlorine; used for disinfection. They are available as liquids (bleach) or solids (powder, granules, and pellets) in barrels, drums, and cans. Salts of hypochlorous acid.

6. *Potable* (POE-tuh-bull) *Water.*   Water that does not contain objectionable pollution, contamination, minerals, or infective agents and is considered satisfactory for drinking.

12.022 Other Methods

ULTRAVIOLET LIGHT (UV)

UV has been used to disinfect potable water supplies and, more frequently, to disinfect wastewater effluents. UV is generally not used for disinfection of effluents other than very high quality tertiary effluents because the equipment used requires a high maintenance routine when disinfecting lower quality effluents. The effluent must flow through the apparatus in a narrow stream, usually through a glass tube with the UV source of quartz lamps radiating through the glass into the effluent. If the effluent has a high suspended solids content (in excess of 8 mg/L) or contains any grease or oil, it will foul the tube and restrict the UV transmittance to the effluent, thus reducing pathogen kill and discharging an unacceptable effluent. In some systems, it is necessary to clean the glass in the UV unit every hour, making it impractical to use except in rare cases. For additional information on disinfection using ultraviolet light, see Section 12.51, "Ultraviolet (UV) Disinfection."

OZONE

Disinfection using ozone has been attempted in limited applications and, again, chlorine has been found to be more practical in these applications. The limitations of using ozone include the requirement that ozone be generated on site and that it cannot be stored due to instability. Also, its energy requirements are high and costly.

HIGH TEMPERATURE

The boiling of water at 220°F (127°C) for 20 minutes at 20 pounds pressure, as in an autoclave, will provide sterilization of the contents. Raising the effluent temperature to 212°F (100°C) for a few minutes has pathogen reduction capability, but again the energy requirement and equipment maintenance favor the use of chlorination systems.

12.03 Reactions of Chlorine Solutions with Impurities in Wastewater

Wastewater contains a great number of complex substances and chemicals and many of these substances have a significant effect on wastewater chlorination. A few of the major impurities are discussed in the following sections.

12.030 Inorganic Reducing Materials

Chlorine reacts rapidly with many inorganic reducing agents and more slowly with others. These reactions complicate the use of chlorine for disinfecting purposes. One of the most widely known inorganic reducing materials is hydrogen sulfide. Hydrogen sulfide reacts with chlorine to form sulfuric acid (sulfate) or elemental sulfur, depending on the concentration of hydrogen sulfide, pH, and temperature.

REACTION 1

Hydrogen Sulfide + Free Chlorine + Water → Sulfuric Acid + Hydrochloric Acid

$$H_2S + 4\,Cl_2 + 4\,H_2O \rightarrow H_2SO_4 + 8\,HCl$$

REACTION 2

Water + Free Chlorine → Hypochlorous Acid + Hydrogen Sulfide → Elemental Sulfur + Hydrochloric Acid + Water

$$H_2O + Cl_2 \rightarrow HOCl + H_2S \rightarrow \underset{\downarrow}{S^0} + HCl + H_2O$$

One part of hydrogen sulfide (H_2S) takes about 8.5 parts of chlorine in Reaction 1 and 2.2 parts of chlorine in Reaction 2. Since these reactions occur before a *CHLORINE RESIDUAL*[7] occurs, the demand caused by the inorganic salts must be satisfied first before any disinfection can take place. Ferrous iron, manganese, and nitrite are examples of other inorganic reducing agents that react with chlorine.

12.031 Reaction with Ammonia (NH_3)

Since ammonia is present in all domestic wastewaters, the reaction of ammonia with chlorine is of great significance. When chlorine is added to waters containing ammonia, the ammonia reacts with hypochlorous acid (HOCl) to form monochloramine, dichloramine, and trichloramine. The formation of these *CHLORAMINES*[8] depends on the pH of the solution and the initial chlorine–ammonia ratio.

Ammonia + Hypochlorous Acid → Chloramine + Water

$$NH_3 + HOCl \rightarrow NH_2Cl + H_2O \qquad \text{monochloramine}$$

$$NH_2Cl + HOCl \rightarrow NHCl_2 + H_2O \qquad \text{dichloramine}$$

$$NHCl_2 + HOCl \rightarrow NCl_3 + H_2O \qquad \text{trichloramine}$$

In general, at the pH levels that are usually found in wastewater (pH 6.5 to 7.5), monochloramine and dichloramine exist together. At pH levels below 5.5, dichloramine exists by itself. Trichloramine is the only compound found below pH 4.0.

The monochloramine and dichloramine forms have definite disinfection powers and are of interest in the measurement of chlorine residuals. Dichloramine has more effective disinfecting power than monochloramine.

---

7. *Chlorine Residual.* The concentration of chlorine present in water after the chlorine demand has been satisfied. The concentration is expressed in terms of the total chlorine residual, which includes both the free and combined or chemically bound chlorine residuals. Also called residual chlorine.

8. *Chloramines* (KLOR-uh-means). Compounds formed by the reaction of hypochlorous acid (or aqueous chlorine) with ammonia.

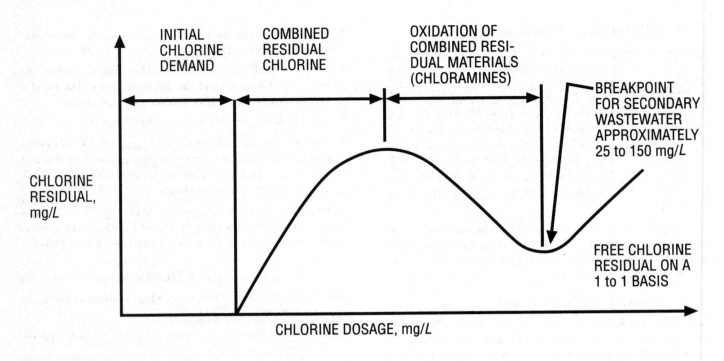

Fig. 12.3 Breakpoint chlorination curve

If enough chlorine is added to react with the inorganic compounds and *NITROGENOUS*[9] compounds, then this chlorine will react with organic matter to produce *CHLORORGANIC*[10] compounds or other combined forms of chlorine, which have slight disinfecting action. Then, if enough chlorine is added to react with all the above compounds, any additional chlorine will exist as a *FREE AVAILABLE CHLORINE RESIDUAL,*[11] which has the highest disinfecting action (Figure 12.3). This situation rarely exists in wastewater that contains nitrogenous compounds. The term *BREAKPOINT CHLORINATION*[12] refers to the breakpoint shown on Figure 12.3.

The exact mechanism of this disinfection action is not fully known. In some theories, chlorine is considered to exert a direct action against the bacterial cell, thus destroying it. Another theory is that the toxic character of chlorine inactivates the *ENZYMES*[13] that the living microorganisms need to digest their food supply. As a result, the organisms die of starvation. From the point of view of wastewater treatment, the mechanism of the action of chlorine is much less important than its effects as a disinfecting agent.

The demand by inorganic and organic materials is referred to as "chlorine demand." The chlorine that remains in combined forms having disinfecting properties plus any free chlorine is referred to as "residual chlorine." The sum of the chlorine demand and the chlorine residual is the chlorine dose.

Chlorine Dose = Chlorine Demand + Chlorine Residual

where

$$\text{Chlorine Residual} = \frac{\text{Combined Chlorine}}{\text{Forms}} + \text{Free Chlorine}$$

9. *Nitrogenous* (nye-TRAH-jen-us). A term used to describe chemical compounds (usually organic) containing nitrogen in combined forms. Proteins and nitrate are nitrogenous compounds.
10. *Chlororganic* (klor-or-GAN-ick). Organic compounds combined with chlorine. These compounds generally originate from, or are associated with, living or dead organic materials, including algae in water.
11. *Free Available Chlorine Residual.* The amount of chlorine available in water at the end of a specified contact period. This chlorine may be in the form of dissolved gas (Cl_2), hypochlorous acid (HOCl), or hypochlorite ion (OCl^-), but does not include chlorine combined with an amine (ammonia or nitrogen) or other organic compound.
12. *Breakpoint Chlorination.* Addition of chlorine to water or wastewater until the chlorine demand has been satisfied. At this point, further additions of chlorine will result in a free chlorine residual that is directly proportional to the amount of chlorine added beyond the breakpoint.
13. *Enzymes* (EN-zimes). Organic substances (produced by living organisms) that cause or speed up chemical reactions. Organic catalysts or biochemical catalysts.

12.04 Hypochlorination[14] of Wastewater

Hypochlorination is a very common method of wastewater disinfection for small systems. Although chlorination with gaseous chlorine can cost less and have a greater disinfection efficiency than hypochlorination, the safety aspects usually make hypochlorination the preferred approach. Hypochlorination is safer because of the ease of handling hypochlorite as compared with the safety hazards of handling chlorine liquid or gas. When the pH of the wastewater is increased, the hypochlorite ion (OCl^-) formation is increased. Hypochlorite ion is a less efficient disinfectant than the hypochlorous acid ($HOCl$) that is formed at lower pH levels.

Hypochlorination will raise the pH of the wastewater being treated. The rise in pH will decrease the effectiveness of the hypochlorite, thereby requiring a higher dosage. This, in turn, will increase the pH even more.

12.05 Factors Influencing Disinfection

Both chlorine addition and contact time are essential for the effective killing or inactivation of pathogenic microorganisms. Experimental determination of the best combination of combined residual chlorine and contact time is necessary to ensure both proper chlorine dose and minimum use of chlorine. Changes in pH affect the disinfection ability of chlorine and the operator must re-examine the best combination of chlorine addition and contact time when the pH fluctuates. Critical factors influencing the effectiveness of disinfection include the following:

1. Injection point and method of mixing to get disinfectant in contact with wastewater being disinfected.

2. Design (shape) of *CHLORINE CONTACT CHAMBERS*.[15] Contact chambers or basins are designed in various sizes and shapes. Rectangular or square contact chambers often allow short-circuiting and consequently reduced contact times. Baffles often are installed in these basins to increase mixing action, to obtain better distribution of disinfectant, and to reduce short-circuiting, which in turn increases contact time. Long, narrow channels or pipelines have proven to be good contact chambers.

3. Contact time. With good initial mixing, the longer the contact time, the better the disinfection. Most chlorine contact basins are designed to provide a contact time of 30 minutes. For most wastewaters, extending the chlorine contact time is more effective than increasing the chlorine dose to improve disinfection.

4. Effectiveness of upstream treatment processes. The lower the suspended solids and dissolved organic content of the wastewater, the better the disinfection.

5. Temperature. The higher the temperature, the more rapid the rate of disinfection.

6. Dose rate and type of chemical. Normally, the higher the dose rate, the more rapid the disinfection rate. The form or type of chemical also influences the disinfection rate.

7. pH. The lower the pH, the better the disinfection.

8. Numbers and types of organisms. The greater the concentration of organisms, the longer the time required for disinfection. Bacterial cells are killed quickly and easily, but bacterial *SPORES*[16] are extremely resistant.

To produce effective disinfecting action, sufficient chlorine must be added after the chlorine demand is satisfied to produce a chlorine residual that will persist through the contact period.

QUESTIONS

Please write your answers to the following questions and compare them with those on page 204.

12.0E   What does chlorine produce when it reacts with organic matter?

12.0F   How is the chlorine dosage determined?

12.0G   How much chlorine must be added to wastewater to produce disinfecting action?

12.06 Estimating Chlorine Requirements

The quantity of chlorine-demanding substances differs from plant to plant, so the amount of chlorine that has to be added to ensure proper disinfection also differs. The chlorine requirement (chlorine dose) also will vary with wastewater flow, time of contact between the wastewater effluent and chlorine, temperature of the effluent, pH, and major waste constituents. The amount of chlorine required to satisfy the chlorine-demanding substances is called the "chlorine demand." This demand is equal to the chlorine dose minus the chlorine residual. Wastewater chlorine demand is caused by waste constituents (reducing compounds) that react with the chlorine. These constituents include hydrogen sulfide, ammonia, ferrous iron, nitrogenous compounds, and organic matter. Chlorine demand is influenced by contact time, effectiveness of upstream processes, temperature, dose rate, type of chemical, pH, and the number and types of organisms.

Chlorine Demand = Chlorine Dose – Chlorine Residual

To determine the chlorine dosage rate for a particular wastewater, start with the suggested chlorine dosage listed in Table 12.1. Increase or decrease this dosage to achieve the desired chlorine residual and level of disinfection (meet effluent coliform discharge requirements).

---

14. *Hypochlorination* (HI-poe-klor-uh-NAY-shun).   The application of hypochlorite compounds to water or wastewater for the purpose of disinfection.

15. *Chlorine Contact Chamber.*   A baffled basin that provides sufficient detention time of chlorine contact with wastewater for disinfection to occur. The minimum contact time is usually 30 minutes. Also commonly referred to as basin or tank.

16. *Spore.*   The reproductive body of certain organisms, which is capable of giving rise to a new organism either directly or indirectly. A viable (able to live and grow) body regarded as the resting stage of an organism. A spore is usually more resistant to disinfectants and heat than most organisms. Gangrene and tetanus bacteria are common spore-forming organisms.

TABLE 12.1   TYPICAL CHLORINE DOSE RANGES FOR VARIOUS APPLICATIONS

Application	Dosage Range, mg/L
Collection Systems	
Slime Control	1–15
Corrosion Control	2–9[a]
Odor Control	2–9[a]
Treatment	
Filter Flies	0.1–0.5
BOD Reduction	0.5–2[b]
Grease Removal	1–10
Filter Ponding	1–10
Inorganic Compounds	2–12
Activated Sludge	
Bulking	1–10
Digester	
Foaming	2–20
Supernatant	20–150
Disinfection	
Recirculating Gravel/Sand Filter Effluent	2–5
Wetlands Effluent	2–10
Rotating Biological Contactor Effluent[c]	3–20
Oxidation Pond Effluent	5–20
Septic Tank Effluent	5–20

a. per mg/L H_2S (Hydrogen sulfide gas reacts with water vapor to form sulfuric acid, which causes corrosion of concrete in manholes, concrete pipe lines, and concrete structures.)
b. per mg/L BOD destroyed
c. after secondary clarification

The chlorine residual is determined by one of several laboratory tests. The method of choice must be one of those approved by state and federal water pollution control agencies, otherwise the results will not be recognized. These tests are discussed in the laboratory section (Chapter 14, "Laboratory Procedures") of this manual. The amount of chlorine residual that one should maintain is determined by the desired microorganism population. The microorganism population is usually specified by state or federal NPDES permit requirements. The microorganism population usually is estimated by determining the *MPN*[17] of *COLIFORM*[18] group organisms present. This determination does not test for individual pathogenic microorganisms, but uses the coliform group of organisms as the indicator organism. Coliform organisms are found in domestic wastewaters. See *Operation of Wastewater Treatment Plants*, Volume II, Chapter 16, for the laboratory test to determine the MPN of coliforms present.

Calculations to determine the chlorine dosage and chlorine demand are shown in the following examples.

EXAMPLE 1

A hypochlorinator is set to feed a dose of 4 pounds of chlorine per 24 hours. The wastewater flow is 0.05 MGD and the chlorine, as measured by the chlorine residual test after 30 minutes of contact time, is 0.5 mg/L. Find the chlorine dosage and the chlorine demand in mg/L.

Known		Unknown
Chlorine Feed, lbs/day	= 4 lbs/day	1. Chlorine Dose, mg/L
Flow, MGD	= 0.05 MGD	
Chlorine Residual, mg/L	= 0.5 mg/L	2. Chlorine Demand, mg/L

1. Calculate the chlorine dose in milligrams per liter.

$$\text{Chlorine Feed or Dose, mg/L} = \frac{\text{Chlorine Feed, lbs/day}}{(\text{Flow, MGD})(8.34 \text{ lbs/gal})}$$

$$= \frac{4 \text{ lbs/day}}{(0.050 \text{ MGD})(8.34 \text{ lbs/gal})}$$

$$= 9.5 \text{ mg/L}$$

2. Determine the chlorine demand in milligrams per liter.

$$\text{Chlorine Demand, mg/L} = \text{Chlorine Dose, mg/L} - \text{Chlorine Residual, mg/L}$$

$$= 9.5 \text{ mg/L} - 0.5 \text{ mg/L}$$

$$= 9.0 \text{ mg/L}$$

The objective of disinfection is the destruction or inactivation of pathogenic organisms. The ultimate measure of its effectiveness is the bacteriological test result. The measurement of residual chlorine does supply a tool for practical control. If the residual chlorine value commonly effective in most wastewater treatment plants does not yield satisfactory bacteriological kills in a particular plant, the amount of residual chlorine that does must be determined and used as a control in that plant. In other words, the 0.5 mg/L residual chlorine, while generally effective, is not a rigid standard but a guide that may be changed to meet local requirements.

One special case would be the use of chlorine in the effluent from a plant serving a tuberculosis hospital. Studies have shown that a chlorine residual of at least 2.0 mg/L should be maintained in the effluent from this type of institution, and that the detention time should be at least two hours at the average rate of flow instead of the thirty minutes normally used as a basis of design. Two-stage chlorination may be particularly effective in this case.

17. *MPN.*   MPN is the Most Probable Number of coliform-group organisms per unit volume of sample water. Expressed as a density or population of organisms per 100 mL of sample water.
18. *Coliform* (KOAL-i-form).   A group of bacteria found in the intestines of warm-blooded animals (including humans) and also in plants, soil, air, and water. The presence of coliform bacteria is an indication that the water is polluted and may contain pathogenic (disease-causing) organisms. Fecal coliforms are those coliforms found in the feces of various warm-blooded animals, whereas the term coliform also includes other environmental sources.

Table 12.1 provides a list of chlorine dosages that are a reasonable guideline to produce chlorine residual adequate for applications indicated for domestic wastewaters. Individual plants may require higher or lower dosages, depending on the type and amount of suspended and dissolved organic compounds in the chlorinated sample. Large amounts of suspended and dissolved organic compounds in an effluent increase the chlorine demand and tend to make the coliform test results more erratic.

QUESTIONS

Please write your answers to the following questions and compare them with those on page 205.

12.0H   What factors influence chlorine demand?

12.0I   A hypochlorinator is set to feed a dose of 5 pounds of chlorine per 24 hours. The wastewater flow is 0.06 MGD and the chlorine, as measured by the chlorine residual test after 30 minutes of contact time, is 0.4 mg/L. Find the chlorine dosage and the chlorine demand in mg/L.

12.0J   What is the objective of disinfection?

12.0K   How is the effectiveness of the chlorination process for a particular plant determined?

END OF LESSON 1 OF 3 LESSONS

on

DISINFECTION AND CHLORINATION

Please answer the discussion and review questions next.

DISCUSSION AND REVIEW QUESTIONS

Chapter 12.   DISINFECTION AND CHLORINATION

(Lesson 1 of 3 Lessons)

At the end of each lesson in this chapter, you will find discussion and review questions. Please write your answers to these questions to determine how well you understand the material in the lesson.

1. Why must wastewater be disinfected?

2. Why is chlorination used to disinfect wastewater?

3. To improve disinfection, which is more effective—increasing the chlorine dose or extending the chlorine contact time?

4. What constituents in wastewater are mainly responsible for the chlorine demand?

5. How do suspended and dissolved organic compounds in an effluent affect disinfection?

CHAPTER 12.   DISINFECTION AND CHLORINATION

(Lesson 2 of 3 Lessons)

12.1   POINTS OF CHLORINE APPLICATION
(Figure 12.4)

12.10   Collection System Chlorination

One of the primary benefits of up-sewer collection system chlo-rination is preventing the deterioration of structures. Other benefits include odor and *SEPTICITY* [19] control, and possibly BOD reduction to decrease the load imposed on the wastewater treatment processes. In some instances, the maximum benefit may result from a single application of chlorine at a point on the main intercepting sewer before the junction of all feeder sewer lines. In others, several applications at more than one point on the main intercepting sewer or at the upper ends of the feeder lines may prove most effective. Due to high costs, chlorination should be considered as a temporary or emergency measure in most cases, with emphasis being placed on proper design of the system. Aeration also is effective in controlling septic conditions in collection systems. Although many problems result from im-proper design or design for future capacity requirements, the need for hydrogen sulfide protection exists under the best of conditions in some locations.

12.11   Prechlorination

"Prechlorination" is defined as the addition of chlorine to waste-water at the entrance to the treatment plant, ahead of settling units and prior to the addition of other chemicals. In addition to its application for aiding disinfection and odor control at this point, prechlorination is applied to reduce plant BOD load, as an aid to settling, to control foaming in Imhoff units, and to help remove oil. Current trends are moving away from prechlo-rination and toward collection system aeration or other chemi-cal treatments for control of odors.

12.12   Plant Chlorination

Chlorine is added to wastewater during treatment by other processes, and the specific point of application is related to the desired outcome. The purpose of plant chlorination is the con-trol and prevention of odors, corrosion, sludge bulking, di-gester foaming, filter ponding, filter flies, and as an aid in sludge thickening. Here again, chlorination should be an emer-gency measure. Extreme care must be exercised when applying chlorine because it can interfere with or inhibit biological treat-ment processes.

12.13   Chlorination Before Filtration

More stringent discharge requirements are causing many agen-cies to provide filtration for effluent solids removal before dis-charge. For additional information, refer to *Advanced Waste Treatment,* Chapter 4, "Solids Removal From Secondary Efflu-ents," in this series of operator training manuals. The better de-signs provide a means of chlorinating the water before applica-tion to the filters to kill algae and large biological organisms. Prechlorination before filtration tends to prevent the develop-ment of biological growths that might cause short-circuiting or excessive backwashing in the filter media. Postchlorination of the effluent from the filter would be done in addition to this application.

12.14   Postchlorination

"Postchlorination" is defined as the addition of chlorine to mu-nicipal or industrial wastewater following other treatment proc-esses. This point of application should occur before a chlorine contact chamber and after the final settling unit in the treat-ment plant. The most effective place for chlorine application for disinfection is after other treatment processes and on a well-clarified effluent. Postchlorination is used primarily for disinfec-tion. As a result of chlorination for disinfection, some reduction in BOD may be observed; however, chlorination is rarely prac-ticed solely for the purpose of BOD reduction.

QUESTIONS

Please write your answers to the following questions and compare them with those on page 205.

12.1A   What is the purpose of up-sewer chlorination?

12.1B   Where should chlorine be applied in sewers?

12.1C   What are the main reasons for prechlorination?

12.1D   Why might chlorine be added to wastewater during treatment by other processes?

12.1E   What is the objective of postchlorination?

---

19. *Septicity* (sep-TIS-uh-tee).   The condition in which organic matter decomposes to form foul-smelling products associated with the absence of free oxygen. If severe, the wastewater produces hydrogen sulfide, turns black, gives off foul odors, contains little or no dissolved oxygen, and the wastewater has a high oxygen demand.

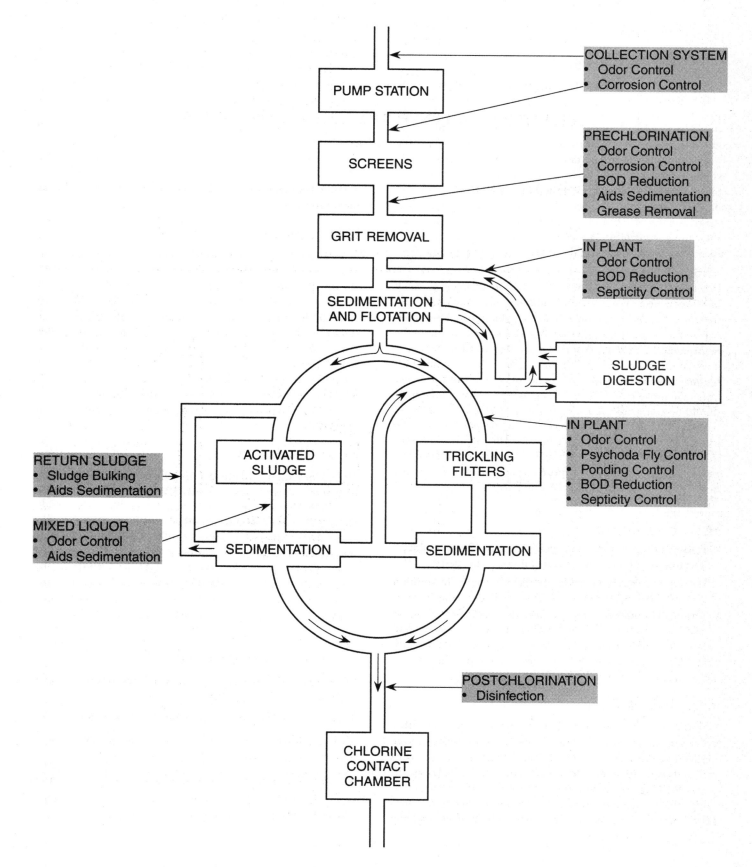

Fig. 12.4   Points of chlorine application

12.2   CHLORINATION PROCESS CONTROL

12.20   Chlorinator Control

The control of chlorine to points of application is accomplished by two basic methods in small wastewater systems, manual control or start–stop control. The selection of control methods should be based on two main factors: treatment costs and the required or desired treatment results. Normally, a waste discharger must meet a disinfection standard. A manual system, which would meet the maximum requirements and overchlorinate at minimum requirement periods, might be used. It is not unusual for a plant to have maximum chlorine residual effluent requirements because of irrigation or marine life tolerances. In these cases, the uncontrolled application of chlorine cannot be considered, no matter how large the added cost for controlling chlorine application or dose.

A specified chlorine residual level may be required at some point downstream from the best residual control sample point. In this case, a *RESIDUAL ANALYZER*[20] should be used to monitor and record residuals at this point. It may also be used to change the control *SET POINT*[21] of the controlling residual analyzer.

12.200   Manual Control

Using the manual control method, the feed-rate adjustment and starting and stopping of equipment are done by hand.

12.201   Start–Stop Control

TABLET HYPOCHLORINATORS[22] (Figure 12.5)

Using this method of start–stop control, tablet hypochlorinators control the feed rate of the chlorine into the effluent in two ways:

1. By the number of tubes that are in service holding the tablets. The number of tubes in use is established by the desired chlorine residual to accomplish the desired disinfection. Once this is determined, only those tubes must be filled with tablets.

2. The actual chlorine dose to the wastewater is controlled automatically by a *V-NOTCH WEIR*[23] to control flow and depth of wastewater through the tablet feeder housing. The flow through the feeder housing may be intermittent or continuous; any time flow passes through the feeder housing chlorine will be applied.

SOLUTION FEED HYPOCHLORINATORS

Using this method of start–stop control, the chlorine feed rate is adjusted by hand through the chemical feed pump (hypochlorinator). The feed pump may be started manually or automatically by a flow or level switch that activates the feed pump motor when effluent flows to the effluent chlorine contact chamber.

12.21   Chlorine Requirements[24] and Residuals

The ultimate control of the chlorine dosage for disinfection depends on the bacterial reduction desired, that is, the bacterial level or concentration acceptable or permissible at the point of discharge. Determination of chlorine requirements and residuals according to the current edition of Hach's *Water Analysis Handbook* (see Section 12.9 for ordering information) is a good method of control. Remember that the chlorine requirement or chlorine dose will vary with wastewater flow, time of contact, temperature, pH, and major waste constituents such as hydrogen sulfide, ammonia, and amount of organic matter.

The chlorine requirement for various flow rates and contact times can be determined on either a plant or laboratory scale. If the determination is made on a laboratory scale, you should expect the plant requirements to be somewhat higher. This is to be expected since the mixing and actual contact times can be more carefully controlled in the laboratory. It is preferable that the determinations be made by both methods and the results compared. If the chlorine requirement, as determined by full plant tests, is significantly greater than that determined by laboratory testing, a wastage of chlorine is indicated. The two major causes of a large discrepancy between laboratory and plant results are poor mixing at the point of chlorine injection and short-circuiting in the contact chamber. Either problem can usually be solved at a relatively small expense as compared to the savings that can be achieved by the reduced use of chlorine.

---

20. *Chlorine Residual Analyzer.*   An instrument used to measure chlorine residual in water or wastewater. This instrument also can be used to control the chlorine dose rate.

21. *Set Point.*   The position at which the control or controller is set. This is the same as the desired value of the process variable. For example, a thermostat is set to maintain a desired temperature.

22. *Hypochlorinators* (HI-poe-KLOR-uh-nay-tors).   Chlorine pumps, chemical feed pumps, or devices used to dispense chlorine solutions made from hypochlorites, such as bleach (sodium hypochlorite) or calcium hypochlorite into the water being treated.

23. *V-Notch Weir.*   A triangular weir with a V-shaped notch calibrated in gallons (liters) per minute readings. The weir can be placed in a pipe or open channel. As the flow passes through the V-notch, the depth of water flowing over the weir can be measured and converted to a flow in gallons (liters) per minute.

24. *Chlorine Requirement.*   The amount of chlorine that is needed for a particular purpose. Some reasons for adding chlorine are reducing the MPN (Most Probable Number) of coliform bacteria, obtaining a particular chlorine residual, or oxidizing some substance in the water. In each case, a definite dosage of chlorine will be necessary. This dosage is the chlorine requirement.

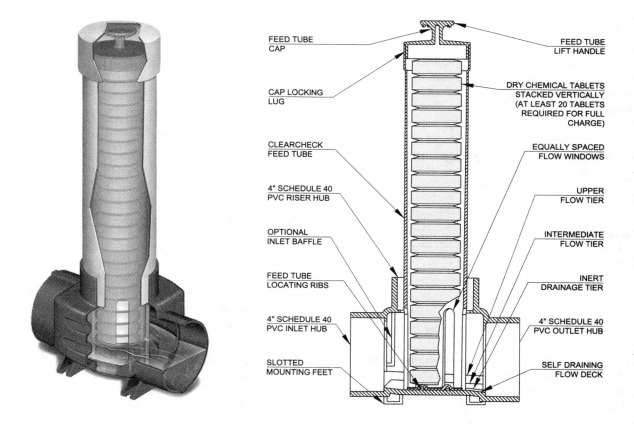

Fig. 12.5 Bio-Dynamic® dry chemical chlorine tablet feeder
(Courtesy of Norweco, Inc.)

QUESTIONS

Please write your answers to the following questions and compare them with those on page 205.

12.2A The selection of a chlorination control method is based on what two main factors?

12.2B The control of the chlorine dosage depends on the bacterial _____ desired.

12.22 Hypochlorinator Feed Rate and Control

Chlorine for disinfection and other purposes is provided in most small plants by the use of hypochlorite compounds. In its dry form (as granules, pellets, or tablets), the amount of chlorine delivered depends on the type of hypochlorite. For example, HTH (high test hypochlorite) contains approximately 65 percent chlorine, by weight, and chlorinated lime contains 34 percent.

Hypochlorite in liquid form is also available and many small plant operators prefer this form of chlorine because it eliminates the need for mixing the dry compounds and storing in a separate solution container for application. For very small systems, hypochlorite is available in the form of laundry bleach and contains 5.25 percent chlorine. For small systems, the liquid hypochlorite is shipped in plastic 30- or 50-gallon drums at a 12.5 percent chlorine concentration. The drums are so designed that the hypochlorinator feed pump sits directly on top of the drum with a suction tube equipped with a flow indicator and level switch attached to the pump to withdraw the premixed solution for direct application.

The useful chlorine concentration in hypochlorite compounds is lower than the concentration in pure gaseous or liquid chlorine and is, therefore, generally more expensive per unit of chlorine. However, the safety, storage, and application equipment costs and the training costs for hypochlorite usage can be much lower in smaller plants and, therefore, hypochlorite use is often cost effective when all costs are considered.

Manufacturers of hypochlorite compounds define "available chlorine" as the amount of gaseous chlorine required to make the equivalent hypochlorite chlorine. If you prepare a hypochlorite solution for disinfection and immediately measure the chlorine residual, the chlorine residual will be about half of the expected value based on the manufacturer's amount of available chlorine. When hypochlorite is mixed with water, approximately half of the chlorine forms hydrochloric acid (HCl) and the other half forms hypochlorite (OCl⁻) in the chlorine residual that you measured.

EXAMPLE 2

A wastewater requires a chlorine feed rate of 17 pounds chlorine per day. How many pounds of chlorinated lime will be required per day to provide the needed chlorine? Assume each pound of chlorinated lime (hypochlorite) contains 0.34 pound of available chlorine.

Known	Unknown
Chlorine Required, lbs/day = 17 lbs/day	Chlorinated Lime
Portion of Chlorine in lb of Hypochlorite, lbs = 0.34 lb	Feed Rate, lbs/day

Calculate pounds of chlorinated lime required per day.

$$\text{Chlorinated Lime Feed Rate, lbs/day} = \frac{\text{Chlorine Required, lbs/day}}{\text{Portion of Chlorine in lb of Hypochlorite, lbs}}$$

$$= \frac{17 \text{ lbs/day}}{0.34 \text{ lb}}$$

$$= 50 \text{ lbs chlorinated lime/day}$$

Hypochlorinators can be installed on either small or large systems. Usually, the larger systems use liquid chlorine because of lower costs. However, some very large cities use hypochlorite because it is safer. Treatment plants serving these cities are located in highly populated areas where escaping chlorine gas could threaten the lives of many people.

The construction and parts of hypochlorinator feed systems are discussed next.

1. Solution storage. The storage container for the solution is made of corrosion-resistant materials such as plastic or other similar materials.

2. Solution piping, which is usually fiberglass or high-density, chemical-resistant polyvinyl chloride (PVC).

3. Diffuser systems, which are made of the same material as piping systems.

4. Valves and *EDUCTORS,* [25] which are made of PVC.

5. Pumps, which are made of corrosion-resistant materials. Epoxy-lined systems have been successful.

6. Flowmeters. These meters can be constructed using a Hastelloy C straight-through metal tube *ROTAMETER* [26] with the float position determined magnetically and the flow rate transmitted either electrically or pneumatically.

7. Chlorine residual analyzers of the *AMPEROMETRIC* [27] type are commonly installed.

8. Automatic controls, which consist of a hypochlorite flow controller, recorder and totalizer, a ratio control station, and the necessary electronic signal converters.

The feed system can either be operated automatically or by manual control. The operation of the hypochlorite system will usually cost twice as much as a gas/liquid chlorine system. Maintenance of a hypochlorinator system requires more operator hours than a gas/liquid chlorine system.

QUESTIONS

Please write your answers to the following questions and compare them with those on page 205.

12.2C If you prepare a hypochlorite solution for disinfection and immediately measure the chlorine residual, why will the residual be about half of the expected value, based on the manufacturer's amount of available chlorine?

12.2D A wastewater requires a chlorine feed rate of 55 pounds chlorine per day. How many pounds of high test hypochlorite (HTH) will be required per day to provide the needed chlorine? Assume each pound of HTH contains 0.65 pound of available chlorine.

12.23 Applying Chlorine Solutions to Discharge Lines, Diffusers, and Mixing

12.230 Solution Discharge Lines

Solution discharge lines are made from a variety of materials depending on the requirements of service. Two primary requirements are that the material must be resistant to the corrosive effects of chlorine solution and it must be of adequate size to carry the required chlorine solution flows. Additional considerations are pressure conditions, flexibility (if required), resistance to external corrosion and stressors when underground or passing through structures, ease and tightness of connection, and the adaptability to field fabrication or alteration.

The development of plastics has contributed greatly to chemical solution transmission. Chlorine discharge lines may be made of PVC pipe or black polyethylene flexible tubing. The use of these materials has all but eliminated the use of rubber hosing for this purpose. Both are generally less expensive and both outlast rubber in normal service. The use of rubber hose is now almost

---

25. *Eductor* (e-DUCK-ter). A hydraulic device used to create a negative pressure (suction) by forcing a liquid through a restriction, such as a Venturi. An eductor or aspirator (the hydraulic device) may be used in the laboratory in place of a vacuum pump. As an injector, it is used to produce vacuum for chlorinators. Sometimes used instead of a suction pump.

26. *Rotameter* (ROTE-uh-ME-ter). A device used to measure the flow rate of gases and liquids. The gas or liquid being measured flows vertically up a tapered, calibrated tube. Inside the tube is a small ball or bullet-shaped float (it may rotate) that rises or falls depending on the flow rate. The flow rate may be read on a scale behind or on the tube by looking at the middle of the ball or at the widest part or top of the float.

27. *Amperometric* (am-purr-o-MET-rick). A method of measurement that records electric current flowing or generated, rather than recording voltage. Amperometric titration is a means of measuring concentrations of certain substances in water.

exclusively limited to applications where flexibility is required or where extremely high back pressures exist.

PVC and polyethylene can be field fabricated and altered. Schedule 80 should be used to limit the tendency of PVC to partially collapse under vacuum conditions. Schedule 80 PVC may be threaded and assembled with ordinary pipe tools or may be installed using solvent-welded fittings. PVC pipe greater than one inch (2.5 cm) in size may be manufactured with bell and spigot joints.

Rubber-lined steel pipe has been used for many years to make discharge lines where resistance to external stresses is required. It cannot be field fabricated or altered and is thus somewhat restricted in application. PVC lining of steel pipe has not yet become economically competitive, but other plastics that can readily compete with rubber lining and are adaptable to field fabrication and alteration have been developed.

Never use neoprene hose to carry chlorine solutions because it will become hard and brittle in a short time. If you have questions regarding the material or equipment needed to handle the chlorine solutions being used, be sure to consult with your chemical supplier.

12.231 Chlorine Solution Diffusers

Chlorine diffusers are normally constructed of the same materials used for solution lines. Their design is an extremely important part of a chlorination program related to the mixing of the chlorine solution with the wastewater being treated; however, strength and flexibility also must be given consideration. In most circular, filled conduits flowing at 0.25 ft/sec (0.08 m/sec) or greater, a solution injected at the center of the pipeline will mix with the entire flow after flowing a distance equal to 10 pipe diameters downstream. Mixing in open channels can be accomplished by using a hydraulic jump (Figure 12.6) or by sizing diffuser orifices so that a high velocity (about 16 ft/sec or 4.8 m/sec) is produced at the diffuser discharge. This accomplishes two things: (1) introducing a pressure drop to get equal discharge from each orifice, and (2) imparting sufficient energy to the surrounding wastewater to complete the mixing. Generally speaking, a diffuser should be supplied for each two to three feet (0.6 to 1 meter) of channel depth. Diffusers should be placed across the width of the channel rather than in the direction of flow. Mixing also can be achieved by using mechanical mixers specifically designed for this purpose.

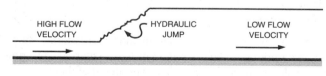

Fig. 12.6   Hydraulic jump

12.232 Mixing

Thorough mixing of the chlorine solution with wastewater is necessary for effective disinfection and consistent results. The process of mixing and the speed of mixing are extremely important ahead of a chlorine contact chamber/basin or a residual

sampling point. Because a contact chamber is usually designed for low velocity, little mixing occurs after wastewater enters it. Mixing must be achieved before the wastewater enters the contact chamber. The same is true for a chlorine residual sampling point. If good mixing does not occur upstream from the sampling point, erratic results will be obtained by the residual chlorine analyzing system.

12.24 Measurement of Chlorine Residual

12.240 Sampling

Samples taken to determine chlorine residuals must be collected at the discharge (effluent) end of the chlorine contact structure, whether it is a specifically built tank or an effluent pipeline that has provided the required chlorine/effluent contact time (usually at least 30 minutes), as established by the regulatory agency. The chlorine residual determination must be made as quickly as possible to prevent chlorine concentration changes in the sample. Samples that are exposed to high temperatures (above 68°F/20°C), sunlight, or waiting periods longer than 15 minutes can degrade and lower the actual chlorine concentration of the sample.

Where appropriate, for example, in larger plants or for critical discharge receiving waters, continuous-flow automatic chlorine residual analyzers are used (Figure 12.7). Chlorinated effluent is continuously pumped to the analyzer for analysis and automatically recorded for the record to satisfy NPDES permit requirements.

Manual (grab) sampling is more common in smaller wastewater treatment plants, where the operator collects the sample, performs the analysis, records the results and, if required, corrects the chlorine dosage feed rate to maintain a chlorine residual necessary to meet disinfection requirements.

12.241 Methods to Measure Chlorine Residual

If you are chlorinating your wastewater plant's effluent, it is because it was mandated in the plant's NPDES permit by the controlling environmental agency. Normally, you must report at some prescribed frequency the chlorine residual concentrations in the effluent being discharged, probably along with several other water quality indicators. The method of determining chlorine residual concentrations will most likely be prescribed by the NPDES permit. There are several methods that can be used to make chlorine measurements analytically and the specific procedures for making those determinations are thoroughly explained in Hach's *Water Analysis Handbook*. It contains the information needed to measure chlorine residuals, including apparatus, reagents, and procedures. The methods listed in this

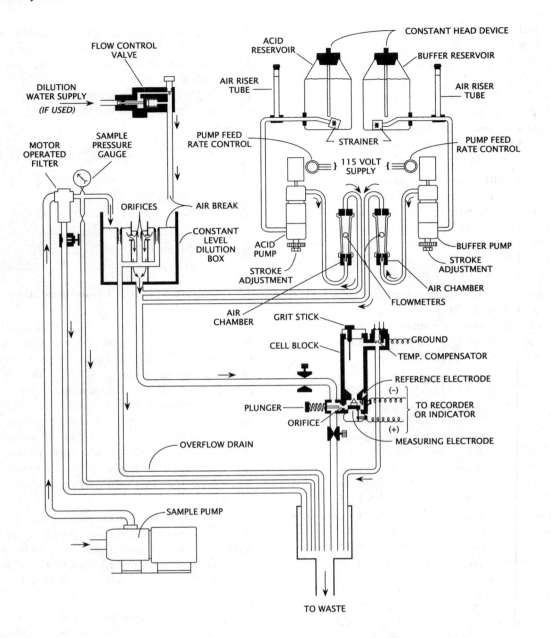

Fig. 12.7   Automatic chlorine residual analyzer
(Permission of Wallace & Tiernan Division, Penwalt Corporation)

book have been reviewed, updated, and approved. This book is a standard reference for examination of potable water and wastewater (see Section 12.9 for ordering information). Chapter 14, "Laboratory Procedures," also lists procedures for measuring chlorine residuals. Your local environmental agency has the historical background data for your area. You may recall from Sections 12.02 through 12.031 that certain substances interfere with the chlorine. This interference can also distort chlorine residual determinations if certain methods and reagents are used. Consequently, one method may provide a more accurate determination than another. In some instances, regulatory agencies permit the use of a less accurate method than another due to local conditions, if slightly higher or lower chlorine concentrations will not adversely affect the environment of that area.

Table 12.2, "How to Select the Right Chlorine Procedure," provides an excellent outline to determine which procedure is best for measuring chlorine residuals in your plant's effluent. In an emergency, a swimming pool or hot tub chlorine residual analyzer can be used, but this method is not considered accurate for treated effluents from a wastewater treatment plant due to waste matter that interferes with the test.

TABLE 12.2   HOW TO SELECT THE RIGHT CHLORINE PROCEDURE[a]

Because there are several chlorine procedures,
the following decision tree will help you select
the appropriate procedure for your application.

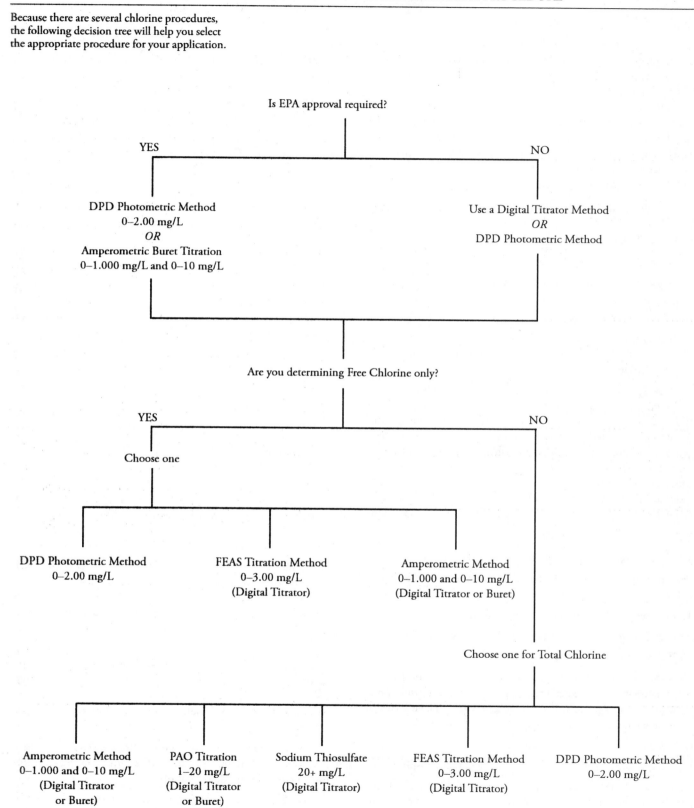

12.3   CHLORINE SAFETY PROGRAM

12.30   Purpose of Safety Program

An effective safety program begins with cooperation between the employee and the employer. The employee must take an active part in the overall program by being responsible and taking all necessary steps to prevent accidents. This begins with the attitude that a good effort must be made by everyone. Safety is everyone's responsibility. The employer must take an active part by developing and supporting effective safety programs. This includes funding the purchase of equipment, providing employee training, and enforcing safety regulations as required by OSHA and state industrial safety programs.

The following elements should be included in all safety programs:

1. Establishment of a formal safety program

2. Written rules and guidelines

3. Periodic hands-on training in the use of safety equipment

4. Establishment of a maintenance and calibration program for safety devices and equipment

5. Tours of the treatment facility for police and fire department personnel to locate hazardous areas and share chlorine safety information

All persons handling chlorine should be thoroughly aware of its hazardous properties. Personnel should know the location and use of the various pieces of protective equipment and should be instructed in safety procedures. In addition, an emergency procedure should be established and each individual should be instructed in how to follow the procedures. An emergency checklist also should be developed and available. For additional information on this topic, see the Chlorine Institute's *Chlorine Basics* (see Section 12.9, "Additional Reading and Contacts," for ordering information).

12.31   Chlorine Safety and First-Aid Management

Prompt action is essential if there is a chlorine spill or leak. If a chlorine spill or leak occurs, take the following first-aid measures:

1. Remove the exposed person(s) to fresh air.

2. Call 911 immediately and notify company safety personnel.

3. If the victim is not breathing, begin artificial respiration.

4. If the victim is breathing, place them in a seated position or lying down with the head and upper body in an upright position. Encourage slow, deep, regular breaths. Have a health professional administer oxygen as soon as possible.

5. Keep the victim warm and quiet.

6. Persons with serious symptoms may need to be hospitalized.

Clothing or the victim's skin that is soaked with chlorine solutions may be caustic and expose rescuers, as well as victims, to vapors. To decontaminate:

1. Remove soaked clothing from the victim and double-bag it immediately.

2. Flush exposed skin and hair with water for 2 to 3 minutes, then wash twice with mild soap. Rinse thoroughly with water.

3. Flush exposed or irritated eyes with water or saline for 15 to 30 minutes. If the person is wearing contact lenses, try to remove them.

12.32   Spill Management

Chlorine spills will become increasingly dangerous if they are not contained promptly. If a spill or leak has occurred, take the following actions:

- Notify trained personnel immediately, such as the company HAZMAT (hazardous material) team or the local fire department. Untrained persons or those without proper personal protective equipment must not enter areas with high concentrations of chlorine.

- Evacuate people from the hazardous area for at least 50 feet (15 meters) in all directions and have them stay upwind from the chlorine release. They should be sheltered in a building with doors and windows shut and air conditioners turned off.

- Stop or control the source of exposure. If the exposure is from a leaking cylinder, take the cylinder outdoors or to an open area until it has completely drained and the contents have evaporated.

- Ventilate potentially explosive atmospheres by opening windows.

- Keep combustibles such as wood, paper, and oil, away from the leak.

- Remove all sources of heat and ignition.

- Refer to the manufacturer's material safety data sheet (MSDS) for more information about chlorine hazards.

12.33   Hypochlorite Safety and First-Aid Management

Hypochlorite does not present the hazards that gaseous chlorine does and therefore is safer to handle, but certain precautions are required when handling it and storing it because hypochlorite is an oxidizing agent. As an oxidizer, it will corrode metals and should not be stored in storage containers not specified to hold the solution.

Dry hypochlorite compounds are normally shipped in pails or drums, depending upon the supplier. Containers of hypochlorite should be stored in a cool, dry area away from other combustibles such as gasoline, solvents, oils, greases, rags, or papers.

Do not inhale container vapors. After use, securely close the hypochlorite container to protect the remaining material from loss of chlorine and deterioration.

In case of fire, apply liberal quantities of water.

Operators handling liquid hypochlorite solutions must wear protective safety gear including safety glasses or a face shield for eye protection as well as a chemical apron and gloves. Without proper protection, operators can suffer severe skin burns from splashing of liquid solutions during handling or spraying due to a leaking pipe, tubing, or pressure fitting. Eyes should be flushed with eyewash and abundant amounts of water and a physician should be consulted. When hypochlorite comes in contact with skin, wash off the chemical, and thoroughly wash the affected area with water. If there is any indication of a chemical burn or rash, consult a physician.

If hypochlorite is spilled, splashed, or sprayed on skin or clothing, the water immediately starts to evaporate. This evaporation causes an increase in the chlorine and caustic concentrations until time of complete loss of water when almost pure sodium hydroxide (caustic) is left on the skin or clothing. Splashed clothing should be removed and washed.

When spills occur, wash the area with large volumes of water. The solution is messy to handle. Hypochlorite causes damage to eyes and skin upon contact. Immediately wash affected areas thoroughly with water. Consult a physician if the area appears burned. Hypochlorite solutions are very corrosive. Hypochlorite compounds are nonflammable; however, they can cause fires when they come in contact with organics or other easily oxidizable substances.

12.34   Operator Safety Training

Training is a concern to everyone, especially when your safety and perhaps your life are concerned. Every utility agency should have a chlorine safety training program that introduces new operators to the program and updates previously trained operators. As soon as a training session ends, obsolescence begins. People will forget what they have learned if they do not use and practice their knowledge and skills. Operator turnover can dilute a well-trained staff. New equipment, techniques, and procedures also can dilute the readiness of trained operators. An ongoing training program can consist of a monthly luncheon seminar, a monthly safety bulletin that is to be read by every operator, and a presentation by outside speakers who can be brought in to reinforce and refresh specific elements of a safety

training program. Safety training should include a scheduled review of all material safety data sheet (MSDS) information.

The American Water Works Association (AWWA) is an excellent source of training materials for safety programs, including a series of 52 lectures on common water utility safety practices (see Section 12.9, "Additional Reading and Contacts," for ordering information).

12.35   CHEMTREC

Safely handling chemicals and other hazardous materials that are used daily in wastewater treatment is one of an operator's primary responsibilities. In the event of a chemical or hazardous materials incident, CHEMTREC (Chemical Transportation Emergency Center) is a free, 24-hour emergency call center that is an immediate resource for first responders to get critical information about the hazardous materials they are dealing with at the scene. CHEMTREC will also contact the manufacturer of the products involved for more detailed assistance and for appropriate follow-up.

CHEMTREC'S emergency toll-free phone number is (800) 424-9300. Post it in a prominent location for easy access during an emergency.

QUESTIONS

Please write your answers to the following questions and compare them with those on page 206.

12.2G   Why must the chlorine residual be measured as quickly as possible?

12.3A   What would you do if hypochlorite came in contact with your hand?

END OF LESSON 2 OF 3 LESSONS

on

DISINFECTION AND CHLORINATION

Please answer the discussion and review questions next.

DISCUSSION AND REVIEW QUESTIONS

Chapter 12.   DISINFECTION AND CHLORINATION

(Lesson 2 of 3 Lessons)

Please write your answers to the following questions to determine how well you understand the material in the lesson. The question numbering continues from Lesson 1.

6. Where might chlorine be applied in the treatment of wastewater?

7. When is chlorine usually applied for disinfection purposes?

8. What is the ultimate control of chlorine dosage for disinfection?

9. Why must the chlorine solution be well mixed with the wastewater?

CHAPTER 12.   DISINFECTION AND CHLORINATION

(Lesson 3 of 3 Lessons)

12.4   OPERATION AND MAINTENANCE OF CHLORINATION EQUIPMENT

12.40   Disinfection Systems

The following discussion of chlorination equipment is based on disinfection systems that would be the most reliable, safest, and easiest for the small system operator to maintain. Consistency of adequate disinfection of an effluent is the major concern.

This manual is not endorsing any particular manufacturer or supplier, only the method used to meet NPDES permit requirements for the small wastewater system. For this reason, the order of discussion about chlorine feeders is tablet chlorinators followed by liquid hypochlorinators.

For those plants with flows greater than what can be effectively disinfected with these methods, information is available in *Operation of Wastewater Treatment Plants,* Volume I, Chapter 10, "Disinfection Processes," which describes the operation and maintenance of systems using 150-pound containers to railroad tank cars containing chlorine.

12.41   Tablet Chlorinators

12.410   Description

The tablet-type chlorinator (Figure 12.5 on page 179) is a relatively maintenance-free method of chlorinating the discharge from a small wastewater treatment plant. Chlorine tablets are added in tubes that extend into a box through which the effluent passes. As the wastewater depth increases, more of the tablet is submerged, allowing more chlorine to dissolve to match the increased flow rate. Using this type of chlorinator is advantageous because it does not have any moving parts or electrical requirements, and it does not require mixing of solutions.

There are several models of tablet chlorinators. Selection of a model should be based on the flow of water to be disinfected. For higher flows, more tubes are provided so that more tablets may dissolve. An adjustable weir is used in four-tube feeders to control depth of flow in the contact box. The bottom of the tablet chlorinator should be four inches above the liquid level in the contact chamber. Shims are sold with some new tablet chlorinators to prevent overchlorination at low flows. Older tablet chlorinators can be shimmed with approximately ¼-inch (0.6 cm) thick circular disks to prevent overchlorination.

The tablet chlorinator may be blocked by sticks, leaves, grease balls, and other matter, which would cause some additional chlorine tablets to dissolve as the liquid depth increases in the box. A blockage of this type may cause the wastewater to back

up in preceding processes. Sometimes, the tablets will also catch on the sides of the tubes and not drop to dissolve, usually due to solids buildup on the walls or improper placement of tablets in the tubes. Therefore, maintenance such as cleaning weirs, adding tablets, and cleaning and checking the tubes for "caught" tablets is necessary. Gloves, aprons, and eye protection should be worn when handling chlorine tablets.

QUESTIONS

Please write your answers to the following questions and compare them with those on page 206.

12.4A   List the different types of chlorine feeders described in this chapter.

12.4B   What type of debris can block flow to tablet chlorinators?

12.411   Tablet Chlorine Feeder

12.4110   OPERATION

Tablet chlorine feeders operate automatically without operator attention for as long as 60 days, depending on plant flow and the required chlorine residual. After the initial adjustment of the feeder, operation is simple, automatic, and continuous.

The entire plant flow of treated wastewater passes into the unit through the inlet adaptor or pipe. As the stream of water flows past the feed tubes containing the chlorine tablets, active chlorine is released into the wastewater by the dissolving action of the wastestream in contact with the tablets.

At the outlet end, a weir (selected to match plant capacity) controls the height of the water level in the feeder, which actually controls the chlorine dose and residual in the wastewater, regardless of surges in wastewater flow entering the feeder. As the incoming wastewater flow rate increases, the wastewater level in the unit rises, immersing a greater number of chlorine tablets. When the incoming flow rate decreases, the wastewater level in the unit drops, exposing fewer tablets to the water. Therefore, chlorine doses are adjusted to flow changes by allowing the frequency and duration of tablet submergence to fluctuate with the flow. Because the amount of chlorine dissolved depends on the number of tablets in the chlorine feeder when it is immersed in the water, the initial chlorine concentration remains constant, regardless of the water level in the feeder.

From the feeder, the chlorinated wastewater flows into the chlorine contact tank where it is held for a sufficient time to permit effective pathogen-killing action. The actual number of tubes to be filled with chlorine tablets and the size of the weir to be used are determined by the average daily flow rate through the plant and the required chlorine residual.

After the system has been in operation for approximately one hour, samples should be taken from the effluent end of the chlorine contact tank for chlorine residual analysis. Although the chlorine residual may be determined by a variety of test methods, the preferred and recommended method is the starch-iodide procedure described in Hach's *Water Analysis Handbook*.

Several samples should be taken for chlorine analysis between the first and second hours of operation. This will allow time for the chlorine residual to reach a state of equilibrium in the detention tank. When two or three chlorine residuals determined 15 minutes apart are consistent with each other, the system may be considered stable. Occasional samples taken in the same way may be used to check the chlorine delivery at any time. If the results of the chlorine residual testing, or any other test, indicate that the unit is not providing sufficient chlorine, the necessary adjustments may be made by changing the weir size or feed tubes. For more information about insufficient chlorination, see Section 12.4112, "Troubleshooting."

If desired, bacteriological analysis for *Escherichia coli* (*E. coli*) may be performed on the effluent from the detention tank. Optional analyses that may be run on the effluent are BOD, COD, pH, suspended solids, nitrite, ammonia, and dissolved oxygen. The tests for all of these are described in Hach's *Water Analysis Handbook*.

12.4111 MAINTENANCE

Tablet chlorine feeders require very little maintenance. Refilling of the feed tubes and an occasional on-stream cleaning of the feeder are the only maintenance procedures required.

Refilling the tubes is done on a schedule based on plant flow and weir size. The refill intervals are at the option of the operator. The tubes should always be filled to the top to ensure the longest possible periods of unattended operation.

Occasional cleaning of the feeder to remove accumulated residues may be required. This will depend on the quality of the influent passing through the unit. Usually, once every six to twelve weeks is sufficient. Solids accumulated around the feed tubes are removed by pulling the tubes a few inches off the bottom of the feeder and raising the weir plate to permit the water to rush out the exit end of the dissolver. This action, if repeated a few times, will flush out most solids. Fibrous materials may occasionally stick to the bottom of the feed tubes. Using a rod or gloved hand, combined with the shearing force of channeled water in the feeder, usually is sufficient to remove these materials.

Once a year, the feed tubes may require removal of internal scale buildup by simply scraping the inside surfaces of the tubes.

When working with chlorine solutions and chlorine handling equipment, always wear gloves, eye protection, and an apron.

12.4112 TROUBLESHOOTING

The delivery of incorrect amounts of chlorine may have several causes. Some of the potential problems and their remedies are outlined next.

Insufficient Chlorine

If the chlorine residual is consistently too low in the chlorine contact tank, use an additional feed tube or a smaller size weir. Samples taken at the effluent end of the unit can be used to determine whether sufficient chlorine is being delivered to the chlorine contact tank.

If the feed tubes do not reach the bottom of the unit, insufficient chlorine will be released into the wastewater, making it important to check the feed tubes for contact with the bottom of the unit.

If improperly loaded, the tablets can jam, causing the stack to remain suspended above the water level. This causes low dosage levels. Check to be sure that all the tablets are flat in the stack. If one or more tablets is jammed, a hard tap on the tube will loosen the bound tablets.

A gross hydraulic overload will cause a lower than desired chlorine residual. Tablet chlorine feeders are designed to accommodate 100 to 200 percent over average flow. This level may be exceeded temporarily by abnormal conditions, such as heavy rains, and no adjustment is needed if this is the case. However, if the overload into the plant becomes permanent due to additional influent from homes, businesses, or other sources, the additional flow must be compensated for by using a larger weir or more feed tubes to maintain the desired chlorine level.

Overchlorination

Overchlorination is caused by excessive consumption of the chlorine tablet. The chlorine residual may be reduced by removing one or more feed tubes, adding spacers, or using a larger weir.

Restricted Flow

If the flow through the weir is restricted, the water level will rise and cause backup in the unit, exposing more tablets to the wastewater flow. The most common obstacles are leaves, sticks, or wastewater solids, which need to be removed whenever found.

Under most flow conditions, there is sufficient shearing action by the water to keep the slots in the feed tubes from becoming blocked. If the slots become obstructed, they may be cleared as described in Section 12.4111, "Maintenance," or by removing the feed tubes from the unit feeder and clearing the obstruction by hand.

12.4113   TABLET HANDLING AND STORAGE

Tablet chlorine is an oxidizing agent. It should be stored in a cool, dry area away from oxidizable material such as rags, paper, kerosene, or other combustible materials. Do not inhale chlorine vapors from the container. After use, securely close the container to protect the remaining material. In case of fire, apply liberal quantities of water.

12.4114   DECHLORINATION

Some effluents must be dechlorinated to protect aquatic life (fish) from toxic chlorine residuals. Sodium sulfite tablets can be used to dechlorinate water using equipment and procedures similar to those used to chlorinate wastewater with chlorine tablets. For additional information on dechlorination, see *Operation of Wastewater Treatment Plants*, Volume I, Chapter 10, Section 10.8, "Dechlorination."

CAUTION: *Never mix or allow contact of chlorine with sulfite tablets. This is a potentially explosive mixture.*

12.4115   ACKNOWLEDGMENT

Material in this section was obtained from ELTECH. Permission to use this material is appreciated.

QUESTIONS

Please write your answers to the following questions and compare them with those on page 206.

12.4C   How is chlorine released into wastewater by tablet chlorinators?

12.4D   How are chlorine doses adjusted to flow changes with a chlorine tablet disinfection system?

12.4E   Why should the tubes always be filled to the top with chlorine tablets?

12.4F   What happens if the feed tubes do not reach the bottom of the unit?

12.42   Liquid Hypochlorinators

12.420   Sources of Hypochlorite

Communities have three options for selecting a hypochlorination system:

1. Generating hypochlorite on site

2. Purchasing prepared (ready-to-use) hypochlorite liquid solutions

3. Purchasing dry chlorine compounds to be mixed into a solution for dispensing

For the small wastewater treatment facility, any of these systems are acceptable; however, there are several advantages to using on-site chlorine/hypochlorite generation, including the following:

1. The hypochlorite solution is generated as needed. This system provides a fresh solution that does not lose its strength when the disinfection system is not in use. This reduces the operational problem of readjusting dosage rates to meet a required chlorine residual.

2. The on-site generation system is safer for operators. They do not have to handle and transfer prepared liquid hypochlorite solutions or mix dry chlorine compounds to prepare a hypochlorite solution to be dispensed into the system. The only raw chemical that the operator is required to handle is salt. The on-site generated hypochlorite is 0.8 percent sodium hypochlorite, which may be only half the strength of premixed or purchased hypochlorite solutions; however, premixed or purchased hypochlorite solutions tend to lose strength during storage.

12.421   On-Site Chlorine/Hypochlorite Generation

On-site chlorine/hypochlorite generation systems are considered safer than other chlorination systems because they eliminate the need to store and handle dangerous gas/liquid chlorine or commercial hypochlorite solutions.

On-site systems use salt, water, and electricity to produce sodium hypochlorite. No liquid or gaseous chlorine is ever produced in the system, eliminating potentially dangerous storage and handling problems. A similar system using salt has been developed to generate chlorine gas on site as needed.

The production of hypochlorite is very simple. Salt is dissolved in water to form a concentrated brine solution. This solution is diluted and passed through a reaction tank containing cells where *ELECTROLYSIS*[28] takes place (Figures 12.8 and 12.9). The final product solution contains approximately 0.8 percent sodium hypochlorite and 2 percent unreacted salt.

Before installing or starting any chlorination system, consult the manufacturer's O&M manual for installation, start-up, operation, maintenance, and troubleshooting procedures.

12.422   Chlorine Chemicals for Hypochlorination Systems

Hypochlorite systems are safer to operate than gaseous/liquid chlorine systems. However, hypochlorite systems can be very time consuming for the operator.

Operation and maintenance of a hypochlorination system requires the handling or preparation of hypochlorite solutions. Dry chlorine products must be mixed with water to produce hypochlorite solutions. Solutions of known strength are prepared by handling dry material that must be accurately weighed and transferred to another container for mixing with a given

---

28. *Electrolysis* (ee-leck-TRAWL-uh-sis).   The decomposition of material by an outside electric current.

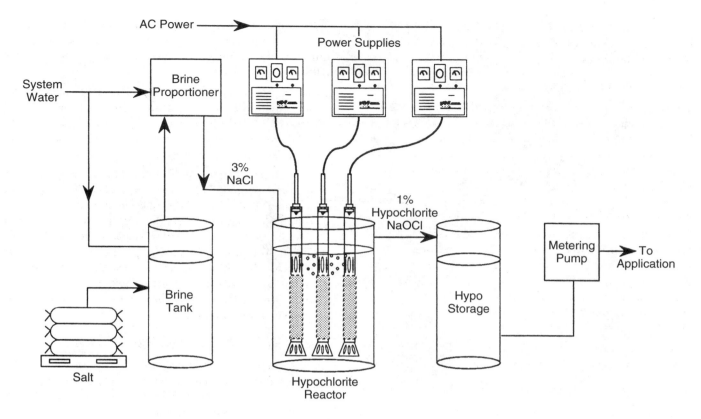

NOTE: Hydrogen gas from the hypochlorite reactor must be vented to the atmosphere in order to prevent flammable/explosive conditions.

Fig. 12.8   On-site generation of hypochlorite
(Permission of Chemical Services Company)

volume of water. The material requires mixing to dissolve the dry chlorine into the water. Some compounds require continuous mixing to keep the product in solution. For this reason, many operators of small wastewater systems prefer to purchase prepared solutions of strengths up to 15 percent chlorine by volume. Hypochlorite liquid may be obtained in 1-gallon, 30-gallon, or 50-gallon plastic containers. Liquid solutions will lose their strength if not stored properly or if stored for long periods—particularly if left open—all of which reduce their effectiveness (chlorine strength).

The loss of hypochlorite solution is a problem in intermittent flow applications such as disinfecting the effluent from a school where the system is shut down for a weekend or a holiday period and the chlorine solution is not used. An even more serious problem with some hypochlorination systems is the unreliable flow of hypochlorite solution, which affects the system's ability to provide continuous, adequate disinfection. This problem has caused a number of wastewater treatment plants to switch to other more reliable disinfection methods.

Determining an adequate pump setting to meet disinfection using the available hypochlorite solution requires running the pump, checking the chlorine residual, and readjusting the pump to achieve the adequate dose rate. This may require several adjustments and testing the chlorine residual in the effluent. There is no established guide available to inform the operator about which pump stroke will meet the required dose rate and desired chlorine residual.

Hypochlorinator pumping units are prone to the following problems:

- losing prime

- plugging of solution tubing lines from deposits of chemicals being pumped

- plugging of filters

- leakage in pump heads that prevents solution flow through the valves of the metering pump

- air lock

- plugged valves

- stripped gear trains

- ruptured diaphragms

Brine Tank   Hypochlorite Reactor   Hypochlorite Storage

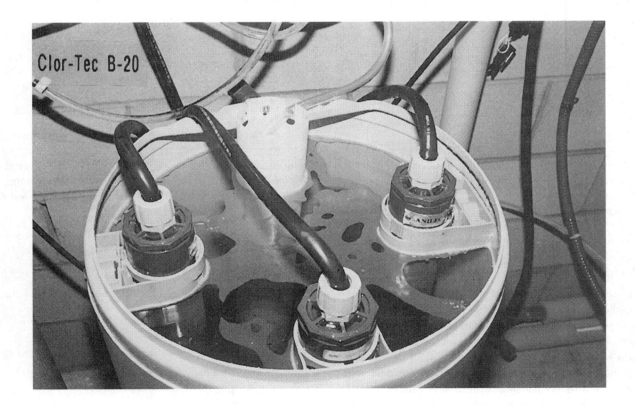

Fig. 12.9   On-site generation of hypochlorite facility
(Permission of Chemical Services Company)

12.423 Hypochlorination System Components

Figure 12.10 is a diagram of a basic hypochlorination system. Hypochlorinators used on small wastewater systems are very simple and relatively easy to install. A typical wall-mounted installation is shown in Figure 12.11. The metering pump can be table mounted, wall mounted, or mounted on a special plastic lid directly over the container holding the solution. On a table- or wall-mounted unit, the rigid suction line is often replaced with a flexible suction line placed into a 30- or 50-gallon prepared hypochlorite solution. This replacement eliminates the need to move the pumping system to a new container and is preferred by many small system operators.

When the solution container is nearly empty, a new, full container is placed beside the used one and the metering pump suction line is transferred over to the new container. When a sufficient amount of solution has been pumped out of the new container, the remaining contents of the old container can be transferred carefully into the container now in service.

NOTE: Calcium hypochlorite (HTH) contains a significant amount of insoluble material that can plug pumps, pipes, and fittings. To reduce this problem, many operators mix the HTH powder or granules in a separate mixing tank and allow the insoluble solids to settle out before using the mixture. The remaining chlorine solution (supernatant) is then used to make up the hypochlorite feed solution in a separate tank.

12.424 Component Functions

The functions of the components in a hypochlorinator are discussed in this section.

1. Storage container for solution. Premixed solutions are delivered in drums or containers and the solution may be pumped directly from them. When mixing dry compounds to make solutions, at least two containers must be available, one to hold solutions previously mixed that are being used, and one available to mix a new batch of solution when required. Containers should be large enough to hold a two or three days'

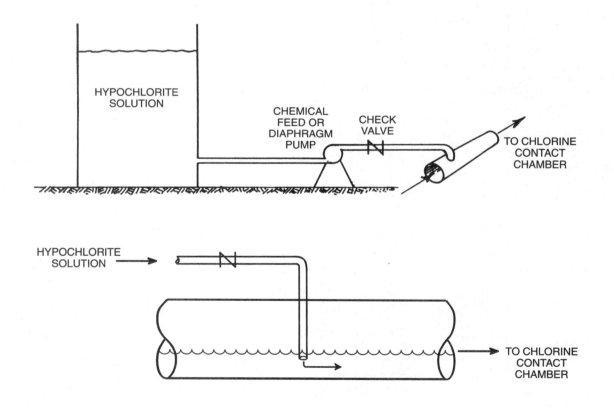

NOTE: The hypochlorite solution application could be a weighted hose or a rigid pipe discharging into an open channel.

Fig. 12.10   Hypochlorinator direct pumping system

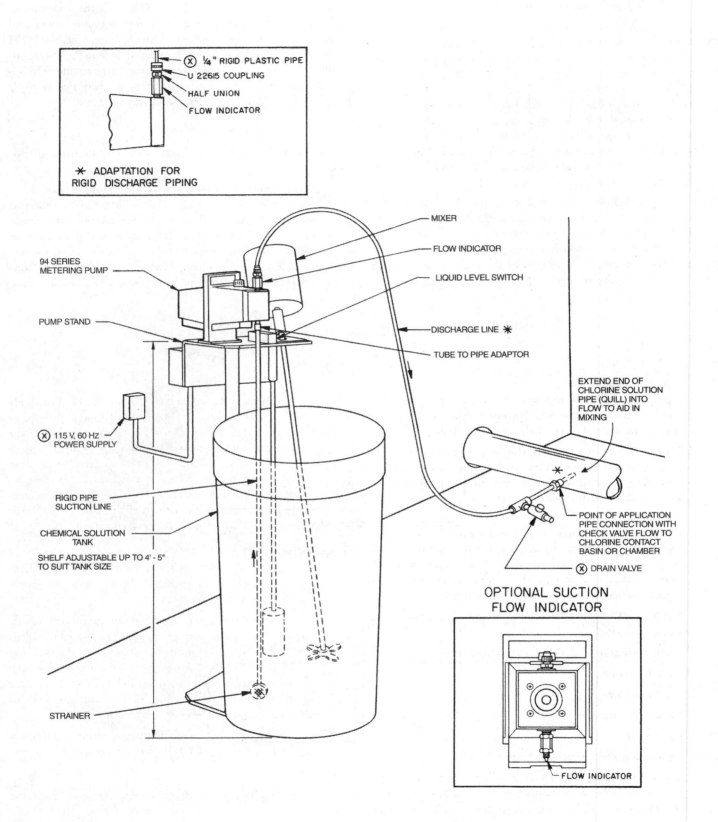

(X) ¼" RIGID PLASTIC PIPE

U 22615 COUPLING

HALF UNION

FLOW INDICATOR

✱ ADAPTATION FOR
RIGID DISCHARGE PIPING

MIXER

FLOW INDICATOR

94 SERIES
METERING PUMP

LIQUID LEVEL SWITCH

PUMP STAND

DISCHARGE LINE ✱

TUBE TO PIPE ADAPTOR

EXTEND END OF
CHLORINE SOLUTION
PIPE (QUILL) INTO
FLOW TO AID IN
MIXING

(X) 115 V, 60 Hz
POWER SUPPLY

RIGID PIPE
SUCTION LINE

CHEMICAL SOLUTION
TANK

SHELF ADJUSTABLE UP TO 4' - 5"
TO SUIT TANK SIZE

POINT OF APPLICATION
PIPE CONNECTION WITH
CHECK VALVE FLOW TO
CHLORINE CONTACT
BASIN OR CHAMBER

(X) DRAIN VALVE

OPTIONAL SUCTION
FLOW INDICATOR

STRAINER

FLOW INDICATOR

Fig. 12.11   Typical hypochlorinator installation
(Permission of Wallace & Tiernan Division, Penwalt Corporation)

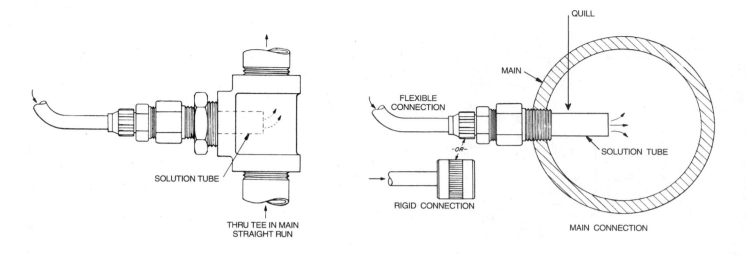

Fig. 12.12   Pipe connection
(Permission of Wallace & Tiernan Division, Penwalt Corporation)

supply of hypochlorite solution. Use only containers made of materials compatible with hypochlorite solution. Never use empty containers that have held other chemicals.

2. Mixers are required when mixing and dissolving dry compounds to prepare a solution, and in some instances other chemical solutions may require continuous mixing to maintain the chemical in suspension. Purchased, prepared solutions usually do not require mixing.

3. Metering pump. There are several suppliers of chemical application pumping units. For the unit that is installed for your facility, the operation and maintenance procedures should be in your plant O&M manual. If not, contact the supplier and request the operation and maintenance manuals for your unit.

The hypochlorinator shown in Figure 12.11 consists of the following components:

1. Suction line equipped with a strainer/foot valve

2. Metering pump

3. Flow indicator

4. Metering pump discharge line

5. Drain valve

6. Connection with check valve (Figure 12.12)

The connection is designed for installation in a ½-inch pipe tap. This may be in the bottom or on the side of a pipe or channel or in a tee in a line too small to tap. The point of application must be where the solution will be thoroughly mixed with all of the wastewater flowing through the pipe or system. For proper dispersion of solution, the end of the solution tube must be extended into the effluent pipe or channel approximately one-third of the diameter of the pipe. In performing routine annual maintenance, the exact replacement of the tube is critical. The chlorine solution mixes with the wastewater being treated as the flow moves to the chlorine contact chamber or basin.

7. Float switch (liquid level switch)

The liquid level switch will activate an alarm. The alarm warns the operator to replenish the hypochlorite supply before the pump loses prime.

NOTE: Sometimes a chemical metering pump will lose its prime and not pump any chemical. This problem is often caused by brittle cracking of the tubing from the chemical container at the point where the tubing connects to the pump. Cutting off about half an inch (1.3 cm) of the cracked tubing and reattaching the tubing to the pump can solve this problem.

Another common cause of the loss of chemical pumping capacity is plugged *FOOT VALVES*[29] and strainers.

8. Spray guard

For operator protection when pumping hazardous fluids, use a spray guard covering the pump head (Figure 12.13).

29. *Foot Valve.* A special type of check valve located at the bottom end of the suction pipe on a pump. This valve opens when the pump operates to allow water to enter the suction pipe but closes when the pump shuts off to prevent water from flowing out of the suction pipe.

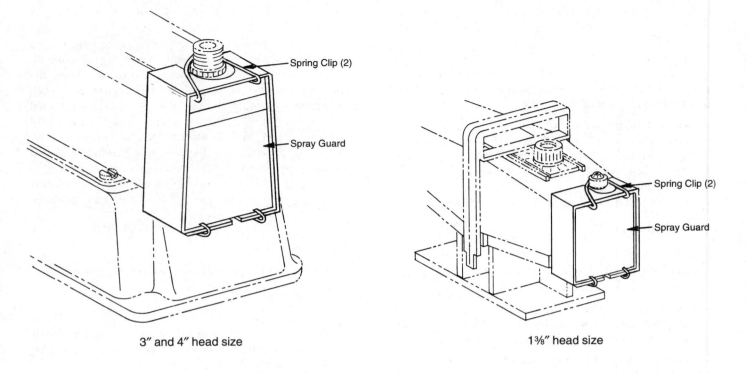

3" and 4" head size   1⅜" head size

Fig. 12.13   Metering pump spray guard

(Permission of Wallace & Tiernan Division, Penwalt Corporation)

12.43   Hypochlorinator Operation

12.430   Description

NOTE: The following material was adapted from "Manual of Instruction for Package Plant Operators," pages 11 and 12, prepared by West Virginia Department of Health, Office of Environmental Health Services, January 1992.

Liquid hypochlorinators are generally positive displacement pumps used to meter a chlorine solution into the influent end of the contact basin. A stock solution is made by mixing either sodium hypochlorite (5.25 to 7 percent by weight *FREE AVAILABLE CHLORINE* [30] (bleach) or 12.5 percent industrial strength) or calcium hypochlorite (67 to 70 percent free available chlorine) with water. Calcium hypochlorite is not recommended for use, but if used, the stock solution should be mixed in a container other than the feed container so the insoluble material can settle out. The solution should be prepared every two or three days. If a greater strength solution is mixed, the solution may lose its strength and thus affect the chlorine feed rate. Normally, a week's supply of hypochlorite should be stored and available for preparing hypochlorite solutions. Store the solution in a cool, dark place. Sodium hypochlorite can lose from two to four percent of its available chlorine content per month when stored at room temperatures. Therefore, manufacturers recommend a maximum shelf life of 60 to 90 days.

The amount of chlorine feed required may be varied by diluting the stock solution, increasing or decreasing the feed rate, or using time clock control. The time clock control should only be used in locations where load patterns are predictable such as at schools.

The hypochlorinator should be installed in a locked, insulated housing. A strip heater or heat lamp is required within the housing to prevent freezing of the stock solution. Verify that the

---

30. *Free Available Chlorine.*   The amount of chlorine available in water. This chlorine may be in the form of dissolved gas (Cl_2), hypochlorous acid ($HOCl$), or hypochlorite ion (OCl^-), but does not include chlorine combined with an amine (ammonia or nitrogen) or other organic compound.

tubing and containers are suitable for use with high chlorine concentrations. Some materials (fiberglass) will deteriorate if exposed to concentrations in excess of one percent or 10,000 mg/L. An inlet strainer on a float should be installed on the suction side of the hypochlorinator. The float allows the strainer to withdraw chlorine from the upper portion of the stock solution container where fewer solids are found.

Calcium hypochlorite must be stored in closed containers, in a cool, dark place, away from any oils, greases, or fuels. Hypochlorites are strong oxidizing agents and will support and accelerate the burning of these substances. In one example, hypochlorite caused the rapid combustion (explosion) of a plastic bucket when ignited by a spark from a nearby refrigerator.

12.431   Start-Up

The start-up of a hypochlorinator is discussed in this section.

1. Solution. Chemical solutions have to be made up, or prepared solutions opened and placed in a container to be pumped.

2. Electrical. Lock out the circuit while making an inspection of an electrical circuit. Look for frayed wires. Normally, no repairs or adjustments are required. Turn the power back on.

3. Priming the pump. If the pump has been off for an extended period, the pump may have to be primed. If the pump discharges to the atmosphere, the pump will be self-priming. Having the valves wet may speed the priming process. Attempting to prime against pressure will prevent priming. Priming is fastest with the pump feed rate selector set at its highest output. *CAUTION: Make any necessary adjustments of the pump output control selector only while the pump is running. Never adjust the pump while it is off because doing so will cause damage to the pump.*

    Prime the pump as follows:

    a. Place a suitable container under the discharge drain valve to catch any liquid when performing Step b.

    b. If a pressure connection with a check valve is used, open the discharge drain valve. Or, if a discharge shut-off valve is used, shut that valve off and open the discharge drain valve. *WARNING: To avoid possible severe personal injury, ensure that the spray guard is in place before starting the pump.*

    c. Start the pump. (Make pump adjustment to maximum output.)

    d. When liquid is seen discharging from the drain valve, the pump is primed.

    e. Return pump output selector to the position at which it previously operated. Stop the pump.

    f. Close the drain valve and open the discharge shut-off valve (if used).

4. Restart the pump for normal operation.

5. Check the chlorine residual in the effluent. Recommended chlorine residual should be 0.5 mg/L at the discharge of the chlorine contact structure. Adjust the chemical feed as required.

NOTE: The feed rate is adjusted by the selector handle or knob on the pump housing. It is suggested that solution strength be adjusted so that the pump may be operated at a dial setting of 6 or 7. This permits adjustment in either direction, if required, by a temporary change in chlorine demand. Remember, the selector may be readjusted only when the pump is running. Because capacity is somewhat affected by operating pressure, the dial reading should be considered a reference point and not an exact indication of pump delivery. Once the dial setting has been identified for the desired residual, an adjustment should not be necessary unless other changes in conditions such as the hypochlorite solution strength, chlorine demand of the effluent, or pump discharge pressure occur.

12.432   Shutdown

The shutdown of a hypochlorinator is discussed in this section.

1. Short duration shutdown.

    a. Turn the pump off.

    b. When making any repairs, lock out and tag the electric circuit supplying the pump, or pull the pump motor plug from its electrical outlet socket.

2. Long duration shutdown, particularly in cold weather.

    a. Disconnect and drain the suction and discharge lines.

    b. If the antisiphon spring is not used, turn the pump upside down to drain.

    c. If the antisiphon spring is used, remove the discharge half union and attached parts before inverting the pump to drain.

3. Failure of hypochlorinator feed pump.

    a. Replace damaged pump with standby unit. Repair damaged pump or obtain replacement.

12.433   Normal Operation and Preventive Maintenance

Normal operation of the hypochlorination process requires routine observation and preventive maintenance.

DAILY

1. Inspect the building to check for trespassing.

2. Read and record the level of the solution in the tank at the same time every day.

3. Check the chlorine residual of the effluent. Adjust the feed rate as required.

WEEKLY

1. Clean the building.

2. Replace the chemicals and wash the chemical storage tank. Have a 15- to 30-day supply of chlorine in storage for future needs. When preparing hypochlorite solutions, prepare only enough for a two- or three-day supply.

MONTHLY

1. Check the operation of the check valve.

2. Perform any required preventive maintenance.

3. Cleaning.

Commercial sodium hypochlorite solutions (such as Clorox) contain an excess of caustic (sodium hydroxide or NaOH). When this solution is diluted with water containing calcium and also carbonate alkalinity, the resulting solution becomes supersaturated with calcium carbonate. This calcium carbonate tends to form a coating on the poppet valves in the solution feeder or in the tubing of finger and diaphragm pumps. The coated valves will not seal properly and the feeder will fail to feed properly.

Use the following procedure to remove carbonate scale from hypochlorinators:

a. Take a one-quart (one-liter) glass jar and fill it half full of tap water.

b. Place one fluid ounce (20 mL) of 30 to 37 percent hydrochloric acid (swimming pool acid) in the jar, or use white vinegar.

ALWAYS ADD ACID TO WATER, NEVER THE REVERSE.[31]

c. Fill the remainder of the jar with tap water.

d. Place the suction hose of the hypochlorinator in the jar and pump the entire contents of the jar through the system.

e. Return the suction hose to the hypochlorite solution tank and resume normal operation.

You can prevent the formation of the calcium carbonate coatings by obtaining the dilution water from an ordinary home water softener.

NORMAL OPERATION CHECKLIST

1. Check chemical usage. Record the solution level and the number of hours of pump operation. Calculate the amount of chemical solution used and compare it with the desired feed rate.

2. Determine if every piece of equipment is operating.

3. Inspect the lubrication of the equipment.

4. Check the building for any possible problems.

5. Clean up the area.

FORMULAS

To determine the actual chlorine dose of wastewater effluent being treated by either a chlorinator or a hypochlorinator, the operator must know the gallons of wastewater treated and the pounds of chlorine used to disinfect the wastewater.

1. Calculate the pounds of wastewater disinfected.

$$\text{Wastewater, lbs} = (\text{Plant Effluent, gallons})(8.34 \text{ lbs/gal})$$

2. To calculate the volume of hypochlorite used from a container, the operator must know the dimensions of the container.

$$\frac{\text{Volume,}}{\text{gallons}} = \frac{(0.785)(\text{Diameter, in})^2(\text{Depth, ft})(7.48 \text{ gal/cu ft})}{144 \text{ sq in/sq ft}}$$

3. To calculate the pounds of chlorine used to disinfect wastewater, the operator must know the gallons of hypochlorite used and the percent available chlorine in the hypochlorite solution.

$$\frac{\text{Chlorine,}}{\text{lbs}} = (\text{Hypochlorite, gal})(8.34 \text{ lbs/gal})\left(\frac{\text{Hypochlorite, \%}}{100\%}\right)$$

4. To determine the actual chlorine dose in milligrams per liter, divide the pounds of chlorine used by the millions of pounds of wastewater treated. Pounds of chlorine per million pounds of water is the same as parts per million or milligrams per liter.

$$\text{Chlorine Dose, mg/L} = \frac{\text{Chlorine Used, lbs}}{\text{Wastewater Treated, Million lbs}}$$

12.434 Abnormal Operation and Troubleshooting

1. Inform your supervisor of the problem.

2. If the hypochlorinator malfunctions, it should be repaired immediately. See the shutdown operation (Section 12.432, Step 3).

3. Determine if the solution tank level is

a. too low: Check the adjustment of the pump.
                            Check the hour meter of the pump.

b. too high: Check the chemical pump.
                            Check the hour meter of the pump.

4. Determine if the chemical pump is not operating.

TROUBLESHOOTING GUIDELINES

a. Check the electrical connection.
b. Check the circuit breaker.
c. Check for stoppages in the flow lines.

CORRECTIVE MEASURES

a. Stop the plant effluent discharge, if possible.
b. Check for a blockage in the solution tank.
c. Check the operation of the check valve.
d. Check the connection of the suction hose to the pump.
e. Check the electric circuits.
f. Replace the chemical feed pump with another pump while repairing the defective unit.

5. Determine if the hypochlorite solution is not being pumped sufficiently to the chlorine contact diffuser.

TROUBLESHOOTING GUIDELINES

a. Check the solution level.
b. Check for blockages in the solution line.

---

31. Water poured into a beaker or container of acid produces a violent chemical reaction generating heat and splattering of the chemical, which can cause serious injury to operators.

12.435 *Routine Maintenance*

Suggested maintenance tasks are outlined in this section. Review the manufacturer's suggestions for additional maintenance requirements.

MONTHLY

1. Check the strainer for clogging, cleaning as needed.
2. Check flex lines for chemical buildup, for cuts and cracks, and for loose connections at fittings.
3. Check vat or container condition. Look for cracks and worn bottoms from moving container.
4. Inventory chemical supplies on hand.
5. Clean and check the check valve.
6. Clean and check the pump head.
7. Check volume of pump output.

SEMIANNUALLY

1. Check and clean the chlorine diffuser.
2. Review the pump spare parts inventory and reorder as necessary to maintain adequate repair stock.

QUESTIONS

Please write your answers to the following questions and compare them with those on page 206.

12.4J   How are hypochlorite solutions prevented from freezing?

12.4K   Why should the hypochlorinator pump spare parts inventory be reviewed?

12.5 DISINFECTION SYSTEMS

12.50 Restricted Applications of Chlorine for Disinfection

All plants are required to disinfect their effluent before discharge to surface waters. To be effective, most effluents must maintain a chlorine residual of at least 0.5 mg/L. Excessive chlorine residual is not required because it is costly and can be toxic to aquatic life in the receiving streams. Chlorination as a form of disinfection may be restricted on certain chlorine-limited streams or trout streams. Dechlorination, usually using sulfur dioxide, may be used in these cases. Sulfur dioxide usually depletes the dissolved oxygen (DO) necessitating post-aeration and may lower the effluent pH. Use of ultraviolet disinfection on chlorine-limited streams may be required. For information on dechlorination, see *Operation of Wastewater Treatment Plants*, Volume I, Chapter 10, "Disinfection Processes," in this series of operator training manuals.

12.51 Ultraviolet (UV) Disinfection (See Figure 12.14)

Common ultraviolet (UV) units either encase UV bulbs in a quartz sleeve or convey wastewater effluent through Teflon tubes surrounded by UV bulbs. UV light is able to pass through both clear Teflon and quartz with a minimal loss of energy. UV light

UV lamp assembly

(Reproduced with permission of Fischer & Porter (Canada) Limited)

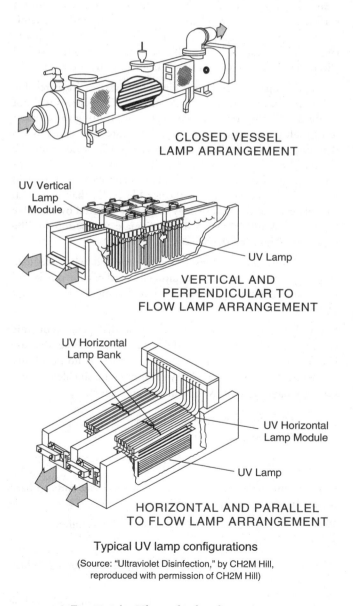

CLOSED VESSEL
LAMP ARRANGEMENT

UV Vertical Lamp Module

UV Lamp

VERTICAL AND
PERPENDICULAR TO
FLOW LAMP ARRANGEMENT

UV Horizontal Lamp Bank

UV Horizontal Lamp Module

UV Lamp

HORIZONTAL AND PARALLEL
TO FLOW LAMP ARRANGEMENT

Typical UV lamp configurations

(Source: "Ultraviolet Disinfection," by CH2M Hill, reproduced with permission of CH2M Hill)

Fig. 12.14   Ultraviolet disinfection systems

disinfects by damaging the *DNA*[32] and *RNA*[33] of organisms. The effectiveness of UV disinfection depends on the amount of UV energy absorbed by the organism. The amount of UV energy available to be absorbed by the organism is diminished by distance from the UV source, UV-absorbing constituents in the wastewater, the reduced level of cleanliness of the UV unit's transmission surface (usually Teflon or quartz), and the suspended solids concentration in the effluent. Organisms covered or shielded by suspended solids may be very resistant to UV energy.

UV disinfection is not suitable for effluents with high turbidity, greases, oils, suspended solids, dyes, organics, or other UV-absorbing constituents. UV disinfection is applicable to most high-quality secondary treatment effluents with total suspended solids (TSS) and biochemical oxygen demand (BOD) of less than 20 mg/L. Disinfection efficiency is related to the degree of treatment of TSS and BOD.

Quartz or Teflon surfaces exposed to wastewater must be kept clean. There are several methods available to clean the fouled surfaces, for example, mechanical wipers, ultrasound, or chemical cleaning with high-pressure spray wash. To protect quartz lamps from breaking, removable screens are usually installed ahead of the unit to prevent debris from entering the system.

Properly operated and maintained UV systems are as reliable as the quality of effluent to be disinfected. UV systems will not perform well if upstream process units fail, causing the effluent to be high in suspended solids or other wastewater constituents that can absorb UV light. Low-intensity UV bulbs must be changed every nine to twelve months. UV systems require constant energy. Improperly matched lamps and ballasts either will not work or will have much shorter life cycles than expected. The life cycle of the ballasts may be greatly reduced by excessive heat, so it is important to have adequate ventilation for the power panel housing the ballasts.

UV disinfection systems must be provided with adequate access and proper handling equipment to allow for quick and safe removal, cleaning, repair, and replacement of parts and equipment.

UV disinfection systems do not produce halogenated organic compounds (such as trihalomethanes) as does chlorine. Because UV disinfection does not use chemicals, there is no possibility of chemical spills or releases. UV light must be shielded to prevent eye damage to O&M staff and plant visitors.

12.52   Chlorination Troubleshooting
(See Table 12.3 and Table 12.4)

QUESTIONS

Please write your answers to the following questions and compare them with those on page 206.

12.5A   Why is excessive chlorine residual not required?

12.5B   How does UV light disinfect?

12.5C   Under what conditions is UV disinfection not suitable?

12.6   ARITHMETIC CALCULATIONS (ENGLISH)

12.60   Computing Chlorine Dose and Chlorine Demand

EXAMPLE 3

A hypochlorinator is set to feed a dose of 30 pounds of chlorine per 24 hours. The wastewater flow is 0.8 MGD and the chlorine, as measured by the chlorine residual test after 30 minutes of contact time, is 0.5 mg/L. Find the chlorine dosage and the chlorine demand in mg/L.

Known		Unknown
Chlorine Feed, lbs/day	= 30 lbs/day	1. Chlorine Dose, mg/L
Flow, MGD	= 0.8 MGD	
Chlorine Residual, mg/L	= 0.5 mg/L	2. Chlorine Demand, mg/L

1. Calculate the chlorine dose in milligrams per liter.

$$\text{Chlorine Feed or Dose, mg/L} = \frac{\text{Chlorine Feed, lbs/day}}{(\text{Flow, MGD})(8.34 \text{ lbs/gal})}$$

$$= \frac{30 \text{ lbs/day}}{(0.8 \text{ MGD})(8.34 \text{ lbs/gal})}$$

$$= 4.5 \text{ mg/L}$$

2. Determine the chlorine demand in milligrams per liter.

$$\text{Chlorine Demand, mg/L} = \text{Chlorine Dose, mg/L} - \text{Chlorine Residual, mg/L}$$

$$= 4.5 \text{ mg/L} - 0.5 \text{ mg/L}$$

$$= 4.0 \text{ mg/L}$$

12.61   Computing Chlorinated Lime Feed Rate

EXAMPLE 4

A wastewater requires a chlorine feed rate of 14 pounds chlorine per day. How many pounds of chlorinated lime will be required per day to provide the needed chlorine? Assume each pound of chlorinated lime (hypochlorite) contains 0.34 pound of available chlorine.

Known		Unknown
Chlorine Required, lbs/day	= 14 lbs/day	Chlorinated Lime Feed Rate, lbs/day
Portion of Chlorine in lb of Hypochlorite, lbs	= 0.34 lb	

Calculate pounds of chlorinated lime required per day.

$$\text{Chlorinated Lime Feed Rate, lbs/day} = \frac{\text{Chlorine Required, lbs/day}}{\text{Portion of Chlorine in lb of Hypochlorite, lbs}}$$

$$= \frac{14 \text{ lbs/day}}{0.34 \text{ lb}}$$

$$= 41.2 \text{ lbs chlorinated lime/day}$$

---

32. *DNA.*   Deoxyribonucleic acid. A chemical that encodes genetic information that is transmitted between generations of cells.

33. *RNA.*   Ribonucleic acid. A chemical that provides the structure for protein synthesis (building up).

TABLE 12.3   TROUBLESHOOTING GUIDE—CHLORINATION

Indicator	Probable Cause	Check or Monitor	Solution		
				Type of Chlorination Equipment	
				Tablet Feeder	Pump System
1. Coliform count fails to meet required NPDES permit	a. Chlorine residual too low	a. 1. Chlorine residual	a. 1. Increase chlorine feed rate	a. 1. Increase chlorine feed rate • Check tubes for tablets • Verify tablets reach bottom of tubes • Weir is holding proper water level in feeder box • Effluent flow higher than usual. Add tablets to additional tube or adjust box water depth by installing higher weir	a. 1. Increase chlorine feed rate • Check hypochlorite solution tank for material to be pumped • If pump functions normally, adjust pump feed selector to a higher rate of feed
		2. 30 minutes after increasing feed rate check chlorine residual in effluent		2. • If chlorine residual is OK, leave as is • If chlorine residual is still low, increase box water depth or add more tubes or tablets	2. If chlorine residual is still low and feed rate is already in high range: • Check pump discharge for correct output (see Section 12.431, item 3) • If pump discharges solution but at low volume, check strainer in hypochlorite solution tank and pump suction line for solids clogging • Check pump delivery system (see Table 12.4 for solutions for different types of pumps)
	b. Chlorine residual too low. Chlorine demand keeps increasing	b. Check for solids buildup in chlorine contact chamber		b. Clean chlorine contact chamber	b. Clean chlorine contact chamber
	c. Short-circuiting in chlorine contact chamber (basin)	c. 1. Detention time through chlorine contact chamber. Run dye test. Dump dye into influent to chlorine contact chamber. Monitor time dye enters until first sign of dye in effluent leaving contact chamber		c. 1. If less than 30 minutes of contact time, install baffling to create a serpentine flow through basin or add mixing to reach outer edge of chamber to achieve even distribution of flow through tank	c. 1. If less than 30 minutes of contact time, install baffling to create a serpentine flow through basin or add mixing to reach outer edge of chamber to achieve even distribution of flow through tank
		2. Hypochlorite solution available			2. Provide solution to be pumped
		3. Suction strainer fouled			3. Clean strainer

TABLE 12.3 TROUBLESHOOTING GUIDE—CHLORINATION (*continued*)

Indicator	Probable Cause	Check or Monitor	Solution — Type of Chlorination Equipment	
			Tablet Feeder	**Pump System**
1. Coliform count fails to meet required NPDES permit (*continued*)	c. Short-circuiting in chlorine contact chamber (basin) (*continued*)	4. Suction line plugged		4. Clean line
		5. Old hypochlorite solution		5. Increase pump feed rate or change to fresh hypochlorite solution
		6. Cold hypochlorite solution that is thick or frozen		6. Heat solution tank; if possible agitate solution tank contents
		7. Discharge of chemical to effluent		7. Check if valve is stuck closed, if discharge line is plugged or broken: • Plugged. Provide high discharge pressure at pump • Broken. No excessive pressure at pump
		8. High chlorine demand of effluent due to upstream influences of spills or dumps of toxic waste or other upstream treatment processes		8. Correct upstream condition, feed maximum chlorine rate, divert effluent flow to temporary tank or storage (for example, an empty tank or pond), or control influent flow by storing in collection system without flooding
	d. No chlorine residual in effluent	Check chlorination feed system; pump not running	d. Check 1a.1	d. Electrical system 1. Circuit breaker tripped 2. Time clock not operating 3. Pump motor burned out or pump not working
2. Chlorine residual in plant effluent too high to meet requirements	a. Chlorine feed rate too high	a. Chlorine residual	a. Reduce chlorine feed rate 1. Check tablet feeder box for debris or grease on weir raising water level on tubes. Clean weir 2. Weir clean, flow normal, more than one tube in service: remove tablets from one tube 3. One tube in service: install smaller weir to increase flow through box, reducing water contact time with tablets	a. Reduce chlorine feed rate 1. Move pump feed rate control knob to a lower setting 2. Reset time clock for shorter pumping cycle 3. Dilute hypochlorite solution to make a weaker chemical dosage

TABLE 12.3   TROUBLESHOOTING GUIDE—CHLORINATION (continued)

Indicator	Probable Cause	Check or Monitor	Solution	
			Type of Chlorination Equipment	
			Tablet Feeder	Pump System
3. Coliform count fails to meet required standards for disinfection	a. Chlorine residual too low	a. Chlorine residual	a. 1. Increase chlorine feed rate 2. Increase chlorine contact time	a. 1. Increase chlorine feed rate 2. Increase chlorine contact time
	b. Inadequate chlorine residual control	b. Continuously record residual in effluent	b. Use chlorine residual analyzer to monitor and control the chlorine dosage automatically	b. Use chlorine residual analyzer to monitor and control the chlorine dosage automatically
	c. Inadequate chlorination equipment capacity	c. Check capacity of equipment	c. Replace equipment as necessary to provide treatment based on maximum flow through plant	c. Replace equipment as necessary to provide treatment based on maximum flow through plant
	d. Solids buildup in contact chamber	d. Visual inspection	d. Clean contact chamber to reduce solids buildup	d. Clean contact chamber to reduce solids buildup
	e. Short-circuiting in contact chamber	e. Contact time	e. 1. Install baffling in contact chamber 2. Install mixing device in contact chamber	e. 1. Install baffling in contact chamber 2. Install mixing device in contact chamber
	f. Coliform regrowth in piping/sample station	f. Effluent coliform sampling station	f. Modify system as needed to provide adequate chlorine residual to prevent growth	f. Modify system as needed to provide adequate chlorine residual to prevent growth
4. Chlorine residual too high in plant effluent to meet requirements	a. Chlorine feed rate too high	a. Chlorine residual	a. Reduce chlorine feed rate	a. Reduce chlorine feed rate
	b. Malfunctioning chlorine residual control	b. Operation of residual control system/analyzer	b. Repair/calibrate chlorine feed/residual control loops	b. Repair/calibrate chlorine feed/residual control loops
5. Breakout (breakaway) of Chlorine[a]	a. Overfeeding chlorine	a. Chlorine feed rate	a. Decrease chlorine feed rate to minimize breakout and maximize application efficiency	a. Decrease chlorine feed rate to minimize breakout and maximize application efficiency

a. *Breakout of Chlorine.* A point at which chlorine leaves solution as a gas because the chlorine feed rate is too high. The solution is saturated and cannot dissolve any more chlorine. The maximum strength a chlorine solution can attain is approximately 3,500 mg/L. Beyond this concentration molecular chlorine will break out of solution and cause off-gassing at the point of application.

TABLE 12.4 TROUBLESHOOTING GUIDE—CHLORINE SOLUTION PUMPS

Problem	Type of Pump	Check/Observe	Solution
Low discharge of chlorine solution	Finger Pump	1. Pump running, fingers activating and forcing solution through tubing but discharge is low	a. Stop pump, disconnect power, tag, and lock out b. Check tubing through pump head for cut, crimped, or plugged tubing. Replace tubing if needed c. Restart pump
		2. Discharge still low	2. Replace pump
	Diaphragm Pump	1. Pump running, diaphragm being activated but discharge is low	a. Stop pump, disconnect power, tag, and lock out b. Open head, check diaphragm. Replace cut, torn, or unseated diaphragm with new diaphragm
		2. Discharge still low	2. Replace pump
	Gear/Lobe Pump	1. Lobes or gears	a. Stop pump, disconnect power, tag, and lock out b. Bleed pressure from pump head c. Open head and manually check operation of lobes or gears; should mesh, rotate freely
		2. Chemical buildup on gears or housing restricting operation	a. Replace worn housing to prevent bypassing of gears or lobes b. Restore clearances or replace pump
	Piston/Poppet Valve Pump	1. Piston or valves for damage or chemical deposits	a. Stop pump, disconnect power, tag, and lock out b. Bleed pressure from pump head c. Remove head and check piston or valves for free rotation, proper meshing
		2. Pistons or valves damaged or chemical deposit blocking actions	2. Remove deposits; pistons/valves should mesh, rotate freely
		3. Head assembly appears OK	3. Replace pump

12.62 Computing Hypochlorinator Setting

EXAMPLE 5

Calculate the chlorine dose in milligrams per liter and the hypochlorinator setting in pounds per 24 hours to treat a wastewater with a chlorine demand of 8 mg/L, when a chlorine residual of 1 mg/L is desired. The flow is 0.6 MGD.

Known	Unknown
Chlorine Demand, mg/L = 8 mg/L	1. Chlorine Dose, mg/L
Chlorine Residual, mg/L = 1 mg/L	
Flow, MGD = 0.6 MGD	2. Hypochlorinator Setting, lbs/day

1. Calculate the chlorine dose in milligrams per liter.

$$\text{Chlorine Dose, mg/L} = \text{Chlorine Demand, mg/L} + \text{Chlorine Residual, mg/L}$$

$$= 8 \text{ mg/L} + 1 \text{ mg/L}$$

$$= 9 \text{ mg/L}$$

2. Calculate the hypochlorinator setting in pounds per day.

$$\text{Hypochlorinator Setting, lbs/day} = (\text{Flow, MGD})(\text{Dose, mg/L})(8.34 \text{ lbs/gal})$$

$$= (0.6 \text{ MGD})(9 \text{ mg/L})(8.34 \text{ lbs/gal})$$

$$= 45 \text{ lbs/day}$$

12.7   ARITHMETIC CALCULATIONS (METRIC)

12.70   Computing Chlorine Dose and Chlorine Demand (Metric)

EXAMPLE 6

A hypochlorinator is set to feed a dose of 25 kilograms of chlorine per 24 hours. The wastewater flow is 3,200 cu m/day and the chlorine, as measured by the chlorine residual test after 30 minutes of contact time, is 0.5 mg/L. Find the chlorine dosage and the chlorine demand in mg/L.

Known		Unknown
Chlorine Feed, kg/day	= 25 kg/day	1. Chlorine Dose, mg/L
Flow, cu m/day	= 3,200 cu m/day	2. Chlorine Demand, mg/L
Chlorine Residual, mg/L	= 0.5 mg/L	

1. Calculate the chlorine dose in milligrams per liter.

$$\text{Chlorine Feed or Dose, mg/L} = \frac{(\text{Chlorine Feed, kg/day})(1,000 \text{ gm/kg})(1,000 \text{ mg/gm})}{(\text{Flow, cu m/day})(1,000 \text{ L/cu m})}$$

$$= \frac{(25 \text{ kg/day})(1,000 \text{ gm/kg})(1,000 \text{ mg/gm})}{(3,200 \text{ cu m/day})(1,000 \text{ L/cu m})}$$

$$= 7.8 \text{ mg/L}$$

2. Determine the chlorine demand in milligrams per liter.

$$\text{Chlorine Demand, mg/L} = \text{Chlorine Dose, mg/L} - \text{Chlorine Residual, mg/L}$$

$$= 7.8 \text{ mg/L} - 0.5 \text{ mg/L}$$

$$= 7.3 \text{ mg/L}$$

12.71   Computing Chlorinated Lime Feed Rate (Metric)

EXAMPLE 7

A wastewater requires a chlorine feed rate of 8 kilograms chlorine per day. How many kilograms of chlorinated lime will be required per day to provide the needed chlorine? Assume each kilogram of chlorinated lime (hypochlorite) contains 0.34 kilogram of available chlorine.

Known		Unknown
Chlorine Required, kg/day	= 8 kg/day	Chlorinated Lime Feed Rate, kg/day
Portion of Chlorine in kg of Hypochlorite, kg	= 0.34 kg	

Calculate kilograms of chlorinated lime required per day.

$$\text{Chlorinated Lime Feed Rate, kg/day} = \frac{\text{Chlorine Required, kg/day}}{\text{Portion of Chlorine in kg of Hypochlorite, kg}}$$

$$= \frac{8 \text{ kg/day}}{0.34 \text{ kg}}$$

$$= 23.5 \text{ kg chlorinated lime/day}$$

12.72   Computing Hypochlorinator Setting (Metric)

EXAMPLE 8

Calculate the chlorine dose in milligrams per liter and the hypochlorinator setting in kilograms per 24 hours to treat a wastewater with a chlorine demand of 10 mg/L, when a chlorine residual of 1 mg/L is desired. The flow is 2,500 cu m per day.

Known		Unknown
Chlorine Demand, mg/L	= 10 mg/L	1. Chlorine Dose, mg/L
Chlorine Residual, mg/L	= 1 mg/L	2. Hypochlorinator Setting, kg/24 hr
Flow, cu m/day	= 2,500 cu m/day	

1. Calculate the chlorine dose in milligrams per liter.

$$\text{Chlorine Dose, mg/L} = \text{Chlorine Demand, mg/L} + \text{Chlorine Residual, mg/L}$$

$$= 10 \text{ mg/L} + 1 \text{ mg/L}$$

$$= 11 \text{ mg/L}$$

2. Calculate the hypochlorinator setting in kilograms per day.

$$\text{Hypochlorinator Setting, kg/day} = \left(\text{Flow, cu m/day}\right)\left(\text{Dose, mg/L}\right)\left(\frac{1 \text{ kg}}{1,000,000 \text{ mg}}\right)\left(\frac{1,000 \text{ L}}{1 \text{ cu m}}\right)$$

$$= (2,500 \text{ cu m/day})(11 \text{ mg/L})\left(\frac{1 \text{ kg}}{1,000,000 \text{ mg}}\right)\left(\frac{1,000 \text{ L}}{1 \text{ cu m}}\right)$$

$$= 27.5 \text{ kg/day}$$

$$= 27.5 \text{ kg/24 hours}$$

12.8   ARITHMETIC ASSIGNMENT

Turn to the Appendix, "How to Solve Wastewater System Arithmetic Problems," at the back of this manual. In Section A.2, "Typical Wastewater System Problems (English System)," read and work the problems in Section A.23, "Disinfection and Chlorination," and check the arithmetic using a calculator.

12.9   ADDITIONAL READING AND CONTACTS

1. *Water Analysis Handbook,* Fifth Edition. Obtain from Hach Company, PO Box 389, Loveland, CO 80539-0389. Order No. 2954700. Price, $108.00.

2. *Chlorine Basics,* Seventh Edition. A PDF of the document is available as a free download from the Chlorine Institute, Inc., at www.chlorineinstitute.org.

3. *Let's Talk Safety—2012 Safety Talks.* A series of 52 lectures on water utility safety practices. Obtain from American Water Works Association (AWWA), Bookstore, 6666 West Quincy Avenue, Denver, CO 80235. Order No. 10123-12. ISBN 978-1-58321-872-3. Price to members, $98.50; nonmembers, $130.50; price includes cost of shipping and handling.

END OF LESSON 3 OF 3 LESSONS

on

DISINFECTION AND CHLORINATION

Please answer the discussion and review questions next.

DISCUSSION AND REVIEW QUESTIONS
Chapter 12.   DISINFECTION AND CHLORINATION
(Lesson 3 of 3 Lessons)

Please write your answers to the following questions to determine how well you understand the material in the lesson. The question numbering continues from Lesson 2.

10. How should chlorine tablets be stored?

11. What precautions would you take before starting any chlorination system?

12. What happens to hypochlorite solutions if they are not stored properly or are kept too long?

13. How can carbonate scale on hypochlorinators be removed?

14. How can fouled quartz or Teflon surfaces in UV systems be cleaned?

15. Calculate the hypochlorinator setting in pounds per 24 hours to treat a waste with a chlorine demand of 8 mg/L, when a chlorine residual of 1 mg/L is desired. The flow is 0.6 MGD.

SUGGESTED ANSWERS
Chapter 12.   DISINFECTION AND CHLORINATION

ANSWERS TO QUESTIONS IN LESSON 1

Answers to questions on page 168.

12.0A   Pathogenic bacteria are destroyed or removed from water by (1) physical removal through sedimentation and filtration, (2) natural die-away or die-off of microorganisms in an unfavorable environment, and (3) destruction through chemical treatment.

12.0B   Disinfection is the destruction or inactivation of pathogenic organisms. Its main objective is to prevent the spread of waterborne diseases.

12.0C   Sterilization of wastewater is impractical and unnecessary and may be detrimental to other treatment processes that are dependent on the activity of non-pathogenic saprophytes.

12.0D   Chlorine is used for disinfection because it is cheap to manufacture and easy to obtain.

Answers to questions on page 173.

12.0E   Chlorine reacts with organic matter to form chlororganic compounds.

12.0F   Chlorine Dose = Chlorine Demand + Chlorine Residual

12.0G   To produce an effective disinfecting action, sufficient chlorine must be added after the chlorine demand is satisfied to produce a chlorine residual that will persist through the contact period.

Answers to questions on page 175.

12.0H Chlorine demand is influenced by contact time, effectiveness of upstream processes, temperature, dose rate, type of chemical, pH, and numbers and types of organisms.

12.0I A hypochlorinator is set to feed a dose of 5 pounds of chlorine per 24 hours. The wastewater flow is 0.06 MGD and the chlorine, as measured by the chlorine residual test after 30 minutes of contact time, is 0.4 mg/L. Find the chlorine dosage and the chlorine demand in mg/L.

Known

Chlorine Feed, lbs/day	= 5 lbs/day
Flow, MGD	= 0.06 MGD
Chlorine Residual, mg/L	= 0.4 mg/L

Unknown

1. Chlorine Dose, mg/L
2. Chlorine Demand, mg/L

1. Calculate the chlorine dose in milligrams per liter.

$$\text{Chlorine Dose, mg/L} = \frac{\text{Chlorine Feed, lbs/day}}{(\text{Flow, MGD})(8.34 \text{ lbs/gal})}$$

$$= \frac{5 \text{ lbs/day}}{(0.06 \text{ MGD})(8.34 \text{ lbs/gal})}$$

$$= 10 \text{ mg/L}$$

2. Determine the chlorine demand in milligrams per liter.

$$\text{Chlorine Demand, mg/L} = \text{Chlorine Dose, mg/L} - \text{Chlorine Residual, mg/L}$$

$$= 10.0 \text{ mg/L} - 0.4 \text{ mg/L}$$

$$= 9.6 \text{ mg/L}$$

12.0J The objective of disinfection is the destruction or inactivation of pathogenic organisms, and the ultimate measure of the effectiveness is the bacteriological test result.

12.0K The ultimate measure of effectiveness is the bacteriological test result. The residual chlorine that yields satisfactory bacteriological test results in a particular plant must be determined and used as a control in that plant.

ANSWERS TO QUESTIONS IN LESSON 2

Answers to questions on page 176.

12.1A The purpose of up-sewer chlorination is to control odors and septicity, prevent deterioration of structures, and decrease BOD load.

12.1B Chlorine should be applied at one or more points in the main intercepting sewer or at the upper ends of feeder lines.

12.1C Prechlorination provides partial disinfection and odor control.

12.1D Plant chlorination provides control of odors, corrosion, sludge bulking, digester foaming, filter ponding, filter flies, or sludge thickening. Be careful that chlorination does not interfere with biological treatment processes.

12.1E Postchlorination is used primarily for disinfection.

Answers to questions on page 179.

12.2A The selection of a chlorination control method is based on the treatment costs and the required or desired treatment results.

12.2B The control of the chlorine dosage for disinfection depends on the bacterial reduction desired; that is, the bacterial level or concentration acceptable or permissible at the point of discharge.

Answers to questions on page 180.

12.2C When hypochlorite is mixed with water, approximately half of the chlorine forms hydrochloric acid (HCl) and the other half forms hypochlorite (OCl⁻) in the chlorine residual that you measured.

12.2D A wastewater requires a chlorine feed rate of 55 pounds chlorine per day. How many pounds of high test hypochlorite (HTH) will be required per day to provide the needed chlorine? Assume each pound of HTH contains 0.65 pound of available chlorine.

Known

Chlorine Required, lbs/day	= 55 lbs/day
Portion of Chlorine in lb of HTH, lbs	= 0.65 lb

Unknown

HTH Feed Rate, lbs/day

Calculate pounds of HTH required per day.

$$\text{HTH Feed Rate, lbs/day} = \frac{\text{Chlorine Require, lbs/day}}{\text{Portion of Chlorine in lb of HTH, lbs}}$$

$$= \frac{55 \text{ lbs/day}}{0.65 \text{ lb}}$$

$$= 85 \text{ lbs HTH/day}$$

Answers to questions on page 181.

12.2E Chlorine discharge lines may be made of PVC or black polyethylene flexible tubing. Rubber hose is rarely used today, but rubber-lined steel pipe has been used.

12.2F Little mixing of chlorine solution with wastewater occurs in a chlorine contact basin because the chamber is designed for low-flow velocity.

Answers to questions on page 185.

12.2G   The chlorine residual must be measured as quickly as possible to prevent chlorine concentration changes in the sample.

12.3A   Whenever hypochlorite comes in contact with your hand, immediately wash off the hypochlorite and thoroughly wash your hand. Consult a physician if the area appears burned.

ANSWERS TO QUESTIONS IN LESSON 3

Answers to questions on page 186.

12.4A   The different types of chlorine feeders described in this chapter are tablet chlorinators and liquid hypochlorinators.

12.4B   Flow to tablet chlorinators can be blocked by sticks, leaves, grease balls, and other matter.

Answers to questions on page 188.

12.4C   Chlorine is released into the wastewater by the dissolving action of the wastewater in contact with the tablets.

12.4D   Chlorine doses are adjusted to flow changes with a chlorine tablet disinfection system by allowing the frequency and duration of tablet submergence to fluctuate with the flows.

12.4E   The tubes should always be filled to the top with chlorine tablets to ensure the longest possible periods of unattended operation.

12.4F   If the feed tubes do not reach the bottom of the unit, insufficient chlorine will be released into the wastewater.

Answers to questions of page 194.

12.4G   Major advantages of on-site chlorine/hypochlorite generation include (1) hypochlorite solution is generated as needed, and (2) the on-site generation system is safer for the operators.

12.4H   On-site hypochlorite systems use salt, water, and electricity to produce sodium hypochlorite.

12.4I   A major problem with sodium hypochlorite systems is the unreliable flow of hypochlorite solution, which affects the system's ability to provide continuous adequate disinfection.

Answers to questions on page 197.

12.4J   Hypochlorite solutions can be prevented from freezing by the use of a strip heater or a heat lamp.

12.4K   Review the spare parts inventory and then reorder as necessary to maintain adequate repair stock.

Answers to questions on page 198.

12.5A   Excessive chlorine residual is not required because it is costly and can be toxic to aquatic life in the receiving streams.

12.5B   UV light disinfects by damaging the DNA and RNA of organisms.

12.5C   UV disinfection is not suitable for effluents with high turbidity, suspended solids, dyes, organics, or other UV-absorbing constituents.

CHAPTER 13

ALTERNATIVE WASTEWATER TREATMENT, DISCHARGE, AND REUSE METHODS

by

John Brady

TABLE OF CONTENTS

Chapter 13. ALTERNATIVE WASTEWATER TREATMENT, DISCHARGE, AND REUSE METHODS

OBJECTIVES

Chapter 13.   ALTERNATIVE WASTEWATER TREATMENT, DISCHARGE, AND REUSE METHODS

1. Describe alternative wastewater treatment, discharge, and reuse methods for small systems.

2. Determine where wetlands can be used effectively.

3. Identify the components of subsurface constructed wetlands.

4. Start, operate, and maintain subsurface constructed wetlands.

5. Safely operate, maintain, and troubleshoot land treatment systems.

6. Monitor a land treatment system.

7. Review plans and specifications for a land treatment system.

WORDS
Chapter 13.   ALTERNATIVE WASTEWATER TREATMENT, DISCHARGE, AND REUSE METHODS

AEROBIC (air-O-bick)

AEROBIC

A condition in which atmospheric or dissolved oxygen is present in the aquatic (water) environment.

ANAEROBIC (AN-air-O-bick)

ANAEROBIC

A condition in which atmospheric or dissolved oxygen (DO) is *NOT* present in the aquatic (water) environment.

ANOXIC (an-OX-ick)

ANOXIC

A condition in which the aquatic (water) environment does not contain dissolved oxygen (DO), which is called an oxygen deficient condition. Generally refers to an environment in which chemically bound oxygen, such as in nitrate, is present. The term is similar to ANAEROBIC.

AVAILABLE MOISTURE CONTENT

AVAILABLE MOISTURE CONTENT

The quantity of water present that is available for use by vegetation.

CATION (KAT-EYE-en) EXCHANGE CAPACITY

CATION EXCHANGE CAPACITY

The ability of a soil or other solid to exchange cations (positive ions such as calcium, Ca^{2+}) with a liquid.

DENITRIFICATION (dee-NYE-truh-fuh-KAY-shun)

DENITRIFICATION

(1) The anoxic biological reduction of nitrate nitrogen to nitrogen gas.

(2) The removal of some nitrogen from a system.

(3) An anoxic process that occurs when nitrite or nitrate ions are reduced to nitrogen gas and nitrogen bubbles are formed as a result of this process. The bubbles attach to the biological floc and float the floc to the surface of the secondary clarifiers. This condition is often the cause of rising sludge observed in secondary clarifiers or gravity thickeners. Also see NITRIFICATION.

DRAIN TILE SYSTEM

DRAIN TILE SYSTEM

A system of tile pipes buried under agricultural fields that collect percolated waters and keep the groundwater table below the ground surface to prevent ponding.

DRAINAGE WELLS

DRAINAGE WELLS

Wells that can be pumped to lower the groundwater table and prevent ponding.

EVAPOTRANSPIRATION (ee-VAP-o-TRANS-purr-A-shun)

EVAPOTRANSPIRATION

(1) The process by which water vapor is released to the atmosphere by living plants. This process is similar to people sweating. Also called transpiration.

(2) The total water removed from an area by transpiration (plants) and by evaporation from soil, snow, and water surfaces.

FREEBOARD

FREEBOARD

(1) The vertical distance from the normal water surface to the top of the confining wall.

(2) The vertical distance from the sand surface to the underside of a trough in a sand filter. This distance is also called available expansion.

GATE

GATE

(1) A movable, watertight barrier for the control of a liquid in a waterway.

(2) A descriptive term used on irrigation distribution piping systems instead of the word valve. Gates cover outlet ports in the pipe segments. Water flows are regulated or distributed by opening the gates by either sliding the gate up or down or by swinging the gate to one side and uncovering an individual port to permit water flow to be discharged or regulated from the pipe at that particular point.

HYDROLOGIC (HI-dro-LOJ-ick) CYCLE

HYDROLOGIC CYCLE

The process of evaporation of water into the air and its return to earth by precipitation (rain or snow). This process also includes transpiration from plants, groundwater movement, and runoff into rivers, streams, and the ocean. Also called the water cycle.

NITRIFICATION (NYE-truh-fuh-KAY-shun)

NITRIFICATION

An aerobic process in which bacteria change the ammonia and organic nitrogen in wastewater into oxidized nitrogen (usually nitrate). The second-stage BOD is sometimes referred to as the nitrogenous BOD (first-stage BOD is called the carbonaceous BOD). Also see DENITRIFICATION.

PLUG FLOW

PLUG FLOW

A type of flow that occurs in tanks, basins, or reactors when a slug of water or wastewater moves through a tank without ever dispersing or mixing with the rest of the water or wastewater flowing through the tank.

DIRECTION OF FLOW

PLUG FLOW

SIDESTREAM

SIDESTREAM

Wastewater flows that develop from other storage or treatment facilities. This wastewater may or may not need additional treatment.

TRANSPIRATION (TRAN-spur-RAY-shun)

TRANSPIRATION

The process by which water vapor is released to the atmosphere by living plants. This process is similar to people sweating. Also called evapotranspiration.

VECTOR

VECTOR

An insect or other organism capable of transmitting germs or other agents of disease.

WATER CYCLE

WATER CYCLE

The process of evaporation of water into the air and its return to earth by precipitation (rain or snow). This process also includes transpiration from plants, groundwater movement, and runoff into rivers, streams, and the ocean. Also called the hydrologic cycle.

CHAPTER 13.   ALTERNATIVE WASTEWATER TREATMENT, DISCHARGE, AND REUSE METHODS

(Lesson 1 of 2 Lessons)

13.0   ALTERNATIVE WASTEWATER TREATMENT METHODS FOR SMALL SYSTEMS

Alternative wastewater treatment methods provide high-quality effluents for disposal or reuse by small systems in comparison to the more expensive and complex conventional secondary wastewater treatment plants. Primary treatment of the wastewater is required and, in most instances, is accomplished by septic tanks prior to the wastewater being discharged to these effluent disposal or reuse systems. The alternative treatment systems discussed in this chapter include constructed wetlands and land application of effluents.

13.1   WETLANDS

Wetlands, though in existence for millennia, are a recent interest to society as a wastewater treatment system. Land application of wastewater over various grasses and feed crop vegetation was just being introduced in the 1970s, and pilot projects reported encouraging results. At that time, NASA was looking for a reasonable, economical alternative to the conventional wastewater treatment systems at one of their facilities, which consisted of a package plant and ponds that had difficulty meeting discharge requirements. Consideration was given to marshes where reeds, cattails, bulrushes, and other water-tolerant plants grew. Engineered systems using vegetation that could function in some of the colder climates such as the Dakotas and Canada were then developed. Today, many communities are using wetlands for wastewater treatment. Wetlands systems, predominantly subsurface constructed wetlands, are discussed in this section.

13.10   Basic Types of Wetlands

Wetlands can be either natural or constructed. Natural wetlands exist in nature as bogs, marshes, swamps, and cypress domes. Early treatment projects were associated with facilities next to existing natural wetlands. Two well-known systems are Cannon Beach, Oregon, and Arcata, California (Figure 13.1). Constructed wetlands can be built almost anywhere and resemble natural wetlands.

Today, three types of wetlands are currently in use:

1. Free water surface (FWS) wetlands have areas of open water and are similar in appearance to natural marshes.

2. Horizontal subsurface flow (HSSF) wetlands typically use a gravel bed planted with wetland vegetation. The water being treated, kept below the surface of the bed, flows horizontally from the inlet to the outlet.

3. Vertical flow (VF) wetlands distribute water across the surface of a sand or gravel bed planted with wetland vegetation. The water is treated as it percolates down through the plant root zone.

All of these types of wetlands can be operated as continuous flow systems or as intermittent flow systems, with a fill and drain mode.

Free Water Surface (FWS) (Figures 13.2 and 13.3)

A free water surface (FWS) system consists of basins or channels with a natural or constructed subsurface barrier made of clay or impervious material that prevents seepage from the wetland. Soil, sand, gravel, rock, or another suitable material is used to support the emergent vegetation and the water is maintained at a relatively shallow depth flowing over the surface. In the FWS wetland, water is visible throughout the bed and maintained at a depth of 18 to 24 inches (45 to 60 cm) with various plants growing in the water and the bed media. Long, narrow channels are used to create *PLUG FLOW*[1] conditions, but wider, shorter basins are also used. FWS systems are predominantly used in larger communities as a secondary or polishing wastewater treatment method. Systems in use today treat wastewater flows as high as 20,000,000 gallons per day (20 MGD or 75 MLD). FWS systems are not recommended for individual home sites because of associated problems such as vector and disease control.

Limitations of FWS systems include the associated potential problems of *VECTOR*[2] and disease control, which must be considered and managed. Short-circuiting may occur in FWS systems due to clumps of grasses and plants that could hinder the

---

1. *Plug Flow.*   A type of flow that occurs in tanks, basins, or reactors when a slug of water or wastewater moves through a tank without ever dispersing or mixing with the rest of the water or wastewater flowing through the tank.

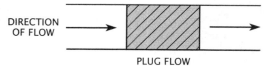

DIRECTION OF FLOW

PLUG FLOW

2. *Vector.*   An insect or other organism capable of transmitting germs or other agents of disease.

Inlet to No. 1 wetland cell from oxidation pond (Cannon Beach, Oregon)

Wetland marsh (Arcata, California)

Fig. 13.1   Natural wetlands

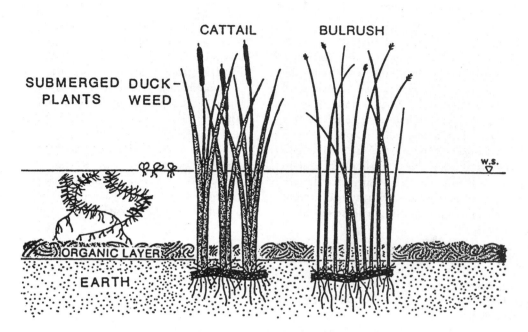

NOTE: Care must be taken to select plants that can survive in local weather conditions.

Fig. 13.2   *Common aquatic plants growing in a free
water surface (FWS) wetland (natural wetland)*

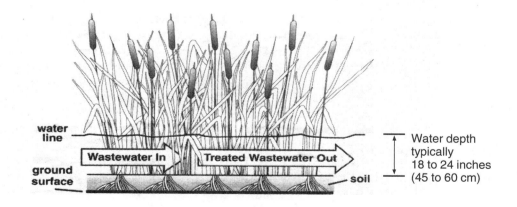

Fig. 13.3   *Free water surface flow constructed wetlands*
(Source: *Small Flows*)

free flow of the water. Solids deposition can occur when trees lose their leaves and plants die. Over time, these deposits may create anaerobic zones that could release obnoxious odors.

Horizontal Subsurface Flow (HSSF) System
(Figures 13.4 and 13.5)

A horizontal subsurface flow wetland is constructed as a trench or bed with an impermeable layer of clay or a synthetic liner. The bed contains media of rock, gravel, or soil that will support the growth of emergent vegetation. The wetland bed is constructed with a slight slope between the inlet and outlet. Influent is applied to the wetland at the high end and is permitted to flow under the media surface, through the media and plant roots to the outlet where it is collected and discharged. Water standing or pooling on the surface is not desirable; you should be able to walk on the dry surface. HSSFs are used at single-family residences to provide a high degree of wastewater treatment without creating nuisance conditions or requiring complex operation

and maintenance skills. The beds often look like flower beds. Figure 13.4 shows cross-sectional views of small HSSF constructed wetlands. Figure 13.5 shows HSSF constructed wetlands in use at two different homes. These systems are becoming more prominent in small communities, treating flows up to several million gallons per day.

The advantages of these wetlands include their lower initial construction cost if land is relatively cheap and available and reduced operation and maintenance costs compared to conventional mechanical plants. The level of operational skill needed to maintain the system is much less than other systems, and yet the system achieves a very high and reliable degree of wastewater treatment for the community. Over time, HSSF systems can become root bound, leading to short-circuiting and channeling. Care must be taken to ensure that the organic loading and seasonal die back of vegetation do not cause the system to operate in an anaerobic condition. If anaerobic conditions develop, obnoxious odors could be released.

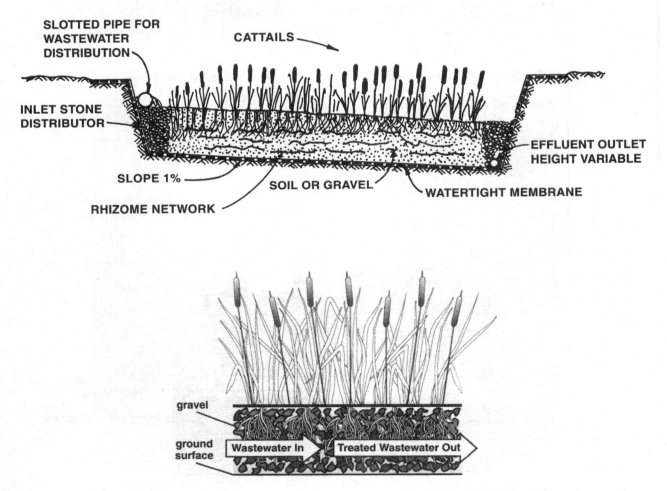

NOTE: Excessive root growth could cause these systems to become root bound.

Fig. 13.4   Horizontal subsurface flow constructed wetland
(Source of lower image: *Small Flows*)

This wetland treatment system is hard to see, but that is the point.
The ferns in the foreground constitute the treatment system.

This treatment system has a small footprint with a relatively manicured garden
look that is achievable in systems designed especially for residential use.

Fig. 13.5   Two examples of single family horizontal subsurface flow (HSSF) wetlands
(Source: Naturally Wallace Consulting)

Vertical Flow (VF) System (Figure 13.6)

Vertical flow wetlands use surface flooding either as a continuous flow loading or as an intermittent flow loading. Many operate as vegetated recirculating gravel filters. The surface layer consists of sharp sand and supports the wetland vegetation. Beneath the sand layer are three layers of gravel of increasing size and then a bottom layer of stones and agricultural drainage pipes.

13.11   Where Can Wetlands Be Used?

A wetland system can be constructed almost anywhere it is needed. When considering the construction of a wetland treatment system, the major controlling factor for communities and individual home sites is land availability. Other controlling factors include zoning and buffer distances. Constructed wetlands that discharge to surface waters require four to ten times more land area than a conventional wastewater treatment plant; for a zero discharge wetland, the land area is ten to one hundred times the area that a conventional treatment plant would need. The advantage of the wetland is that it can be built and used in places where other discharge systems (such as leach fields, seepage pits, and mounds) have failed. Failure was due to tight clay soils that bind, or loose sandy soils that percolate water too fast, or locations where rock outcroppings or a high groundwater table prevent safe wastewater treatment and discharge. The cost to construct an HSSF wetland for an individual home is no more expensive than the long-used septic tank/leach field wastewater

discharge system. HSSF wetlands are predominantly used for homes and small communities. There are two main reasons for their acceptance:

1. The wastewater is not exposed to the surface until it has been treated and stabilized at the time of discharge, if a discharge does exist. This prevents contact of wastewater with children or pets, and the HSSF minimizes the potential odor, insect vector, and public health problems. Studies have indicated that the possibility of contracting a disease or virus from an HSSF wetland is very remote. The HSSF further reduces this possibility as compared to the FWS wetland system.

2. The HSSF bed can be constructed to fit the available land area. Some designers prefer long, narrow beds; others prefer a shorter, wider bed. Wetland beds have even been constructed on the side of a hill as multiple beds or terraces; therefore, the configuration of the wetland system is flexible and depends on the site. Once constructed, the wetland is planted with plants that can grow and survive in the area's climate. Figure 13.7 shows a cold weather constructed wetland containing plants that are suitable to the climate. There is a wide variety of plants that do well in a wetland habitat and some are very attractive, creating colorful flower beds or a different type of planting area that would not normally have been available at that site. As a result, wastewater treatment systems are becoming a visual asset to their communities rather than a necessary and unattractive liability.

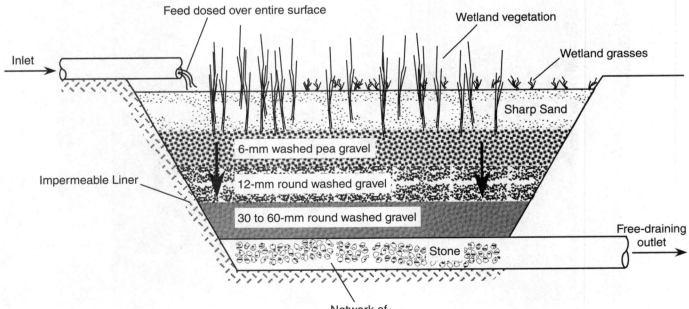

Fig. 13.6   Vertical flow constructed wetland
(Adapted from *Reed Beds and Constructed Wetlands for Wastewater Treatment*,
P. F. Cooper et al., WRC Publications. Swindon, UK)

This constructed wetland is marked by the twigs and tall reeds seen growing in front of the log cabin.

Fig. 13.7   Cold weather constructed wetland
(Source: Laura Downing, Mount Elbert Lodge, Twin Lakes, Colorado, 2011)

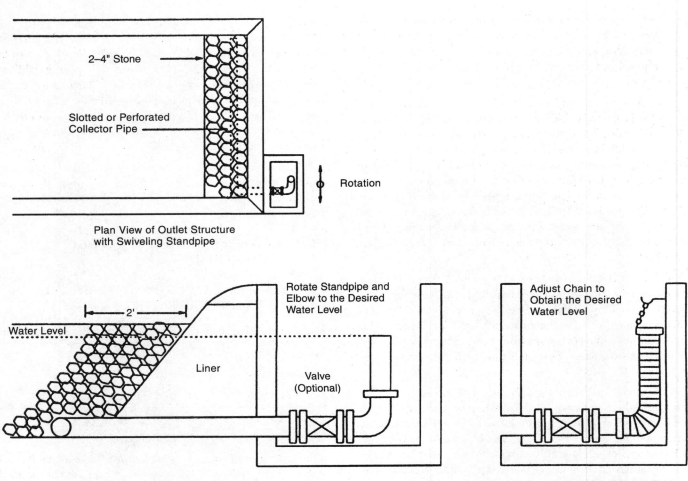

2–4" Stone

Slotted or Perforated
Collector Pipe

Rotation

Plan View of Outlet Structure
with Swiveling Standpipe

Rotate Standpipe and
Elbow to the Desired
Water Level

2'

Water Level

Liner

Valve
(Optional)

Adjust Chain to
Obtain the Desired
Water Level

Control Structure with Swiveling Standpipe

Control Structure with Collapsible Tubing

Fig. 13.8   Water level control structures
(Source: *General Design, Construction, and Operation
Guidelines: Constructed Wetlands*, TVA Technical Report)

13.12   Components of Subsurface Constructed Wetlands

Subsurface wetland systems can be constructed as a single bed or as multiple beds. The first bed or cell is made watertight by using clay or an impermeable liner. Liners are normally covered at the top of the berms to protect the plastic or rubber liner from ultraviolet sunlight, which will deteriorate those materials over time. The bed is constructed with a berm or wall around the perimeter (outer edge) to contain the wastewater and control hydraulic loading by preventing the entry of surface water runoff from adjoining land. The walls or berms are constructed with at least 6 inches (15 cm) *FREEBOARD* [3] above adjoining

land and at least 9 to 12 inches (22.5 to 30 cm) freeboard inside the bed and above the top surface of the bed media. Bed media depth is normally 12 inches (30 cm), making the total inside bed depth 21 to 24 inches (52 to 60 cm).

Inlet distribution piping and outlet collection piping are placed at opposite ends of the cell with a minimum of 2 feet (60 cm) of round stone, 2 to 4 inches (5 to 10 cm) in diameter, that forms a barrier between the piping and the smaller gravel media filling the bed (Figure 13.8). The stone helps distribute flows evenly and prevents plugging or clogging of the distribution and collection piping. The inlet piping to the first cell should have a

---

3. *Freeboard.*   (1) The vertical distance from the normal water surface to the top of the confining wall.
(2) The vertical distance from the sand surface to the underside of a trough in a sand filter. This
distance is also called available expansion.

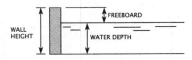

WALL
HEIGHT

FREEBOARD

WATER DEPTH

cleanout or small chamber to enable visual inspection for solids accumulation that may be carried over from the primary treatment unit. Algae and other biological growths can coat the walls of the pipes and eventually could plug the pipes. Accessible cleanouts must be provided so the pipes can be cleaned.

Round ¾- to 1-inch (2- to 2.5-cm) gravel was originally used as media. Later studies indicated that using smaller pea gravel as media provided for better plant growth. In some instances, coarse sand has been used. In systems using two cells, media in the second (percolation) cell is coarse sand with a layer of mulch on the surface. The sand provides a good media for the plants, ensuring their survival in fluctuating water levels in the cell during dry periods (Figure 13.9). Cells with depths up to 30 inches (75 cm) have been constructed in colder climates to facilitate root growth for plants predominant in those areas (Figure 13.10).

The media surface is covered with a mulch (litter) material approximately 3 inches (7.5 cm) deep. The mulch on top of the gravel prevents sun scalding of the plants, provides visual aesthetics, and helps control potential odors. In colder climates, the mulch insulates the plant roots from freezing temperatures. Reports have been made of wetlands functioning at temperatures as low as −30°F (−34°C).

Constructed wetlands are designed to be dosed by gravity flow but, where necessary, low-head pump systems are installed for communities or individual residences. Water level control is important for plant survival and wetlands health (see Figure 13.8). Roots of emergent plants must be kept wet, but plants will not survive if completely covered with water. In addition, if water is permitted to stand above the mulch layer, surface odors may develop. The water level in a horizontal subsurface flow wetland is controlled through the outlet structure regulating device, which lowers and raises the overflow discharge level.

13.13   How a Subsurface Wetland Functions

Primary treated wastewater (septic tank effluent) is applied across the inlet end of the wetland bed at the bottom. The wastewater flows through the gravel or sand media and around the plant roots, just below the surface of the gravel media in the wetland bed. Bacteria growing in the voids of the gravel use the nutrients in the wastewater for cell growth and reproduction. The major role of the plants is to transfer oxygen to the root zone. The plant stalks, roots, and rhizomes penetrate the soil or support media and transport oxygen deeper than it would naturally travel by diffusion alone. The plants also take up nutrients in the media for plant growth and remove some of the water by *TRANSPIRATION*.[4]

In areas with high groundwater or where the soil has no ability to absorb water, the effluent from the wetland is collected and discharged to a small percolation field or a small sand filter trench for irrigation. In areas of soil conditions that have some percolation capacity, a wetland of two-cell or bed configuration is constructed (Figure 13.11). The first cell is lined or sealed to prevent water loss into the soil to accommodate waste stabilization without contaminating groundwater aquifers. The second cell is not lined or sealed; its purpose is to discharge the effluent from the first cell through percolation into the surrounding soil and to permit water losses due to evaporation and transpiration (see Figure 13.11).

Wetlands have been shown to significantly reduce BOD, suspended solids, nitrogen, and coliform organisms from the wastewater but have not been shown to be efficient in the removal of phosphate.

13.14   Constructed Wetlands Loading Criteria (FWS and HSSF)

TABLE 13.1   CONSTRUCTED WETLANDS LOADING CRITERIA (FWS and HSSF)

Pretreatment Required	Minimum of Primary Sedimentation
Hydraulic Loading	0.67–0.77 gal/day/sq ft
Hydraulic Detention Time	7 days+
Organic Loading	0.05 lb BOD/sq ft/day

13.15   Starting a New HSSF Wetland

The wetland bed is built at the same time the septic tank is installed. If the types of vegetation to be planted are not specified by the design, the local agricultural extension agent or a local nursery may be consulted to recommend suitable plants adapted to the local climate. Most wetland plants require six hours of full sunlight for good growth. For shady locations, ferns or other shade-tolerant species should be selected. Decorative species such as water iris, canna lily, elephant's ear, and sweet flag can be used in warmer climates. All will grow over the entire United States (except in parts of Alaska) but some will die out in the colder regions during winter months; these areas must rely on bulrushes, cattails, and other marsh plants.

Place the plants no less than one foot (0.3 m) apart in a grid pattern; this spacing usually produces a uniform cover of vegetation over the entire surface of the cell in one to two years. Irrigate the new plants with domestic water for the first six to eight weeks, if possible, before applying wastewater to the bed. After the initial six weeks, add makeup water to control the water level, along with liquid fertilizer to stimulate plant growth (domestic wastewater may not have all of the nutrients required for good plant growth). In facilities that have two cells, the second cell should be given a light application of liquid fertilizer to stimulate growth.

---

4. *Transpiration* (TRAN-spur-RAY-shun).   The process by which water vapor is released to the atmosphere by living plants. This process is similar to people sweating. Also called evapotranspiration.

Constructed wetlands like this one are being built

throughout the nation to handle wastewater from mostly small rural communities and homes where traditional treatment systems are a problem.

Wastewater flows into the constructed wetland from a septic tank or other type of primary treatment system. Here the wastewater is evenly distributed among the plants where microorganisms and chemical reactions break down organic materials and pollutants.

Constructed wetlands provide simple, effective, and low-cost wastewater treatment when compared with conventional systems.

Wastewater enters the constructed wetland (1) where it is distributed evenly across the width of the first cell by a series of plastic valves or PVC tees (2). The first cell contains gravel (3). A waterproof liner is used on the sides and the bottom of the first cell to conserve water and provide more effective treatment (4). Cattails and bulrushes are usually planted in the first cell (5). The roots of these marsh plants form a dense mat among the gravel (6). Here chemical, biological, and physical processes take place that purify the water. Water from the first cell passes into the second cell through a perforated pipe embedded in large stone (7). The water level within each cell is regulated by swivel standpipes located in concrete tanks at the end of each cell (8). Wastewater in the second cell is distributed evenly across this cell through another perforated pipe (9). Cell 2 has a layer of gravel (10) covered with topsoil (11) and then mulch (12). This cell is planted with a variety of ornamental wetland plants such as iris, elephant ear, and arrowhead (13). The water in cell 2 eventually seeps into the soil below (14) or passes into another perforated pipe (15) where it is released into a drainfield similar to those used with conventional septic tanks (16).

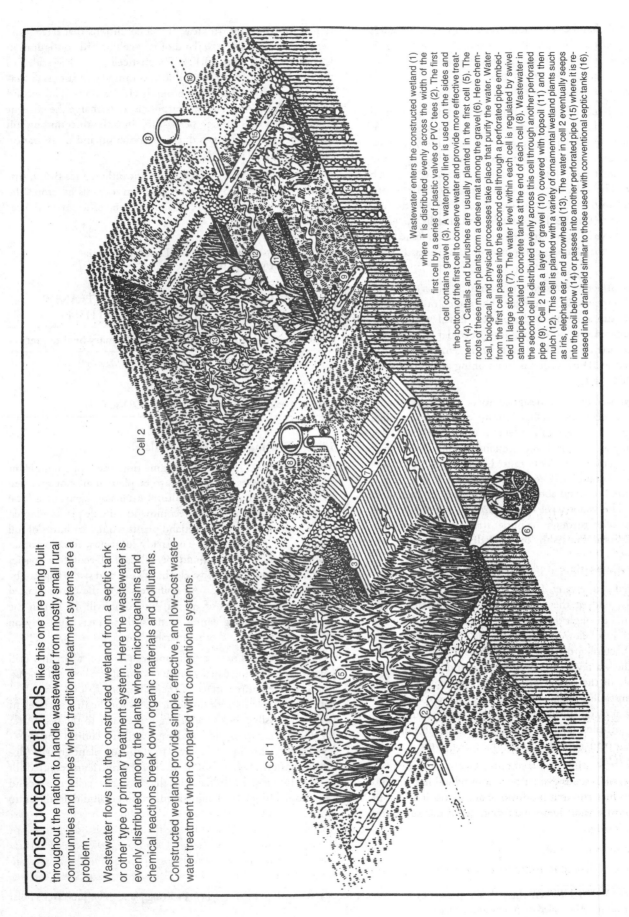

Cell 2

Cell 1

Fig. 13.9   Cut-away view of a constructed wetlands system

(Source: *General Design, Construction, and Operation Guidelines: Constructed Wetlands*, TVA Technical Report)

*Fig. 13.10   The inlet distribution piping at American
Crystal Sugar Company's constructed wetlands system*
(Source: *Small Flows,* January 1993; photo courtesy of
Donald Hammer, Ph.D., Tennessee Valley Authority)

QUESTIONS

Please write your answers to the following questions and compare them with those on page 248.

13.1A   List the three types of wetlands.

13.1B   Why are free water surface (FWS) wetlands not recommended for individual home sites?

13.1C   An advantage of wetlands is that they can be built where what other types of wastewater treatment systems have failed?

13.16   Operation and Maintenance

Constructed wetlands are "natural" systems, meaning that their operation is mostly passive and requires little operator intervention. Operation involves simple procedures similar to the requirements for operation of a facultative lagoon. The operator must be observant, take appropriate actions when problems develop, and conduct required monitoring and operational monitoring. The most critical items requiring operator intervention include the following:

- Adjustment of water levels
- Maintenance of flow uniformity (inlet and outlet structures)
- Management of vegetation
- Odor control
- Control of nuisance pests and insects
- Maintenance of berms and dikes
- Cultivate maturation of wetlands
- Mechanical equipment maintenance
- Ensuring health and safety

13.160   Water Level and Flow Control

Water level and flow control are usually the only operational variables that have a significant impact on the performance of a well-designed constructed wetland. Changes in water levels affect the hydraulic residence time, atmospheric oxygen diffusion into the water phase, and the amount of plant cover. Significant changes in water levels should be investigated immediately, as they may be due to leaks, clogged outlets, breached berms, stormwater drainage, or other causes.

Seasonal water level adjustment helps prevent freezing in the winter. In cold climates, water levels should be raised approximately 18 inches (45 cm) in late fall until an ice sheet develops. Once the water surface is completely frozen, the water levels can be lowered to create an insulating air pocket under the ice and snow cover to maintain higher water temperatures in the wetland.

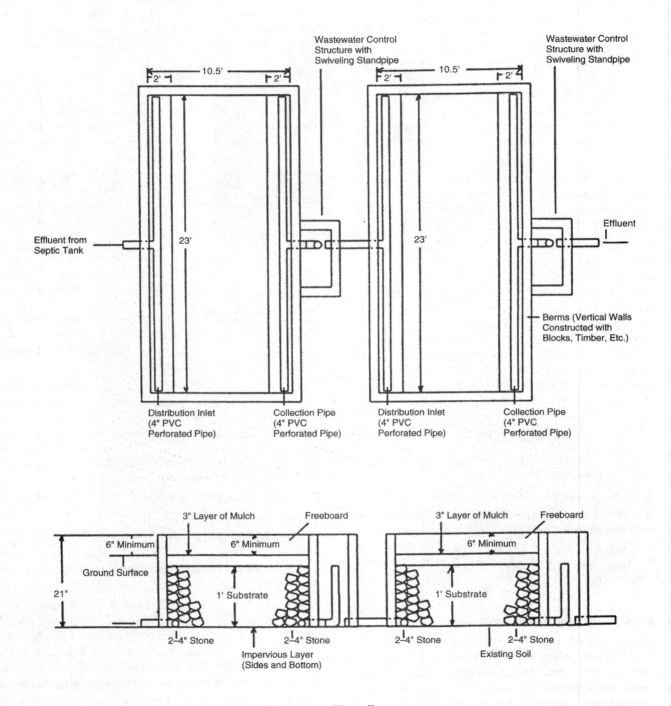

Fig. 13.11   Two-cell system

(Source: *General Design, Construction, and Operation
Guidelines: Constructed Wetlands*, TVA Technical Report)

Operational variables to consider include the following:

1. The main operational control is proper water level. In subsurface flow wetlands, water level in the first cell is maintained at one inch below the top of the gravel media at the inlet end. Water levels can be monitored easily by installing 4-inch diameter pipes standing vertically at random locations throughout the cell. The pipes will act as stilling wells and provide an instant view of the water level. Immediately after an application of wastewater to a small bed, an increase in water level may result from surging. In larger community beds this is less noticeable. Water level control through the outlet structure regulating device occurs by lowering or raising the overflow discharge level (see Figure 13.8 on page 222).

2. Low water level. In periods of low flows or during hot weather, water in the first cell may evaporate or be used by the plants. During severe cold periods, it could freeze solid, damaging the roots and tubers. Makeup water may have to be added to the cell during those periods, as needed.

3. Surface pooling or ponding in the wetland bed that cannot be controlled by water level adjustment is probably caused by either excessive flows above the design hydraulic loading rate or clogging of the media by excessive solids in the influent being applied. Check the cleanout in front of the wetland. Solids accumulated in the pipe indicate excessive solids from the pretreatment unit (septic tank or tanks from the community), which must be located and pumped. If there are no solids in the cleanout, ponding is probably caused by excessive water flow that exceeds the hydraulic capacity of the media. Employ water conservation practices to decrease flows. If ponding is not corrected, a portion of the media may have to be replaced, starting with the area near the inlet.

An example of operational control is the City of Davis, California, which regulates water levels in its FWS wetlands seasonally. In March and April, water levels are drawn down for the summer to expose the benches along the shorelines. This creates a foraging habitat for shorebirds. Also, by keeping water off the benches, mosquitoes are discouraged from breeding. In the fall and winter months, increased stormwater flows are directed to the wetlands. The higher water levels flood the benches, creating a shallow habitat for waterfowl and wading birds. This is but one example of how operators effectively regulate water levels to prepare for seasonal changes and to serve the needs of local waterfowl and wildlife as appropriate.

13.161 Maintenance of Flow Uniformity

Maintaining uniform flow across the wetland using inlet and outlet adjustments is extremely important to achieve the expected treatment performance. The inlet and outlet manifolds should be inspected routinely, regularly adjusted, and cleaned of debris that may clog the inlets and outlets. Debris removal and bacterial slime removal from weir and screen surfaces is necessary. Submerged inlet and outlet manifolds should be flushed periodically. Additional cleaning with a high-pressure water spray or by mechanical means also may become necessary.

Wastewater suspended solids will settle and accumulate near the inlets to the wetland. These accumulations can decrease hydraulic detention times. Over time, these accumulations of solids will require removal. HSSF systems cannot be desludged easily without draining and removing the media. Therefore, HSSFs should not be considered for treating wastewaters with high suspended solids loads, such as facultative pond effluents, which have high algal concentrations. Also, suspended solids from septic tanks should not be allowed to enter HSSFs.

13.162 Vegetation Management

Routine maintenance of wetland vegetation is not required for systems operating within their design guidelines and with precise bottom-depth control of vegetation. Wetland plant communities are self-maintaining and will grow, die, and regenerate each year. Plants will naturally spread to unvegetated areas with suitable environments, within the plants' depth range, and be displaced from areas that are environmentally stressful. Operators must control plants in FWS systems by harvesting to prevent them from spreading into open water areas that are intended by design to be aerobic zones.

The primary objective in vegetation management is to maintain the desired plant communities where they are intended to grow within the wetland. This is achieved through consistent pretreatment process operation, small, infrequent changes in water levels, and harvesting plants when and where necessary. Where plant cover is deficient, activities designed to improve cover may include water level adjustment, reduced loadings, pesticide application, and replanting.

Harvesting and litter removal may be necessary depending on the design of the system. Plant removal from some free water surface wetlands may be required to meet the treatment goals, but a well-designed and well-operated horizontal subsurface flow system should not require routine harvesting. Harvesting of vegetation from wetlands at the height of and just before the end of the growing season helps remove some nitrogen from the system, but phosphorus removal is limited. Winter burning of vegetation, where allowed, can be used to control pests.

Vegetation management also includes the following:

1. Periodically, check vegetation for signs of disease or other stress such as yellowing, spots, withering, browning, and insects. Vegetation will show natural seasonal changes with temperature and day length changes. First, check the water level in the bed. If the water level is appropriate, obtain guidance from a local agricultural extension agent or a knowledgeable garden center for plant enhancement or correct chemical use and application rate of chemicals for insect control.

2. If vegetation does not appear healthy although correct water levels are maintained, add a balanced liquid fertilizer periodically (three times per growing season) to the wastewater. Pour the fertilizer into an upstream toilet or into the cleanout inspection pipe just prior to the bed inlet.

3. Replace dead plants to fill open spaces in the bed, as necessary.

4. Remove weeds, trees, and shrubs from the beds along with unwanted growth adjacent to the wetlands because these unwanted species will shade and crowd out the desirable plants.

5. Remove mature wetland vegetation after the plants have browned in the fall for visual aesthetics, if desirable. Only cut approximately two-thirds of the height of the plants. The removed material may be left on the bed as mulch.

6. Divide and replant decorative flowering species such as irises to enhance the system's appearance.

7. Encourage deep root growth by slowly lowering the water level over several weeks during the dormant period. Do not lower the water level too quickly because doing so will leave the roots stranded without water.

8. Maintain a three-inch layer of mulch on top of the media, either with litter from the wetland vegetation, pine straw, bark, or other suitable material on HSSFs.

9. Native, noninvasive vegetation species are preferred to invasive species that can spread from the treatment wetlands to local wetlands, replacing native species. In addition, the plant life selected should not create a habitat for unwanted animals and invertebrates, such as rats and mosquitoes, or for threatened or endangered species that could limit or eliminate the use of the treatment wetland.

13.163   Odor Control

Odors are seldom a nuisance in properly loaded wetlands. Odorous compounds emitted from open water areas are typically associated with anaerobic conditions, which can be created by excessive BOD and ammonia loadings. Therefore, reducing the organic and nitrogen loadings by pretreatment can control odors. If odors persist, drain each cell, one cell at a time, and then refill it to restore aerobic conditions in the cell. Alternatively, aerobic open water zones interspersed in areas between fully vegetated zones introduce oxygen to the system. Turbulent flow structures such as cascading outfall structures and channels with hydraulic jumps, used to introduce oxygen into the system effluent, can generate serious odor problems by stripping volatile compounds such as hydrogen sulfide if the constructed wetland has failed to remove these constituents.

Standing water on the HSSF media surface (mulch) is the probable cause of objectionable odors. Level any low and high spots that create small standing pools on the media surface by raking or filling with additional gravel and mulch. If the water level is too high, causing standing water on most of the media surface, lower the water level to about one inch below the surface of the gravel in the bed.

13.164   Nitrogen Control

Wetlands can be an effective treatment process for removing ammonia by converting it to nitrate. Wetlands require aerobic conditions to convert ammonia to nitrate (*NITRIFICATION*[5]). Aerobic conditions can be achieved if the water being treated is near the surface of submerged flow systems. After nitrification occurs, wetlands require *ANOXIC*[6] conditions for *DENITRIFICATION*[7] to occur.

13.165   Control of Nuisance Pests and Insects

Wetland plant life should not create a habitat for unwanted animals or invertebrates, such as rats and mosquitoes, or for threatened or endangered species that could limit or eliminate the use of the treatment wetland.

Potential nuisances and vectors that may occur in FWS wetlands include burrowing animals, dangerous reptiles, mosquitoes, and odors. An infestation of burrowing animals such as muskrats and nutria can seriously damage vegetation and berms in a system. These animals use both cattails and bulrushes as food and nesting materials. These animals can be controlled during the design phase by decreasing the slope on berms to 5:1 or using a coarse riprap. Temporarily raising the operating water level may also discourage the animals. Live trap and release programs may be successful, but in most cases it has been necessary to eliminate the animals. Fencing has had little success.

Dangerous reptiles are common in the southeastern states. The most common are snakes, particularly water moccasins, and alligators. It is difficult to control these animals directly. Warning signs, fencing, raised boardwalks, and mowed hiking trails can be used to minimize human contact with the animals. Operators should be made aware of the dangers and preventive actions that can be taken to avoid dangerous situations.

Mosquito control is a critical issue in FWS wetlands. In warm climates, wetlands have been seeded with mosquito fish (gambusia) and dragonfly larvae to control mosquitoes. Mosquito fish also can be used in northern climates but they need to be restocked each year. However, these fish have difficulty reaching all parts of the wetland when the accumulation of litter is too dense, particularly if cattails are grown. Other natural control methods have included erecting bat and bird houses. Desirable birds include purple martins and swallows. Bacterial larvicides, *Bacillus thuringiensis israelensis* (BTI), and *Bacillus sphaericus* (BS) have been used successfully in a number of wetlands.

---

5. *Nitrification* (NYE-truh-fuh-KAY-shun).   An aerobic process in which bacteria change the ammonia and organic nitrogen in wastewater into oxidized nitrogen (usually nitrate). The second-stage BOD is sometimes referred to as the nitrogenous BOD (first-stage BOD is called the carbonaceous BOD). Also see DENITRIFICATION.

6. *Anoxic* (an-OX-ick).   A condition in which the aquatic (water) environment does not contain dissolved oxygen (DO), which is called an oxygen deficient condition. Generally refers to an environment in which chemically bound oxygen, such as in nitrate, is present. The term is similar to ANAEROBIC.

7. *Denitrification* (dee-NYE-truh-fuh-KAY-shun).   (1) The anoxic biological reduction of nitrate nitrogen to nitrogen gas. (2) The removal of some nitrogen from a system. (3) An anoxic process that occurs when nitrite or nitrate ions are reduced to nitrogen gas and nitrogen bubbles are formed as a result of this process. The bubbles attach to the biological floc and float the floc to the surface of the secondary clarifiers. This condition is often the cause of rising sludge observed in secondary clarifiers or gravity thickeners. Also see NITRIFICATION.

13.166 Maintenance of Berms and Dikes

Berms and dikes require mowing, controlling erosion, and preventing animal burrows and tree growth. When the wetland is operated at a shallow depth, periodically removing tree seedlings from the wetland bed may be necessary. If the trees are allowed to reach maturity, they may shade out the emergent vegetation and with it the necessary conditions to enhance flocculation, sedimentation, and denitrification.

13.167 Maturation of Wetlands

Wetlands may require up to three years to mature and achieve expected performance. This amount of time is required for plant roots and aquatic organisms to develop and become effective in treating wastewater.

13.168 General Maintenance

These general maintenance tasks should be completed on a regular basis:

1. Pumps, if used in the system, must be lubricated and cleaned. Control systems must be checked on a routine basis.

2. Berms must be repaired immediately if erosion is evident; leaks from the beds should not be permitted. Earthen berms or dikes should be mowed to maintain an attractive site.

3. Polyethylene, PVC, neoprene, or rubber liners that extend above the media or water level should be covered with mulch or soil to prevent ultraviolet degradation.

4. Surface drainage should be routed away from the wetland beds so as not to affect water level and hydraulic loading.

13.169 Health and Safety

For health and safety reasons, prevent children and pets from playing or digging in the wetland to avoid contact with potentially infectious microorganisms and to avoid destroying vegetation or disturbing media and mulch.

13.17 Monitoring

Routine monitoring is essential in managing a wetland system. In addition to monitoring done to meet regulatory requirements, inflow and outflow rates, water quality, water levels, and indicators of biological conditions should be regularly monitored and evaluated. Monitoring of biological conditions includes measuring microbial populations and monitoring changes in water quality, percent of cover provided by dominant plant species, and benthic macroinvertebrate and fish populations at representative stations. Over time, these data help the operator to predict potential problems and select appropriate corrective actions.

13.18 Acknowledgment

The author wishes to acknowledge that a major portion of the material in this lesson on wetlands was adapted from these EPA manuals: *Wastewater Collection, Treatment and Disposal Options for Small Communities* and *Constructed Wetlands Treatment of Municipal Wastewaters.*

QUESTIONS

Please write your answers to the following questions and compare them with those on page 248.

13.1D   Which operational variables can have a significant impact on the performance of a well-designed constructed wetland?

13.1E   What maintenance should be performed on the inlets and outlets to wetlands?

13.1F   Why is vegetation harvested from wetlands?

13.1G   Why should wetland vegetation be checked periodically?

13.1H   What type of habitat should not be created by wetland plant life?

13.1I   What is the probable cause for odors from a horizontal subsurface flow system (HSSF) wetland?

END OF LESSON 1 OF 2 LESSONS

on

ALTERNATIVE WASTEWATER TREATMENT, DISCHARGE, AND REUSE METHODS

Please answer the discussion and review questions next.

DISCUSSION AND REVIEW QUESTIONS

Chapter 13.   ALTERNATIVE WASTEWATER TREATMENT, DISCHARGE, AND REUSE METHODS

(Lesson 1 of 2 Lessons)

At the end of each lesson in this chapter, you will find discussion and review questions. Please write your answers to these questions to determine how well you understand the material in the lesson.

1. How is emergent vegetation supported in a free water surface (FWS) wetland?

2. How is the water level controlled in a horizontal subsurface flow wetland?

3. What basic tasks must an operator perform when operating and maintaining a treatment wetland?

4. What is the primary objective of vegetation management in a wetland?

CHAPTER 13.   ALTERNATIVE WASTEWATER TREATMENT, DISCHARGE, AND REUSE METHODS

(Lesson 2 of 2 Lessons)

13.2   LAND TREATMENT SYSTEMS

13.20   Description of Treatment Systems

When a high-quality effluent or no discharge is required, land treatment offers an effective means of wastewater reclamation or ultimate disposal. Land treatment systems use soil, plants, and bacteria to treat and reclaim wastewaters. They can be designed and operated for the sole purpose of wastewater disposal, for crop production, or for both purposes. In land treatment systems, effluent is pretreated and applied to land by conventional irrigation methods. When systems are designed for crop production, the wastewater and its nutrients (nitrogen and phosphorus) are used as a resource. Treatment is provided by natural processes (physical, chemical, and biological) as the effluent flows through the soil and plants. Part of the wastewater returns to the *HYDRO-LOGIC CYCLE*[8] by *EVAPOTRANSPIRATION*[9] and part of it returns by surface runoff or percolation through the soil to groundwater (Figure 13.12). Land disposal of wastewater may be done by one of the three methods shown in Figure 13.13: irrigation, infiltration–percolation, or overland flow.

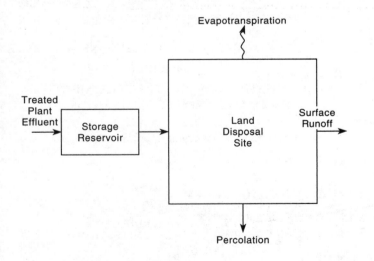

Fig. 13.12   Land disposal system schematic

The method of irrigation depends on the type of crop being grown. Irrigation methods include traveling sprinklers, fixed sprinklers, furrows, and flooding. Infiltration–percolation systems are not suited for crop growth because the water is not held in the soil for crop uptake. Overland flow systems are similar to other treatment processes and have runoff that must either be discharged or recycled in the system. The other systems usually do not have a significant surface runoff. Typical loading rates for these systems are shown in Table 13.2.

Land application systems include the following components:

1. Treatment before application
2. Transmission to the land treatment site
3. Storage
4. Distribution over site
5. Runoff recovery system, if needed
6. Crop systems

EXAMPLE 1

A plot of land 1,000 feet long by 800 feet wide is used for a land disposal system. If 0.4 million gallons of water is applied to the land during a 24-hour period, calculate the land area in acres and the hydraulic loading in MGD per acre and also in inches per day.

Known	Unknown
Length, ft = 1,000 ft	1. Area, acres
Width, ft = 800 ft	2. Hydraulic Loading, MGD/ac
Flow, MGD = 0.4 MGD	3. Hydraulic Loading, in/day

1. Determine the surface area in acres.

$$\text{Area, acres} = \frac{\text{Length, ft} \times \text{Width, ft}}{43,560 \text{ sq ft/acre}}$$

$$= \frac{1,000 \text{ ft} \times 800 \text{ ft}}{43,560 \text{ sq ft/acre}}$$

$$= 18.4 \text{ acres}$$

---

8. *Hydrologic* (HI-dro-LOJ-ick) *Cycle.*   The process of evaporation of water into the air and its return to earth by precipitation (rain or snow). This process also includes transpiration from plants, groundwater movement, and runoff into rivers, streams, and the ocean. Also called the water cycle.

9. *Evapotranspiration* (ee-VAP-o-TRANS-purr-A-shun).   (1) The process by which water vapor is released to the atmosphere by living plants. This process is similar to people sweating. Also called transpiration. (2) The total water removed from an area by transpiration (plants) and by evaporation from soil, snow, and water surfaces.

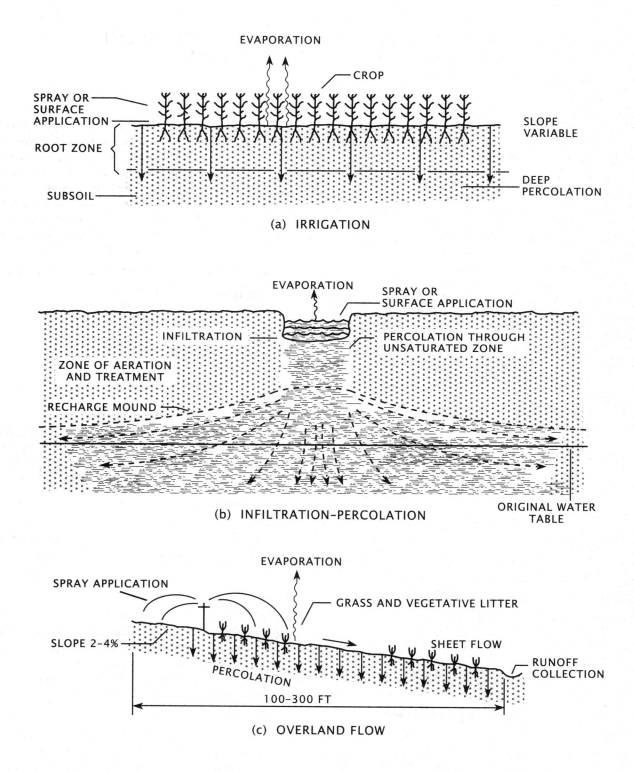

Fig. 13.13   Methods of land application

(*Costs of Wastewater Treatment by Land Application*, C. E. Pound et al.,
US Environmental Protection Agency. EPA No. 430-9-75-003.)

TABLE 13.2   TYPICAL LOADINGS FOR IRRIGATION, INFILTRATION–PERCOLATION, AND OVERLAND FLOW SYSTEMS[a]

Factor	Irrigation		Infiltration–Percolation	Overland Flow
	Low-Rate	High-Rate		
Liquid loading rate, in/wk[b]	0.5 to 1.5	1.5 to 4.0	4 to 120	2 to 9
Annual application, ft/yr[c]	2 to 4	4 to 18	18 to 500	8 to 40
Land required for 1 MGD flow rate, acres [d, e]	280 to 560	62 to 280	6 to 62	28 to 140
Application techniques	Spray or surface		Usually surface	Usually surface
Vegetation required	Yes	Yes	No	Yes
Crop production	Excellent	Good/fair	Poor/none	Fair/poor
Soils	Moderately permeable soils with good productivity when irrigated		Rapidly permeable soils such as sands, loamy sands, and sandy loams	Slowly permeable soils such as clay loams and clays
Climatic constraints	Storage often needed		Reduce loadings in freezing weather	Storage often needed
Wastewater loss due to	Evaporation and percolation		Percolation	Surface runoff and evaporation with some percolation
Expected treatment performance				
BOD and SS removal	98+%		85 to 99%	92+%
Nitrogen removal	85+%[d]		0 to 50%	70 to 90%
Phosphorus removal	80 to 99%		60 to 95%	40 to 80%

a *Costs of Wastewater Treatment by Land Application,* C. E. Pound et al., US Environmental Protection Agency. EPA No. 430-9-75-003, June 1975.
b in/wk × 2.54 = cm/wk.
c ft/yr × 0.3 = m/yr.
d Dependent on crop uptake.
e acres × 0.00107 = hectares for 1 cu m/day
   or
   acres × 9.24 = hectares for 1 cu m/sec.

2. Determine the hydraulic loading in million gallons per day per acre.

$$\text{Hydraulic Loading, MGD/ac} = \frac{\text{Flow, MGD}}{\text{Area, acres}}$$

$$= \frac{0.4 \text{ MGD}}{18.4 \text{ acres}}$$

$$= 0.022 \text{ MGD/acre}$$

3. Determine the hydraulic loading in inches per day.

$$\text{Hydraulic Loading, in/day} = \frac{\text{Flow, MGD} \times 1,000,000/\text{M} \times 12 \text{ in/ft}}{\text{Length, ft} \times \text{Width, ft} \times 7.48 \text{ gal/cu ft}}$$

$$= \frac{0.4 \text{ MGD} \times 1,000,000/\text{M} \times 12 \text{ in/ft}}{1,000 \text{ ft} \times 800 \text{ ft} \times 7.48 \text{ gal/cu ft}}$$

$$= 0.8 \text{ in/day}$$

13.200   Equipment Requirements

Irrigation systems apply water by sprinkling or by surface spreading (Figure 13.14). Sprinkler systems may be fixed or movable. Fixed systems are permanently installed on the ground or buried with sprinklers set on risers that are spaced along pipelines. These systems have been used in all types of terrain. Movable systems include center pivot, side wheel roll, and traveling gun sprinklers.

Surface irrigation systems include flooding, border-check, and ridge and furrow systems. Flooding systems are very similar to overland flow systems except the slopes are nearly level. Border-check systems are simply a controlled flooding system. Ridge and furrow systems are used for row crops such as corn where the water flows through furrows between the rows and seeps into the root zone of the crop.

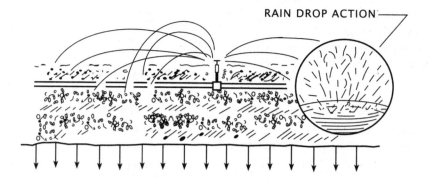

(a) SPRINKLER

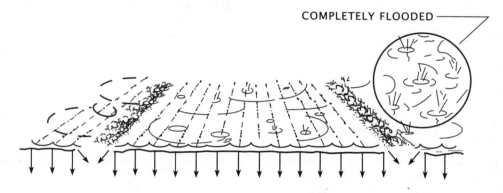

(b) FLOODING

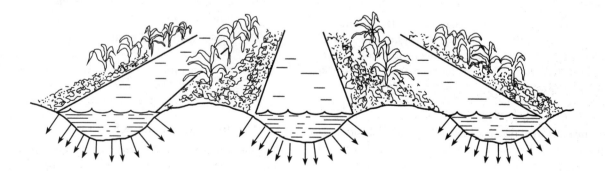

(c) RIDGE AND FURROW

Fig. 13.14   Irrigation techniques

(*Evaluation of Land Application Systems*, C. E. Pound et al.,
US Environmental Protection Agency. EPA No. 430-9-74-015.)

An overland flow system consists of effluent being sprayed or diverted over sloping terraces where it flows downslope through the vegetation. The vegetation provides a filtering action, removing suspended solids and insoluble BOD. This system can be operated as a treatment system or, with a recycle system included, as a discharge system. When operated as a discharge system, the overland flow process is nearly the same as the flood irrigation process, so it will be discussed with the flood systems in the following sections.

13.201   Sidestreams and Their Treatment

There are two possible sidestreams with land disposal systems: percolation from storage reservoirs and runoff from irrigated fields. Unlined storage reservoirs will result in wastewater percolating down to groundwater. If the water stored in reservoirs is the final effluent from a treatment plant, percolation down to the groundwater probably is acceptable. However, if the water is untreated or partially treated (primary effluent), the reservoir should be lined or an underground collection system should be installed to collect any percolation or seepage from the reservoir.

In some areas, percolation may cause a rise in the area's groundwater level. To prevent groundwater problems, a seepage ditch may be built around the outside of the reservoir. This ditch should be positioned somewhat lower than the bottom of the reservoir. Wastewater that percolates through the reservoir bottom is collected in the ditches and can be pumped back into the reservoir. The groundwater table also can be lowered by a series of shallow wells with water being pumped out as necessary. Lowering of the groundwater can result in increased percolation rates.

Wastewater that seeps out of storage reservoirs or runs off from a land treatment system may be pumped back to a storage reservoir, recycled through the land treatment system again, or released to surface waters, depending on local conditions.

The other major sidestream is surface runoff. Runoff quantities vary depending on the type of irrigation system used. In all systems, provisions should be made for collecting runoff and returning it for reapplication. In some locations, runoff water can be discharged to surface water. Discharge is the preferred approach due to cost savings and minimizing operational problems. For information on discharge to surface waters, see *Operation of Wastewater Treatment Plants,* Volume II, Chapter 13, "Effluent Discharge, Reclamation, and Reuse," in this series of operator training manuals.

13.202   Limitations of Land Treatment

Limitations or problems encountered when using land treatment systems usually involve soil problems and weather conditions. If proper care is not taken, soils can lose their ability to percolate applied water. During the cold winter season, plants and crops will not grow. Under these conditions, no water will be treated by transpiration processes. Also, precipitation can soak the soil to the point that no wastewater can be treated;

therefore, provisions must be made to store wastewater during cold and wet weather.

One of the most common land treatment problems is the sealing of the soil by suspended solids in the final effluent. These solids are deposited on the surface of the soil and form a mat that prevents the percolation of water down through the soil. There are three possible solutions to this problem:

1. Remove the suspended solids from the effluent.
2. Apply water intermittently and allow a long enough drying period for the solids mat to dry and crack.
3. Disk or plow the field to break the mat of solids.

Another serious problem associated with land treatment is the buildup of salts in the soil (salinity). If the effluent has a high chloride content, within one year enough salts can build up in the soil to create a toxic condition for most grasses and plants. To overcome salinity problems, leach out the salts by applying fresh water (not effluent), or till the field and turn it over to a depth of 4 to 5 feet (1.2 to 1.5 meters).

The severity of soil sealing due to suspended solids and salinity problems due to dissolved solids depends on the type of soil in the discharge area. These problems are more common and more difficult to correct in clay soils than in sandy soils.

QUESTIONS

Please write your answers to the following questions and compare them with those on pages 248 and 249.

13.2A   What are the three methods by which land disposal of wastewater is accomplished?

13.2B   What are the major components of a land application system?

13.2C   What are the definitions of the terms "evapotranspiration" and "hydrologic cycle"?

13.2D   A plot of land 2,000 feet long by 1,000 feet wide is used for a land disposal system. If one million gallons of water is applied to the land during a 24-hour period, calculate the land area in acres and the hydraulic loading in MGD per acre and also in inches per day.

13.21   Operating Procedures for Crops

The operating procedures described in this section apply to a spray irrigation system for an area where crops are grown. The procedures for this type of system are more complex than for other systems.

13.210   Pre-Start Checklist

Table 13.3 lists items that should be checked before starting a land irrigation system. The table includes a checklist for aluminum pipe. A similar checklist can be developed for other pipe materials. Using a combination of inspections and corrective maintenance can save considerable time and money.

TABLE 13.3 SPRAY IRRIGATION PRE-START CHECKLIST [a]

Check Every Time Pump Is Started	Check at the Beginning of the Season	
		Electric Motors
	_____	Replace winter lubricant.
	_____	Oil bath bearings. Drain oil and replace with proper weight of clean oil.
	_____	Grease lube bearings. If grease gun is used, be sure old grease is purged through outlet hole.
_____		Before turning on switch, have power company check voltage.
	_____	Check for proper rotation of motor and pump.
_____	_____	Check fuses to make sure they are still working.
_____	_____	Check electrical contact points for excessive corrosion.
_____	_____	Physically inspect for rodent and insect invasion.
		Pumps
	_____	Replace oil or grease with proper weight-bearing lubricant.
_____	_____	Tighten packing gland to proper setting.
_____	_____	Check discharge head, discharge check valve, and suction screen thoroughly for foreign matter.
_____		Pump shaft should turn freely without noticeable dragging.
		Aluminum Pipe
	_____	If you did not properly handle and store aluminum pipe or tubing last fall, make a mental note to do that at the end of this growing season.
		Always carefully drain aluminum tubing or pipe when you are finished using it. Aluminum pipe bends very easily. If you pick up a length of pipe that is half-full of water and has one end plugged, the pipe will bend. Flush out the pipe to clean it, drain the pipe, place the pipe on a long-bed trailer for transport, and then store the pipe on racks until next season.
	_____	Inspect pipe ends to make certain that no damage has occurred. Ends should be round for best operation. A slightly tapered wooden plug of the proper diameter can be used to round out the ends. The diameter of aluminum pipe varies from 2 to 12 inches (50 to 300 mm).

Check Every Time Pump Is Started	Check at the Beginning of the Season	
		Aluminum Pipe (continued)
	_____	Check pipe for pit corrosion. If the pipe has corrosion spots, contact the aluminum pipe supplier for advice.
	_____	Inspect pipe gaskets, couplers (irrigation pipe couplings), and gates to find those in need of replacement.
_____	_____	Check pipeline to see that all couplers are still fastened, pipe supports have not fallen over, and gates and valves are still open.
_____		Pipe makes an excellent nesting area for small animals. Flush out the pipeline before installing the end plug. Make sure you are not near overhead power lines before you raise the aluminum pipe to drain water from it.
		Sprinkler Systems
	_____	Sprinkler bearing washers should be replaced if there is indication of wear.
_____	_____	Visually check all moving parts, seals, bearings, and flexible hose for replacement or repair.
_____	_____	Check to see that hose is laid out straight or on a long radius for turns. Be sure there are no kinks in the hose. There should be sufficient hose at the end of the sprinkler to act as a brake or to hold back the sprinkler system initially as it drags the hose through the field. Also check earth anchors.
_____	_____	If possible, operate the system to check speed adjustment, alignment, and safety switch mechanisms.
_____	_____	Check sprinkler oscillating arm for proper adjustment. If damage has occurred to the sprinkler oscillating arm, the arm should be replaced or bent back to the correct angle. Your dealer can help in correcting a damaged arm. The angle of water-contact surface, if not correct, will change the turning characteristics of the sprinkler. Excessive wear of sprinkler nozzles can be checked with proper size drill bit.

a From *Wastewater Resources Manual,* Edward Norum, ed., by permission of The Irrigation Association.

TABLE 13.4   GUIDE FOR DETERMINING SOIL MOISTURE BY FEEL AND APPEARANCE [a]

This chart is very useful in estimating how much available moisture is in your soil. Although the plant's daily moisture use may range from 0.1 inch to 0.4 inch per day (0.25 to 1.0 cm), it will average about 0.20 inch (0.5 cm) per day, and 0.25 inch (0.6 cm) per day during hot days.

Moisture Condition	Percent of Available Moisture Remaining in Soil, %	Soil Texture		
		Sands–Sandy Loams	Loams–Silt Loams	Clay Loams–Clay
Dry	Wilting point	Dry, loose, flows through fingers.	Powdery, sometimes slightly crusted but easily broken down into powdery condition.	Hard, baked, cracked; difficult to break down into powdery condition.
Low	50% or less		Will form a weak ball when squeezed but will not stick to tools.	Pliable, but not slick; will ball under pressure—sticks to tools.
Time to Irrigate When Available Moisture is 50 Percent or Less				
Fair	50 to 75%	Tends to ball under pressure but seldom will hold together when bounced in the hand.	Forms a ball somewhat plastic, will stick slightly with pressure. Does not stick to tools.	Forms a ball, will ribbon out between thumb and forefinger, has a slick feeling.
Good	75 to 100%	Forms a weak ball, breaks easily when bounced in the hand, can feel moistness in soil.	Forms a ball, very pliable, sticks readily, clings slightly to tools.	Easily ribbons out between thumb and forefinger, has a slick feeling, very sticky.
Ideal	Field capacity 100%	Soil mass will cling together. Upon squeezing, outline of ball is left on hand.	Wet outline of ball is left on hand when soil is squeezed. Sticks to tools.	Wet outline of ball is left on hand when soil is squeezed. Sticky enough to cling to fingers.

a   From *Wastewater Resources Manual,* Edward Norum, ed., by permission of The Irrigation Association.

13.211   Start-Up

The procedures in this section are prepared to help you determine when you should irrigate and how much water should be applied. These procedures are based on soil conditions and may require adjustment based on the type of crop being irrigated. Complete the following steps as part of the start-up procedures for a spray irrigation system:

1. Determine the need to irrigate. The amount of water that should be applied depends on the type of crop. Some crops require a lot of water while other crops will not tolerate any excess water. Table 13.4 will aid in determining the need to irrigate based on soil conditions. To use this table you must first determine the type of soil you are irrigating. Walk around the field and identify the different soil types if more than one type of soil is present. Pick up a handful of soil and examine the grains or particles. Very small, hard particles indicate a sandy soil. Very fine, soft, smooth particles indicate a clay soil.

   Once you have identified the type of soil, the *AVAILABLE MOISTURE CONTENT* [10] can be estimated by squeezing the soil together in your hand. The wetter the soil, the more likely the soil will stick or cling together. Based on the way the soil sticks together and having identified the type of soil, you can estimate when the available moisture content drops to 50 percent or less. Irrigate when the moisture content is 50 percent or less.

   Soil types vary throughout an area of land and some areas need irrigating before others. Many farmers have specific areas in their fields that they call "hot spots." Due to soil characteristics or other reasons, these areas dry out faster than the rest of the field. Whenever the soil will not form a ball when squeezed together or crops start to show signs of stress due to lack of water, this is the time to irrigate.

2. Determine the amount of moisture to apply using Table 13.5. To use this table, determine the soil type by examining a handful of soil as described in Step 1. The root zone depth is determined by the type of crop. If necessary, dig down to determine how far down the roots are growing. Using Table 13.4, you can estimate the percent of available moisture remaining in the soil before irrigation. Knowing the percent available moisture in the soil before irrigation, you can use Table 13.5 to determine the net inches of water to apply when you irrigate.

3. Use Figure 13.15 or 13.16 to determine the time required to apply one inch (2.5 cm) of water. Figure 13.15 is used for small areas (up to 60 acres) and Figure 13.16 is used for larger areas (over 60 acres). To use the tables, determine (1) the area of land you wish to irrigate and (2) the capacity of

10.   *Available Moisture Content.*   The quantity of water present that is available for use by vegetation.

TABLE 13.5   AMOUNT OF MOISTURE TO APPLY TO VARIOUS SOILS UNDER DIFFERENT MOISTURE RETENTION CONDITIONS[a]

Soil Type	Root Zone Depth	Available Moisture Plants Will Use	Net Inches to Apply Per Irrigation with Various Percents Available Moisture Retained in the Soil Before Irrigation		
			Percent Available Moisture Before Irrigation		
	Feet	Inches	67%	50%	33%
Light Sandy	1	1.00	0.33	0.50	0.67
	1½	1.50	0.50	0.75	1.00
	2	2.00	0.56	1.00	1.33
	2½	2.50	0.83	1.25	1.67
	3	3.00	0.99	1.50	2.00
Medium	1	1.69	0.57	0.85	1.13
	1½	2.53	0.84	1.26	1.70
	2	3.38	1.11	1.69	2.26
	2½	4.21	1.39	2.11	2.82
	3	5.06	1.67	2.53	3.38
Heavy	1	2.39	0.79	1.20	1.59
	1½	3.58	1.18	1.79	2.38
	2	4.78	1.58	2.39	3.25
	2½	5.97	1.97	2.98	3.97
	3	7.17	2.36	3.58	4.77

a From *WASTEWATER RESOURCES MANUAL*, Edward Norum, ed., by permission of The Irrigation Association.

your irrigation system in gallons per minute. By starting at the bottom of the figure with the known area, draw a line vertically upward. Next, draw a line from the system capacity on the left, horizontally to the right. Where these two lines intersect is the time required to apply one inch of water to the area being irrigated.

4. Determine the total run time for the irrigation system. After you have determined the net inches of water you wish to apply from Table 13.5 (Step 2), multiply the net inches by the time required (Figures 13.15 or 13.16) to determine the total irrigation time.

   If you need help determining soil types, irrigation requirements, soil moisture content, types of crops to plant, salt tolerance of plants, depth of root zone, and fertilizer needs, contact your local farm adviser. In many areas, an expert adviser is available free of charge through an agency of the federal, state, or local government.

5. Check the irrigation system's pump discharge check valve and suction screen for foreign matter.

6. Check pipelines to see that all couplings are still fastened, blocks or pipe supports have not fallen over, and gates and valves are open.

7. Start the pump.

8. Inspect the irrigation system to be sure everything is working properly.

EXAMPLE 2

The need to irrigate a plot of land has been determined. Using the procedures outlined above, calculate the total time it will take to irrigate 40 acres of land with a light sandy soil, a root zone depth of 3 feet, and 50 percent available moisture retained in the soil at irrigation. The pumping system capacity is 1,000 GPM.

Known		Unknown
Area, ac	= 40 acres	Total Time to
Soil Type	= Light Sandy	Irrigate, hr
Root Zone Depth, ft	= 3 ft	
Moisture Retention, %	= 50%	
Pump, GPM	= 1,000 GPM	

1. Determine the amount of water to be applied in inches. Table 13.5 indicates that a light sandy soil with a root zone depth of 3 feet and 50 percent available moisture requires 1.50 inches of water to be applied.

2. Determine the time it will take to apply one inch of water to 40 acres of land when the pumping system capacity is 1,000 GPM. Figure 13.15 shows that approximately18 hours are needed per inch.

3. Determine the total time to irrigate in hours.

$$\text{Total Time to Irrigate, hr} = \frac{\text{Time to Irrigate One Inch, hr/in}}{} \times \text{Amount to Apply, in}$$

$$= 18 \text{ hours/in} \times 1.50 \text{ in}$$

$$= 27 \text{ hours}$$

13.212   Normal Operation

The following steps are part of the normal operating procedure for the pump:

1. When irrigating a plot of land, run the pump or pumps for the amount of time determined in Step 4 of the start-up procedure (Section 13.211) and as shown in the example in the same section.

2. Turn the sprinkler irrigation pump off when water begins to pond on the fields.

13.213   Shutdown

This shutdown procedure applies to the end-of-season shutdown.

1. Drain all lines.

2. Plug open ends of pipelines.

3. Lubricate motors and pumps for winter.

4. Store small, movable materials and equipment.

5. Store aluminum tubing or piping. See Table 13.3 (page 235) for guidelines for handling aluminum pipe.

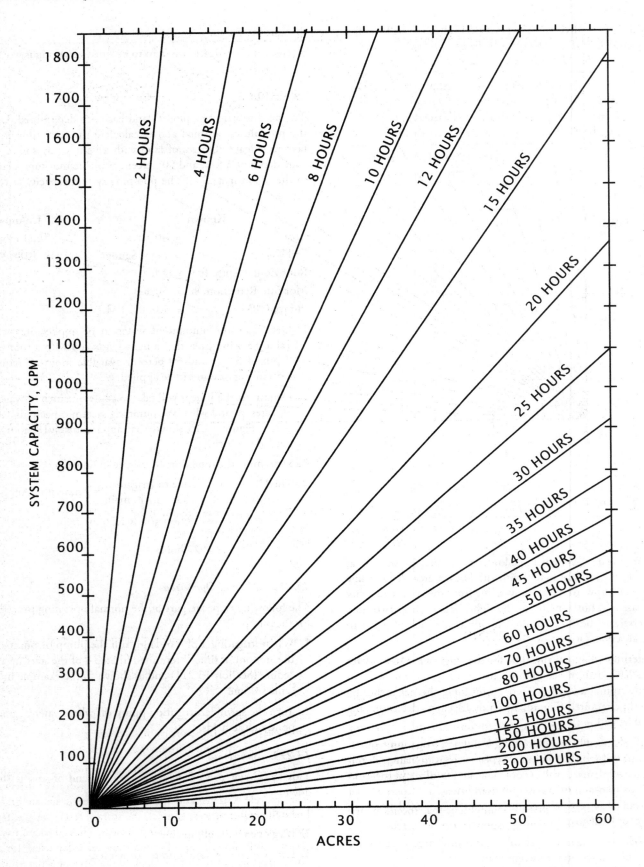

Fig. 13.15   Time required to apply one inch (2.5 cm) of water on small acreages
(From *Wastewater Resources Manual,* Edward Norum, ed.,
by permission of The Irrigation Association)

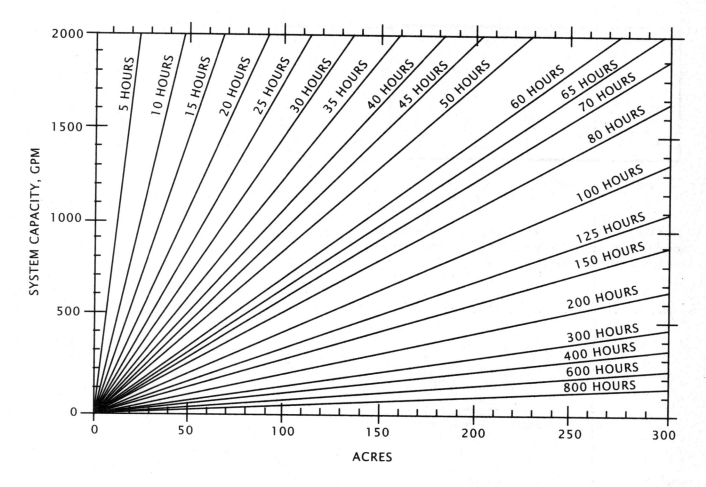

Fig. 13.16   Time required to apply one inch (2.5 cm) of water on large acreages
(From *Wastewater Resources Manual*, Edward Norum, ed.,
by permission of The Irrigation Association)

QUESTIONS

Please write your answers to the following questions and compare them with those on page 249.

13.2E   The need to irrigate a plot of land has been determined. Calculate the total time it will take to irrigate 30 acres of land with a medium soil type, a root zone depth of 2 feet, and 50 percent available moisture retained in the soil at irrigation. The pumping system capacity is 1,200 GPM. Use the figures and tables in this lesson to answer this question.

13.2F   List the major items of equipment that should be inspected before starting a spray irrigation system.

13.214   Operational Strategy

The operational strategy for a land disposal system is based on the physical control and process control of the system. The first strategy involves physical control. The main objective of a land disposal system is to discharge effluent without harming surface waters or creating nuisance conditions. An irrigation system can be designed and operated to produce a crop. The sale of this crop then can help reduce treatment costs. The purpose of physical control is to discharge effluent at the highest rate possible without damaging the crop. For all types of land discharge systems, physical controls consist of valves or gates that are used to direct treated effluent to different discharge areas.

The second strategy involves process control. There are three areas of process control: storage reservoirs, runoff and seepage water recycle systems, and soil systems where crops are grown. The first area of process control, the storage reservoir, is usually equipped with aeration or mixing devices. These devices may be operated full-time or for a limited time each day using timers. The correct length of time to operate the devices is determined by measuring the dissolved oxygen (DO) content at several points in the reservoir. A minimum of four points should be sampled. An example showing six sampling locations is shown in Figure 13.17.

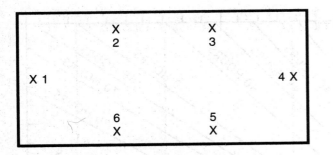

Fig. 13.17   Possible reservoir sampling locations

Each sample should have at least 0.4 mg/L of DO and the average of all samples should be at least 0.8 mg/L. Two sample sets of test results are shown below:

Set 1

Sample No.	mg/L DO
1	1.2
2	1.8
3	2.0
4	1.6
5	0.2
6	0.4

$$\text{Average} = \frac{7.2}{6} = 1.2 \text{ mg/L}$$

Set 2

Sample No.	mg/L DO
1	1.0
2	0.4
3	0.8
4	1.4
5	0.6
6	1.2

$$\text{Average} = \frac{5.4}{6} = 0.90 \text{ mg/L}$$

Set 1 does not meet the requirements. The average DO is 1.2 mg/L which is good, but one sample (#5) is 0.2 mg/L, which does not meet the minimum requirement. This result indicates that there is adequate aeration but either a portion of the system is not operating properly or the reservoir is not being adequately mixed.

Even though Set 2 has a lower average DO than Set 1, the average is greater than 0.8 and all samples are 0.4 mg/L or greater. Therefore, Set 2 is acceptable.

The second area of process control is the runoff and seepage water recycle systems. These recycle systems may not be necessary, depending on the particular application. For example, the storage reservoir may be lined so there would be no seepage water to recycle. Sprinkler systems that are carefully controlled will have no significant runoff. Your goal is to dispose of as much water as possible without causing runoff and seepage

from the disposal area. This is done to reduce recycle pumping costs. The reduction in seepage and runoff is accomplished by taking more care in applying effluent and turning off sprinklers or closing gates when water begins to stand in the field.

The third area of process control applies to those systems where crops are grown. In dry climates such as those found in the southwestern states and western mountainous states, farmers who irrigate are concerned with saline and alkaline soils. Some minerals found in the effluent may cause a decrease in crop production in soils of this type. Analyses and irrigation practices for these areas are described in detail in the US Department of Agriculture's Handbook No. 60, *Diagnosis and Improvement of Saline and Alkaline Soils.* A diagram for the classification of irrigation waters (taken from Handbook No. 60) is shown in Figure 13.18.

A simplified version of this diagram with other critical constituents (substances) is shown in Table 13.6.

TABLE 13.6   CLASSIFICATION OF IRRIGATION WATERS [a]

Chemical Properties	Class I Excellent to Good — Suitable under most conditions	Class II Good to Injurious — Suitability dependent on soil crop, climate, and other factors	Class III Injurious to Unsatisfactory — Unsuitable under most conditions
Total dissolved solids (mg/L)	Less than 700	700–2,000	More than 2,000
Chloride (mg/L)	Less than 175	175–350	More than 350
Sodium (percent of base constituents)	Less than 60	60–75	More than 75
Boron (mg/L)	Less than 0.5	0.5–2.0	More than 2.0

a   From *Diagnosis and Improvement of Saline and Alkaline Soils,* L. A. Richards, ed., Agricultural Handbook No. 60, US Department of Agriculture, Washington, DC.

The local US Department of Agriculture, Soil Conservation Service office can provide a list of crops that can be irrigated with Class II and Class III water.

There is not an economically feasible method to control the concentration of these constituents. If one of the constituents exceeds the Class I limit, the crop should be changed to one that is more tolerant. Special agricultural practices can be used to minimize the effects of these constituents. These practices vary from one local area to another. The local soil conservation service office and farm adviser can assist you by providing information appropriate to your area.

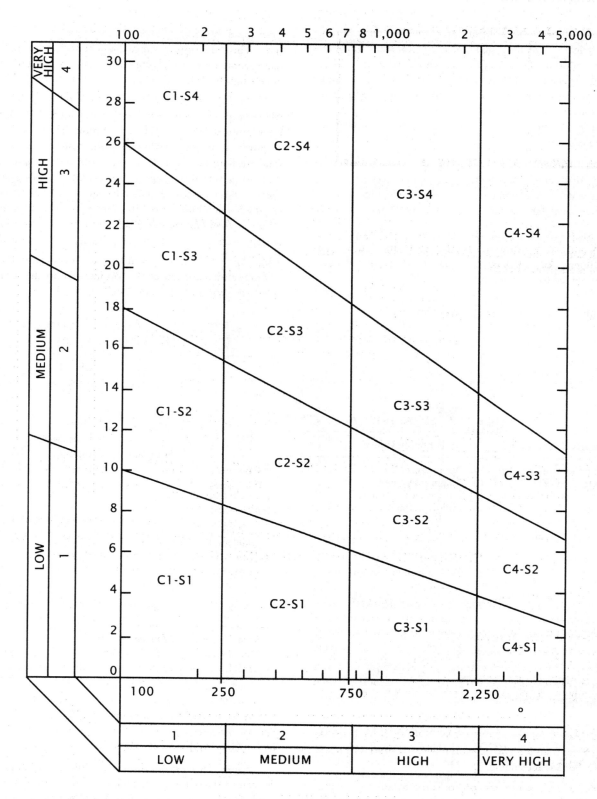

Fig. 13.18 Diagram for the classification of irrigation waters

(From Diagnosis and Improvement of Saline and Alkaline Soils, L. A. Richards, ed., Agricultural Handbook No. 60, US Department of Agriculture, Washington, DC)

Important observations and interpretations for determining when to irrigate were discussed previously (Section 13.211). The most important observations in a system where crops are grown and surface runoff is not allowed are observing ponding or runoff. When one of those occurs, either the application rate is excessive (too many gates open in segments of the pipe), the application rate is normal but greater than the soil infiltration rate, or there is a broken pipe in the system.

The visual appearance of the crop being grown is extremely important to observe. Discoloration in plant leaves can indicate excess water (poor drainage) or a nutrient or mineral deficiency. Your local farm adviser can assist in diagnosing such problems.

Observations are critical in storage reservoirs. Odors can result if effluent treatment was inadequate or if insufficient aeration was provided to the reservoir.

13.215 Emergency Operating Procedures

It is important to have an emergency operating procedure in place to respond to power outages. The loss of power will disrupt sprinkler systems (assuming electric motors are used for pumps). If power is lost, the effluent will be retained in the storage reservoir. Gravity-flow flood irrigation systems will not be affected during power outages.

The loss of other treatment units for a few days due to a power outage is generally not a problem. Longer downtimes may result in an overloaded and an odorous storage reservoir and possible odors at the disposal area.

13.216 Troubleshooting Guide

If your land disposal system is not performing well, use Table 13.7, "Spray Irrigation System Troubleshooting Guide," to identify the cause and select an appropriate solution.

TABLE 13.7   SPRAY IRRIGATION SYSTEM TROUBLESHOOTING GUIDE

Indicator/Observation	Probable Cause	Check or Monitor	Solution
1. Water ponding in irrigated area where ponding normally has not been observed.	1a. Application rate is excessive.	1a. Application rate.	1a. Reduce rate to normal value.
	1b. If application rate is normal but greater than the soil infiltration rate.	1b. (1) Seasonal variation in groundwater level.	1b. (1) Irrigate portions of the site where groundwater is not a problem or store wastewater until level has dropped.
		(2) Operability of any drainage wells.	(2) Repair drainage wells or increase pumping rate.
		(3) Condition of drain tiles.	(3) Repair drain tiles.
	1c. Broken pipe in distribution system.	1c. Leaks in system.	1c. Repair pipe.
2. Lateral aluminum distribution piping deteriorating.	2a. Effluent permitted to remain in aluminum pipe too long causing electrochemical corrosion.	2a. Operating techniques.	2a. Drain aluminum lateral lines except when in use.
	2b. Dissimilar metals (steel valves and aluminum pipe).	2b. Pipe and valve specifications.	2b. Coat steel valves or install cathodic or anodic protection.
3. No flow from some sprinkler nozzles.	3. Nozzle clogged with particles from wastewater due to lack of screening at inlet side of irrigation pumps.	3. Screen may have developed hole due to partial plugging of screen.	3. Repair or replace screen.
4. Wastewater is running off of irrigated area.	4a. Sodium adsorption ratio (SAR) of wastewater is too high, causing the clay soil to become impermeable.	4a. SAR should be less than 9.	4a. Feed calcium and magnesium to adjust SAR.
	4b. Soil surface sealed by solids.	4b. Soil surface.	4b. Strip crop area.
	4c. Application rate exceeds infiltration rate of soil.	4c. Application rate.	4c. Reduce application rate until compatible with infiltration rate.
	4d. Break in distribution piping.	4d. Leaks in distribution piping.	4d. Repair breaks.
	4e. Soil permeability has decreased due to continuous application of wastewater.	4e. Duration of continuous operation on the given area.	4e. Each area should be allowed to rest (2–3 days) between applications of wastewater to allow soil to drain.
	4f. Rain has saturated soil.	4f. Rainfall records.	4f. Store wastewater until soil has drained.

TABLE 13.7   SPRAY IRRIGATION SYSTEM TROUBLESHOOTING GUIDE (*continued*)

Indicator/Observation	Probable Cause	Check or Monitor	Solution
5. Irrigated crop is dead.	5a. Too much (or not enough) water has been applied.	5a. Water needs of specific crop versus application rate.	5a. Reduce (or increase) application rate.
	5b. Wastewater contains excessive amount of toxic elements.	5b. Analyze wastewater and consult with county agricultural agent.	5b. Eliminate industrial discharges of toxic materials.
	5c. Too much insecticide or weed killer applied.	5c. Application of insecticide or weed killer.	5c. Proper control of application of insecticide or weed killer.
	5d. Inadequate drainage has flooded root zone of crop.	5d. Water ponding.	5d. See Item 1.
6. Growth of irrigated crop is poor.	6a. Too little nitrogen (N) or phosphorus (P) applied.	6a. N and P quantities applied— check with county agricultural agent.	6a. If increased wastewater application rates are not practical, supplement wastewater N or P with commercial fertilizer.
	6b. Timing of nutrient application not consistent with crop needs. Also, see 5a–5c.	6b. Consult with county agricultural agent.	6b. Adjust application schedule to meet crop needs.
7. Irrigation pumping station shows normal pressure but above-normal flow.	7a. Broken main, lateral, riser, or gasket.	7a. Inspect distribution system for leaks.	7a. Repair leak.
	7b. Missing sprinkler head or end plug.	7b. Inspect distribution system for leaks.	7b. Repair leak.
	7c. Too many laterals on at one time.	7c. Number of laterals in service.	7c. Make appropriate valving changes.
8. Irrigation pumping station shows above-average pressure but below-average flow.	8. Blockage in distribution system due to plugging sprinklers, valves, screens, or frozen water.		8. Locate blockage and eliminate.
9. Irrigation pumping station shows below-normal flow and pressure.	9a. Pump impeller is worn.	9a. Pump impeller.	9a. Replace impeller. See Section 13.28, "Review of Plans and Specifications," No. 6.
	9b. Partially clogged inlet screen.	9b. Screen.	9b. Clean screen.
10. Excessive erosion occurring.	10a. Excessive application rates.	10a. Application rate.	10a. Reduce application rate.
	10b. Inadequate crop cover.	10b. Condition of crop cover.	10b. See Items 5 and 6.
11. Odor complaints.	11a. Wastewater turning septic during transmission to irrigated site, and odors are being released as it is discharged to pretreatment.	11a. Sample wastewater as it leaves transmission system.	11a. Contain and treat off-gases from discharge point of transmission system by covering inlet with building and passing off-gases through deodorizing system.
	11b. Odors from storage reservoirs.	11b. DO in storage reservoirs.	11b. Improve pretreatment or aerate reservoirs.
12. Center pivot irrigation rigs stuck in mud.	12a. Excessive application rates.		12a. Reduce application rates.
	12b. Improper tires or rigs.		12b. Install tires with higher flotation capabilities.
	12c. Poor drainage.		12c. Improve drainage. See Item 1b.
13. Nitrate concentration of groundwater in vicinity of irrigation site is increasing.	13a. Application of nitrogen is not in balance with crop needs.	13a. Check lbs/acre/yr of nitrogen being applied with needs of crops.	13a. Change crop to one with higher nitrogen needs.
	13b. Nitrogen being applied during periods when crops are dormant.	13b. Application schedules.	13b. Apply wastewater only during periods of active crop growth.
	13c. Crop is not being harvested and removed.	13c. Farming management.	13c. Harvest and remove crop.

QUESTIONS

Please write your answers to the following questions and compare them with those on page 249.

13.2G   What is the main objective of a land disposal system?

13.2H   List the three main areas of process control in a land disposal system.

13.2I   How many points in a storage reservoir should be sampled for DO?

13.2J   What are the minimum recommended DO requirements for a storage reservoir?

13.2K   What are the probable causes of water ponding in an irrigated area where ponding normally has not been observed?

13.22   Monitoring Program for an Irrigation System

13.220   Monitoring Frequency

The four areas requiring monitoring in an irrigation system in which crops are grown are effluent, vegetation, soils, and groundwater (or collected seepage). The monitoring areas are reduced to the two areas of effluent and groundwater for systems in which crops are not grown. Wells located near a land disposal irrigation system should be monitored to identify any adverse effects on groundwater. Testing requirements and monitoring frequencies are shown in Table 13.8.

TABLE 13.8   TESTING REQUIREMENTS AND MONITORING FREQUENCY

Area	Test	Frequency
Effluent and groundwater or seepage	BOD	Two times per week
	Fecal coliform	Weekly
	Total coliform	Weekly
	Flow	Continuous
	Nitrogen	Weekly
	Phosphorus	Weekly
	Suspended solids	Two times per week
	pH	Daily
	Total dissolved solids (TDS)	Monthly
	Boron	Monthly
	Chloride	Monthly
Vegetation	—Variable depending on crop—	
Soils	Conductivity	Two times per month
	pH	Two times per month
	Cation exchange capacity[a]	Two times per month

a   Cation (KAT-EYE-en) Exchange Capacity.   The ability of a soil or other solid to exchange cations (positive ions such as calcium, Ca^{2+}) with a liquid.

The water quality indicators that should be monitored related to groundwater supply include salinity, chemical buildup (especially nitrate), toxicity, and organics (Table 13.9).

TABLE 13.9   SAMPLING WELL TESTS AND MONITORING FREQUENCY

Area	Test	Frequency
Wells	Salinity	
	Conductivity	Monthly
	Chloride	Quarterly
	TDS	Quarterly
	Chemical Buildup	
	Nitrate	Monthly
	Calcium	Semiannually
	Magnesium	Semiannually
	Toxicity (Heavy Metals)	
	Cadmium	Monthly
	Lead	Annually
	Zinc	Annually
	Mercury	Annually
	Molybdenum	Annually
	Selenium	Annually
	Organics	
	Trihalomethanes	Quarterly
	Pesticides (depends on local application)	Quarterly

Sampling wells should be located within the irrigation site as well as on all sides of the site to identify any changes or trends in water quality. Typical sampling well tests and monitoring frequencies are listed in Table 13.9.

13.221   Interpretation of Test Results and Follow-Up Actions

Excessive levels and concentrations greater than desired for effluent BOD, fecal and total coliforms, nitrogen, phosphorus, and suspended solids typically are not a concern for crop-growing operations. The total dissolved solids (TDS), boron, chloride, and pH are important during long periods of land treatment, but not for treatment times of less than two to three weeks. Excessive nitrogen is a potential problem in spreading basins as nitrate in water supplies can be harmful to infants. If TDS, boron, or chloride levels increase and do not return to previous levels, a change in farming practices may be necessary.

Increased levels in any of the constituents in the groundwater are unacceptable. Most likely, the only constituent that will increase is nitrate–nitrogen. If this occurs, then a nitrogen removal system (partial or complete) should be added to the treatment plant.

13.23   Safety

Safe operating procedures should be practiced at all times. The major cause of accidents and fatalities to operators while working with sprinkler irrigation systems is contact with electricity used either to power the pumping plant or to transmit electricity associated with the area being irrigated.

Moving portable sprinkler lateral pipelines is an extremely hazardous activity. Raising a pipeline into the air to dislodge a small animal or debris and contacting overhead electrical transmission lines has resulted in severe electric shock or death to the person holding the pipe.

A sprinkler throwing a stream of water into a power line has shorted the power to ground through the sprinkler system and resulted in severe electrical injuries to anyone touching the sprinkler system parts.

Always have the electric motor well bonded to a good ground with suitably sized conductors. Injuries have occurred from touching an ungrounded motor or pump frame having shorted electrical windings in electrically powered pumping plants.

Electric shocks have occurred from faulty starting equipment and from working on energized circuits. Always pull the line disconnect switch; lock out and tag it when making repairs or checks on electrical equipment of any kind.

Surface spreading systems can be hazardous due to wet surfaces and muddy areas. Look over each sprinkler system, mark the potential safety hazards, and then avoid the hazards.

13.24   Operation and Maintenance

Maintenance of land treatment systems requires operators to keep the wastewater distribution piping, pumps, valves, and sprinklers in good working condition. Pump and valve maintenance is discussed in *Operation of Wastewater Treatment Plants*, Volume II, Chapter 15, "Maintenance." Storage reservoir maintenance is similar to pond maintenance as outlined in Chapter 9 of this manual and in *Operation of Wastewater Treatment Plants*, Volume I, Chapter 9.

The timing and application rate of treated wastewater for land treatment must be scheduled so the plants benefit as much as possible from the nutrients and other constituents in the wastewater without being overwhelmed by them. In addition, there is the potential that certain wastewater constituents may accumulate in the soil and plants over time, becoming toxic to the plants, clogging the soil, or altering the soil structure. For example, too much nitrogen can result in nitrate accumulation in crops, but too little can result in reduced yields. If evaporation regularly exceeds precipitation, too much salt may remain in the soil, damaging roots. In general, rapid infiltration of the applied water will not increase salts in the soil or the groundwater.

The particular characteristics of the wastewater must be considered in relation to such factors as climate and the individual nutrient requirements of the crops, grass, and landscape plants selected. In addition, the need to dispose of the wastewater has to be balanced with the needs of the plants during the various stages of growth and the hydraulic capacity of the soil and its ability to effectively provide treatment.

Proper operation of a land treatment system requires management of the applied wastewater, the crop, and the soil profile. Applied wastewater must be rotated around the site through the application cycle, allowing time for drying, maintenance, cultivation, and crop harvest. The soil profile must be managed to maintain infiltration rates, avoid soil compaction, and maintain soil chemical balance.

Compaction and surface sealing can reduce soil infiltration and runoff. The compaction, solids accumulation, and crusting

of surface soils may be broken up by cultivation, plowing, or disking when the soil surface is dry. A check of the soil chemical balance is required periodically to determine if the soil pH and percent exchangeable sodium are in the acceptable range. Soil pH can be adjusted by adding lime (to increase pH) or gypsum (to decrease pH). Exchangeable sodium can be reduced by adding sulfur or gypsum followed by leaching to remove the displaced sodium.

The spray headers in the spray irrigation system require regular maintenance. Open pipes and spray headers can become damaged, plugged, or frozen. Any changes in pressure in the system can alter spray patterns in the field. Therefore, spray patterns should be tested to ensure that the system still complies with all setback requirements and also covers the entire area as evenly as possible. To test whether the entire area is being covered evenly, place some pans (like an oil drain pan) in various locations and, after a specific time period, measure the volume of water in each pan using a graduated cylinder or similar device.

Erosion control is an important maintenance activity. Inspect the area around the land treatment system to ensure that there are no bare spots that are eroding. Erosion is usually not a problem on slopes of less than two percent while erosion problems can be expected on slopes greater than six to eight percent. Also, erosion can be a problem if the spray headers do not distribute the spray uniformly over the area. If sprayed water concentrates in one area, high runoff flows can result and cause erosion. Land treatment systems that do not use spray irrigation must ensure that the water enters the area and is applied uniformly over the area or erosion can be a serious problem.

Use of native grasses in land treatment systems is recommended because native grasses can thrive under local variables such as soils, moisture, and seasons. Competition from non-native grasses and broadleaf weeds can reduce expected treatment levels. Weed control consists of pulling weeds or the use of selective herbicide spraying.

13.25 Monitoring for Land Treatment

Monitoring requirements vary depending on state and local regulations, public access to the site, and system size. In some systems, regular daily or weekly monitoring is needed to check influent and effluent quality, system storage capacity, wind speed and direction, signs of ponding or runoff in the spray field, and depth to water table. Cumulative levels of nutrients, heavy metals, fecal coliforms, and wastewater constituents must be monitored in the soils (and groundwater) at some sites once or twice per year.

13.26 Drip and Subsurface Drip Irrigation Systems

Drip irrigation systems are another type of land treatment. They are also known as "trickle systems" and are an efficient and proven technology used by many small communities to recycle and dispose of wastewater. Subsurface drip systems deliver treated effluent directly to plant roots.

Drip systems apply treated wastewater to soil slowly and uniformly from a network of narrow tubing, usually plastic or polyethylene, placed either on the ground surface or below the ground surface at shallow depths of 6 to 12 inches (15 to 30 centimeters) in the plant root zone. Wastewater is pumped through tubes under pressure and drips out slowly from a series of evenly spaced openings in the tubing.

Because subsurface drip systems release wastewater below ground directly to plant roots, they irrigate more efficiently and have advantages over surface irrigation systems. For example, the soil surface tends to stay dry, meaning less water is lost to evaporation and there is almost no opportunity for the wastewater to come in contact with plant foliage, humans, or animals. Also, percolation losses are reduced because the wastewater is applied directly to the plant roots over a wide area of soil at a slow rate. Drip systems also deliver wastewater to the most biologically active part of the soil, enhancing treatment and minimizing the possibility of groundwater contamination. The constant moisture in the root zone also may increase the availability of nutrients to plants, reducing the delivery of nitrogen to groundwater.

13.27 Drip System Operation, Maintenance, and Scheduling

As with spray systems, drip irrigation must be scheduled so the plants benefit from the nutrients and other constituents in the wastewater without being overwhelmed by them. The needs of the plants must be balanced with the capacity of the soil to treat the most restrictive components in the wastewater. These concerns must be balanced, in turn, with climate and other site factors. Less labor is usually required to operate and maintain fixed subsurface drip system components compared to spray systems and subsurface drip systems with movable components. For small and individual home systems, the pattern of flow may be fixed or adjusted manually or automatically by the homeowner or operator, depending on the system design and sophistication. In general, the best care for subsurface drip systems is provided by following the manufacturer's recommendations.

13.28 Review of Plans and Specifications

Many operational and maintenance problems can be avoided by a careful review of the plans and specifications for a land treatment system. Once familiar with the plans and specifications, examine or address the following items in your land system.

1. Ponding. Ponding problems can be avoided if the proper site is selected and provided with proper drainage. Soils at the site must be suitable for percolation and for planned crops. Adequate drainage that prevents ponding can be provided by leveling or sloping the land surface so the water will flow evenly over all of the land. *DRAINAGE WELLS*[11] or *DRAIN TILE SYSTEMS*[12] may be necessary to remove excess water and prevent ponding.

2. Plastic pipe laterals. Plastic pipe laterals installed above ground may break because of cold weather or deteriorate due to sunlight. Install plastic pipe laterals below ground.

3. Screens. Install screens on the inlet side of irrigation pumps to prevent spray nozzles from becoming plugged.

4. Buffer area. Be sure a sufficient buffer area is provided around spray areas to prevent mist from drifting onto nearby homes and yards. If necessary, do not schedule spraying during days when the wind is blowing toward treatment system neighbors.

5. Potential for odor problems. If odors are a problem, consider furrow or flood irrigation rather than spraying. Spraying can cause odor problems by releasing odors to the atmosphere.

6. Protection of pumps. Excessive wear on pumps can result from sand in the water being pumped. If sand is a problem, improve pretreatment or install a sand trap ahead of the pumps. Remember to drain out-of-service pumps before freezing weather occurs in the fall or winter.

7. Alternate effluent disposal site. Ensure that an alternate location is provided to pump or dispose of effluent in case of system failure. This pre-planning allows operators to respond quickly.

8. Use specially designed effluent spray irrigation heads that will not plug like conventional sprinkler heads will under these conditions.

---

11. *Drainage Wells.* Wells that can be pumped to lower the groundwater table and prevent ponding.
12. *Drain Tile System.* A system of tile pipes buried under agricultural fields that collect percolated waters and keep the groundwater table below the ground surface to prevent ponding.

13.29   References and Additional Reading

13.290   References

1. *Costs of Wastewater Treatment by Land Application,* C. E. Pound et al., US Environmental Protection Agency. EPA No. 430-9-75-003.

2. *Evaluation of Land Application Systems,* C. E. Pound et al., US Environmental Protection Agency. EPA No. 430-9-74-015.

3. *Wastewater Resources Manual,* Edward Norum, ed., by permission of The Irrigation Association.

4. *Diagnosis and Improvement of Saline and Alkaline Soils,* L. A. Richards, ed., Agricultural Handbook No. 60, US Department of Agriculture.

13.291   Additional Reading

1. *Wastewater Reclamation and Reuse,* Water Quality Management Library, Volume 10, edited by Takashi Asano. Obtain from CRC Press, Attn: Order Entry, 6000 Broken Sound Parkway NW, Suite 300, Boca Raton, FL 33487. ISBN 978-1-56676-620-3. Price, $279.95.

2. *Manual of Wastewater Treatment,* "Texas Manual," Chapters 3 and 22,* published by Texas Water Utilities Association. No longer in print.

3. *Western Fertilizer Handbook,* Ninth Edition, Western Plant Health Association. Published by Waveland Press, Inc. ISBN 978-1-57766-679-0.

4. *Land Treatment Systems for Municipal and Industrial Wastes,* Ronald W. Crites, Sherwood C. Reed, and Robert K. Bastian. Published by McGraw-Hill. ISBN 978-0-07-061040-8.

*Depends on edition.

QUESTIONS

Please write your answers to the following questions and compare them with those on page 249.

13.2L   What are the four monitoring areas for an irrigation system where crops are grown?

13.2M   What is the major cause of accidents to operators while working with sprinkler irrigation systems?

13.2N   What equipment needs to be maintained in a land treatment system?

13.2O   List the items that should be examined when reviewing plans and specifications for a land disposal system.

END OF LESSON 2 OF 2 LESSONS
on
ALTERNATIVE WASTEWATER TREATMENT, DISCHARGE, AND REUSE METHODS

Please answer the discussion and review questions next.

DISCUSSION AND REVIEW QUESTIONS
Chapter 13.   ALTERNATIVE WASTEWATER TREATMENT, DISCHARGE, AND REUSE METHODS

(Lesson 2 of 2 Lessons)

Please write your answers to the following questions to determine how well you understand the material in the lesson. The question numbering continues from Lesson 1.

5. How does land treatment work?

6. What should be done with wastewater that seeps out of storage reservoirs and runs off from a land treatment system?

7. Describe the limitations of land treatment systems.

8. How long should the pumps be run while irrigating a plot of land?

9. What water quality indicators should be monitored to ensure that a land disposal system does not adversely affect a groundwater supply?

10. Why should safe operating procedures be practiced at all times when working in a sprinkler irrigation system?

SUGGESTED ANSWERS
Chapter 13.   ALTERNATIVE WASTEWATER TREATMENT, DISCHARGE, AND REUSE METHODS

ANSWERS TO QUESTIONS IN LESSON 1

Answers to questions on page 225.

13.1A   The three types of wetlands are (1) free water surface (FWS), (2) horizontal subsurface flow (HSSF), and (3) vertical flow (VF).

13.1B   Free water surface (FWS) wetland systems are not recommended for individual home sites because of the associated problems of vector and disease control.

13.1C   An advantage of wetlands is that they can be built where other wastewater disposal systems such as leach fields, seepage pits, and mounds have failed.

Answers to questions on page 229.

13.1D   Water level and flow control are usually the only operational variables that have a significant impact on a well-designed constructed wetland.

13.1E   Inlet and outlet maintenance includes cleaning of debris that may clog the inlets and outlets. Debris removal and removal of bacterial slimes from weir and screen surfaces are necessary.

13.1F   Harvesting vegetation from wetlands at the height of and just before the end of the growing season helps remove some nitrogen from the system.

13.1G   Wetland vegetation should be periodically checked for signs of disease or other stress such as yellowing, spots, withering, browning, and insects.

13.1H   Wetland plant life should not create a habitat for unwanted animals or invertebrates, such as rats and mosquitoes, or for threatened or endangered species that could limit or eliminate the use of the treatment wetland.

13.1I   Standing water on the media surface of an HSSF wetland is the probable cause of odors.

ANSWERS TO QUESTIONS IN LESSON 2

Answers to questions on page 234.

13.2A   Land disposal of wastewater is accomplished by irrigation, infiltration–percolation, and overland flow

13.2B   The major components of land application systems include the following:

1. Treatment before application
2. Transmission to the land treatment site
3. Storage
4. Distribution over site
5. Runoff recovery system, if needed
6. Crop systems

13.2C   Evapotranspiration.   (1) The process by which water vapor is released to the atmosphere by living plants. This process is similar to people sweating. Also called transpiration. (2) The total water removed from an area by transpiration (plants) and by evaporation from soil, snow, and water surfaces.

Hydrologic cycle is the process of evaporation of water into the air and its return to earth by precipitation (rain or snow). This process also includes transpiration from plants, groundwater movement, and runoff into rivers, streams and the ocean. Also called the water cycle.

13.2D   A plot of land 2,000 feet long by 1,000 feet wide is used for a land disposal system. If one million gallons of water is applied to the land during a 24-hour period, calculate the land area in acres and the hydraulic loading in MGD per acre and also in inches per day.

Known	Unknown
Length, ft = 2,000 ft	1. Area, acres
Width, ft = 1,000 ft	2. Hydraulic Loading, MGD/ac
Flow, MGD = 1 MGD	3. Hydraulic Loading, in/day

1. Determine the surface area in acres.

$$\text{Area, acres} = \frac{\text{Length, ft} \times \text{Width, ft}}{43,560 \text{ sq ft/acre}}$$

$$= \frac{2,000 \text{ ft} \times 1,000 \text{ ft}}{43,560 \text{ sq ft/acre}}$$

$$= 45.9 \text{ acres}$$

2. Determine the hydraulic loading in million gallons per day per acre.

$$\text{Hydraulic Loading, MGD/ac} = \frac{\text{Flow, MGD}}{\text{Area, acres}}$$

$$= \frac{1 \text{ MGD}}{45.9 \text{ acres}}$$

$$= 0.02 \text{ MGD/ac}$$

3. Determine the hydraulic loading in inches per day.

$$\text{Hydraulic Loading, in/day} = \frac{\text{Flow, MGD} \times 1{,}000{,}000/M \times 12 \text{ in/ft}}{\text{Length, ft} \times \text{Width, ft} \times 7.48 \text{ gal/cu ft}}$$

$$= \frac{1 \text{ MGD} \times 1{,}000{,}000/M \times 12 \text{ in/ft}}{2{,}000 \text{ ft} \times 1{,}000 \times 7.48 \text{ gal/cu ft}}$$

$$= 0.8 \text{ in/day}$$

Answers to questions on page 239.

13.2E The need to irrigate a plot of land has been determined. Calculate the total time it will take to irrigate 30 acres of land with a medium soil type, a root zone depth of 2 feet, and 50 percent available moisture retained in the soil at irrigation. The pumping system capacity is 1,200 GPM.

Known		**Unknown**
Area, ac	= 30 acres	Total Time to
Soil Type	= Medium	Irrigate, hr
Root Zone Depth, ft	= 2 ft	
Moisture Retention, %	= 50%	
Pump, GPM	= 1,200 GPM	

1. Determine the amount of water to be applied in inches. Table 13.5 (page 237) indicates that a medium soil with a root zone depth of 2 feet and 50 percent available moisture requires 1.69 inches of water to be applied.

2. Determine the time it will take to apply one inch of water to 30 acres of land when the pumping system capacity is 1,200 GPM. Figure 13.15 (page 238) shows that approximately 11 hours are needed per inch.

3. Determine the total time to irrigate in hours.

$$\text{Total Time to Irrigate, hr} = \frac{\text{Time to Irrigate One Inch, hr/in}}{} \times \frac{\text{Amount to Apply, in}}{}$$

$$= 11 \text{ hours/in} \times 1.69 \text{ in}$$

$$= 18.6 \text{ hours}$$

13.2F The major items of equipment that should be inspected before starting a spray irrigation system include electric motors, pumps, aluminum tubing or piping, and sprinkler systems.

Answers to questions on page 244.

13.2G The main objective of a land disposal system is to discharge effluent without harming surface waters or creating nuisance conditions.

13.2H The three main areas of process control in a land disposal system are storage reservoirs, runoff and seepage water recycle systems, and soil systems where crops are grown.

13.2I A minimum of four points in a storage reservoir should be sampled for DO.

13.2J The minimum DO requirements for a storage reservoir are a minimum DO of 0.4 mg/L for all samples and the average of all samples should be at least 0.8 mg/L.

13.2K Probable causes of ponding include the following:

1. Application rate is excessive (too many gates open in the segments of pipe in the area)
2. If application rate is normal but greater than the soil infiltration rate
3. A broken pipe in the distribution system

Answers to questions on page 247.

13.2L The four monitoring areas for an irrigation system where crops are grown are effluent, vegetation, soils, and groundwater (or collected seepage).

13.2M The major cause of accidents to operators while working with sprinkler irrigation systems is contact with electricity used either to power the pumping plant or to transmit electricity associated with the area being irrigated.

13.2N Equipment requiring maintenance in a land treatment system includes distribution piping, pumps, valves, and sprinklers.

13.2O Items to be examined when reviewing the plans and specifications for a land disposal system include the following:

1. Ponding
2. Plastic pipe laterals
3. Screens
4. Buffer area
5. Potential for odor problems
6. Protection of pumps
7. Alternate effluent disposal

CHAPTER 14

LABORATORY PROCEDURES

by

John Brady

and

Ken Kerri

Much of this text is taken from Chapter 16, "Laboratory Procedures and Chemistry," *Operation of Wastewater Treatment Plants,* Volume II, by James Paterson, Joe Nagano, and James Sequeira.

TABLE OF CONTENTS
Chapter 14.   LABORATORY PROCEDURES

OBJECTIVES
Chapter 14.   LABORATORY PROCEDURES

1. Work safely in a laboratory.

2. Operate laboratory equipment.

3. Collect representative samples of influents to and effluents from a treatment process as well as sample the process.

4. Prepare samples for analysis in the plant lab.

5. Perform plant control tests.

6. Record laboratory test results.

7. Collect and prepare samples of plant effluent for analysis in accordance with NPDES permit requirements.

WORDS

Chapter 14.   LABORATORY PROCEDURES

>GREATER THAN

DO >5 mg/L would be read as DO GREATER THAN 5 mg/L.

<LESS THAN

DO <5 mg/L would be read as DO LESS THAN 5 mg/L.

ALIQUOT (AL-uh-kwot)

Representative portion of a sample. Often, an equally divided portion of a sample.

AMBIENT (AM-bee-ent) TEMPERATURE

Temperature of the surroundings.

AMPEROMETRIC (am-purr-o-MET-rick)

A method of measurement that records electric current flowing or generated, rather than recording voltage. Amperometric titration is a means of measuring concentrations of certain substances in water.

ANAEROBIC (AN-air-O-bick)

A condition in which atmospheric or dissolved oxygen (DO) is *NOT* present in the aquatic (water) environment.

ASEPTIC (a-SEP-tick)

Free from the living germs of disease, fermentation, or putrefaction. Sterile.

BIOASSAY (BUY-o-AS-say)

(1) A way of showing or measuring the effect of biological treatment on a particular substance or waste.

(2) A method of determining the relative toxicity of a test sample of industrial wastes or other wastes by using live test organisms, such as fish.

BIOMONITORING

A term used to describe methods of evaluating or measuring the effects of toxic substances in effluents on aquatic organisms in receiving waters. There are two types of biomonitoring, the BIOASSAY and the BIOSURVEY.

BIOSURVEY

A survey of the types and numbers of organisms naturally present in the receiving waters upstream and downstream from plant effluents. Comparisons are made between the aquatic organisms upstream and those organisms downstream of the discharge.

BLANK

A bottle containing only dilution water or distilled water; the sample being tested is not added. Tests are frequently run on a sample and a blank and the differences are compared. The procedure helps to eliminate or reduce test result errors that could be caused when the dilution water or distilled water used is contaminated.

BUFFER

A solution or liquid whose chemical makeup neutralizes acids or bases without a great change in pH.

>GREATER THAN

<LESS THAN

ALIQUOT

AMBIENT TEMPERATURE

AMPEROMETRIC

ANAEROBIC

ASEPTIC

BIOASSAY

BIOMONITORING

BIOSURVEY

BLANK

BUFFER

BUFFER CAPACITY

BUFFER CAPACITY

A measure of the capacity of a solution or liquid to neutralize acids or bases. This is a measure of the capacity of water or wastewater for offering a resistance to changes in pH.

CHAIN OF CUSTODY

CHAIN OF CUSTODY

A record of each person involved in the handling and possession of a sample from the person who collected the sample to the person who analyzed the sample in the laboratory and to the person who witnessed disposal of the sample.

COLORIMETRIC MEASUREMENT

COLORIMETRIC MEASUREMENT

A means of measuring unknown chemical concentrations in water by measuring a sample's color intensity. The specific color of the sample, developed by addition of chemical reagents, is measured with a photoelectric colorimeter or is compared with color standards using, or corresponding with, known concentrations of the chemical.

COMPOSITE (PROPORTIONAL) SAMPLE

COMPOSITE (PROPORTIONAL) SAMPLE

A composite sample is a collection of individual samples obtained at regular intervals, usually every one or two hours during a 24-hour time span. Each individual sample is combined with the others in proportion to the rate of flow when the sample was collected. Equal volume individual samples also may be collected at intervals after a specific volume of flow passes the sampling point or after equal time intervals and still be referred to as a composite sample. The resulting mixture (composite sample) forms a representative sample and is analyzed to determine the average conditions during the sampling period.

COMPOUND

COMPOUND

A pure substance composed of two or more elements whose composition is constant. For example, table salt (sodium chloride, NaCl) is a compound.

DESICCATOR (DESS-uh-kay-tor)

DESICCATOR

A closed container into which heated weighing or drying dishes are placed to cool in a dry environment in preparation for weighing. The dishes may be empty or they may contain a sample. Desiccators contain a substance (DESICCANT), such as anhydrous calcium chloride, that absorbs moisture and keeps the relative humidity near zero so that the dish or sample will not gain weight from absorbed moisture.

DISTILLATE (DIS-tuh-late)

DISTILLATE

In the distillation of a sample, a portion is collected by evaporation and recondensation; the part that is recondensed is the distillate.

ELEMENT

ELEMENT

A substance that cannot be separated into its constituent parts and still retain its chemical identity. For example, sodium (Na) is an element.

END POINT

END POINT

The completion of a desired chemical reaction. Samples of water or wastewater are titrated to the end point. This means that a chemical is added, drop by drop, to a sample until a certain color change (blue to clear, for example) occurs. This is called the end point of the titration. In addition to a color change, an end point may be reached by the formation of a precipitate or the reaching of a specified pH. An end point may be detected by the use of an electronic device, such as a pH meter.

FACULTATIVE (FACK-ul-tay-tive) BACTERIA

FACULTATIVE BACTERIA

Facultative bacteria can use either dissolved oxygen or oxygen obtained from food materials such as sulfate or nitrate ions. In other words, facultative bacteria can live under aerobic, anoxic, or anaerobic conditions.

FLAME POLISHED

FLAME POLISHED

Melted by a flame to smooth out irregularities. Sharp or broken edges of glass (such as the end of a glass tube) are rotated in a flame until the edge melts slightly and becomes smooth.

GRAB SAMPLE

GRAB SAMPLE

A single sample of water collected at a particular time and place that represents the composition of the water only at that time and place.

GRAVIMETRIC

GRAVIMETRIC

A means of measuring unknown concentrations of water quality indicators in a sample by weighing a precipitate or residue of the sample.

INDICATOR

INDICATOR

(1) (Chemical indicator) A substance that gives a visible change, usually of color, at a desired point in a chemical reaction, generally at a specified end point.

(2) (Instrument indicator) A device that indicates the result of a measurement, usually using either a fixed scale and movable indicator (pointer), such as a pressure gauge, or a moving chart with a movable pen like those used on a circular flow-recording chart. Also called a receiver.

INTEGRATOR

INTEGRATOR

A device or meter that continuously measures and sums a process rate variable in cumulative fashion over a given time period. For example, total flows displayed in gallons per minute, million gallons per day, cubic feet per second, or some other unit of volume per time period. Also called a totalizer.

M or MOLAR

M or MOLAR

A molar solution consists of one gram molecular weight of a compound dissolved in enough water to make one liter of solution. A gram molecular weight is the molecular weight of a compound in grams. For example, the molecular weight of sulfuric acid (H_2SO_4) is 98. A one M solution of sulfuric acid would consist of 98 grams of H_2SO_4 dissolved in enough distilled water to make one liter of solution.

MPN

MPN

MPN is the Most Probable Number of coliform-group organisms per unit volume of sample water. Expressed as a density or population of organisms per 100 mL of sample water.

MSDS

MSDS

See MATERIAL SAFETY DATA SHEET (MSDS).

MATERIAL SAFETY DATA SHEET (MSDS)

MATERIAL SAFETY DATA SHEET (MSDS)

A document that provides pertinent information and a profile of a particular hazardous substance or mixture. An MSDS is normally developed by the manufacturer or formulator of the hazardous substance or mixture. The MSDS is required to be made available to employees and operators or inspectors whenever there is the likelihood of the hazardous substance or mixture being introduced into the workplace. Some manufacturers are preparing MSDSs for products that are not considered to be hazardous to show that the product or substance is not hazardous.

MENISCUS (meh-NIS-cuss)

MENISCUS

The curved surface of a column of liquid (water, oil, mercury) in a small tube. When the liquid wets the sides of the container (as with water), the curve forms a valley. When the confining sides are not wetted (as with mercury), the curve forms a hill or upward bulge.

MOLECULAR WEIGHT

The molecular weight of a compound in grams per mole is the sum of the atomic weights of the elements in the compound. The molecular weight of sulfuric acid (H_2SO_4) in grams is 98.

Element	Atomic Weight	Number of Atoms	Molecular Weight
H	1	2	2
S	32	1	32
O	16	4	64
			98

MOLECULE

The smallest division of a compound that still retains or exhibits all the properties of the substance.

N or NORMAL

A normal solution contains one gram equivalent weight of reactant (compound) per liter of solution. The equivalent weight of an acid is that weight that contains one gram atom of ionizable hydrogen or its chemical equivalent. For example, the equivalent weight of sulfuric acid (H_2SO_4) is 49 (98 divided by 2 because there are two replaceable hydrogen ions). A one N solution of sulfuric acid would consist of 49 grams of H_2SO_4 dissolved in enough water to make one liter of solution.

NPDES PERMIT

National Pollutant Discharge Elimination System permit is the regulatory agency document issued by either a federal or state agency that is designed to control all discharges of potential pollutants from point sources and stormwater runoff into US waterways. NPDES permits regulate discharges into US waterways from all point sources of pollution, including industries, municipal wastewater treatment plants, sanitary landfills, large animal feedlots, and return irrigation flows.

NONVOLATILE MATTER

Material such as sand, salt, iron, calcium, and other mineral materials that are only slightly affected by the actions of organisms and are not lost on ignition of the dry solids at 550°C (1,022°F). Volatile materials are chemical substances usually of animal or plant origin. Also see INORGANIC WASTE and VOLATILE SOLIDS.

OXIDATION

Oxidation is the addition of oxygen, removal of hydrogen, or the removal of electrons from an element or compound; in the environment and in wastewater treatment processes, organic matter is oxidized to more stable substances. The opposite of REDUCTION.

OXIDATION STATE/OXIDATION NUMBER

In a chemical formula, a number accompanied by a polarity indication (+ or −) that together indicate the charge of an ion as well as the extent to which the ion has been oxidized or reduced in a REDOX REACTION.

Due to the loss of electrons, the charge of an ion that has been oxidized would go from negative toward or to neutral, from neutral to positive, or from positive to more positive. As an example, an oxidation number of 2+ would indicate that an ion has lost two electrons and that its charge has become positive (that it now has an excess of two protons).

Due to the gain of electrons, the charge of the ion that has been reduced would go from positive toward or to neutral, from neutral to negative, or from negative to more negative. As an example, an oxidation number of 2− would indicate that an ion has gained two electrons and that its charge has become negative (that it now has an excess of two electrons). As an ion gains electrons, its oxidation state (or the extent to which it is oxidized) lowers; that is, its oxidation state is reduced. Also see REDOX REACTION.

OXIDATION-REDUCTION POTENTIAL (ORP)

The electrical potential required to transfer electrons from one compound or element (the oxidant) to another compound or element (the reductant); used as a qualitative measure of the state of oxidation in water and wastewater treatment systems. ORP is measured in millivolts, with negative values indicating a tendency to reduce compounds or elements and positive values indicating a tendency to oxidize compounds or elements.

OXIDATION-REDUCTION (REDOX) REACTION

See REDOX REACTION.

PERCENT SATURATION PERCENT SATURATION

The amount of a substance that is dissolved in a solution compared with the amount dissolved in the solution at saturation, expressed as a percent.

$$\text{Percent Saturation, \%} = \frac{\text{Amount of Substance That Is Dissolved} \times 100\%}{\text{Amount Dissolved in Solution at Saturation}}$$

pH (pronounce as separate letters) pH

pH is an expression of the intensity of the basic or acidic condition of a liquid. Mathematically, pH is the logarithm (base 10) of the reciprocal of the hydrogen ion activity.

$$pH = \text{Log} \frac{1}{\{H^+\}}$$

If $\{H^+\} = 10^{-6.5}$, then pH = 6.5. The pH may range from 0 to 14, where 0 is most acidic, 14 most basic, and 7 neutral.

PILLOWS PILLOWS

Plastic tubes shaped like pillows that contain exact amounts of chemicals or reagents. Cut open the pillow, pour the reagents into the sample being tested, mix thoroughly, and follow test procedures.

PLUG FLOW PLUG FLOW

A type of flow that occurs in tanks, basins, or reactors when a slug of water or wastewater moves through a tank without ever dispersing or mixing with the rest of the water or wastewater flowing through the tank.

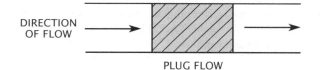

PRECIPITATE (pre-SIP-uh-TATE) PRECIPITATE

(1) An insoluble, finely divided substance that is a product of a chemical reaction within a liquid.

(2) The separation from solution of an insoluble substance.

REAGENT (re-A-gent) REAGENT

A pure, chemical substance that is used to make new products or is used in chemical tests to measure, detect, or examine other substances.

RECEIVER RECEIVER

A device that indicates the result of a measurement, usually using either a fixed scale and movable indicator (pointer), such as a pressure gauge, or a moving chart with a movable pen like those used on a circular flow-recording chart. Also called an INDICATOR.

REDOX (REE-docks) REACTION REDOX REACTION

A two-part reaction between two ions involving a transfer of electrons from one ion to the other. Oxidation is the loss of electrons by one ion, and reduction is the acceptance of electrons by the other ion. Reduction refers to the lowering of the OXIDATION STATE/ OXIDATION NUMBER of the ion accepting the electrons.

In a redox reaction, the ion that gives up the electrons (that is oxidized) is called the reductant because it causes a reduction in the oxidation state or number of the ion that accepts the transferred electrons. The ion that receives the electrons (that is reduced) is called the oxidant because it causes oxidation of the other ion. Oxidation and reduction always occur simultaneously.

REDUCTION (re-DUCK-shun) REDUCTION

Reduction is the addition of hydrogen, removal of oxygen, or the addition of electrons to an element or compound. Under anaerobic conditions (no dissolved oxygen present), sulfur compounds are reduced to odor-producing hydrogen sulfide (H_2S) and other compounds. In the treatment of metal finishing wastewaters, hexavalent chromium (Cr^{6+}) is reduced to the trivalent form (Cr^{3+}). The opposite of OXIDATION.

REFLUX REFLUX

Flow back. A sample is heated, evaporates, cools, condenses, and flows back to the flask.

REPRESENTATIVE SAMPLE REPRESENTATIVE SAMPLE

A sample portion of material, water, or wastestream that is as nearly identical in content and consistency as possible to that in the larger body being sampled.

SOLUTION SOLUTION

A liquid mixture of dissolved substances. In a solution it is impossible to see all the separate parts.

SPECIFIC GRAVITY SPECIFIC GRAVITY

(1) Weight of a particle, substance, or chemical solution in relation to the weight of an equal volume of water. Water has a specific gravity of 1.000 at 4°C (39°F). Wastewater particles or substances usually have a specific gravity of 0.5 to 2.5. Particulates with specific gravity less than 1.0 float to the surface and particulates with specific gravity greater than 1.0 sink.

(2) Weight of a particular gas in relation to the weight of an equal volume of air at the same temperature and pressure (air has a specific gravity of 1.0). Chlorine gas has a specific gravity of 2.5.

STANDARD METHODS STANDARD METHODS

STANDARD METHODS FOR THE EXAMINATION OF WATER AND WASTEWATER, 21st Edition. A joint publication of the American Public Health Association (APHA), American Water Works Association (AWWA), and the Water Environment Federation (WEF) that outlines the accepted laboratory procedures used to analyze the impurities in water and wastewater. Available from: American Water Works Association, Bookstore, 6666 West Quincy Avenue, Denver, CO 80235. Order No. 10084. Price to members, $198.50; nonmembers, $266.00; price includes cost of shipping and handling. Also available from Water Environment Federation, Publications Order Department, PO Box 18044, Merrifield, VA 22118-0045. Order No. S82011. Price to members, $203.00; nonmembers, $268.00; price includes cost of shipping and handling.

STANDARD SOLUTION STANDARD SOLUTION

A solution in which the exact concentration of a chemical or compound is known.

STANDARDIZE STANDARDIZE

To compare with a standard.

(1) In wet chemistry, to find out the exact strength of a solution by comparing it with a standard of known strength. This information is used to adjust the strength by adding more water or more of the substance dissolved.

(2) To set up an instrument or device to read a standard. This allows you to adjust the instrument so that it reads accurately, or enables you to apply a correction factor to the readings.

SURFACTANT (sir-FAC-tent) SURFACTANT

Abbreviation for surface-active agent. The active agent in detergents that possesses a high cleaning ability.

THIEF HOLE THIEF HOLE

A digester sampling well that allows sampling of the digester contents without venting digester gas.

TITRATE (TIE-trate) TITRATE

To titrate a sample, a chemical solution of known strength is added drop by drop until a certain color change, precipitate, or pH change in the sample is observed (end point). Titration is the process of adding the chemical reagent in small increments (0.1–1.0 milliliter) until completion of the reaction, as signaled by the end point.

TOTALIZER TOTALIZER

A device or meter that continuously measures and sums a process rate variable in cumulative fashion over a given time period. For example, total flows displayed in gallons per minute, million gallons per day, cubic feet per second, or some other unit of volume per time period. Also called an integrator.

TURBIDITY (ter-BID-it-tee)

TURBIDITY

The cloudy appearance of water caused by the presence of suspended and colloidal matter. In the waterworks field, a turbidity measurement is used to indicate the clarity of water. Technically, turbidity is an optical property of the water based on the amount of light reflected by suspended particles. Turbidity cannot be directly equated to suspended solids because white particles reflect more light than dark-colored particles and many small particles will reflect more light than an equivalent large particle.

TURBIDITY (ter-BID-it-tee) UNITS (TU)

TURBIDITY UNITS (TU)

Turbidity units are a measure of the cloudiness of water. If measured by a nephelometric (deflected light) instrumental procedure, turbidity units are expressed in nephelometric turbidity units (NTU) or simply TU. Those turbidity units obtained by visual methods are expressed in Jackson turbidity units (JTU), which are a measure of the cloudiness of water; they are used to indicate the clarity of water. There is no real connection between NTUs and JTUs. The Jackson turbidimeter is a visual method and the nephelometer is an instrumental method based on deflected light.

VOLATILE (VOL-uh-tull)

VOLATILE

(1) A volatile substance is one that is capable of being evaporated or changed to a vapor at relatively low temperatures. Volatile substances can be partially removed from water or wastewater by the air stripping process.

(2) In terms of solids analysis, volatile refers to materials lost (including most organic matter) upon ignition in a muffle furnace for 60 minutes at 550°C (1,022°F). Natural volatile materials are chemical substances usually of animal or plant origin. Manufactured or synthetic volatile materials, such as plastics, ether, acetone, and carbon tetrachloride, are highly volatile and not of plant or animal origin. Also see NONVOLATILE MATTER.

VOLATILE ACIDS

VOLATILE ACIDS

Fatty acids produced during digestion that are soluble in water and can be steam-distilled at atmospheric pressure. Also called organic acids. Volatile acids are commonly reported as equivalent to acetic acid.

VOLATILE LIQUIDS

VOLATILE LIQUIDS

Liquids that easily vaporize or evaporate at room temperature.

VOLATILE SOLIDS

VOLATILE SOLIDS

Those solids in water, wastewater, or other liquids that are lost on ignition of the dry solids at 550°C (1,022°F). Also called organic solids and volatile matter.

VOLUMETRIC

VOLUMETRIC

A measurement based on the volume of some factor. Volumetric titration is a means of measuring unknown concentrations of water quality indicators in a sample by determining the volume of titrant or liquid reagent needed to complete particular reactions.

CHAPTER 14. LABORATORY PROCEDURES

(Lesson 1 of 7 Lessons)

14.0 LABORATORY BASICS

Operators conduct laboratory control tests to collect data used to control the efficiency of wastewater treatment processes. By relating laboratory results to operations, the plant operator can select the most effective operational procedures, determine the efficiency of treatment unit processes, and identify treatment problems before they seriously affect effluent quality. For these reasons, a clear understanding of laboratory procedures is a must to any operator. The lab procedures outlined in this chapter are simple and easy procedures used by many operators to control wastewater treatment processes. See Section 14.4, "Laboratory Procedures for Plant Control," for procedures for successful plant control and operation. When lab data must be submitted to regulatory agencies for monitoring and enforcement purposes, you must use approved test procedures. See Section 14.5, "Laboratory Procedures for *NPDES*[1] Monitoring." Most small wastewater system operators collect samples for NPDES monitoring and submit these samples to an approved laboratory for analysis.

Each test section contains the following information:

1. Discussion of test
2. What is tested
3. Apparatus
4. Reagents
5. Procedures
6. Precautions
7. Examples
8. Calculations

14.00 Metric System

The metric system is based on the decimal system. All units of length, volume, and weight (mass) use factors of 10 to express larger or smaller quantities of these units. The metric system is used exclusively in the wastewater plant laboratory. The following is a summary of metric and English unit names and abbreviations.

Type of Measurement	English System Unit	Metric Unit	Metric Abbreviation
Length	inch foot yard	meter	m
Temperature	Fahrenheit	Celsius	°C
Volume	quart gallon	liter	L
Weight	ounce pound	gram	gm
Concentration	lbs/gal strength, %	milligrams per liter	mg/L

In the laboratory, we sometimes use smaller amounts than a meter, a liter, or a gram. To express these smaller amounts, prefixes are added to the names of the metric units. There are many prefixes in use; however, we commonly use the following two prefixes more than any others in the laboratory.

Prefix	Abbreviation	Meaning
milli-	m	1/1000 of; or 0.001 times
centi-	c	1/100 of; or 0.01 times

One milliliter (mL) is 1/1,000 of a liter and likewise one centimeter (cm) is 1/100 of a meter.

EXAMPLE

Convert 2 grams into milligrams.

$$1 \text{ milligram} = 1 \text{ mg} = 1/1{,}000 \text{ gm}$$
$$\text{therefore, } 1 \text{ gram} = 1{,}000 \text{ milligrams}$$
$$2 \text{ grams} \times 1{,}000 \text{ mg/gram} = 2{,}000 \text{ mg}$$

Convert 500 mL to liters.

$$1 \text{ mL} = 1/1{,}000 \text{ liter}$$
$$\text{therefore, } 1 \text{ liter} = 1{,}000 \text{ mL}$$
$$500 \text{ mL} \times 1 \text{ liter}/1{,}000 \text{ mL} = 0.500 \text{ liters}$$

---

1. *NPDES Permit.* National Pollutant Discharge Elimination System permit is the regulatory agency document issued by either a federal or state agency that is designed to control all discharges of potential pollutants from point sources and stormwater runoff into US waterways. NPDES permits regulate discharges into US waterways from all point sources of pollution, including industries, municipal wastewater treatment plants, sanitary landfills, large animal feedlots, and return irrigation flows.

The Celsius (or centigrade) temperature scale is used in the laboratory rather than the more familiar Fahrenheit scale.

	Fahrenheit (°F)	Celsius (°C)
Freezing point of water	32	0
Boiling point of water	212	100

To convert Fahrenheit to Celsius, use the following formula:

$$\text{Temperature, °C} = (°F - 32°F) \times \frac{5}{9}$$

EXAMPLE: Convert 68°F to °C.

$$\text{Temperature, °C} = (°F - 32°) \times \frac{5}{9}$$

$$= (68° - 32°) \times \frac{5}{9}$$

$$= 36° \times \frac{5}{9}$$

$$= 4° \times 5$$

$$= 20°C$$

To convert Celsius to Fahrenheit, use the following formula:

$$\text{Temperature, °F} = \left(C° \times \frac{9}{5}\right) + 32°F$$

EXAMPLE: Convert 35°C to °F.

$$\text{Temperature, °F} = \left(C° \times \frac{9}{5}\right) + 32°$$

$$= \left(35° \times \frac{9}{5}\right) + 32°$$

$$= \frac{315°}{5} + 32°$$

$$= 63° + 32°$$

$$= 95°F$$

14.01   Chemical Symbols and Formulas

Chemical symbols are shorthand for the names of elements. Names and symbols for some of these elements are listed below.

Chemical Name	Symbol
Calcium	Ca
Carbon	C
Chlorine	Cl
Hydrogen	H
Iron	Fe
Oxygen	O
Potassium	K
Sodium	Na
Sulfur	S

Many different compounds can be made from the same two or three elements, therefore, read the formula and name carefully to prevent errors and accidents. A chemical formula is a shorthand or an abbreviated way to write the name of a compound. For example, "sodium chloride" (table salt) can be written "NaCl." Table 14.1 lists commonly used chemicals found in the wastewater treatment plant laboratory.

The following procedures show the uses of some of the chemicals whose names and formulas will be seen in wastewater tests:

Dissolved Oxygen Procedure

"… To 300 milliliter (mL) sample placed in a BOD bottle, add 1 mL of manganous sulfate solution. Now, add 1 mL of alkaline iodine-sodium azide solution. Shake well and add 1 mL of concentrated sulfuric acid. Titrate with sodium thiosulfate solution until …"

Preparation of Manganous Sulfate Solution

"… Weigh 480 grams (gm) $MnSO_4 \cdot H_2O$ and dissolve in distilled water. Dilute to 1 liter …"

Chemicals must be properly labeled to avoid poor results and safety hazards in the laboratory. These issues are often caused by laboratory personnel using a chemical from the shelf that is not the same chemical called for in the procedure. This mistake usually occurs when the chemicals have similar names or formulas. This problem can be eliminated if you use both the chemical name and formula as a double check. As you can see in Table 14.1, the spellings of many chemical names are very similar. These slight differences are critical because the chemicals do not behave alike. For example, the chemicals potassium nitrate (KNO_3) and potassium nitrite (KNO_2) are just as different in meaning chemically as the words fat and fit are to your doctor.

TABLE 14.1   CHEMICAL NAMES AND FORMULAS COMMONLY USED IN WASTEWATER ANALYSES

Chemical Name	Chemical Formula
Ammonium chloride	NH_4Cl
Calcium chloride (heptahydrate)*	$CaCl_2 \cdot 7H_2O$
Dipotassium hydrogen phosphate	K_2HPO_4
Disodium hydrogen phosphate (heptahydrate)*	$Na_2HPO_4 \cdot 7H_2O$
Ferric chloride	$FeCl_3$
Magnesium sulfate (heptahydrate)*	$MgSO_4 \cdot 7H_2O$
Manganous sulfate (tetrahydrate)*	$MnSO_4 \cdot 4H_2O$
Phenylarsine oxide	C_6H_5AsO
Potassium iodide	KI
Sodium azide	NaN_3
Sodium chloride	$NaCl$
Sodium hydroxide	$NaOH$
Sodium iodide	NaI
Sodium thiosulfate (pentahydrate)*	$Na_2S_2O_3 \cdot 5H_2O$
Sulfuric acid	H_2SO_4

*Note that: tetra = 4, penta = 5, and hepta = 7, thus heptahydrate = $7H_2O$

14.02   References

For information about laboratory procedures used by approved laboratories that are not included in this chapter, see the following references:

1. *Operation of Wastewater Treatment Plants,* Volume II, Chapter 16, "Laboratory Procedures and Chemistry." Obtain from Office of Water Programs, California State University, Sacramento, 6000 J Street, Sacramento, CA 95819-6025. Price, $49.00.

2. *Laboratory Procedures and Chemistry for Operators of Water Pollution Control Plants.* This publication is a reproduction of *Operation of Wastewater Treatment Plants,* Volume II, Chapter 16, "Laboratory Procedures and Chemistry." Obtain from California Water Environment Association (CWEA), 7677 Oakport Street, Suite 600, Oakland, CA 94621. Price to members, $10.00; nonmembers, $14.85; plus shipping and handling.

3. *Water Analysis Handbook,* Fifth Edition. Obtain from HACH Company, PO Box 389, Loveland, CO 80539-0389. Order No. 2954700. Price, $108.00.

4. *Standard Methods for the Examination of Water and Wastewater,* 21st Edition, 2005. Obtain from Water Environment Federation, Publications Order Department, PO Box 18044, Merrifield, VA 22118-0045. Order No. S82011. Price to members, $203.00; nonmembers, $268.00; price includes cost of shipping and handling.

14.03   Acknowledgments

Many of the illustrated laboratory procedures in this chapter were provided by Joe Nagano, Laboratory Director, Hyperion Treatment Plant, City of Los Angeles, California. These procedures originally appeared in *Laboratory Procedures for Operators of Water Pollution Control Plants,* prepared by Mr. Nagano and published by the California Water Pollution Control Association. Information in this chapter is a condensed version of Chapter 16, "Laboratory Procedures and Chemistry," by James Paterson and revised by James Sequeira, in *Operation of Wastewater Treatment Plants,* Volume II.

14.1   LABORATORY EQUIPMENT AND TECHNIQUES

14.10   Equipment

In every wastewater laboratory, there are certain pieces of equipment that are used routinely by the lab analyst to perform chemical tests such as those in wastewater analyses. Pictures and names of such equipment and a description of the use of each piece follow.

Volumetric glassware is calibrated either "to contain" (TC) or "to deliver" (TD). Glassware designed to contain (graduated cylinders, volumetric flasks) is usually marked with "TC." When liquid is poured from TC glassware, a small amount remains behind. Glassware designed to deliver (pipets, burets) is usually marked with "TD." This glassware is designed such that when a certain amount of liquid is poured from it, the TD glassware empties completely. It will do so accurately only when the inner surface of the glassware is so scrupulously clean that water wets the surface immediately and forms a uniform film on the surface upon emptying.

BEAKERS. Beakers are the most common pieces of laboratory equipment. They are short, wide cylinders that come in sizes from 1 mL to 4,000 mL. They are used mainly to mix chemicals and to measure approximate volumes.

Beaker

GRADUATED CYLINDERS. Graduated cylinders also are basic to any laboratory. These long, narrow cylinders come in sizes from 5 mL to 4,000 mL. They are used to measure volumes more accurately than beakers.

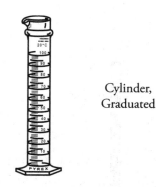

Cylinder,
Graduated

PIPETS. Pipets are used for accurate volume measurement and transfer. They are very thin tubes with a pointed tip that come in sizes from 0.1 mL to 100 mL.

Pipet, Volumetric

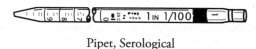

Pipet, Serological

BURETS. Burets are also used for accurate measurement of liquid volumes. They are especially useful in a procedure called titration. These long tubes with graduated walls and a stopcock come in sizes from 10 to 1,000 mL.

FLASKS. Flasks are used for mixing, heating, and weighing chemicals. There are many different sizes and shapes.

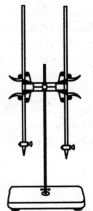

Support, Buret, and Buret Clamp

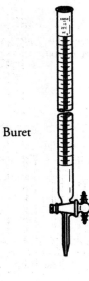

Buret

Automatic Buret

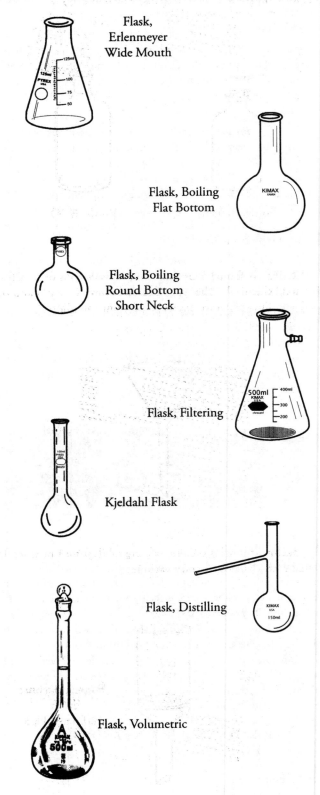

Flask, Erlenmeyer Wide Mouth

Flask, Boiling Flat Bottom

Flask, Boiling Round Bottom Short Neck

Flask, Filtering

Kjeldahl Flask

Flask, Distilling

Flask, Volumetric

BOTTLES. Bottles are used to store chemicals, to collect samples for testing purposes, and to dispense liquids.

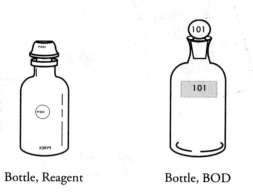

Bottle, Reagent   Bottle, BOD

FUNNELS. A funnel is used for pouring solutions or transferring solid chemicals. The type of funnel shown here can also be used with filter paper to remove solids from a solution.

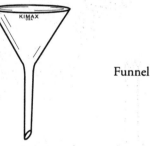

Funnel

A Buchner funnel is used to separate solids from a mixture. It is used with a filter flask and a vacuum.

Funnel, Buchner
with
Perforated Plate

Separatory funnels are used to separate one chemical mixture from another. The separated chemical usually is dissolved in one or two layers of liquid.

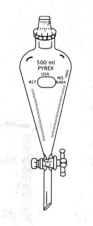

Separatory Funnel

TUBES. Test tubes are used for mixing small quantities of chemicals. They are also used as containers for bacterial testing (culture tubes).

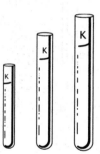

Test Tube   Culture Tube
                   without Lip

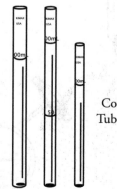

Color
Comparison
Tubes, Nessler

OTHER LABWARE AND EQUIPMENT

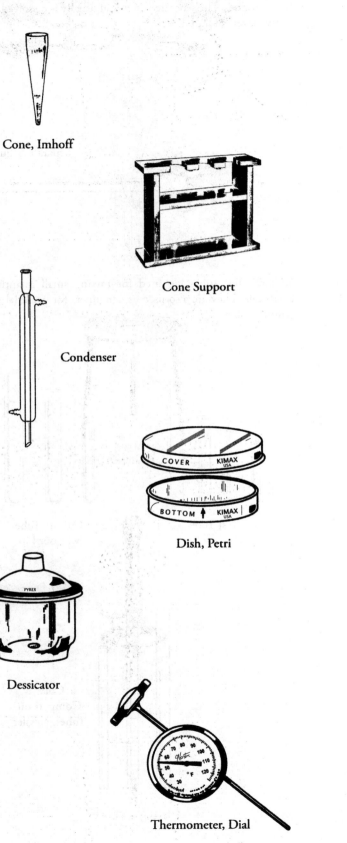

Cone, Imhoff

Cone Support

Condenser

Dish, Petri

Dessicator

Thermometer, Dial

Hot Plate

Oven, Mechanical Convection

Muffle Furnace, Electric

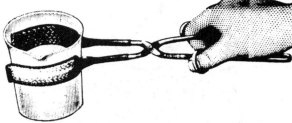

Clamp, Beaker, Safety Tongs

Clamp, Utility

Clamp, Dish, Safety Tongs

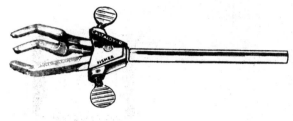

Clamp

Clamp, Flask, Safety Tongs

Tripod, Concentric Ring

Clamp, Test Tube

Burner, Bunsen

Clamp Holder

Triangle, Fused

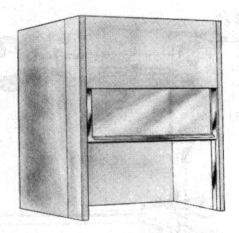

Fume Hood

Portable Dissolved Oxygen Meter
(with computer docking station)
(Courtesy of HACH Company)

Portable pH Meter
(Courtesy of HACH Company)

Crucible, Porcelain

Crucible, Gooch
Porcelain

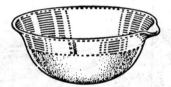

Dish, Evaporating

Test Paper, pH 1–11

Pump, Air Pressure and Vacuum

BOD Incubator

Weight = 95.5580 gm.

Balance, Analytical

(Permission of Mettler)

14.11 Using Laboratory Glassware

BURETS

A buret is used to give accurate measurements of liquid volumes. The stopcock controls the amount of liquid that flows from the buret. A glass stopcock must be lubricated (stopcock grease) and should not be used with alkaline solutions. A Teflon stopcock never needs to be lubricated.

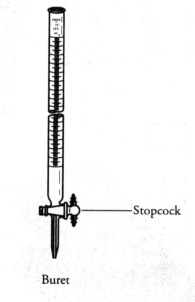

—Stopcock

Buret

Burets come in several sizes, with those holding 10 or 25 milliliters used most frequently.

When a buret is filled with liquid, the surface of the liquid is curved. This curve of the surface is called the meniscus (meh-NIS-cuss). Depending on the liquid, the curve forms a valley, as with water, or forms a hill, as with mercury. Since most solutions used in the laboratory are water-based, always read the bottom of the meniscus with your eye at the same level (Figure 14.1). If you have the meniscus at eye level, the closest marks that go all the way around the buret will appear as straight lines, not circles.

GRADUATED CYLINDERS

The graduated cylinder or "graduate" is one of the most-used pieces of laboratory equipment. This cylinder is made either of glass or plastic and ranges in size from 10 mL to 4 liters. The graduate is used to measure volumes of liquid with an accuracy less than burets but greater than beakers or flasks. Never heat graduated cylinders in an open flame because they will break.

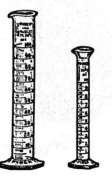

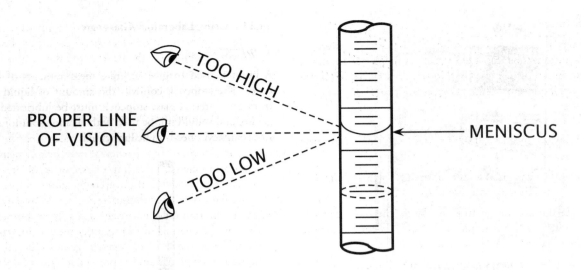

Fig. 14.1   How to read meniscus

FLASKS AND BEAKERS

Beakers and flasks are used for mixing, heating, and weighing chemicals. Most beakers and flasks are not calibrated with exact volume lines; however, they are sometimes marked with approximate volumes and, therefore, can be used to estimate volumes.

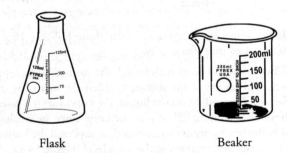

Flask   Beaker

VOLUMETRIC FLASKS

Volumetric flasks are used to prepare solutions and come in sizes from 10 to 2,000 mL. Volumetric flasks should never be heated. Rather than store liquid chemicals in volumetric flasks, the chemicals should be transferred to a storage bottle. Volumetric flasks are more accurate than graduated cylinders.

PIPETS

Pipets are used for accurate volume measurements and transfer. There are three types of pipets commonly used in the laboratory: volumetric pipets, graduated (measuring) or Mohr pipets, and serological pipets.

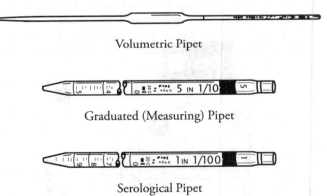

Volumetric Pipet

Graduated (Measuring) Pipet

Serological Pipet

Volumetric pipets are available in sizes such as 1, 10, 25, 50, and 100 mL. They are used to deliver a single volume. Measuring (graduated) and serological pipets deliver fractions of the total volume indicated on the pipet.

Volumetric pipets should be held in a vertical position when emptying and the outflow should be unrestricted. The tip should be touched to the wet surface of the receiving vessel and kept in contact with it until the emptying is complete. Under no circumstance should the small amount remaining in the tip be blown out.

Graduated (measuring) and serological pipets should be held in the vertical position. After outflow has stopped, the tip should be touched to the wet surface of the receiving vessel. No drainage period is allowed. Where the small amount remaining in the tip is to be blown out and added, indication is made by a frosted band near the top of the pipet.

14.12 Solutions

Many laboratory procedures do not give the concentrations of standard solutions in grams/liter or milligrams/liter. Instead, the concentrations are usually given as normality (N), which is the standard designation for solution strengths in chemistry.

EXAMPLES

$0.025\,N\,H_2SO_4$ means a 0.025 normal solution of sulfuric acid

$2\,N\,NaOH$ means that the normality of a sodium hydroxide solution is 2

The larger the number in front of the N, the more concentrated the solution. For example, 1 N NaOH solution is more concentrated than a 0.2 N NaOH solution.

When the exact concentration of a chemical or compound in a solution is known, it is referred to as a "standard solution." To standardize a solution means to determine its concentration accurately, thereby making it a standard solution. "Standardization" is the process of using one solution of known concentration to determine the concentration of another solution. This action often involves a procedure called a "titration." Standard solutions can be ordered already prepared. Once a standard has been prepared, it can then be used to standardize other solutions.

14.13 Titration

Titration involves the addition of one solution, which is usually in a buret, to another solution in a flask or beaker. The solution in the buret is referred to as the "titrant" and is added to the other solution until there is a measurable change in the solution in the flask or beaker. This change is frequently a color change as a result of the addition of a special chemical called an "indicator" to the solution in the flask before the titration begins. The solution in the buret is added slowly to the flask until the change, which is called the "end point," is reached.

14.14 Using a Spectrophotometer

In the field of wastewater analysis, measuring the intensity of color in a sample at a particular wavelength helps determine many components, including phosphorus, nitrite, and nitrate. The color is formed in the sample by adding a specific developing reagent to it. The intensity of the color formed is directly related to the amount of material (such as phosphorus) in the sample. For the analysis of phosphorus present in wastewater, for example, ammonium molybdate reagent is added as the developing reagent; if phosphorus is present, a blue color develops. The more phosphorus there is, the deeper and darker the blue color.

The human eye can detect some differences in color intensity; however, for very precise measurements an instrument called a spectrophotometer is used.

The *SPECTROPHOTOMETER* (Figure 14.2) is an instrument generally used to measure the color intensity of a chemical solution. In its simplest form, it consists of a light source that is focused on a prism or other suitable light dispersion device to separate the light into its separate bands of energy. Each different wavelength or color may be selectively focused through a narrow slit. This beam of light then passes through the sample to be measured. The sample is usually contained in a glass tube called a cuvette (kyoo-VET). Most cuvettes are standardized to have a 1.0-cm light path length, however, other sizes are available.

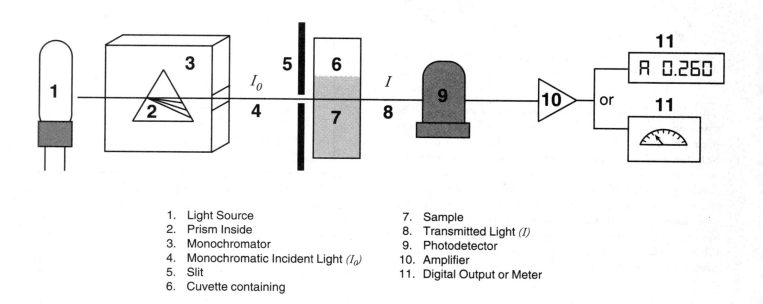

1. Light Source
2. Prism Inside
3. Monochromator
4. Monochromatic Incident Light (I_0)
5. Slit
6. Cuvette containing
7. Sample
8. Transmitted Light (I)
9. Photodetector
10. Amplifier
11. Digital Output or Meter

Fig. 14.2 Working parts of a spectrophotometer

After the selected beam of light has passed through the sample, it emerges and strikes the photodetector. If the solution in the sample cell has absorbed any of the light, the total energy content will be reduced. If the solution in the sample cell does not absorb the light, then there will be no change in energy. When the transmitted light beam strikes the photodetector, it generates an electric current that is proportional to the intensity of light energy striking it. By connecting the photodetector to a device for measuring electric current that produces a reading either digitally or using a scale, a means of measuring the intensity of the transmitted beam is achieved. This is a general description of how this instrument works. The operator should always follow the manufacturer instructions provided with the instrument.

UNITS OF SPECTROSCOPIC MEASUREMENT. The scale on a spectrophotometer is generally graduated in two ways:

1. In units of percent transmittance (%T), an arithmetic scale with units graded from 0 to 100%

2. In units of absorbance (A), a logarithmic scale of nonequal divisions graduated from 0.0 to 2.0

Some specialized spectrophotometers also contain a scale that directly reads the concentration of one chemical constituent in milligrams per liter.

Both the units, percent transmittance and absorbance, are associated with the term "color intensity." That is, a sample that has a low color intensity will have a high percent transmittance but a low absorbance.

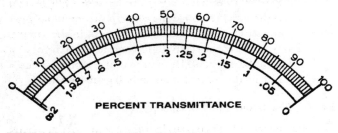

PERCENT TRANSMITTANCE

Absorbance Scale

As illustrated above, the absorbance scale is ordinarily calibrated on the same scale as the percent transmittance on a spectrophotometer. The chief usefulness of absorbance lies in the fact that it is a logarithmic function rather than linear (arithmetic). Beer's Law states that the concentration of a light-absorbing, colored solution is directly proportional to absorbance over a given range of concentrations. If one were to plot a graph showing %T (percent transmittance) versus concentration on straight graph or line paper, and another showing absorbance versus concentration on the same paper, the following curves (graphs) would result:

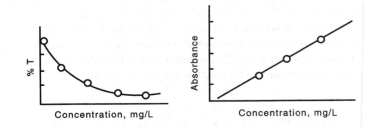

CALIBRATION CURVES. The calibration curve is used to determine the concentration of the water quality indicator (phosphorus, nitrite) contained in a sample. Three steps must be completed in order to make a calibration graph:

1. Prepare a series of standards. A standard is a solution that contains a known amount of the same chemical constituent that is being determined in the sample.

2. Treat the standard solutions and a sample (containing none of the constituent being tested for) with the developing reagent, in the same manner as the sample would be treated. The sample in this step usually consists of distilled water and is generally referred to as a blank.

3. Use a spectrophotometer to determine the absorbance or transmittance of the standards and blank at the specified wavelength. From the values obtained, a calibration curve of absorbance versus concentration can be plotted. After these points are plotted, you can then extend the plotted points by connecting the known points with a straight line. For example, with the data given below one could construct the following calibration curve.

Absorbance	Concentration, mg/L
0.0	0.0
0.30	0.25
0.55	0.50
0.80	0.75

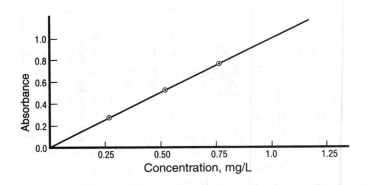

Once you have established a calibration curve for the water quality indicator in question, you can easily determine the amount of that substance contained in a solution of unknown concentration. Do this by taking an absorbance reading on the color developed by the unknown and locate it on the vertical axis. Then a straight line is drawn to the right on the graph until it intersects with the experimental standard curve. A line is then dropped to the horizontal axis and this value identifies the concentration of your unknown water quality indicator.

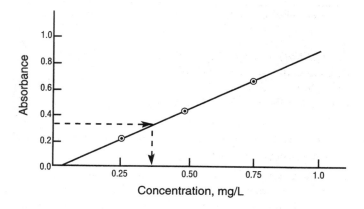

In this example, an absorbance reading of 0.32 was read on the unknown solution or sample, which indicates a concentration of about 0.37 mg/L.

14.15 Indicators

Indicators used in water testing are chemicals that will change color in the presence of specific chemical ions. Even though the indicator changes color as the test is run, it is neither involved with nor does it interfere with the reaction that takes place during the test.

Many different indicators are used in the testing of water samples. Three different uses for indicators are discussed in this section. The person who sets up the test procedure will select the indicator that will give the most recognizable and reproducible results. Generally, the test kits and procedures that are used in small plants will specify or include the best indicator for each test.

Indicators are used in three different types of testing.

1. Qualitative Analysis. In this type of test, the indicator is added to an unmeasured amount of sample. The only purpose of the test is to find out if a specific chemical, one that will cause the color change, is present in the sample.

2. Quantitative Titration. In this test, a reagent solution that has a very accurately known concentration is added to an accurately measured volume of a test sample that contains the selected indicator. The reagent can be added by using a standardized dropper bottle or a buret. The purpose of the test is to measure how much reagent it takes to react with all of the test chemical in the sample. When the indicator changes color, it signals that the reaction is complete. Then, by using a calculation, the amount of reagent used in the test is converted to the amount of test chemical present in the sample. Charts provided in the test kits or laboratory manuals show the equivalent concentrations of the chemical being tested in milligrams per liter.

3. Quantitative Colorimetric Analysis. In this test, an accurately measured volume of sample is placed in a glass tube called a cuvette and the indicator is added to the sample to develop the color. The intensity of the color indicates the concentration of the test chemical that is in the sample. The color intensity is measured in one of two ways:

   a. Color comparator: This method enables the tester to match the color of the test sample to the color of one of the standards provided with the test kit.

   b. Photodetector: The prepared test sample is placed in a device that directs a light source through the sample. A photodetector then measures the amount of light that passes through the sample and converts the amount of light to the concentration of the test chemical in the sample. The test devices either report the concentration directly on a dial on the meter, or they report a value that can be used to calculate the concentration in milligrams per liter.

14.16 Notebooks and Worksheets

The plant operator has two goals in using a laboratory notebook and worksheets: to record data and to arrange data in an orderly manner. Often, days of work can be wasted if data are written down on a scrap of paper, which can be misplaced or thrown away. Worksheets and notebooks help prevent error and provide a record of the work. The routine use of worksheets and notebooks is one of the best ways an operator can be sure that all important information for a test is properly recorded.

There is no standard laboratory worksheet form. Most treatment plants usually develop their own worksheets for recording laboratory results and other important plant data. The data sheets are prepared in a manner that makes it easy to record the data, review it, and recover it when necessary. Each plant may have different needs for collecting and recording data and may use five to eight different types of worksheets. Figures 14.3 and 14.4 illustrate typical laboratory worksheets (sometimes called bench sheets).

Laboratory worksheets provide an organized method for recording data. They are used to review effluent quality, identify problems, and search for the cause of problems. Worksheets also provide the information needed to complete NPDES permit reporting forms.

Acknowledgment

Pictures of laboratory glassware and equipment in this manual are reproduced with the permission of VWR Scientific, San Francisco, California.

QUESTIONS

Please write your answers to the following questions and compare them with those on page 337.

14.1A   For each piece of laboratory equipment listed, describe the item and its use.

1. Beakers
2. Graduated Cylinders
3. Pipets
4. Burets

14.1B   Why should graduated cylinders never be heated in an open flame?

14.1C   What is a bench sheet?

END OF LESSON 1 OF 7 LESSONS

on

LABORATORY PROCEDURES

Please answer the discussion and review questions next.

DISCUSSION AND REVIEW QUESTIONS

Chapter 14.   LABORATORY PROCEDURES

(Lesson 1 of 7 Lessons)

At the end of each lesson in this chapter, you will find discussion and review questions. Please write your answers to these questions to determine how well you understand the material in the lesson.

1. Why must chemicals be properly labeled?

2. How are pipets emptied or drained? Why are these procedures important?

3. How would you titrate a solution?

PLANT _____

DATE _____

SUSPENDED SOLIDS AND DISSOLVED SOLIDS

SAMPLE							
Crucible							
Sample, mL							
Wt Dry + Dish, gm Wt Dish, gm							
Wt Dry, gm							
$mg/L = \dfrac{Wt\ Dry,\ gm \times 1{,}000{,}000}{Sample,\ mL}$							
Wt Dish + Dry, gm Wt Dish + Ash, gm							
Wt Volatile, gm							
$\% = \dfrac{Wt\ Vol}{Wt\ Dry} \times 100\%$.							

BOD

Blank _____

SAMPLE										
DO Sample										
Bottle #										
% Sample										
Blank or adjusted blank DO after incubation										
Depletion, 5 days										
Depletion, %										

SETTLEABLE SOLIDS

Sample, mL _____   _____

Direct mL/L _____   _____

COD

Sample _____   _____

Blank Titration _____   _____

Sample Titration _____   _____

Depletion _____   _____

$mg/L = \dfrac{Dep \times N\ FAS \times 8{,}000}{Sample,\ mL}$ _____   _____

Fig. 14.3   Typical laboratory worksheet

TOTAL SOLIDS

SAMPLE

Dish No.

Wt Dish + Wet, gm
Wt Dish, gm

Wt Wet, gm

Wt Dish + Dry, gm
Wt Dish, gm

Wt Dry, gm

% Solids = $\dfrac{Wt\ Dry}{Wt\ Wet} \times 100\%$

Wt Dish + Dry, gm
Wt Dish + Ash, gm

Wt Volatile, gm

% Volatile = $\dfrac{Wt\ Vol \times 100\%}{Wt\ Dry}$

pH

Volatile Acid, mg/L

Alkalinity as CaCO$_3$, mg/L

Grease

Sample

Sample, mL

Wt Flask + Grease, mg

Wt Flask, mg

Wt Grease, mg

mg/L = $\dfrac{Wt\ Grease,\ mg \times 1,000}{Sample,\ mL}$

Fig. 14.4   Typical laboratory worksheet

CHAPTER 14.   LABORATORY PROCEDURES

(Lesson 2 of 7 Lessons)

14.2   SAFETY AND HYGIENE IN THE LABORATORY

Safety is just as important in the laboratory as in the rest of the treatment plant. State laws and the Occupational Safety and Health Act (OSHA) demand proper safety procedures to be exercised in the laboratory at all times. OSHA specifically deals with "safety at the place of work." The Act states that "each employer shall furnish to each of his employees employment and a place of employment which are free from recognized hazards that are causing or are likely to cause death or serious physical harm to his employees."

Personnel working in a wastewater treatment plant laboratory must recognize that a number of hazardous materials and conditions exist that require them to be alert, careful, and aware of potential dangers at all times. Safe practices in the laboratory prevent injuries to lab personnel.

Each laboratory must develop and implement a written hazard communication program to ensure that personnel working with chemicals are appropriately informed of any health and safety considerations related to the chemicals. Personnel must also be trained in methods to safely handle and store chemicals. As a minimum, the written program must describe how the OSHA criteria for labels and other forms of warnings, material safety data sheets (MSDSs), and employee information and training will be met. Plants that have a laboratory must also develop and implement a written chemical hygiene plan.

14.20   Laboratory Hazards

Working with chemicals and other materials in a wastewater treatment plant laboratory can be hazardous. Exposure to such hazards can be minimized by employing safe practices and by using proper techniques and equipment. Laboratory hazards include the following:

- Infectious Materials
- Hazardous Materials
- Toxic Fumes
- Burns (heat and chemical)
- Electric Shock
- Poisons
- Explosions
- Fires

The above dangers to yourself and others can be minimized, however, by using proper techniques and equipment.

14.200   Infectious Materials

Wastewater and sludge contain millions of biological organisms. Some of these are infectious and dangerous and can cause diseases such as giardiasis, cryptosporidiosis, tetanus, typhoid, dysentery, and hepatitis. Personnel should wear gloves when handling wastewater, and then thoroughly wash their hands with soap and water afterward, particularly before handling food.

Though not mandatory, inoculations by your local health department are recommended for each employee to reduce the possibility of contracting diseases.

Do not pipet anything by mouth. Pipet all samples and reagents by mechanical means (rubber pipet bulbs) to prevent severe illness or death. Never drink from a beaker or other laboratory glassware. A beaker used specifically for drinking is a dangerous practice in a laboratory setting.

14.201   Hazardous Materials

Hazardous materials include corrosive, toxic, and explosive or flammable materials.

1. CORROSIVE MATERIALS

   Corrosive materials include acids, bases (caustics), and miscellaneous chemicals.

   Acids

   a. Examples: Sulfuric (H_2SO_4), hydrochloric or muriatic (HCl), nitric (NHO_3), glacial acetic ($H_4C_2O_2$), and chromic acid.

   b. Acids are extremely corrosive and hazardous to human tissue, clothing, wood, metal, cement, stone, and concrete.

   c. Commercially available spill cleanup kits should be kept on hand to neutralize the acid in the event of an accidental spill. Baking soda (bicarbonate, not laundry soda) is a nontoxic acid neutralizer, so it can be used effectively on skin and lab surfaces.

   Bases (Caustics)

   a. Examples: Sodium hydroxide (caustic soda or lye (NaOH)), potassium hydroxide (KOH), and alkaline iodine-sodium azide solution (used in the dissolved oxygen test).

   b. Bases are extremely corrosive to skin, clothing, and leather. Caustics can quickly and permanently cloud vision if not immediately flushed out of eyes. Determine the location of the safety showers and eye wash station in the plant before starting to work with dangerous chemicals.

   c. Commercially available spill cleanup materials should be kept on hand for use in the event of an accidental spill. A jug of ordinary vinegar can be kept on hand to neutralize bases and it will not harm your skin.

   Miscellaneous Chemicals

   Examples: Alum, chlorine, ferric salts (ferric chloride), and other strong oxidants

2. TOXIC MATERIALS

Examples

- Solids: Cyanide compounds, chromium, orthotolidine, cadmium, mercury, and other heavy metals

- Liquids: Carbon tetrachloride, ammonium hydroxide, nitric acid, bromine, chlorine water, aniline dyes, formaldehyde, and carbon disulfide. Carbon tetrachloride is absorbed into the skin on contact. Its vapors will damage the lungs, and it will build up in your body to a dangerous level.

- Gases: Hydrogen sulfide, chlorine, ammonia, sulfur dioxide, and chlorine dioxide

3. EXPLOSIVE OR FLAMMABLE MATERIALS

Examples

- Liquids: Acetone, ethers, and gasoline
- Gases: Propane and hydrogen

14.21   Personal Safety and Hygiene

14.210   Laboratory Safety

Laboratory work can be dangerous if proper precautions and techniques are not taken. Always follow these basic laboratory safety rules:

1. Never work alone in the laboratory. Someone should always be available to help in case you should have an accident that blinds you, leaves you unconscious, or starts a fire you cannot handle alone. If necessary, have someone check on you regularly to be sure you are OK.

2. Wear protective goggles or safety glasses at all times in the laboratory. Contact lenses may be worn under safety goggles, but contact lenses are not eye protective devices, and wearing them does not reduce the requirement for eye and face protection. Because fumes can seep between the lens and the eyeball, irritating the eye, the National Institute of Occupational Safety and Health (NIOSH) recommends an eye injury hazard evaluation be conducted in accordance with the safety guidelines listed in Current Intelligence Bulletin (CIB) 59 2005-139 (http://www.cdc.gov/niosh/docs/2005-139/).

Safety Glasses

3. Wear a face shield if there is any possibility of a hot liquid erupting from a container or pieces of glassware flying from an exploded apparatus. If in doubt as to its need, wear it.

4. Wear protective or insulated gloves when handling hot equipment or very cold objects, or when handling liquids or solids that are skin irritants.

5. Always wear a lab coat or apron in the laboratory to protect your skin and clothes.

6. Never pipet hazardous materials by mouth.

7. Never eat or smoke in the laboratory. Never use laboratory glassware for serving food or drinks.

8. Do not keep food in a refrigerator that is used for chemical or sample storage.

9. Use a ventilated laboratory fume hood whenever handling toxic chemicals.

10. Maintain clear access to emergency eye wash stations and deluge showers.

11. Practice good housekeeping in the laboratory as an effective way to prevent accidents.

12. Never pour volatile liquids down the sink.

14.211   Personal Hygiene

Although it is highly unlikely that personnel will contract diseases by working in wastewater treatment plants, such a possibility does exist with certain diseases, which can be contracted in the following ways:

1. Some diseases may be contracted through breaks in skin, cuts, or puncture wounds. In such cases, the bacteria causing the disease may be covered over and trapped by flesh, creating an *ANAEROBIC*[2] environment in which the bacteria may thrive and produce toxins that can spread throughout the body.

For protection against diseases contracted in this manner, anyone working in or around wastewater should receive immunization from tetanus. Immunization must be received

---

2. *Anaerobic* (AN-air-O-bick).   A condition in which atmospheric or dissolved oxygen (DO) is *NOT* present in the aquatic (water) environment.

before the infection occurs in order to be effective. To prevent diseases from entering open wounds, care must be taken to keep wounds protected either with adhesive bandages or, if necessary, with rubber gloves or waterproof protective clothing. The use of protective clothing is very important, particularly gloves and boots for safety and injury prevention or protection.

Changing from your work clothes into street clothes before leaving work is highly recommended to prevent transferring unsanitary materials into your home. Store your work clothes and street clothes in separate lockers.

2. Diseases that may be contracted through the gastrointestinal system or the mouth are giardiasis, cryptosporidiosis, typhoid, cholera, dysentery, amebiasis, worms, salmonella, infectious hepatitis, and polio virus. These diseases are transmitted when infected wastewater materials are ingested or swallowed. The best protection against these diseases is achieved by thorough cleansing. Hands, face, and body should be thoroughly washed with soap and water, particularly the hands, in order to prevent the transfer of any unsanitary materials or germs to the mouth. The kind of soap used is less important than the thorough use of soap. (Special disinfectant soaps are not essential.)

Immunization is available for typhoid, polio, and hepatitis A and B. However, little is known about how infectious hepatitis A is acquired except that it can be transmitted by wastewater and dried sludge solids. Hepatitis is frequently associated with gross wastewater pollution.

3. Diseases that may be contracted by breathing contaminated air include tuberculosis, infectious hepatitis, and San Joaquin fever. There is no past evidence of tuberculosis being transmitted through the air at wastewater treatment plants. However, there is one reported case of tuberculosis being contracted by an employee who fell into wastewater, inhaling it into his lungs. San Joaquin fever is caused by a fungus that may be present in wastewater. However, there is no record of operators contracting the disease while on the job.

The best way to prevent these diseases is with proper personal hygiene and immunization. Your plant should have an immunization program for tetanus, polio, and hepatitis. Immunization against typhoid should be made available if typhoid could be present in your area. Check with your local or state health department for recommendations regarding immunization.

There is no absolute insurance against contracting disease in a wastewater treatment plant. However, the likelihood of transmission is practically negligible, and there appears to be no special risk associated with working in treatment plants. In fact, operators may develop a natural immunity by working in this environment.

The possibility of contracting AIDS from exposure to raw wastewater has been discounted by researchers. Researchers have found that, although the AIDS virus is present in waste from AIDS victims, the raw wastewater environment is hostile to the virus itself and has not been identified as a mode of transmission

to date. Needle sticks from potentially contaminated syringes should remain a concern to operators, maintenance personnel, and laboratory analysts. Fluids in or on syringes may provide a less severe environment for pathogens than raw wastewater where dilution and chlorination significantly reduce infection potential.

14.22 Preventing Laboratory Accidents

Preventing laboratory accidents is best accomplished by using proper laboratory techniques, by understanding how to avoid accidents related to electric shock, cuts, toxic fumes, etc., and by practicing the safe storage and transfer of chemicals in the laboratory.

14.220 Proper Laboratory Techniques

Using faulty techniques is one of the chief causes of accidents in the laboratory, and it is one of the most difficult to correct. Proper laboratory techniques will help you prevent accidents and avoid injuries. Because of their nature and prevalence in the laboratory, acids and other corrosive materials constitute a series of hazards ranging from poisoning, burning, and gassing through explosion when mishandled. As a precaution, always flush the outsides of acid bottles before opening them. Do not place the bottle stopper on a countertop where someone might rest an arm or hand on it. Keep all acids tightly stoppered when not in use and make sure no spilled acid remains on the floor, table, or bottle after use. To avoid splashing, always pour acid into water never water into acid—as you would add chemicals to the water in a swimming pool.

Mercury requires special care. Keep all mercury containers tightly stoppered. Even a small amount left in the bottom of a drawer can poison the atmosphere in a room. After a laboratory accident or spill involving mercury, repeatedly clean the entire area carefully until there are no globules remaining. Also, follow all of the safety requirements specified on the material safety data sheet.

14.221 Accident Prevention and Response

Accident prevention comes primarily from understanding the hazards and how to avoid situations that lead to accidents in a laboratory setting. Although prevention is the goal, being prepared for accidents is also important, so have a well-stocked first-aid kit available in the laboratory at all times.

ELECTRIC SHOCK. Wherever there are electrical outlets, plugs, and wiring connections, there is a danger of electric shock. The usual do's and don'ts of protection against shock in the home are applicable in the laboratory. Do not use equipment with worn or frayed wiring. Replace connections when there is any sign of thinning insulation. Ground all apparatus using plugs or pigtail adapters. Install ground-fault circuit interrupters (GFCIs) on all electric circuits located near laboratory sinks or anywhere activities involving liquids occur. Immediately shut off a motor after any liquid has spilled on it, then clean and dry the inside thoroughly before attempting to operate it again.

Electrical units that are operated in an area exposed to flammable vapors should be explosion-proof. All permanent wiring should be installed by an electrician using proper conduit or BX cable to eliminate any danger of circuit overloading.

CUTS. Some of the pieces of glass used in the laboratory, such as glass tubing, thermometers, and funnels, must be inserted through rubber stoppers. If the glass is forced through the hole in the stopper by applying too much pressure, the glass usually breaks. This is one of the most common sources of cuts in the laboratory.

Use care in making rubber-to-glass connections. Lengths of glass tubing should be supported while being inserted into rubber. The ends of the glass should be *FLAME POLISHED*[3] and either wetted or covered with a lubricating jelly for ease of connections. Never use oil or grease to lubricate the glass. Gloves should be worn when making such connections and the tubing should be held as close to the insertion end as possible to prevent bending or breaking. Never force rubber tubing or stoppers from glassware. Cut off the rubber or other material instead.

BURNS. Heating glassware and porcelain usually produces a red color that disappears in seconds, however the glass remains hot enough to burn your skin for several minutes. After heating a piece of glass or porcelain, put it aside to let it cool. Many safeguards against burns are available. Special gloves, safety tongs, and aprons are but a few examples. Make it a habit to wear a pair of gloves or to use a pair of tongs to handle a dish or flask that has been heated.

Spattering from acids, caustic materials, and strong oxidizing solutions on skin should be washed off immediately with large quantities of water. Neutralize an acid with sodium carbonate or bicarbonate. Every worker in the wastewater laboratory should have access to a sink, an emergency deluge shower, and an eye wash station.

Perhaps the most harmful and painful chemical burn occurs when small objects, chemicals, or fumes get into your eyes. If this happens, immediately flood your eyes with water or a special eyewash solution from a safety kit, eye wash station, or fountain. Washing the eyes with large amounts of water for at least 15 minutes is recommended.

TOXIC FUMES. Certain chemicals are dangerous to breathe; therefore, use a ventilated laboratory fume hood for routine reagent preparation or whenever handling substances that may generate harmful atmospheric contaminants. Select a hood that has adequate air displacement and expels harmful vapors and gases at their source. The face velocity of laboratory fume hoods must average from 100 to 150 feet per minute (FPM) or 30 to 45 meters per minute (MPM), depending on the substance handled. A device that continuously indicates that air is flowing should be provided. Noxious fumes can be spread by the heating and cooling system of the building. Check with your local safety regulatory agency for specific requirements for your facility. An annual check should be made of the entire laboratory building.

WASTE DISPOSAL. A good safety program requires the regular disposal of laboratory waste. Hazardous chemicals or substances must be disposed of by using methods that comply with local environmental regulations. Confirm the local requirements before starting disposal. To protect maintenance personnel from injuries, use separate, clearly marked, covered containers to dispose of broken glass.

FIRE. The laboratory should be equipped with a fire blanket, which is used to smother clothing fires. Small fires that occur in an evaporating dish or beaker may be put out by covering the container with a glass plate, wet towel, or wet blanket. Do not use a fire extinguisher on small beaker fires because the force of the spray may knock over the beaker and spread the fire. Use a fire extinguisher for larger fires or those that may spread rapidly. Become familiar with the operation and use of your fire extinguisher before an emergency occurs.

Using the proper type of fire extinguisher for each class of fire provides the best control of the situation and avoids compounding the problem. The class of fire is based on the type of material being consumed (burned), and the class of extinguisher corresponds with the class of fire it will extinguish.

Fire classifications are important for determining the type of fire extinguisher needed to control the fire. Classifications also aid in recordkeeping. Fires are classified as A, B, C, or D fires based on the type of material being consumed. Fire extinguishers are also classified as A, B, C, or D to correspond with the class of fire each will extinguish:

---

3. *Flame Polished.*   Melted by a flame to smooth out irregularities. Sharp or broken edges of glass (such as the end of a glass tube) are rotated in a flame until the edge melts slightly and becomes smooth.

Class A fires consume ordinary combustibles such as wood, paper, cloth, rubber, many plastics, dried grass, hay, and stubble. Use foam, water, soda-acid, carbon dioxide gas, or almost any type of extinguisher.

Class B fires consume flammable and combustible liquids such as gasoline, oil, grease, tar, oil-based paint, lacquer, and solvents, and also flammable gases. Use foam, carbon dioxide, or dry chemical extinguishers.

Class C fires consume energized electrical equipment such as starters, breakers, and motors. Use carbon dioxide or dry chemical extinguishers to smother the fire; both types are nonconductors of electricity.

Class D fires consume combustible metals such as magnesium, sodium, zinc, and potassium. Operators rarely encounter this type of fire. Use a Class D extinguisher or fine dry soda ash, sand, or graphite to smother the fire. Consult with your local fire department about the best methods to use for specific hazards that exist at your facility.

Multipurpose extinguishers are also available, such as a Class BC carbon dioxide extinguisher that can be used to smother both Class B and Class C fires. A multipurpose ABC carbon dioxide extinguisher will handle most laboratory fire situations. (When using carbon dioxide extinguishers, remember that carbon dioxide can displace oxygen, so take appropriate precautions.)

There is no single type of fire extinguisher that is effective for all fires so it is important that you identify the class of fire you are trying to control. You must be trained in the use of the different types of extinguishers, and the proper type should be located near the area where that class of fire may occur.

14.222 Chemical Storage and Transfer

An adequate storeroom is essential for the safe storage of chemicals in a wastewater laboratory. The storeroom should be properly ventilated and lighted and be laid out to segregate incompatible chemicals. Flammable liquids, acids, bases, and oxidizing agents should be separated from each other by distances, partitions, or other means so as to prevent accidental contact between them. Order and cleanliness must be maintained. All chemicals and bottles or reagents must be clearly labeled and dated. Never handle chemicals with bare hands. Use a spoon, spatula, or tongs.

Heavy items should be stored on or as near to the floor as possible. VOLATILE LIQUIDS[4] that may escape as a gas, such as ether, must be kept away from heat sources, sunlight, and electric switches. Do not store ether in the refrigerator because laboratory explosions have occurred when the gas ignited as the refrigerator light came on.

Cylinders of gas in storage should be capped and secured to prevent rolling or tipping. They should also be placed away from any possible sources of heat or open flames.

Flammable gases must be stored separately. The storage room should be fitted with explosion-proof wiring and lighting fixtures. Appropriate warning signs prohibiting sources of ignition from being brought into the area must be posted in conspicuous locations.

Also of concern in the prevention of laboratory accidents is the safe transfer of chemicals, apparatus, gases, or other hazardous materials from the storeroom to the laboratory for use. Use cradles or tilters to facilitate handling carboys or other larger chemical vessels. Use a trussed hand truck to transport cylinders of compressed gases. Never move or transport a cylinder without the valve protection hood in place. Never roll a cylinder by its valve. Immediately after they are positioned for use, cylinders should be clamped securely into place to prevent shifting or toppling. Carry flammable liquids in safety cans or, in the case of reagent-grade chemicals, protect the bottle with a carrier. Always wear protective gloves, safety shoes, and rubber aprons in case of accidental spilling of chemical containers.

Drum Tilter

---

4. Volatile Liquids. Liquids that easily vaporize or evaporate at room temperature.

The usual common sense rules and industry standards for chemical storage and transfer should be followed. Good housekeeping practices make significant contributions to an effective safety program in an active chemical storage room.

14.23   Acknowledgments

Portions of this section were taken from material written by A. E. Greenberg, "Safety and Hygiene," which appeared in the California Water Pollution Control Association's *Operators' Laboratory Manual.* Some of the ideas and material also came from the *Fisher Safety Manual.*

14.24   Additional Reading

1. *Fisher Safety Catalog.* Obtain from Literature Fulfillment at the Fisher Scientific website, www.fishersci.com, or phone (800) 772-6733.

2. *General Industry, OSHA Safety and Health Standards* (CFR, Title 29, Labor Pt. 1900-1910.999 (most recent edition)). Obtain from the US Government Printing Office, PO Box 979050, St. Louis, MO 63197-9000. Order No. 869-074-00110-8. Price, $67.00.

END OF LESSON 2 OF 7 LESSONS

on

LABORATORY PROCEDURES

Please answer the discussion and review questions next.

DISCUSSION AND REVIEW QUESTIONS

Chapter 14.   LABORATORY PROCEDURES

(Lesson 2 of 7 Lessons)

Please write your answers to the following questions to determine how well you understand the material in the lesson. The question numbering continues from Lesson 1.

4. What precautions should personnel take to protect themselves from diseases when working in a wastewater treatment plant?

5. What are the basic rules for working in a laboratory?

6. Why should work with certain chemicals be conducted under a ventilated laboratory fume hood?

CHAPTER 14.   LABORATORY PROCEDURES

(Lesson 3 of 7 Lessons)

14.3   SAMPLING

14.30   Importance

The basis for any plant monitoring program is the information obtained by sampling. Decisions based on incorrect data may be made if sampling is performed in a careless manner. Obtaining good results will depend to a great extent on the following factors:

1. Ensuring that the samples taken are truly representative of the wastestream

2. Using proper sampling techniques

3. Protecting and preserving the samples until they are analyzed

4. Following a proper *CHAIN OF CUSTODY*[5] procedure if samples are sent off site for analysis

The greatest errors found in laboratory results are usually caused by improper sampling, poor preservation, or lack of enough mixing during *COMPOSITING*[6] and testing.

14.31   Representative Sampling

Wastewater flows can vary widely in quantity and composition over a 24-hour period. Also, composition can vary within a given stream at any single time due to the partial settling of solids or floating of light materials. To efficiently operate a wastewater treatment plant, an operator must rely on test results to indicate what is happening in the treatment process. This requires samples to be taken from the material, water, or wastestream where it is well mixed. Obtaining a *REPRESENTATIVE SAMPLE*[7] should be a major priority in any sampling and monitoring program. If a representative sample is not collected, the test results will not have any significant meaning.

Laboratory equipment is generally quite accurate. Analytical balances weigh to 0.1 milligram. Graduated cylinders, pipets, and burets usually measure to 1 percent accuracy, so that the errors introduced by these items should total less than 5 percent, and under the worst possible conditions only 10 percent.

Under ideal conditions, assume that a test of raw wastewater for suspended solids should run about 300 mg/L. Because of the previously mentioned equipment or apparatus variables, the value may actually range from 270 to 330 mg/L. Results in this range are reasonable for operation. Other less obvious factors are usually present that make it quite possible to obtain results that are 25, 50, or even 100 percent in error, unless certain precautions are taken. The following example illustrates how these errors are produced.

The Dumpmore Wastewater Treatment Plant is a secondary treatment facility with a flow of 8 million gallons per day (30 ML/day). The plant has an aerated grit basin, two 750,000-gallon (2.8-ML) capacity, circular primary clarifiers, two digesters, two aeration basins, two secondary clarifiers, four chlorinators, and two chlorine contact basins.

Monthly summary calculations based on the suspended solids test showed that about 8,000 pounds (3,640 kg) of suspended solids were captured per day during primary sedimentation, assuming 200 mg/L for the influent and 100 mg/L for the effluent. However, 12,000 pounds (5,450 kg) per day of raw sludge solids were being pumped out of the primary clarifiers into the digester. If sampling and analyses were perfect, these weights would have been balanced, provided the waste activated sludge was not returned to the primary clarifiers. The capture should equal the removal of solids. A study was made to determine why the variance in these values was so great. The problem could be due to incorrect testing procedures, poor sampling, incorrect metering of the wastewater or sludge flow, or any combination of these reasons.

In the first case, the equipment was in excellent condition. The operator carried out the laboratory procedures carefully and had previously run successful tests on comparative samples. It was concluded that the equipment and test procedures were completely satisfactory.

A survey was then made to determine if the sampling stations were in need of relocation. By using Imhoff cones and running

---

5. *Chain of Custody.*   A record of each person involved in the handling and possession of a sample from the person who collected the sample to the person who analyzed the sample in the laboratory and to the person who witnessed disposal of the sample.

6. *Composite (Proportional) Sample.*   A composite sample is a collection of individual samples obtained at regular intervals, usually every one or two hours during a 24-hour time span. Each individual sample is combined with the others in proportion to the rate of flow when the sample was collected. Equal volume individual samples also may be collected at intervals after a specific volume of flow passes the sampling point or after equal time intervals and still be referred to as a composite sample. The resulting mixture (composite sample) forms a representative sample and is analyzed to determine the average conditions during the sampling period.

7. *Representative Sample.*   A sample portion of material, water, or wastestream that is as nearly identical in content and consistency as possible to that in the larger body being sampled.

settleable solids tests along the influent channel and in the aerated grit chamber, it was quickly recognized that the best mixed and most representative samples would be taken from the aerated grit chamber rather than from the influent channel.

The settleable solids ran 13 mL/L in the aerated grit chamber against 10 mL/L in the channel. By the simple process of determining the best sampling station, the suspended solids value in the influent was corrected from 200 mg/L to the more representative 300 mg/L. Calculations made using the correct figures changed the solids capture from 8,000 pounds to 12,000 pounds per day (3,640 kg to 5,450 kg/day), and a balance was obtained.

This example illustrates the importance of selecting a good sampling point in order to collect a truly representative sample. It also emphasizes the point that even though a test is accurately performed, the results may be entirely erroneous and meaningless insofar as their usability for process control is concerned, unless a good representative sample is taken. Furthermore, a good sample is highly dependent on the sampling station. Whenever possible, select a place where mixing is thorough and the wastewater quality is uniform. As the solids concentration increases above about 200 mg/L, mixing becomes even more significant because the wastewater solids will tend to separate rapidly with the heavier solids settling toward the bottom, the lighter solids in the middle, and the floatables rising toward the surface. If, as is usual, a one-gallon portion is taken as representative of a million gallon flow, the job of sample location and sampling must be taken seriously.

14.32 Time of Sampling

The time and frequency of sampling are also important factors in the sampling process. Samples should be taken to represent typical weekdays or even varied from day to day within the week to obtain a good indication of the characteristics of the wastewater. In carrying out a testing program, particularly where personnel and time are limited due to operational responsibilities, testing may necessarily be restricted to about one test day per week. If you decide to start your tests early in the week by taking samples early on Monday morning, you may wind up with odd results.

For example, during a test for SURFACTANTS[8] at one plant, samples were taken early on Monday morning and rushed to the laboratory for testing. Due to the detention time in the sewers, these wastewater samples actually represented Sunday flow on the graveyard shift, the weakest wastewater obtainable. The surfactant content was only 1 mg/L, whereas it would usually run 8 to 10 mg/L, so, the time and day of sampling is quite important.

14.33 Types of Samples

The two types of samples collected in treatment plants are grab samples and composite samples, either of which may be obtained manually or automatically.

GRAB SAMPLES

A grab sample is a single sample of wastewater taken at neither a set time nor flow. The grab samples show the waste characteristics only at the time the sample is taken. A grab sample may be preferred over a composite sample under these conditions:

1. The wastewater to be sampled does not flow on a continuous basis.

2. The wastewater characteristics are relatively constant.

3. The operator wants to determine whether or not a composite sample obscures extreme conditions of the waste.

4. The wastewater is to be analyzed for dissolved oxygen (DO), coliforms, chlorine residual, temperature, and pH. The analyses for these water quality indicators must be performed immediately. (*NOTE:* Grab samples for these water quality indicators may be collected at set times or specific time intervals.)

COMPOSITE SAMPLES

As wastewater quality changes from moment to moment and hour to hour, the best results would be obtained by using some type of continuous sampler-analyzer. However, because operators are usually the sampler-analyzer, continuous analysis would leave little time for any other tasks. Except for tests that cannot wait due to rapid chemical or biological change of the sample, such as tests for dissolved oxygen and sulfide, a fair compromise may be reached by taking samples throughout the day at hourly or two-hour intervals.

Samples should be refrigerated immediately after collection to preserve them from continued bacterial decomposition. When all of the samples have been collected for a 24-hour period, the samples from a specific location should be combined or composited according to flow to form a single 24-hour composite sample.

If an equal volume of sample was collected each hour and mixed, this would simply be a composite sample. A proportional composite sample may be prepared by collecting a sample every hour. The rate of wastewater flow must be known and each grab sample must then be taken and measured out in direct proportion to the volume of flow at that time. Table 14.2 illustrates the hourly flow and sample volume to be measured for a 12-hour proportional composite sample.

Large wastewater solids should be excluded from a sample, particularly those greater than one-quarter inch (6 mm) in diameter.

During compositing and at the exact moment of testing, the samples must be vigorously remixed[9] so that they will be of the same composition and as well mixed as when they were originally sampled. Such remixing may become lax, so that all the solids are not uniformly suspended. Lack of mixing can cause low results in samples of solids that settle out rapidly, such as those in activated

---

8. *Surfactant* (sir-FAC-tent). Abbreviation for surface-active agent. The active agent in detergents that possesses a high cleaning ability.

9. *NOTE:* If the sample has a low buffer capacity and the real pH is 6.5 or less, vigorous shaking can cause a significant change in pH level.

TABLE 14.2   DATA COLLECTED TO PREPARE PROPORTIONAL COMPOSITE SAMPLE

Time	Flow, MGD	Factor	Sample Vol, mL
6 AM	0.2	100	20
7 AM	0.4	100	40
8 AM	0.6	100	60
9 AM	1.0	100	100
10 AM	1.2	100	120
11 AM	1.4	100	140
12 N	1.5	100	150
1 PM	1.2	100	120
2 PM	1.0	100	100
3 PM	1.0	100	100
4 PM	1.0	100	100
5 PM	0.9	100	90
			1,140

A sample composited in this manner would total 1,140 mL.

TABLE 14.3   DECREASE IN PERCENT TOTAL SOLIDS DURING PUMPING

Pumping Time, min	Total Solids, percent	Cumulative Solids Average
0.5	7.0	7.0
1.0	7.1	7.1
1.5	7.4	7.2
2.0	7.3	7.2
2.5	6.7	7.1
3.0	5.3	6.8
3.5	4.0	6.4
4.0	2.3	5.9
4.5	2.0	5.5
5.0	1.5	5.1

sludge or raw wastewater. Samples must therefore be mixed thoroughly and poured quickly before any settling occurs. If this is not done, errors of 25 to 50 percent may easily occur. For example, on the same mixed liquor sample, one person may find 3,000 mg/L suspended solids while another person may find only 2,000 mg/L due to poor mixing. When a well-mixed, proportional composite sample is tested, a reasonably accurate measurement of the quality of the day's flow can be obtained.

If a 24-hour sampling program is not possible, single representative samples should be taken at a time when typical characteristic qualities are present in the wastewater. The samples should be taken in accordance with the detention time required for treatment. For example, this period may exist between 10 am and 5 pm for the sampling of raw influent. If a sample is taken at 12 noon, other samples should be taken in accordance with the detention periods of the serial processes of treatment in order to follow this slug of wastewater or *PLUG FLOW*.[10] In primary settling, if the detention time in the primaries is two hours, the primary effluent should be sampled at 2 pm. If the detention time in the succeeding secondary treatment process required three hours, this sample should be taken at 5 pm.

14.34   Sludge Sampling

In sampling raw sludge and feeding a digester, keep the following important points in mind:

1. When raw sludge is pumped from a primary clarifier, sludge solids vary considerably with pumping time for samples withdrawn every one-half minute, as shown in Table 14.3. The examples listed in the table show that solids were heavy during the first 2.5 minutes, and thereafter rapidly became thin and watery. Because sludge solids fed to a digester should be as heavy as possible and contain a minimum amount of water, the pumping should probably have been stopped at about 3 minutes. After 3 minutes, the water content became greater than desirable.

2. When sampling sludge from a primary clarifier, prepare a composite sample by mixing small, equal portions of sludge that are taken every 0.5 minute during pumping. If only a single portion of sludge is taken for the sample, there is a chance that the sludge sample will be too thick or too thin, compared to the actual sludge pumped. A composite sample prevents this possibility.

10. *Plug Flow.* A type of flow that occurs in tanks, basins, or reactors when a slug of water or wastewater moves through a tank without ever dispersing or mixing with the rest of the water or wastewater flowing through the tank.

DIRECTION OF FLOW

PLUG FLOW

3. When a sludge sample is left standing, the solids and liquid separate due to gasification and flotation or to settling of the solids. This makes it absolutely necessary to thoroughly remix the sample before pouring it for a test.

4. When individual samples are taken at regular intervals, they should be carefully preserved to prevent sample deterioration by bacterial action. Refrigeration is an excellent method of preservation and is generally preferable to chemical preservation, which may interfere with tests such as biochemical oxygen demand (BOD) and chemical oxygen demand (COD) tests. Because sludge digestion can even occur under refrigeration, do not store refrigerated sludge samples in sealed containers so that the gases produced do not cause sludge to spray out of the container when opened.

14.35   Sampling Devices

Automatic sampling devices are effective timesavers and should be used where possible. However, as with anything automatic, problems do arise and the operator should be aware of potential difficulties. Sample lines to auto-samplers may build up growths that may periodically slough off and contaminate the sample with a high solids content. Cleaning out the intake line is required on a regular basis to prevent this problem.

Manual sampling equipment includes dippers, weighted bottles, hand-operated pumps, and cross-section samplers. Dippers consist of wide-mouth, corrosion-resistant containers (such as cans or jars) on long handles that collect samples for testing. A weighted bottle is a collection container that is lowered to a desired depth. At this location a cord or wire removes the bottle stopper so the bottle can be filled. Sampling pumps allow the inlet to the suction hose to be lowered to the sampling depth. Cross-sectional samplers are used to sample where the wastewater and sludge may be in layers, such as in a digester or clarifier. The sampler consists of a tube, open at both ends, that is lowered at the sampling location. When the tube is at the proper depth, the ends of the tube are closed and a sample is obtained from different layers.

Many operators build their own samplers (Figure 14.5) using the material described below:

1. *SAMPLING BUCKET.* A coffee can attached to an 8-foot (2.5-m) length of ½-inch (1.2-cm) electrical conduit or a wooden broom handle with a ¼-inch (0.6-cm) diameter spring in a 4-inch (10-cm) loop. A section of thin-walled PVC pipe (Schedule 40) of appropriate diameter with a cap on one end will be a more durable sampling bucket than a coffee can.

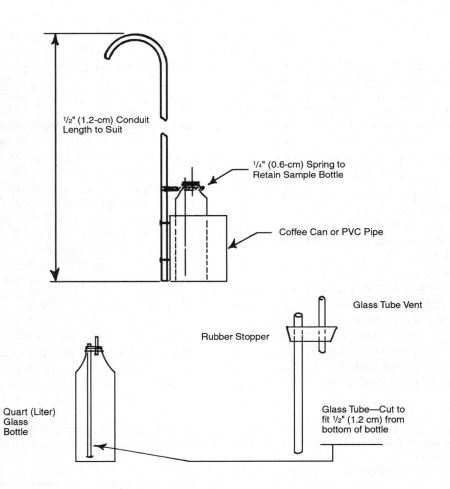

Fig. 14.5   Dissolved oxygen sampling bottle

Surface samples also can be collected using a plastic bucket or a gallon milk jug on a rope with an opening cut out opposite from the handle. Placing a large bolt and nut near one edge of the sampler will cause the bucket to tip so that the sample will fill the bucket, rather than the bucket floating on the water surface.

2. *SAMPLING BOTTLE.* Glass bottle with rubber stopper equipped with two ⅜-inch (1-cm) glass tubes, one ending near the bottom of bottle to allow sample to enter and the other ending at the bottom of the stopper to allow the air in the bottle to escape while the sample is filling the bottle.

For sample containers, wide-mouth glass bottles are recommended. Glass bottles, though somewhat expensive initially, greatly reduce the problem of metal contamination. The wide-mouth bottles ease the washing problem. For regular samples, sets of glass bottles bearing identification labels should be used.

14.36   Preservation of Samples

Sample deterioration starts immediately after collection for most wastewaters. The shorter the time that elapses between collection and analysis, the more reliable will be the analytical results.

Samples should be preserved if they are not going to be analyzed immediately due to operator workload or the remoteness of the laboratory. Preservation is essential to prevent the deterioration of samples. Some water quality constituents (such as temperature, pH, and DO) must be analyzed immediately on site because they cannot be preserved. A summary of acceptable EPA (US Environmental Protection Agency) methods of preservation are shown in Table 14.4.

The information shown in Table 14.4 is adapted from "Table II—Required Containers, Preservation Techniques, and Holding Times," which is part of 40 CFR 136.3 in the Code of Federal

TABLE 14.4   US EPA RECOMMENDED PRESERVATION METHODS FOR WATER AND WASTEWATER SAMPLES [a]

Test	Container[b]	Preservation Method	Max. Holding Time[c]
Acidity/Alkalinity	P, FP, G	Cool to ≤6°C	14 days
Ammonia	P, FP, G	Cool to ≤6°C, Add H_2SO_4 to pH <2	28 days
BOD	P, FP, G	Cool to ≤6°C	48 hours
COD	P, FP, G	Cool to ≤6°C, Add H_2SO_4 to pH <2	28 days
Chloride	P, FP, G	None required	28 days
Chlorine, Total Residual	P, G	None required	Analyze within 15 minutes
Cyanide, Total or Available (CATC)	P, FP, G	Cool to ≤6°C, Add NaOH to pH >12	14 days
Dissolved Oxygen	G, Bottle and top	None required	Analyze within 15 minutes
Fluoride	P	None required	28 days
Mercury (CVAA)	P, FP, G	Add HNO_3 to pH <2	28 days
Metals	P, FP, G	Add HNO_3 to pH <2 at least 24 hours prior to analysis	6 months
Nitrate	P, FP, G	Cool to ≤6°C	48 hours
Nitrate-Nitrite	P, FP, G	Cool to ≤6°C, H_2SO_4 to pH <2	28 days
Nitrite	P, FP, G	Cool to ≤6°C	48 hours
Oil & Grease	G	Cool to ≤6°C, HCl or H_2SO_4 to pH <2	28 days
Organic Carbon	P, FP, G	Cool to ≤6°C, HCl, H_2SO_4, or H_3PO_4 to pH <2	28 days
pH	P, FP, G	None required	Analyze within 15 minutes
Phenols	G	Cool to ≤6°C, H_2SO_4 to pH <2	28 days
Phosphorus:			
Orthophosphate	P, FP, G	Cool to ≤6°C	Filter within 15 minutes; Analyze within 48 hours
Total	P, FP, G	Cool to ≤6°C, H_2SO_4 to pH <2	28 days
Solids	G	Cool to ≤6°C	7 days
Specific Conductance	P, FP, G	Cool to ≤6°C	28 days
Sulfate	P, FP, G	Cool to ≤6°C	28 days
Sulfide	P, FP, G	Cool to ≤6°C, Add zinc acetate plus NaOH to pH >9	7 days
Temperature	P, FP, G	None required	No holding
Total Kjeldahl and Organic Nitrogen	P, FP, G	Cool to ≤6°C, H_2SO_4 to pH <2	28 days
Turbidity	P, FP, G	Cool to ≤6°C	48 hours

a Adapted from 40 CFR 136.3, July 1, 2010, "Table II—Required Containers, Preservation Techniques, and Holding Times." US Government Printing Office, Superintendent of Documents, PO Box 979050, St. Louis, MO 63197-9000. To find the current version of "Table II" in the Code of Federal Regulations (CFR), see the instructions in Section 14.36.

b Polyethylene (P), Fluoropolymer (FP), or Glass (G). For metals, polyethylene with a polypropylene cap (no liner) is preferred.

c Holding times listed above are recommended for properly preserved samples based on currently available data. It is recognized that for some sample types, extension of these times may be possible while for other types, these times may be too long.

Regulations (CFR). (To find the current version of "Table II" in the CFR, go to www.fdsys.gov. In the search box type "40 CFR 136.3." The search results display a list of relevant documents. Click the link for "40 CFR 136.3 - Identification of Test Procedures." When the document opens, scroll through the pages to locate "Table II.")

14.37   Quality Control in the Wastewater Laboratory

Quality control in the wastewater laboratory enables operators to obtain good analytical results. Practices that improve these results include the following:

1. Taking representative samples before any tests are made

2. Selecting a good sampling location

3. Collecting samples and, if necessary, properly preserving them

4. Mixing samples thoroughly before compositing, and at the time of testing

Using the correct methods and having good equipment are not enough to ensure correct analytical results. Each operator must be alert to factors in the laboratory that can cause poor data quality, including sloppy laboratory technique, deteriorated reagents, poorly operating instruments, and calculation mistakes.

14.38   Additional Reading

1. *Handbook for Monitoring Industrial Wastewater,* US Environmental Protection Agency. EPA No. 625-6-73-002. Obtain from National Technical Information Service (NTIS), 5301 Shawnee Road, Alexandria, VA 22312. Order No. PB-259146. Price, $60.00, plus $6.00 shipping and handling.

2. *Handbook for Analytical Quality Control in Water and Wastewater Laboratories,* US Environmental Protection Agency. EPA No. 600-4-79-019. Obtain from National Technical Information Service (NTIS), 5301 Shawnee Road, Alexandria, VA 22312. Order No. PB-297451. Price, $60.00, plus $6.00 shipping and handling.

QUESTIONS

Please write your answers to the following questions and compare them with those on page 338.

14.3A   What causes most errors found in laboratory results?

14.3B   Why must a representative sample be collected?

14.3C   How would you prepare a proportional composite sample?

END OF LESSON 3 OF 7 LESSONS

on

LABORATORY PROCEDURES

Please answer the discussion and review questions next.

DISCUSSION AND REVIEW QUESTIONS

Chapter 14.   LABORATORY PROCEDURES

(Lesson 3 of 7 Lessons)

Please write your answers to the following questions to determine how well you understand the material in the lesson. The question numbering continues from Lesson 2.

7. What is meant by representative sample?

8. Where would you obtain a representative sample?

9. Why would you preserve a sample?

CHAPTER 14.   LABORATORY PROCEDURES

(Lesson 4 of 7 Lessons)

14.4   LABORATORY PROCEDURES FOR PLANT CONTROL

Lesson 4 outlines laboratory procedures used by operators to analyze samples for plant process control. These tests are frequently performed by operators to quickly obtain the results and make any necessary process adjustments. Lessons 6 and 7 outline laboratory procedures for NPDES monitoring. These test procedures may be performed by operators, but are also performed by laboratory analysts, especially in larger plants.

14.40   Clarity

A.   *DISCUSSION*

All plant effluents should be examined daily to observe the clarity of the effluent. The clarity test is an indication of the quality of the effluent with respect to color, solids, and turbidity. The purpose of this test is to determine whether the plant effluent is staying the same, improving, or deteriorating.

The clarity test can be performed in the lab by looking down through the effluent in a graduated cylinder, or it can be performed in the field by looking down through the effluent in a clarifier or chlorine contact basin. The clarity of the effluent should be examined at the same time each day and under the same conditions, so that the results will be comparable from day to day. By noting how far down you can see into the cylinder or basin and comparing each day's test results with the previous day's results, you will be able to see whether the general quality of the effluent is changing.

Sometimes, this test is referred to as a turbidity measurement; however, for our purpose, you are interested in the *clarity of the effluent* rather than the *quantity of solids* in the effluent. Section 14.5, Procedure 6, "Turbidity," on page 335 discusses the testing procedure for turbidity.

B.   *WHAT IS TESTED?*

Sample	Common Range (Field Test)	
	Poor	Good
Secondary Clarifiers:		
Trickling Filter	1 ft (0.3 m)	3 ft (0.9 m)
Activated Sludge	3 ft (0.9 m)	6 ft (1.8 m)
Activated Sludge Blanket in Secondary Clarifier	1 ft (0.3 m)	4 ft (1.2 m)
Chlorine Contact Basins	3 ft (0.9 m)	6 ft (1.8 m)

C.   *APPARATUS*

1. One clarity unit, also known as a Secchi (SECK-key) disk, with an attached rope marked in one-foot units

2. One 1,000-mL graduated cylinder

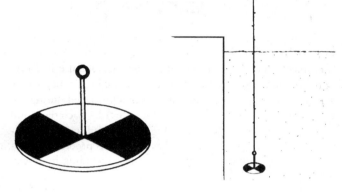

Secchi Disk

D.   *REAGENTS*

None

E.   *PROCEDURES*

1. *FIELD TEST.* Tie the end of the marked rope to the handrail where the tests will be run, for example, in final sedimentation unit. Always take tests at the same time each day for comparable results. Lower the disk slowly until you just lose sight of it, then stop. Bring it up slowly until just visible, then stop again. Look at the marks on the rope to see the depth of water that you can see the disk through. Bring up the disk, then clean and store it. Record the results.

2. *LAB TEST.* Use a clean 1,000-mL graduate. Fill with a well-mixed sample up to the 1,000-mL mark. During every test the same lighting conditions in the lab should be maintained. Look down through the liquid in the cylinder and read the last visible number etched on the side of the graduate and record the results.

Whether you use one or both of these tests, run each test at the same time every day and under similar conditions for comparable results.

F.   *TEST RESULTS*

1. Each foot of depth is better clarity with Secchi disk.

2. Each 100-mL mark seen in depth is better clarity.

(Hydrogen Sulfide)

QUESTIONS

Please write your answers to the following questions and compare them with those on page 338.

14.4A   What does the clarity test indicate?

14.4B   What happens when you attempt to measure clarity under different conditions such as lighting and clarifier loadings?

14.41   Hydrogen Sulfide (H₂S)

Hydrogen sulfide is measured in wastewater because it causes corrosion and odors that should be controlled. The major portion of the sulfide content in wastewater is produced during the conversion of sulfate (SO_4^{2-}) to sulfide (S^{2-}) by bacteria found in the wastewater. Oxygen-reducing bacteria will use any available sulfur-containing compound as food. This process can produce odorous, reduced-sulfur compounds, including hydrogen sulfide (H_2S).

H_2S production in wastewater can be controlled by up-sewer maintenance, which reduces H_2S formation. In some cases, it can be controlled by the application of chemicals such as chlorine, oxygen, hydrogen peroxide, iron salts, or bases that increase the pH.

14.410   H₂S in the Atmosphere

A. DISCUSSION

The rate of corrosion in the sewer collection system and treatment plant is often directly related to the rate of H_2S production or the amount of H_2S in the atmosphere. Hydrogen sulfide gas is toxic to your respiratory system and is both flammable and explosive under certain conditions. The explosive limits of hydrogen sulfide range from 4.3% (lower explosive limit or LEL) to 46% (upper explosive limit or UEL). The parts per million (ppm) concentration of hydrogen sulfide in the sewer atmosphere is quite different from that in wastewater. A concentration of 1 mg/L (ppm) in turbulent wastewater can quickly produce a concentration of 300 ppm in an unventilated, enclosed atmospheric space. The minimum concentration of H_2S in the atmosphere known to cause death is 300 ppm.

B. METHODS

Methods available for testing H_2S in the atmosphere are listed in Section D "Apparatus" that follows. Method 1, using paper tape or tiles with lead acetate, will give a rough estimate of H_2S. Method 2 is a faster and more accurate instrumental measurement of a hazardous level or of the actual concentration of sulfide present.

C. WHAT IS TESTED?

Sample	Common Range
Atmosphere in sewers, outlets from force mains, wet pits, pumping stations, and influent areas to treatment plant.	1. Lead Acetate Method (1) Not black in 24 hours = Good, 24+ hr Black in less than 1 hour = Bad, <1 hour
	2. H₂S Detector Method (2) <3 ppm = good >10 ppm = toxic

D. APPARATUS

Method 1: Lead acetate paper or unglazed tile soaked in lead acetate solution

Method 2: H_2S Detectors or H_2S Detector Tubes

These devices are available in three types:

(1) Personal. Attach to your belt or use a shoulder strap. These devices have an alarm signal; however, some units also have a readout for the level of hydrogen sulfide. You should rely on the alarm signal rather than the readout.

(2) Portable. Carry this device by hand and place it near your work site. These devices have an alarm signal and a meter that indicates the level of H_2S.

(3) Stationary. Install this device permanently in lift stations, gas compressor rooms, and other potentially dangerous areas. These devices have an alarm signal and a meter that indicates the level of H_2S.

Contact your local safety equipment supplier for information on atmospheric monitors.

E. REAGENTS

Method 1: Lead Acetate Solution, saturated. Dissolve 50 grams lead acetate in 80 mL distilled water.

Method 2: No reagents are required.

F. PROCEDURE

Method 1:

1. Obtain pieces of unglazed tile or use lead acetate paper. Using a hacksaw, cut the tile into ½-inch (1.2-cm) strips.

2. Soak the tile pieces or paper strips in the lead acetate solution.

3. Dry the tile in a drying oven or let it air dry.

4. An open manhole or any point where wastewater is exposed to the atmosphere is a good test site. Drive a nail between the metal crown ring of the manhole, concrete, or in another convenient place. Tie the tile or paper to the nail with cotton string.

Replace the manhole cover and return in half an hour or less. If the tile is not black or substantially colored, return periodically to check its progress until it turns black. If H_2S is present as indicated by a color change, then measure flow, temperature, pH, and BOD for further evaluation of the problem.

Method 2: The instructions are included with the instrument.

G. CALCULATIONS

Method 1: None required.

Method 2: None required.

14.411   H₂S in Wastewater

A. DISCUSSION

In sewers, when there is no longer any dissolved oxygen, H_2S tests are run to determine the rate of H_2S increase as the wastewater travels to a pumping station or treatment plant. If the wastewater is exposed to the atmosphere, H_2S will be released and a typical rotten egg odor will be detected. This type of odor is indicative of the anaerobic decomposition of organics in wastewater, which occurs in the absence of oxygen. The process of the anaerobic decomposition of solids can release H_2S into sewers. When the gas leaves the wastewater stream, some of it dissolves in the condensed moisture on the concrete. Sulfur-oxidizing bacteria convert the hydrogen sulfide to sulfuric acid, which is very corrosive to concrete. Only the concrete exposed to the atmosphere is corroded while the concrete below the low waterline is not attacked.

Not all odors in wastewater are from H_2S, and there is no correlation between H_2S and other odors. The total H_2S procedure is good up to 18 mg/L, and higher concentrations must be diluted before testing. H_2S production can be controlled by up-sewer maintenance, which reduces H_2S formation in the wastewater and protects the collection system. In some severe cases, chemicals are applied to flows in the collection system for H_2S control. These chemicals include chlorine, oxygen, hydrogen peroxide, iron salts, or bases that increase the pH.

B. WHAT IS TESTED?

Sample Wastewater Locations	H₂S Concentrations Possible Results, mg/L	
	Good	Bad
Sewers	0.1	1
Outlets from force mains	0.1	1
Wet pits, pumping stations	0.1	0.5
Influents to treatment plants	0	0.5

All of these locations should be sampled, if pertinent, when using upstream H_2S control in the collection system.

C. APPARATUS

Use one of the following apparatus for this test:

1. LaMotte-Pomeroy Sulfide Testing Kit to test:
   a. Total Sulfide
   b. Dissolved Sulfide
   c. Hydrogen Sulfide in Solution

Obtain from LaMotte Company, PO Box 329, Chestertown, MD 21620. Order Code #4630.

2. Hach Hydrogen Sulfide Test Kit

   Obtain from HACH Company, PO Box 389, Loveland, CO 80539-0389. Catalog No. 2238-01.

D. REAGENTS

The reagents are included in the kits.

E. PROCEDURE

The instructions are included in the kits.

F. EXAMPLE

The instructions are included in the kits.

G. CALCULATIONS

The instructions are included in the kits.

H. ADDITIONAL READING

1. *Standard Methods for the Examination of Water and Wastewater,* 21st Edition. Obtain from Water Environment Federation (WEF), Publications Order Department, PO Box 18044, Merrifield, VA 22118-0045. Order No. S82011. Price to members, $203.00; nonmembers, $268.00; price includes cost of shipping and handling.

2. *Operation and Maintenance of Wastewater Collection Systems,* Volumes I and II, prepared by the Office of Water Programs, California State University, Sacramento, 6000 J Street, Sacramento, CA 95819-6025. Price: $49.00 for each volume.

QUESTION

Please write your answers to the following questions and compare them with those on page 338.

14.4C   Why is H_2S concentration measured in:

   1. the atmosphere?
   2. wastewater?

14.42   Settleable Solids

A. DISCUSSION

The settleable solids test measures the volume of settleable solids in one liter of sample that will settle to the bottom of an Imhoff cone within one hour. The test is an indication of the volume of solids removed by sedimentation in sedimentation tanks, clarifiers, or ponds. The results are read directly from the Imhoff cone in milliliters. The actual volume of solids pumped may disagree with the calculated volume from the settleable solids test due to compacting of the sludge in the clarifier, the sludge settling differently in Imhoff cones than in the clarifier, or the clarifier not capturing all of the solids that should settle.

(Settleable Solids)

B. *WHAT IS TESTED?*

Sample	Common Ranges Found
Influent	6 mL/L weak wastewater
	12 mL/L medium wastewater
	20 mL/L strong wastewater
Primary Effluent	0.1 mL/L to 3 mL/L
	Over 3 mL/L poor
Secondary Effluent	Trace to 0.5 mL/L
	Over 0.5 mL/L poor

C. *APPARATUS*

1. Imhoff cones

2. Rack for holding Imhoff cones

3. Glass stirring rod or wire

4. Time clock or watch

D. *PROCEDURE*

1. Thoroughly mix the wastewater sample by shaking and immediately fill an Imhoff cone to the liter mark.

2. Record the time of day that the cone was filled. T = _____.

3. Allow the wastewater sample to settle for 45 minutes.

4. Gently stir sides of the cone to facilitate settling of material adhering to the side of the cone.

5. After one hour, record the number of milliliters of settleable solids in the Imhoff cone. Make allowance for voids among the settled material. For example, if you read a sludge volume of 3.0 mL and voids or spaces in the sludge occupy approximately 0.2 mL, record a sludge volume of 2.8 mL.

6. Record the settleable solids as mL/L or milliliters per liter.

Settleable Solids, Influent = _____ mL/L

Settleable Solids, Effluent = _____ mL/L

Settleable Solids, Removal = _____ mL/L

OUTLINE OF PROCEDURE FOR SETTLEABLE SOLIDS

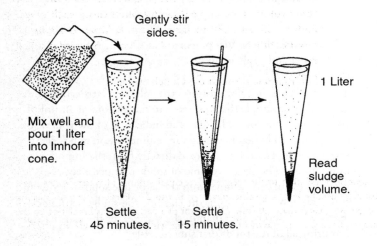

Gently stir sides.

Mix well and pour 1 liter into Imhoff cone.

1 Liter

Read sludge volume.

Settle 45 minutes.   Settle 15 minutes.

E. *EXAMPLE*

Samples were collected from the influent and effluent of a primary clarifier. After one hour, the following results were recorded:

Settleable Solids, mL/L	
Influent	12.0
Effluent	0.2

F. *CALCULATIONS*

1. Calculate the efficiency or percent removal of settleable solids from the primary clarifier discussed in the above example.

$$\begin{aligned} \text{\% Removal of Set Sol} &= \frac{(\text{Infl Set Sol, mL/L} - \text{Effl Set Sol, mL/L})}{\text{Influent Set Sol, mL/L}} \times 100\% \\ &= \frac{(12 \text{ mL/L} - 0.2 \text{ mL/L})}{12 \text{ mL/L}} \times 100\% \\ &= \frac{11.8}{12} \times 100\% \\ &= 98\% \end{aligned}$$

2. Estimate the gallons per day of sludge that should be pumped from the same primary clarifier if the flow is 1 MGD (1 million gallons per day). In your plant, the Imhoff cone may not measure or indicate the exact performance of your clarifier or sedimentation tank, but with some experience you should be able to relate or compare your lab tests with actual performance.

$$\begin{aligned} \text{Sludge Removed by Clarifier, mL/L} &= \text{Influent Set Sol, mL/L} - \text{Effluent Set Sol, mL/L} \\ &= 12 \text{ mL/L} - 0.2 \text{ mL/L} \\ &= 11.8 \text{ mL/L} \end{aligned}$$

To estimate the GPD (gallons per day) of sludge pumped to a digester, use the following formula:

$$\begin{aligned} \text{Sludge to Digester, GPD} &= \frac{\text{Total Set Sol Rem, mL/L}}{} \times 1{,}000 \text{ mg/mL} \times \text{Flow, MGD} \\ &= \frac{11.8 \text{ mL}}{\text{M mg}} \times \frac{1{,}000 \text{ mg}}{\text{mL}} \times \frac{1 \text{ M gal}}{\text{day}} \\ &= 11{,}800 \text{ GPD} \end{aligned}$$

This value may be reduced by 30 to 75 percent due to compaction of the sludge in the clarifier.

If you figure sludge removed as a percentage (1.18%), the sludge pumped to the digester would be calculated as follows:

$$\frac{1.18\%}{100\%} = \frac{\text{Sludge to Digester, GPD}}{\text{Flow of 1,000,000 GPD}}$$

$$\begin{aligned} \text{Sludge to Digester, GPD} &= \frac{(1.18\% \times 1{,}000{,}000 \text{ GPD})}{100\%} \\ &= 11{,}800 \text{ GPD} \end{aligned}$$

G. CLINICAL CENTRIFUGE

Settleable solids may also be measured by a small clinical centrifuge. This method should be used for plant control only and not for NPDES monitoring. A mixed sample is placed in 15-mL graduated centrifuge tubes and spun for 15 minutes. The solid deposition in the tip of the tube is compared with a curve prepared by plotting settleable solids versus centrifuge solid deposition. This test provides a quick estimate of the settleable solids.

QUESTION

Please write your answers to the following questions and compare them with those on page 338.

14.4D Estimate the volume of solids pumped to a digester in gallons per day (GPD) if the flow is 1 MGD, the influent settleable solids are 10 mL/L, and the effluent settleable solids are 0.4 mL/L for a primary clarifier.

14.43 Suspended Solids

A. DISCUSSION

The suspended solids test run on wastewater is used to determine the amount of material suspended within the sample. The result obtained from this test does not mean that all of the suspended solids settle out in the primary clarifier or, for that matter, in the final clarifier. Some suspended material in wastewater will not be removed because the *SPECIFIC GRAVITY* [11] is very near that of water, and the material is not light enough to float or heavy enough to settle. Therefore, suspended solids are a combination of settleable solids and those solids that remain in suspension.

B. WHAT IS TESTED?

Sample	Common Ranges, mg/L
Influent	Weak 150 to 400+ Strong
Primary Effluent	Weak 60 to 150+ Strong
Secondary Effluent	Good 10 to 60+ Bad
Activated Sludge Tests:	Depending on Type of Process
Mixed Liquor	1,000 to <5,000
Return or Waste Sludge	2,000 to <12,000
Digester Tests:	
Supernatant	3,000 to <10,000

When supernatant suspended solids are greater than 10,000 mg/L, the total solids test is usually performed.

C. APPARATUS

1. Glass-fiber filter disks without organic binder (Whatman grade 934H; Gelman type A/E; Millipore type AP40; or equivalent).

2. Filtration apparatus. Use one of the following, which must be suitable for the filter disk selected above:

   a. Membrane filter funnel,

   b. Gooch crucible, 25-mL to 40-mL capacity with Gooch crucible adapter, or

   c. Filtration apparatus with reservoir and coarse 40- to 60-μm/fritted (high density porcelain) disk as filter support.

3. Flask, suction.

4. Vacuum source.

5. Drying oven, 103° to 105°C.

6. Desiccator.

7. Analytical balance.

D. PROCEDURE

The "Outline of Procedure for Suspended Solids" is shown on the next page.

Preparing the Gooch Crucible

1. Insert a 2.2-cm glass-fiber filter into 25-mL Gooch crucible and center it.

2. Apply vacuum by washing the filter with 20 mL of distilled or deionized water three times to seat filter properly.

3. Dry the crucible at 103°C for one hour. If volatile suspended solids are to be determined, ignite crucible in muffle furnace for one hour at 550°C.

4. Cool in desiccator.

5. Obtain the dry weight of the crucible and filter (also known as the tare weight). Record weight.

Sample Analysis

6. Depending on the suspended solids content, measure out a 25-, 50-, or 100-mL portion of a well-mixed sample into a graduated cylinder. Use 25 mL if sample filters slowly. Use larger volumes of sample if sample filters easily, such as secondary effluent. Try to limit filtration time to about 15 minutes or less. Wet prepared Gooch crucible with distilled water and apply vacuum.

7. Filter sample through the Gooch crucible.

---

11. *Specific Gravity.* (1) Weight of a particle, substance, or chemical solution in relation to the weight of an equal volume of water. Water has a specific gravity of 1.000 at 4°C (39°F). Wastewater particles or substances usually have a specific gravity of 0.5 to 2.5. Particulates with specific gravity less than 1.0 float to the surface and particulates with specific gravity greater than 1.0 sink. (2) Weight of a particular gas in relation to the weight of an equal volume of air at the same temperature and pressure (air has a specific gravity of 1.0). Chlorine gas has a specific gravity of 2.5.

(Suspended Solids)

OUTLINE OF PROCEDURE FOR SUSPENDED SOLIDS

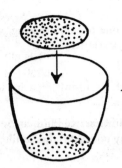

1. Insert glass-fiber filter into Gooch crucible and center it.

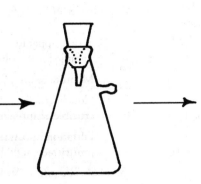

2. Apply vacuum by washing filter with distilled water to seat filter properly.

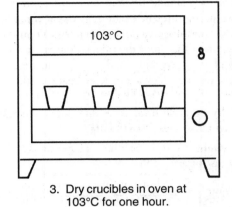

3. Dry crucibles in oven at 103°C for one hour.

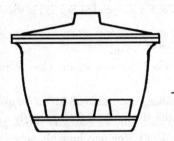

4. Cool in desiccator for 20–30 minutes.

5. Obtain dry weight of crucible and filter.

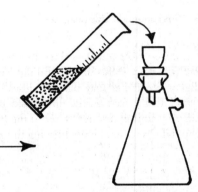

6. Pour measured volume of sample into Gooch crucible.

7. Filter out suspended solids with vacuum.

8. Wash flask, crucible, and filter with distilled water to complete solids transfer.

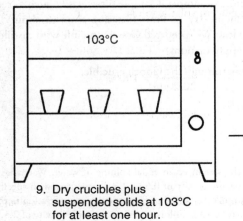

9. Dry crucibles plus suspended solids at 103°C for at least one hour.

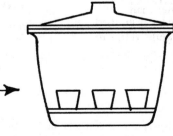

10. Cool in desiccator for 20–30 minutes.

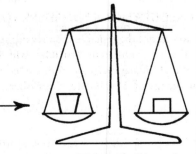

11. Weigh crucible and suspended solids.

(Suspended Solids)

8. Wash dissolved solids from the flask, crucible, and filter with about 20 mL of distilled water. Use two 10-mL portions to complete solids transfer.

9. Dry crucible and suspended solids at 103°C for at least one hour. Some samples may require up to three hours to dry if the residue is thick.

10. Cool in desiccator for 20 to 30 minutes.

11. Weigh crucible and suspended solids.

12. Repeat drying cycle until constant weight is attained or until weight loss is less than 0.5 mg.

13. Record weight:

Total Weight = _____ gm

Tare Weight = _____ gm

Solids Weight = _____ gm

E. PRECAUTIONS

1. Check and regulate the oven temperature between 103° to 105°C.

2. Check the crucible and glass-fiber filter for any possible leaks. A leak will cause solids to pass through and give low results. The glass-fiber filter may become unseated and leaky when the crucible is placed on the filter flask. The filter should be reseated by adding distilled water to the filter in the crucible and applying vacuum before filtering the sample.

3. Mix the sample thoroughly so that it is completely uniform in suspended solids when measured into a graduated cylinder before the sample can settle out. This is especially true of samples heavy in suspended solids, such as raw wastewater and mixed liquor in activated sludge, which settle rapidly. The test can be no better than the mix.

4. A good practice is to prepare a number of extra Gooch crucibles for additional tests if the need arises. If a test result appears faulty or questionable, the test should be repeated. Check the filtration rate and clarity of water passing through the filter.

F. EXAMPLE AND CALCULATIONS

This section provides you with examples and detailed calculations. Practice using the lab worksheet shown in Figure 14.6 to calculate suspended solids and dissolved solids.

CALCULATIONS FOR SUSPENDED SOLIDS TEST

Always record the crucible number of the crucible (or dish) used for the test. Different crucibles will have slightly different weights. Calculations 1–3 are included in Figure 14.6 to illustrate the use of the laboratory worksheet.

EXAMPLE: Assume the following data.

Crucible #015

Volume of Sample = 50 mL

	Recorded Weights
Crucible (dish) Weight	21.6329 gm
Crucible (dish) Plus Dry Solids	21.6531 gm
Crucible (dish) Plus Ash[12]	21.6360 gm

1. Compute total suspended solids.

$$\begin{array}{ll} 21.6531\ gm & \text{Weight of Crucible Plus Dry Solids, gm} \\ -\ 21.6329\ gm & -\ \text{Weight of Crucible, gm} \\ =\ \ 0.0202\ gm & -\ \text{Weight of Dry Solids, gm} \end{array}$$

or

$$=\ 20.2\ mg$$

1,000 milligrams (mg) = 1 gram (gm)

or

$$20.2\ mg\ =\ 0.0202\ gm$$

$$\text{Total Suspended Solids, mg/L} = \frac{\text{Weight of Solids, mg} \times 1{,}000\ \text{mL/L}}{\text{Sample Volume, mL}}$$

$$= \frac{20.2\ mg \times 1{,}000\ mL/L}{50\ mL}$$

$$=\ 404\ mg/L$$

2. Compute volatile or organic suspended solids.

$$\begin{array}{ll} 21.6531\ gm & \text{Weight of Crucible Plus Dry Solids, gm} \\ -\ 21.6360\ gm & -\ \text{Weight of Crucible Plus Ash, gm} \\ =\ \ 0.0171\ gm & -\ \text{Weight of Volatile Solids, gm} \end{array}$$

or

$$=\ 17.1\ mg$$

$$\text{Vol Suspended Solids, mg/L} = \frac{\text{Weight of Vol Solids, mg} \times 1{,}000\ \text{mL/L}}{\text{Sample Volume, mL}}$$

$$= \frac{17.1\ mg \times 1{,}000\ mL/L}{50\ mL}$$

$$=\ 342\ mg/L$$

3. Compute the percent volatile solids.

$$\text{Volatile Solids, \%} = \frac{\text{Weight Volatile, mg} \times 100\%}{\text{Weight Total Dry Solids, mg}}$$

$$= \frac{17.1\ mg}{20.2\ mg} \times 100\%$$

$$=\ 84.7\%$$

---

12. Obtained by placing the crucible plus dry solids in a muffle furnace at 550°C for one hour. The crucible plus remaining ash are cooled and weighed.

(Suspended Solids)

PLANT _____CLEAN WATER_____

DATE _____

SUSPENDED SOLIDS AND DISSOLVED SOLIDS

SAMPLE	INFL.					
Crucible	#015					
Sample, mL	50					
Wt Dry + Dish, gm	21.6531					
Wt Dish, gm	21.6329					
Wt Dry, gm	0.0202					
$mg/L = \dfrac{Wt\ Dry,\ gm \times 1{,}000{,}000}{Sample,\ mL}$	404 mg/L					
Wt Dish + Dry, gm	21.6531					
Wt Dish + Ash, gm	21.6360					
Wt Volatile, gm	0.0171					
$\% = \dfrac{Wt\ Vol}{Wt\ Dry} \times 100\%$	84.7%					

BOD

\# Blank _____

SAMPLE							
DO Sample							
Bottle #							
% Sample							
Blank or adjusted blank							
DO after incubation							
Depletion, 5 days							
Depletion, %							

SETTLEABLE SOLIDS

Sample, mL _____   _____

Direct mL/L _____   _____

COD

Sample _____   _____

Blank Titration _____   _____

Sample Titration _____   _____

Depletion _____   _____

$mg/L = \dfrac{Dep \times N\ FAS \times 8{,}000}{Sample,\ mL}$ _____   _____

Fig. 14.6   Calculating suspended solids and dissolved solids on a laboratory worksheet

(Suspended Solids)

4. Compute fixed or inorganic suspended solids.

21.6360 gm	Weight of Crucible Plus Ash, gm
− 21.6329 gm	− Weight of Crucible, gm
= 0.0031 gm	= Weight of Fixed Solids, gm

or

= 3.1 mg

$$\text{Fixed Suspended Solids, mg/L} = \frac{\text{Weight of Fixed Solids, mg} \times 1,000 \text{ mL/L}}{\text{Sample Volume, mL}}$$

$$= \frac{3.1 \text{ mg} \times 1,000 \text{ mL/L}}{50 \text{ mL}}$$

$$= 62 \text{ mg/L}$$

To check your work:

$$\text{Fixed Susp Solids, mg/L} = \frac{\text{Total Susp}}{\text{Solids, mg/L}} - \frac{\text{Volatile Susp}}{\text{Solids, mg/L}}$$

$$= 404 \text{ mg/L} - 342 \text{ mg/L}$$

$$= 62 \text{ mg/L (check)}$$

5. Compute the percent fixed solids.

$$\text{Fixed Solids, \%} = \frac{\text{Weight Fixed, mg}}{\text{Weight Total, mg}} \times 100\%$$

$$= \frac{3.1 \text{ mg}}{20.2 \text{ mg}} \times 100\%$$

$$= 15.3\%$$

CALCULATIONS FOR OVERALL PLANT REMOVAL OF SUSPENDED SOLIDS IN PERCENT

EXAMPLE: Assume the following data.

Influent Suspended Solids	202 mg/L
Primary Effluent Suspended Solids	110 mg/L
Secondary Effluent Suspended Solids	52 mg/L
Final Effluent Suspended Solids	12 mg/L

To calculate the percent removal or treatment efficiency for a particular process or the overall plant, use the following formula:

$$\text{Removal, \%} = \frac{(\text{In} - \text{Out})}{\text{In}} \times 100\%$$

Compute percentage removed between influent and primary effluent:

$$\text{Removal, \%} = \frac{(\text{In} - \text{Out})}{\text{In}} \times 100\%$$

$$= \frac{(202 \text{ mg/L} - 110 \text{ mg/L})}{202 \text{ mg/L}} \times 100\%$$

$$= \frac{92}{202} \times 100\%$$

$$= 45.5\%$$

Compute percentage removed between influent and secondary effluent:

$$\text{Removal, \%} = \frac{(\text{In} - \text{Out})}{\text{In}} \times 100\%$$

$$= \frac{(202 \text{ mg/L} - 52 \text{ mg/L})}{202 \text{ mg/L}} \times 100\%$$

$$= \frac{150}{202} \times 100\%$$

$$= 74.3\%$$

Compute percentage removed between influent and final effluent (overall plant percentage removed):

$$\text{Removal, \%} = \frac{(\text{In} - \text{Out})}{\text{In}} \times 100\%$$

$$= \frac{(202 \text{ mg/L} - 12 \text{ mg/L})}{202 \text{ mg/L}} \times 100\%$$

$$= \frac{190}{202} \times 100\%$$

$$= 94.1\% \text{ Suspended Solids Removal for the Plant}$$

CALCULATIONS FOR POUNDS SUSPENDED SOLIDS REMOVED PER DAY

EXAMPLE: Assume the following data.

Influent Suspended Solids	= 200 mg/L
Effluent Suspended Solids	= 10 mg/L
Flow in million gallons/day	= 2 MGD
1 gallon of water weighs	= 8.34 lbs

Compute pounds suspended solids removed:

The general formula for computing pounds removed is:

$$\text{Material Removed, lbs/day} = \left(\frac{\text{Conc In,}}{\text{mg/L}} - \frac{\text{Conc Out,}}{\text{mg/L}} \right) \times \text{Flow, MGD} \times 8.34 \text{ lbs/gal}$$

$$= (200 \text{ mg/L} - 10 \text{ mg/L}) \times 2 \text{ MGD} \times 8.34 \text{ lbs/gal}$$

$$= 190 \times 2 \times 8.34$$

$$= 3,169 \text{ lbs/day of Suspended Solids Removed by Plant}$$

DERIVATION

Although not essential to efficient plant operation, this section may provide you with a better understanding of the calculation.

1 mg/L = 1 ppm or 1 part per million

or   = 1 mg/million mg, because 1 liter = 1,000,000 mg

Therefore:

$$\frac{\text{lbs}}{\text{day}} = \frac{\text{mg}}{\text{M mg}} \times \frac{\text{M gal}}{\text{day}} \times \frac{\text{lbs}}{\text{gal}}$$

$$= \text{lbs/day}$$

(Total Sludge Solids (Volatile and Fixed))

QUESTIONS

Please write your answers to the following questions and compare them with those on pages 338 and 339.

14.4E   Why does some of the suspended material in wastewater fail to be removed by settling or flotation?

14.4F   Given the following data:

Volume of Sample   = 100 mL
Crucible Weight   = 19.3241 gm
Crucible Plus Dry Solids = 19.3902 gm
Crucible Plus Ash   = 19.3469 gm

Compute:

1. Total Suspended Solids, mg/L
2. Volatile Suspended Solids, mg/L
3. Volatile Solids, %
4. Fixed Suspended Solids, mg/L
5. Fixed Solids, %

14.4G   Suspended solids data from a wastewater treatment plant are given below:

Influent Suspended Solids = 221 mg/L
Primary Effluent SS   = 159 mg/L
Final Effluent SS   = 33 mg/L

Compute the percent removal of suspended solids by the:

1. Primary clarifier,
2. Secondary process (removal between primary effluent and secondary effluent), and
3. Overall plant.

14.4H   If the data in 14.4G above are from a 1.5 MGD plant, calculate the pounds of suspended solids removed by the:

1. Primary unit,
2. Secondary unit, and
3. Overall plant.

14.44   Total Sludge Solids (Volatile and Fixed)

A. DISCUSSION

The analysis of total solids is used to access wastewater treatment processes such as digester efficiencies. Total solids are the combined amounts of suspended and dissolved solids in a sample.

This test is used for wastewater sludges or where the solids can be expressed in percentages by weight. The weight can be measured on an inexpensive beam balance to the nearest 0.01 of a gram. The total solids are composed of two components, volatile and fixed solids. Volatile solids are composed of organic compounds that are of either plant or animal origin. In a treatment plant, volatile solids indicate the portion of waste material that can be treated by biological processes. Fixed solids are inorganic compounds such as sand, gravel, and salts.

Volatile solids are those solids lost on ignition (heating to 550°C). The remaining sample after ignition represents the fixed solids in the sample. Determining the amount of volatile solids lost enables the operator to calculate the difference between the total solids and fixed solids in the sample. See Section F for the Outline of Procedure for Volatile Solids.

B. WHAT IS TESTED?

Sample	Common Range, % by Weight		
	Total	Volatile	Fixed
Raw Sludge	6% to 9%	75%	25% ± 6%
Raw Sludge Plus Waste Activated Sludge	2% to 5%	80%	20% ± 5%
Recirculated Sludge	1.5% to 3%	75%	25% ± 5%
Supernatant:			
Good Quality, has Suspended Solids	<1%	50%	50% ± 10%
Poor Quality	>1%	50%	50% ± 10%
Digested Sludge to Air Dry	3% too thin to <8%	50%	50% ± 10%

C. APPARATUS

1. Evaporating dish
2. Analytical balance
3. Drying oven, set between 103° and 105°C
4. Measuring device: graduated cylinder
5. Muffle furnace set to 550°C

D. PROCEDURE

1. Dry the dish by ignition in a muffle furnace at 550°C for one hour.
2. Cool dish in desiccator for 20 to 30 minutes.
3. Tare the dish to the nearest 10 milligrams, or 0.01 gm on a single pan balance. Record the weight as:
   Tare Weight = _____ gm.
4. Weigh dish plus 50 to 100 mL of WELL-MIXED sludge sample. Record total weight to nearest 0.01 gram as:
   Gross Weight = _____ gm.
5. Measure sludge into a dish for evaporation in drying oven.
6. Evaporate the sludge sample to dryness in drying oven set to between 103° and 105°C.
7. Cool dish and sample in desiccator for 20 to 30 minutes.
8. Weigh the dried residue in the evaporating dish to the nearest 10 milligrams, or 0.01 gm (step not shown in diagram). Record the weight as:
   Dry Sample and Dish = _____ gm.
9. Compute the net weight of the residue by subtracting the tare weight (obtained in Step 3) of the dish from the dry sample and dish.

(Total Sludge Solids (Volatile and Fixed))

OUTLINE OF PROCEDURE FOR TOTAL SOLIDS

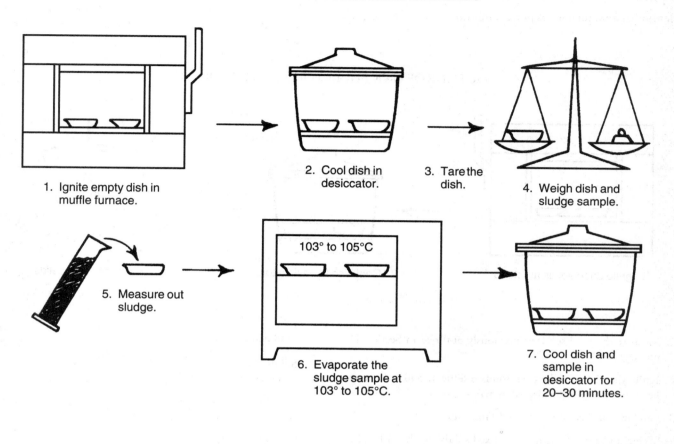

1. Ignite empty dish in muffle furnace.

2. Cool dish in desiccator.

3. Tare the dish.

4. Weigh dish and sludge sample.

5. Measure out sludge.

6. Evaporate the sludge sample at 103° to 105°C.

7. Cool dish and sample in desiccator for 20–30 minutes.

E. *PRECAUTIONS*

1. Be sure that the sample is thoroughly mixed and is representative of the sludge being pumped. Generally, where sludge pumping is intermittent, sludge is much denser at the beginning and is less dense toward the end of pumping. To balance this difference, collect several equal portions of sludge at regular intervals and mix for a good sample.

2. Take a large enough sample. Measuring a 50- or 100-mL sample, which is closely equal to 50 or 100 grams, is recommended. Since this material is so heterogeneous (nonuniform), it is difficult to obtain a good representative sample with less volume. Smaller volumes will show greater variations in results, due to the uneven and lumpy nature of the material.

3. Control the drying oven temperature to remain constant between 103° and 105°C. Some solids are lost at any drying temperature. Close control of the oven temperature is necessary because higher temperatures increase the loss of volatile solids, in addition to the evaporated water.

4. Heat the evaporating dish long enough to ensure evaporation of all water. This usually takes about 3 to 4 hours. If heat drying and weighing are repeated, stop the procedure when the weight change becomes less than 4 percent of the previous weight, or 0.5 mg, whichever is less. The oxidation, dehydration, and degradation of the volatile fraction will not completely stabilize until it is carbonized or becomes ash.

5. Because of the non-uniformity of sludge and the difficulty in obtaining a representative sample, weighing on an analytical balance should be made only to the nearest 0.01 gram (or 10 milligrams).

(Total Sludge Solids (Volatile and Fixed))

F. *PROCEDURE*

(continued from total solids procedure)

OUTLINE OF PROCEDURE FOR VOLATILE SOLIDS

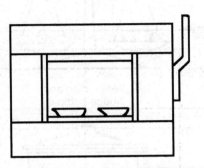

1. Ignite dried solids at 550°C.

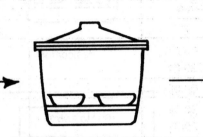

2. Cool dish in desiccator.

3. Weigh fixed solids.

1. Measure the total solids as previously outlined in Section D on page 302.

2. Ignite the dish and the total solids residue at 550°C for one hour or until only a white ash remains.

3. Cool in a desiccator for about 30 minutes.

4. Weigh and record the residue (fixed solids) in the dish. The amount of volatile solids = Total Solids − Fixed Solids.

5. Record your results.

G. *EXAMPLE*

Dish No. 7

Weight of Dish (Tare)	= 20.31 gm
Weight of Dish Plus Wet Solids (Gross)	= 70.31 gm
Weight of Dish Plus Dry Solids	= 22.81 gm
Weight of Dish Plus Ash	= 20.93 gm

H. *CALCULATIONS*

Figure 14.7 includes the following total solids calculations to illustrate the use of the laboratory worksheet.

1. Find weight of sample.

Weight of Dish Plus Wet Solids (Gross)	= 70.31 gm
Weight of Dish (Tare)	= 20.31 gm
Weight of Sample	= 50.00 gm

2. Find weight of total solids.

Weight of Dish Plus Dry Solids	= 22.81 gm
Weight of Dish (Tare)	= 20.31 gm
Weight of Total Solids	= 2.50 gm

3. Find percent sludge.

$$\text{Solids, \%} = \frac{(\text{Weight of Solids, gm})100\%}{\text{Weight of Sample, gm}}$$

$$= \frac{(2.50 \text{ gm})100\%}{50.00 \text{ gm}}$$

$$= 5\%$$

4. Find weight of volatile solids.

Weight of Dish Plus Dry Solids	= 22.81 gm
Weight of Dish Plus Ash	= 20.93 gm
Weight of Volatile Solids	= 1.88 gm

5. Find percent volatile solids.

$$\text{Volatile Solids, \%} = \frac{(\text{Weight of Volatile Solids, gm})100\%}{\text{Weight of Total Solids, gm}}$$

$$= \frac{(1.88 \text{ gm})100\%}{2.50 \text{ gm}}$$

$$= 75\%$$

QUESTIONS

Please write your answers to the following questions and compare them with those on page 339.

14.4I   What are volatile solids composed of?

14.4J   What do volatile solids in a treatment plant indicate?

END OF LESSON 4 OF 7 LESSONS

on

LABORATORY PROCEDURES

Please answer the discussion and review questions next.

(Total Sludge Solids (Volatile and Fixed))

TOTAL SOLIDS

SAMPLE	RAW				
Dish No.	7				
Wt Dish + Wet, gm	70.31				
Wt Dish, gm	20.31				
Wt Wet, gm	50.00				
Wt Dish + Dry, gm	22.81				
Wt Dish, gm	20.31				
Wt Dry, gm	2.50				
% Solids = $\dfrac{\text{Wt Dry}}{\text{Wt Wet}} \times 100\%$	5.0%				
Wt Dish + Dry, gm	22.81				
Wt Dish + Ash, gm	20.93				
Wt Volatile, gm	1.88				
% Volatile = $\dfrac{\text{Wt Vol} \times 100\%}{\text{Wt Dry}}$	75%				
pH					
Volatile Acid, mg/L					
Alkalinity as $CaCO_3$, mg/L					

Grease

Sample

Sample, mL

Wt Flask + Grease, mg

Wt Flask, mg

Wt Grease, mg

$$\text{mg/L} = \frac{\text{Wt Grease, mg} \times 1{,}000}{\text{Sample, mL}}$$

Fig. 14.7   *Calculating total solids on a laboratory worksheet*

DISCUSSION AND REVIEW QUESTIONS
Chapter 14.   LABORATORY PROCEDURES
(Lesson 4 of 7 Lessons)

Please write your answers to the following questions to determine how well you understand the material in the lesson. The question numbering continues from Lesson 3.

10. Why must the clarity test always be run under the same conditions?

11. Why is hydrogen sulfide measured?

12. What process releases H_2S in sewers?

13. Why does the actual volume of solids pumped from a clarifier not agree exactly with calculations based on the settleable solids test?

14. Calculate the efficiency or percent removal by a primary clarifier when the influent settleable solids are 10 mL/L and the effluent settleable solids are 0.3 mL/L.

15. Given the following data:

    Volume of Sample   = 100 mL
    Crucible Weight   = 19.9850 gm
    Crucible Plus Dry Solids = 20.0503 gm
    Crucible Plus Ash   = 20.0068 gm

    Compute:

    1. Total Suspended Solids in mg and mg/L
    2. Volatile or Organic Suspended Solids in mg and mg/L
    3. Percent Volatile Solids

16. Estimate the pounds of solids removed per day by a primary clarifier if the influent suspended solids are 220 mg/L and the effluent suspended solids are 120 mg/L when the flow is 1.5 MGD.

17. Why are solids only weighed to the nearest 0.01 gram when determining the total volatile solids content of digesters?

CHAPTER 14.   LABORATORY PROCEDURES

(Lesson 5 of 7 Lessons)

14.45   Tests for Activated Sludge Control

14.450   Settleability

A. *DISCUSSION*

The settleability of activated sludge solids is an important factor in determining the ability of the solids to separate from the liquid in the final clarifier. The settleability test is run on mixed liquor or return sludge to indicate what will happen to the mixed liquor in the final clarifier—the rate of sludge settling, turbidity, color, and volume of sludge after 60 minutes of settling. The results are plotted on a graph (see sample in Figure 14.8). The settleability test should be run on the same sample of mixed liquor that is used for the suspended solids test. This will allow you to calculate the sludge volume index (SVI) or the sludge density index (SDI), which are explained in the following sections.

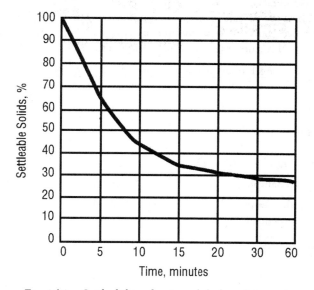

Fig. 14.8   Settleability of activated sludge solids graph

B. *WHAT IS TESTED?*

Sample	Working Range
Mixed Liquor or Return Sludge	Depends on desirable mixed liquor concentration

C. *APPARATUS*

2,000 mL graduated cylinder[13]

D. *REAGENTS*

None

E. *PROCEDURE*

The "Outline of Procedure for Settleable Solids" is shown on page 308. The procedure includes the following steps:

1. Collect a sample of mixed liquor or return sludge.

2. Carefully mix sample and pour into 2,000-mL graduate. Avoid vigorously shaking or mixing the sample as this tends to break up floc, produce slower settling floc, or poorer separation.

3. Record settleable solids percentage at regular intervals.

   NOTE: 1. If a 1,000-mL graduate is used, the percent settleable solids is easier to record.

   2. Some plants use a video recording in the lab to observe and record settleability.

F. *EXAMPLE AND CALCULATION*

The percent settling rate can be compared for the various days of the week and with other measurements—suspended solids, sludge volume index, percent sludge solids returned, aeration rate, and plant inflow. A very slow-settling mixed liquor usually requires air and solids adjustment to encourage increased stabilization during aeration. Both very rapidly settling and very slowly settling mixed liquors can give poor effluent clarification.

14.451   Sludge Volume Index (SVI)

A. *DISCUSSION*

The SVI is the volume in mL occupied by one gram of mixed liquor suspended solids after 30 minutes of settling. The SVI test is used to indicate the condition of sludge (aeration solids or suspended solids) for settleability in a secondary or final clarifier. It is a useful test to indicate changes in sludge characteristics. Determine the proper SVI range for your plant when your final effluent is in the best condition regarding solids and BOD removals and clarity.

---

13. *Mallory Direct Reading Settlometer* (a 2-liter graduated cylinder approximately 5 inches (12.5 cm) in diameter and 7 inches (17.5 cm) high). Obtain from Customer Service, Wilmad Glass/Lab Glass, 1172 NW Boulevard, Vineland, NJ 08360. Catalog No. LG5601-100. Price, $183.94 each.

(Settleability—SDI)

OUTLINE OF PROCEDURE FOR SETTLEABLE SOLIDS

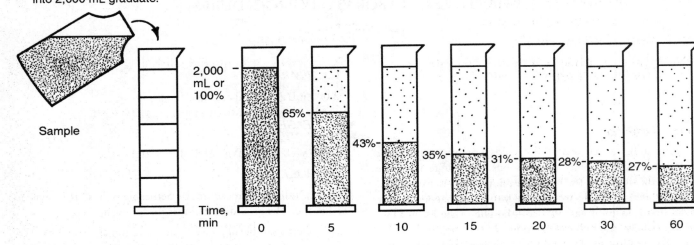

1. Mix sample and pour into 2,000-mL graduate.

2. Record settleable solids, %, at regular intervals.

Sample

2,000 mL or 100%

65%
43%
35%
31%
28%
27%

Time, min   0   5   10   15   20   30   60

B. *WHAT IS TESTED?*

Sample	Preferable Range, SVI
Aeration Tank Suspended Solids	50 mL/gm to 150 mL/gm

C. *APPARATUS*

See Section 14.43, "Suspended Solids," (C. "Apparatus") on page 297, and Section 14.450, "Settleability," (C. "Apparatus") in this lesson.

D. *REAGENTS*

None

E. *PROCEDURE*

See Section 14.43 (D. "Procedure") on page 297, and Section 14.450 (E. "Procedure") on page 307.

F. EXAMPLE

30-minute Settleable Solids Test = 360 mL in 2,000 mL graduate or 18%.

Mixed Liquor Suspended Solids = 1,500 mg/L

G. *CALCULATIONS*

$$\text{Sludge Volume Index, SVI, mL/gm} = \frac{\% \text{ Settleable Solids} \times 1,000 \text{ mg/gm} \times 1,000 \text{ mL/L}}{\text{Mixed Liquor Suspended Solids, mg/L} \times 100\%}$$

$$= \frac{\% \text{ Settleable Solids} \times 10,000}{\text{Mixed Liquor Suspended Solids, mg/L}}$$

$$= \frac{18 \times 10,000}{1,500}$$

$$= \frac{1,600}{15}$$

$$= 120 \text{ mL/gm}$$

14.452   *Sludge Density Index (SDI)*

A. *DISCUSSION*

The SDI is used in a way similar to the SVI to indicate the settleability of a sludge in a secondary clarifier or effluent. The SDI is the weight in mg of one milliliter of mixed liquor suspended solids after 30 minutes of settling. The calculation of the SDI requires the same information as the SVI test.

$$\text{SDI} = \frac{\text{mg/L of Suspended Solids in Mixed Liquor}}{\text{mL/L of Settled Mixed Liquor Solids} \times 10}$$

or

$$\text{SDI} = 100/\text{SVI}$$

B. *WHAT IS TESTED?*

Sample	Preferable Range, SDI
Aeration Tank Suspended Solids	0.6 gm/100 mL to 2.0 gm/100 mL

C. *APPARATUS*

See Section 14.451 (C. "Apparatus").

(Sludge Age)

14.4K   Why is the settleability test run on mixed liquor?

14.4L   What is the sludge volume index (SVI)?

14.4M   Why is the SVI test run?

14.4N   What is the relationship between the sludge density index (SDI) and SVI?

14.453   Sludge Age

A. DISCUSSION

Sludge age is a measure of the length of time a pound of suspended solids is retained under aeration in the activated sludge process. It is a relatively easy operational guide to monitor. The basis for calculating the sludge age is weight of suspended solids in the mixed liquor in the aeration tank divided by weight of suspended solids added per day to the aerator.

$$\text{Sludge Age, days} = \frac{\text{Suspended Solids in Mixed Liquor, mg/L} \times \text{Aerator Vol in MG} \times 8.34 \text{ lbs/gal}}{\text{SS in Primary Effluent, mg/L}^* \times \text{Daily Flow, MGD} \times 8.34 \text{ lbs/gal}}$$

*NOTE: Sludge age is calculated by three different methods:
1. Suspended solids in primary effluent, mg/L
2. Suspended solids removed from primary effluent, mg/L, or primary effluent suspended solids, mg/L – final effluent suspended solids, mg/L
3. BOD or COD in primary effluent, mg/L

Any significant additional loading placed on the aerator by the digester supernatant liquor must be considered when determining the sludge age. Also consider the additional flow (MGD) and concentration (mg/L) if the supernatant is returned following the primary clarifiers. If supernatant is returned ahead of the primary clarifiers, only additional flow must be considered because the solids concentration will be represented in the primary effluent analysis. The available methods for determining sludge age are discussed in Chapter 10, "Activated Sludge." Because the method discussed in this section is based on suspended solids, it should not be used if the soluble organic (BOD) portion of the wastewater is more than 50 percent of the total organic (BOD) component. Also, this method is not recommended for plants that are nutrifying.

B. WHAT IS TESTED?

Sample	Common Range
Suspended Solids in Aerator and Suspended Solids in Primary Effluent	mg/L, Depends on process
Sludge Age	Conventional process, 2.5 to 6 days

C. APPARATUS

See Section 14.43, "Suspended Solids."

D. REAGENTS

None

E. PROCEDURE

See Section 14.43, "Suspended Solids."

F. EXAMPLE

Suspended Solids in Mixed Liquor	= 1,500 mg/L
Aeration Tank Volume	= 0.50 MG
Suspended Solids in Primary Effl	= 100 mg/L
Daily Flow	= 2.0 MGD

G. CALCULATIONS

$$\text{Sludge Age, days} = \frac{\text{Suspended Solids in Mixed Liquor, mg/L} \times \text{Aerator Vol in MG} \times 8.34 \text{ lbs/gal}}{\text{Susp Solids in Primary Effluent, mg/L} \times \text{Flow, MGD} \times 8.34 \text{ lbs/gal}}$$

$$= \frac{\text{Mixed Liquor Susp Solids, lbs}}{\text{Primary Effluent SS, lbs/day}}$$

$$= \frac{1,500 \text{ mg/L} \times 0.50 \text{ MG} \times 8.34 \text{ lbs/gal}}{100 \text{ mg/L} \times 2.0 \text{ MGD} \times 8.34 \text{ lbs/gal}}$$

$$= \frac{1,500 \times 0.50}{100 \times 2.0}$$

$$= \frac{7.5}{2.0}$$

$$= 3.75 \text{ days}$$

14.4O   Determine the sludge age in an activated sludge process if the volume of the aeration tank is 200,000 gallons and the suspended solids in the mixed liquor equals 2,000 mg/L. The primary effluent SS are 115 mg/L, and the average daily flow is 1.8 MGD.

(DO in Aerator)

14.454 Dissolved Oxygen in Aerator

A. DISCUSSION

This test is a modification that is used for biological flocs that have high oxygen utilization rates in the activated sludge process, and when a DO probe is not available. It is very important that some oxygen be present in aeration tanks at all times to maintain aerobic conditions.

This test is similar to the regular DO test (Section 14.5, Procedure 2) except that copper sulfate is added to kill oxygen-consuming organisms, and sulfamic acid is added to combat nitrite before the regular DO test is run.

If the results indicate a DO of less than 1 mg/L, it is possible that the DO in the aeration tank is at or near zero. When the DO in the aeration tank is this low, the copper sulfate-sulfamic acid test procedure can give DO results higher than the actual DO in the aerator. These results may be high because oxygen can enter the sample from the air when the sample is collected, when the copper sulfate-sulfamic acid inhibitor is added, while the solids are settling, or when the sample is transferred to a BOD bottle for the DO test. As a precaution when performing this test, collect the sample in a tall bottle and avoid stirring it.

B. WHAT IS TESTED?

Sample	Common DO Range, mg/L
Aerator Mixed Liquor	0.1 to 3.0

C. APPARATUS

1. One tall bottle, approximately 1,000 mL

2. Regular DO apparatus

D. REAGENTS

1. Copper sulfate-sulfamic acid inhibitor solution. Dissolve 32 gm technical grade sulfamic acid (NH_2SO_2OH) without heat in 475 mL distilled water. Dissolve 50 gm copper sulfate, $CuSO_4 \cdot 5H_2O$, in 500 mL water. Mix the two solutions together and add 25 mL concentrated acetic acid.

2. Regular DO reagents (See Section 14.5, "Procedure 2" on page 319).

OUTLINE OF PROCEDURE FOR DISSOLVED OXYGEN

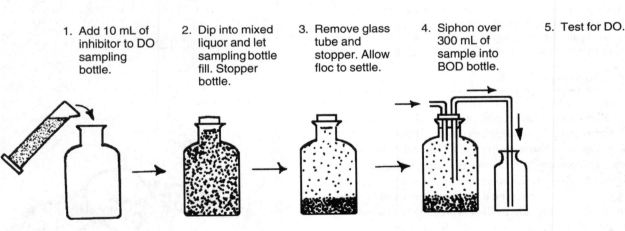

1. Add 10 mL of inhibitor to DO sampling bottle.

2. Dip into mixed liquor and let sampling bottle fill. Stopper bottle.

3. Remove glass tube and stopper. Allow floc to settle.

4. Siphon over 300 mL of sample into BOD bottle.

5. Test for DO.

E. PROCEDURE

1. Add at least 10 mL of inhibitor (5 mL copper sulfate and 5 mL sulfamic acid) to any tall bottle (1-quart or 1-liter bottle) with an approximate volume of 1,000 mL. Place filling tube near the bottom. An emptying tube is placed approximately ¼ inch (6 mm) from the top of the bottle stopper. See Figure 14.5 on page 290 for an example of this step. Attach bottle to rod or aluminum conduit and lower into aeration tank.

2. Submerge bottle 1.5 to 2.0 feet (0.45 to 0.60 m) below the surface of the aerator and allow bottle to fill with mixed liquor sample. Remove bottle from aeration tank.

3. Remove glass tube and stopper from bottle. Insert lid in bottle. Allow bottle to stand until solids (floc) settle and leave a clear supernatant liquor.

4. Siphon the supernatant liquor into a 300-mL BOD bottle. Keep the outlet of the siphon below the water surface. Do not aerate in transfer.

5. Perform regular DO test as outlined on page 320.

F. and G. EXAMPLE AND CALCULATIONS

See DO Test (F. "Example" and G. "Calculation," on page 321.

(Suspended Solids (Centrifuge))

14.4P What are the limitations of the copper sulfate-sulfamic acid procedure for measuring DO in the aeration tank when the DO in the tank is very low?

14.455 Suspended Solids in Aerator (Centrifuge Method)

A. DISCUSSION

This procedure is frequently used in plants as a quick and easy method to estimate the suspended solids concentration of the mixed liquor in the aeration tank instead of using the regular suspended solids test. This method also provides reliable results when the suspended solids concentration is below 1,500 mg/L. If you do not have a centrifuge or if your solids content is over 1,500 mg/L, use the regular method to determine the amount of suspended solids. Many operators control the solids in their aerator on the basis of centrifuge readings. Others prefer to control solids using graphed results (see example in Figure 14.9) to estimate the suspended solids. In either case, the operator should periodically compare centrifuge readings with values obtained from suspended solids tests. If the solids are in a good settling condition, a one percent centrifuge solids reading could have a suspended solids concentration of 1,000 mg/L. However, if the sludge is feathery, a one percent centrifuge solids reading could have a suspended solids concentration of 600 mg/L.

The centrifuge reading versus mg/L suspended solids chart (Figure 14.9) must be developed for each plant by comparing centrifuge readings with suspended solids determined by the standard Gooch crucible method. The points are plotted and a line of best fit is drawn as shown in Figure 14.9. This line must be periodically checked by comparing centrifuge readings with the results of regular suspended solids tests because of the large number of variables influencing the relationship. These variables include influent waste, mixing in the aerator, and organisms in the aerator. If you do not have a centrifuge or if your solids content is over 1,500 mg/L, determine the amount of suspended solids by using the regular method.

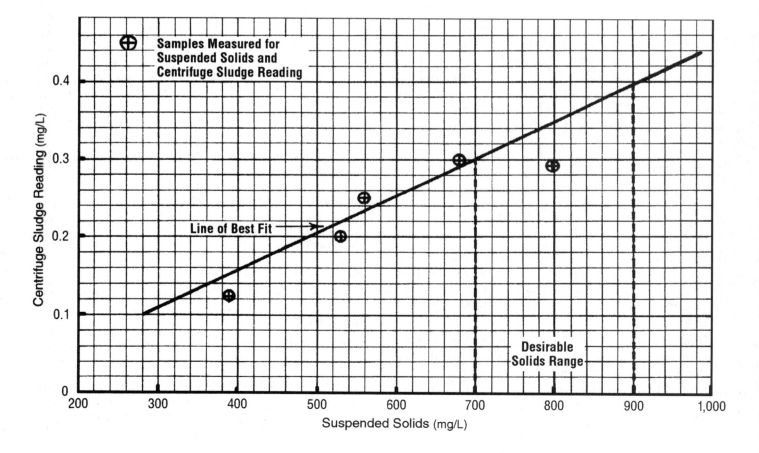

Fig. 14.9 Graph of plant control by centrifuge solids in aeration tank, centrifuge speed, 1,750 RPM

(Suspended Solids (Centrifuge)—Mean Cell Residence Time (MCRT))

B. *WHAT IS TESTED?*

Sample	Common Range
Suspended Solids in Aeration Tanks	800 to 5,000 mg/L

C. *APPARATUS*

1. Centrifuge

2. Graduated centrifuge tubes, 15 mL

D. *REAGENTS*

None

E. *PROCEDURE*

1. Collect sample.

2. Mix sample well and fill each centrifuge tube to the 15-mL line with sample.

3. Place filled sample tubes in centrifuge holders.

4. Crank the centrifuge at a fast speed as you count slowly to 60. Be sure to count and crank at the same speed for all tests. It is extremely important to perform each step exactly the same way every time.

5. Remove one tube and read the amount (in tenths of a milliliter) of suspended solids concentrated in the bottom of the tube. Results in other tubes should be compared.

6. Refer to the conversion curve to determine suspended solids in mg/L.

NOTE: The reason for filling tubes to the 15-mL mark is that the curve on the graph (Figure 14.9) is computed for samples of this size. This curve was developed for a specific mechanically aerated activated sludge plant. The best range for a plant such as this one is 700 to 900 mg/L MLSS. Each plant must develop its own curve based on actual data.

F. *EXAMPLE*

Suspended solids concentration on bottom of centrifuge tube is 0.4 mL.

G. *CALCULATIONS*

From Figure 14.9, find 0.4 mL on the centrifuge reading side (vertical axis). Follow the horizontal line across the chart from 0.4 mL until it intersects the line of best fit. Follow the line downward to the horizontal axis and read the result—900 mg/L suspended solids.

If the suspended solids concentration is above or below the desired range, adjust the pumping rate of the waste and return sludge. For details on controlling the solids concentration, refer to Chapter 10, "Activated Sludge."

H. *DEVELOPMENT OF GRAPH*

To develop the graph like the one shown in Figure 14.9, take a sample from the aeration tank, measure suspended solids, and centrifuge a portion of the sample to obtain the centrifuge

sludge reading in mL of sludge at the bottom of the tube. Obtain other samples of different solids concentrations to obtain the points on the graph. Draw a line of best fit through the points. Periodically, the points should be checked because the influent characteristics and conditions in the aeration tank change.

I. *PRECAUTIONS*

This test works best for low mixed liquor suspended solids (MLSS) concentrations below 1,500 mg/L. Above 1,500 mg/L, the centrifuge results might not produce an accurate estimate of the MLSS.

QUESTION

Please write your answers to the following questions and compare them with those on page 340.

14.4Q   What is the advantage of the centrifuge test for determining suspended solids in an aeration tank in comparison with other methods of measuring suspended solids?

14.456   Mean Cell Residence Time (MCRT)

A. *DISCUSSION*

The MCRT is considered a precise sludge age calculation. MCRT describes the mean (average) residence time of an activated sludge particle in the activated sludge system and is a true measure of the age of the activated sludge. The MCRT considers the solids removed from the process by wasting and the solids removed in the effluent. This calculation also considers the solids in the system. To use MCRT for the control of an activated sludge process, several measurements are required. Representative composite samples of mixed liquor suspended solids, effluent suspended solids, and waste sludge suspended solids are essential measurements. Influent flow and waste sludge flow measurements are also required.

B. *WHAT IS TESTED?*

Sample	Common Range, days
Mean Cell Residence Time (MCRT)	5 to 15 days

C. *APPARATUS AND REAGENTS*

See Section 14.44 for volatile suspended solids test procedures.

D. *PROCEDURE*

The following data are required to calculate the MCRT:

1. Influent flow, MGD

2. Waste sludge flow, MGD

3. Volume of aeration basins in million gallons

4. Mixed liquor volatile suspended solids concentration in mg/L

5. Effluent volatile suspended solids in mg/L

6. Waste sludge volatile suspended solids in mg/L

(Mean Cell Residence Time (MCRT))

E. *EXAMPLE*

Influent Flow, Q	= 3 MGD
Waste Sludge Flow, W	= 0.040 MGD
Volume of Aeration Basins, V	= 1.0 MG
Mixed Liquor Suspended Solids Concentration, X_1	= 1,600 mg/L
Effluent Suspended Solids, X_2	= 8 mg/L
Waste Sludge Suspended Solids, X_W	= 4,700 mg/L

F. *CALCULATIONS*

$$\text{Mean Cell Residence Time, days} = \frac{(V, MG)(X_1, mg/L)}{(Q, MGD)(X_2, mg/L) + (W, MGD)(X_W, mg/L)}$$

$$= \frac{(1.0\ MG)(1,600\ mg/L)}{(3\ MGD)(8\ mg/L) + (0.040\ MGD)(4,700\ mg/L)}$$

$$= \frac{1,600}{24 + 188}$$

$$= 7.5\ \text{days}$$

NOTE: The mixed liquor, effluent and waste sludge volatile suspended solids may be used instead of suspended solids to calculate the MCRT. See Chapter 11, "Activated Sludge," Section 11.552, "Mean Cell Residence Time (MCRT)," *Operation of Wastewater Treatment Plants,* Volume II, for a more information about calculating MCRT.

QUESTION

Please write your answers to the following questions and compare them with those on page 340.

14.4R   What measurements are required to calculate the mean cell residence time (MCRT)?

END OF LESSON 5 OF 7 LESSONS

on

LABORATORY PROCEDURES

Please answer the discussion and review questions next.

DISCUSSION AND REVIEW QUESTIONS

Chapter 14.   LABORATORY PROCEDURES

(Lesson 5 of 7 Lessons)

Please write your answers to the following questions to determine how well you understand the material in the lesson. The question numbering continues from Lesson 4.

18. Calculate the SVI if the mixed liquor suspended solids are 2,000 mg/L and the 30-minute settleable solids test is 500 mL in 2 liters or 25 percent.

19. Calculate the SDI if the SVI is 125.

20. What does sludge age measure?

21. What is the difference between sludge age and mean cell residence time (MCRT)?

CHAPTER 14.   LABORATORY PROCEDURES

(Lesson 6 of 7 Lessons)

14.5   LABORATORY PROCEDURES FOR NPDES MONITORING

14.50   Need for Approved Procedures

The tests discussed in this section are designed for the treatment plant operator or laboratory analyst who is required to monitor effluent discharges under a National Pollutant Discharge Elimination System (NPDES) permit. Principal tests include:

1. Chlorine Residual (Total), below

2. Dissolved Oxygen (DO) and Biochemical Oxygen Demand (BOD), page 319

3. pH (Hydrogen Ion), page 329

4. Total Solids (Residue), page 330

   Additional tests include:

5. Temperature, page 334

6. Turbidity, page 335

   Test procedures for ammonia nitrogen are described in *Operation of Wastewater Treatment Plants,* Volume II, Chapter 16, "Laboratory Procedures and Chemistry."

   Monitoring data required by an NPDES permit must be obtained by using approved test procedures. A list of approved test procedures can be found in *40 CFR 136,* December 13, 2011, "Guidelines Establishing Test Procedures for the Analysis of Pollutants." US Government Printing Office, Superintendent of Documents, PO Box 979050, St. Louis, MO 63197-9000.

14.51   Test Procedures

1.   Chlorine Residual (Total)

A.   DISCUSSION

The many uses of chlorine in wastewater treatment include disinfecting effluent, reducing BOD, controlling odor, improving scum and grease removal, controlling activated sludge bulking, controlling foam, and aiding in chemical coagulation. The most important use of chlorine in the treatment of wastewater is for disinfection, which protects the bacteriological quality of receiving waters. The amount of chlorine residual remaining in treated wastewater after it passes through a contact basin or chamber may be related to the numbers of coliform bacteria allowed in the effluent by regulatory agencies. The effluent from a treatment plant is chlorinated for disinfection purposes. At most plants, the chlorine residual must be removed before final discharge to a receiving stream for the protection of fish and other aquatic life.

Chlorine reacts quickly (within minutes) and completely with ammonia in wastewater to produce monochloramine and dichloramine. Chlorine residual in the monochloramine or dichloramine state is called combined chlorine residual. With the amount of ammonia usually found in wastewater, the chlorine residual will contain all combined chlorine and no free chlorine. Because chlorine residual in wastewater is in a combined state, the determination of chlorine residual presents special challenges.

B.   *WHAT IS TESTED?*

Sample	Common Range, mg/L (After 30 minutes)
Effluent	0.5 to 2.0 mg/L

C.   *METHODS*

The Iodometric Method for measuring chlorine residual gives good results for samples containing wastewater, such as plant effluents or receiving waters. The DPD Titrimetric Method is applicable to wastewaters that do not contain iodine-reducing substances. This method also has the advantage that it can be modified to determine free chlorine residual, monochloramine, and dichloramine. Colorimetric tests for chlorine residual have special limitations and should generally be avoided in wastewater. The *AMPEROMETRIC*[14] Titration Method gives the best results for wastewater samples, but the required titrator instrument is expensive. The Amperometric Titration Method can detect and measure chlorine residual below 0.2 mg/L, but it is not possible to distinguish between free and combined chlorine forms at this low level. However, such differentiation may not be necessary for wastewater.

D.   *APPARATUS REQUIRED*

Iodometric Method

1. Graduated cylinder, 250 mL

2. Pipets, 5 and 10 mL

---

14. *Amperometric* (am-purr-o-MET-rick).   A method of measurement that records electric current flowing or generated, rather than recording voltage. Amperometric titration is a means of measuring concentrations of certain substances in water.

3. Erlenmeyer flask, 500 mL

4. Buret, 5 mL

5. Magnetic stirrer

DPD Titrimetric Method

1. Graduated cylinder, 100 mL

2. Pipets, 1 and 10 mL

3. Erlenmeyer flask, 250 mL

4. Buret, 10 mL

5. Magnetic stirrer

Amperometric Titration Method (All Residual Levels)

1. Amperometric titrator

2. End point detection apparatus

3. Agitator

4. Buret

For more information about this method and the equipment used, see page 4-62, *Standard Methods,* 21st Edition (2005), and the amperometric titrator manufacturer's instruction manual.

E. REAGENTS

Standardized solutions may be purchased from chemical suppliers.

F. VOLUME OF SAMPLE

For chlorine residual concentrations of 10 mg/L or less, take a 200-mL sample for titration. For concentrations greater than 10 mg/L, use proportionately less sample.

G. PROCEDURE

Iodometric Method

1. Pipet 5.00 mL 0.00564 *N* PAO solution into an Erlenmeyer flask.

2. Add excess KI (approximately 1 gm).

3. Add 4 mL acetate buffer solution, or enough solution to lower the pH to between 3.5 and 4.2.

4. Pour in 200 mL of sample.

5. Mix with magnetic stirrer.

6. Just before titration, add 1 mL starch solution.

7. Titrate with 0.0282 *N* Iodine just until the sample turns blue. This color will remain after complete mixing.

DPD Titrimetric Method

1. Place 5 mL each of buffer reagent and DPD indicator in a 250-mL flask. Mix.

2. Add 2 drops (0.1 mL) KI solution. Mix.

3. Add 100 mL of sample. Mix.

4. Let stand for 2 minutes and then titrate with FAS titrant.

Amperometric Titration Method

FREE CHLORINE

1. Unless sample pH is known to be between 6.5 and 7.5, add 1 mL pH 7 phosphate buffer solution to produce a pH of 6.5 to 7.5.

2. Titrate with standard phenylarsine oxide (PAO), observing current changes on microammeter. Add titrant (PAO) in smaller amounts until all needle movement stops. Record buret readings when needle action becomes sluggish, signaling the approach of the end point. Subtract last small increment of titrant that causes no needle movement because of overtitration.

COMBINED CHLORINE

3. To sample remaining from free-chlorine titration, in order, add 1.00 mL KI solution and 1 mL acetate buffer solution.

4. Titrate with phenylarsine oxide (PAO) titrant to the end point, as in Step 2, above. Do not refill buret, but continue titration after recording buret reading for free chlorine. Again, subtract last small increment of titrant to give amount of titrant actually used in reaction with chlorine. If titrant was continued without refilling buret, this buret reading represents total chlorine. Subtracting free chlorine from total gives combined chlorine.

5. Wash apparatus and sample cell thoroughly to remove iodide ion to avoid inaccuracies when the titrator is used subsequently for a free chlorine determination.

MONOCHLORAMINE

6. After titrating for free chlorine, add 0.2 mL KI solution to the same sample and, without refilling buret, continue to titrate with phenylarsine oxide (PAO) to the end point. Subtract last increment of PAO to obtain net volume of PAO consumed by monochloramine.

DICHLORAMINE

7. Add 1 mL acetate buffer solution and 1 mL KI solution to same sample and titrate final dichloramine fraction by titrating with PAO to the end point. Subtract last increment of PAO to obtain net volume of PAO consumed by dichloramine.

(Chlorine Residual)

OUTLINE OF PROCEDURE FOR CHLORINE RESIDUAL
IODOMETRIC METHOD

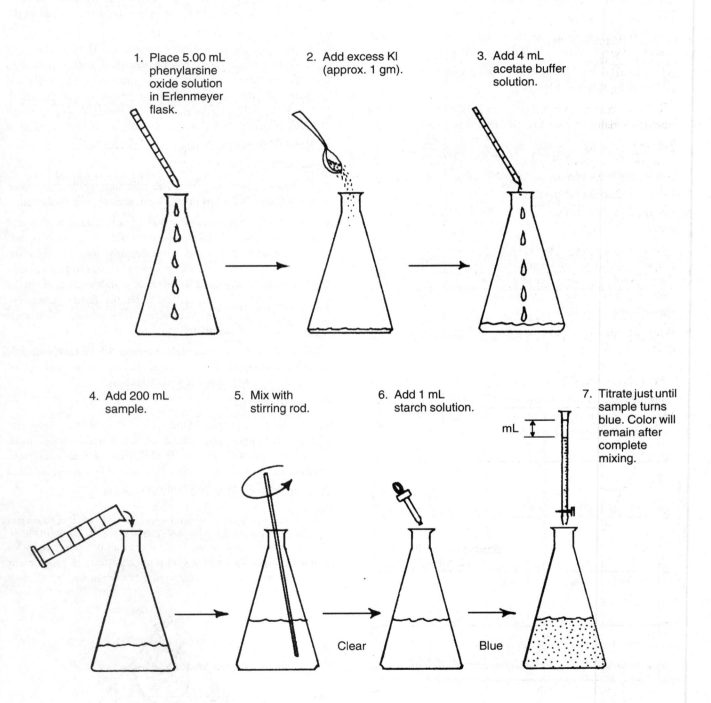

1. Place 5.00 mL phenylarsine oxide solution in Erlenmeyer flask.

2. Add excess KI (approx. 1 gm).

3. Add 4 mL acetate buffer solution.

4. Add 200 mL sample.

5. Mix with stirring rod.

6. Add 1 mL starch solution.

7. Titrate just until sample turns blue. Color will remain after complete mixing.

Clear

Blue

mL

(Chlorine Residual)

Amperometric Titration Method (Low Residual Levels)

1. Select a sample volume requiring no more than 2 mL PAO titrant. A 200-mL sample will be adequate for sample containing less than 0.2 mg total chlorine/L (see Section I, Note 11).

2. Add 200 mL sample to sample container and add approximately 1.5 gm KI crystal. Dissolve using a stirrer or mixer.

3. Add 1.0 mL acetate buffer and place container in end point detection apparatus.

4. When the current (amperage) signal stabilizes, record the reading. Initially, adjust meter to a near full-scale deflection.

5. Titrate by adding a small, known volume of titrant. After each addition, record the cumulative volume added and the current reading when the signal stabilizes (see Section I, Note 12).

6. Continue adding titrant until no further meter deflection occurs.

7. Determine equivalence point by plotting total meter deflection against titrant volume added. Draw straight line through the first several points in the plot and a second, horizontal line corresponding to the final total deflection in the meter as shown in Figure 14.10.

8. Read equivalence point as the volume of titrant added at the intersection of these two lines.

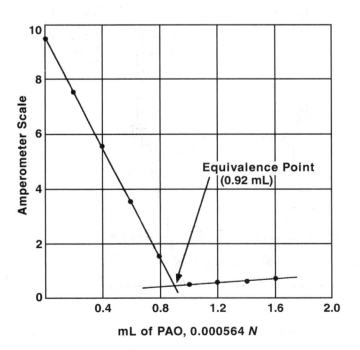

Fig. 14.10   Determination of low-level total chlorine
(Source: Leonard Ashack)

H. CALCULATIONS

Iodometric Method

Total Chlorine Residual, mg/L $= \dfrac{(A - 5B) \times 200}{C}$

A = mL 0.00564 N PAO

B = mL 0.0282 N I$_2$

C = mL of Sample Used

EXAMPLE: Titration of a 200-mL sample requires 0.3 mL I$_2$ solution and 5 mL of PAO.

Total Chlorine Residual, mg/L $= \dfrac{(A - 5B) \times 200}{C}$

$= \dfrac{[5 - (5 \times 0.3)](200)}{200 \text{ mL}}$

$= (5 - 1.5)$

$= 3.5$ mg/L

DPD Titrimetric Method

For a 100-mL sample, 1.00 mL standard FAS titrant = 1.0 mg/L chlorine residual.

EXAMPLE: 100 mL of sample required 3.4 mL standard FAS titrant, therefore

Total Chlorine Residual, mg/L = 3.4 mg/L

Amperometric Titration Method

Chlorine Residual, mg/L $= \dfrac{A \times 200}{\text{mL Sample}}$

Where:

A = mL phenylarsine oxide (PAO) titration

EXAMPLE: 100 mL of sample required 2.4 mL PAO for titration of free and combined chlorine.

Total Chlorine Residual, mg/L $= \dfrac{A \times 200}{\text{mL Sample}}$

$= \dfrac{2.4 \times 200}{100 \text{ mL}}$

$= 4.8$ mg/L

Amperometric Titration Method (Low Residual Levels)

mg Cl as Cl$_2$/L $= \dfrac{A \times 200 \times N}{B \times 0.000564 \times N}$

where: A = mL Titrant at Equivalence Point

B = Sample Volume, mL

N = Phenylarsine (PAO) Normality

(Chlorine Residual)

EXAMPLE: The results of PAO titration on a 200-mL sample of chlorinated and dechlorinated effluent sample are as follows:

Volume of PAO, mL	Current Reading on Amperometer
0	9.5
0.2	7.5
0.4	5.6
0.6	3.6
0.8	1.6
1.0	0.5
1.2	0.6
1.4	0.6
1.6	0.7

From Figure 14.10, the equivalence point was determined at 0.92 mL.

$$\text{mg/L of Cl} = \frac{0.92 \times 200 \times 0.000564}{200 \times 0.000564}$$

$$= 0.92 \text{ mg/L}$$

I. NOTES

1. All chlorine residuals are reported as milligrams of chlorine as Cl_2 per liter (mg Cl as Cl_2/L).

2. Some organic chloramines can interfere with the test. Monochloramine can intrude into the free chlorine fraction and dichloramine can interfere in the monochloramine fraction, especially at high temperatures and prolonged titration times.

3. Free halogens (iodine, bromine) other than chlorine also will titrate as free chlorine.

4. Combined chlorine reacts with iodide ions to produce iodine.

5. When titration for free chlorine follows a combined chlorine titration that requires addition of KI, erroneous results may occur unless the measuring cell is rinsed thoroughly with distilled water between titrations.

6. Interference from copper has been observed in samples taken from copper pipe or after heavy copper sulfate treatment of reservoirs.

7. Silver ions poison the electrode.

8. Interference occurs in some highly colored waters and in waters containing surface-active agents.

9. Vigorous stirring can lower chlorine values by volatilization.

10. Very low temperatures slow the response of the measuring cell and a longer time period is required for titration, but precision is not affected.

11. It is possible to use the same method noted above and measure effluent chlorine concentration greater than 0.2 mg/L by using higher strength of PAO. According to the

formula in Section H, "Calculations," varying the concentration of PAO would allow flexibility as shown in the following table.

Strength of PAO	Dilution from 0.00564 N	Chlorine Concentration in mg/L per mL Titrant	Maximum* Concentration Measured
0.000564	1 mL into 10 mL	0.1 mg/L	0.2 mg/L
0.001128	1 mL into 5 mL	0.2 mg/L	0.4 mg/L
0.00282	1 mL into 2 mL	1.0 mg/L	1.0 mg/L
0.00564	No dilution	1.0 mg/L	2.0 mg/L

NOTE: Assumed the maximum titrant volume is 2.0 mg/L.

12. Keep the tip of the titration needle in contact with the surface of the sample being titrated to avoid drop formation. This will hasten the meter response.

13. Operators using the DPD colorimetric method to test water for a free chlorine residual need to be aware of a potential error that may occur. If the DPD test is run on water containing a combined chlorine residual, a precipitate may form during the test. The particles of precipitated material will give the sample a turbid appearance or the appearance of having color. This turbidity can produce a positive test result for free chlorine residual when there is actually no chlorine present. Operators call this error a "false positive" chlorine residual reading.

14. Home swimming pool chlorine residual analyzers are not approved for commercial use, but can be used in an emergency for a short period of time. However, the results will not be accurate.

J. REFERENCES

Iodometric Method: See page 4-55, *Standard Methods,* 21st Edition.

DPD Titrimetric Method: See page 4-61, *Standard Methods,* 21st Edition.

Amperometric Titration Method: See page 4-58, *Standard Methods,* 21st Edition.

Low-Level Amperometric Titration Method: See page 4-60, *Standard Methods,* 21st Edition.

QUESTIONS

Please write your answers to the following questions and compare them with those on page 340.

14.5A   Why should plant effluents be chlorinated?

14.5B   What are the advantages and disadvantages of the iodometric titration and amperometric titration methods of measuring chlorine residual?

2. Dissolved Oxygen (DO) and Biochemical Oxygen Demand (BOD)

Dissolved Oxygen (DO)

A. *DISCUSSION*

The DO test is the testing procedure used to determine the amount of oxygen dissolved in samples of water or wastewater. The analysis for DO is a key test in water pollution control activities and the waste treatment process control. The generalized principle is that iodine will be released in proportion to the amount of dissolved oxygen present in the sample tested. By using sodium thiosulfate with starch as the indicator, one can titrate the sample and determine the amount of dissolved oxygen.

There are various types of testing procedures that can be run to obtain the amount of dissolved oxygen. Two of these procedures are discussed in this section. The Sodium Azide Modification of the Winkler Method, a "wet method" (industry term for wet chemistry analysis), is best suited for relatively clean waters. Substances that can interfere with this test include color, organics, suspended solids, sulfide, chlorine, and ferrous and ferric iron. Nitrite will not interfere with the test if fresh azide is used. The other testing procedure discussed in this section is the DO Probe method on page 321. The BOD test discussion starts on page 322. The "Outline of Procedure for BOD" on page 324 shows the steps in this testing procedure.

B. *WHAT IS TESTED?*

Sample	Common Range, mg/L
Influent	Usually 0. Greater than 1 is very good.
Primary Clarifier Effluent	Usually 0. Recirculated from filters. Greater than 2 is good.
Secondary Effluent	50% to 95% saturation. 3 to greater than 8 is good.
Oxidation Ponds	1 to 25+ (supersaturated with oxygen).
Activated Sludge— Aeration Tank Outlet	Greater than 2 is desirable.

C. *APPARATUS*

Method A (Sodium Azide Modification of Winkler Method)

1. Buret, graduated to 0.1 mL
2. Three 300-mL glass-stoppered BOD bottles
3. Wide-mouth Erlenmeyer flask, 500 mL
4. One 10-mL measuring pipet
5. One 1-liter reagent bottle to collect activated sludge

Method B (DO Probe)

Follow manufacturer's instructions. Also see Section H, "DO Probe," on page 321.

D. *REAGENTS*

Standardized solutions may be purchased from chemical suppliers.

E. *PROCEDURE* (page 321)

Sodium Azide Modification of the Winkler Method

Sodium azide destroys nitrate, which will interfere with this test. Add the reagents using the order, methods, and quantities listed as follows:

1. Collect a sample to be tested in a 300-mL (BOD) bottle, taking special care to avoid aeration of the liquid being collected. Fill bottle completely and add cap.

2. Remove cap and add 1 mL of manganous sulfate solution below the surface of the liquid.

3. Add 1 mL of alkaline iodide-sodium azide solution below the surface of the liquid.

4. Replace the stopper, avoid trapping air bubbles, and mix the solution well by inverting the bottle several times. Repeat this mixing after the floc has settled halfway. Allow the floc to settle halfway a second time.

5. Wearing protective gloves, acidify with 1 mL of concentrated sulfuric acid by allowing the acid to run down the neck of the bottle above the surface of the liquid. Handle the bottle carefully to avoid acid burns.

6. Restopper the bottle and mix the solution well until the precipitate has dissolved. The solution will then be ready to titrate.

7. Pour 200 mL of the original sample from the bottle into an Erlenmeyer flask.

8. If the solution is brown in color, titrate with 0.025 *N* PAO until the solution is a pale yellow color. Add a small quantity of starch indicator and proceed to Step 10. (Note: Either PAO or 0.025 *N* sodium thiosulfate can be used.) If the solution is not brown, proceed to Step 9.

9. If the solution has no brown color, or is only slightly colored, add a small quantity of starch indicator. If no blue color develops, the samples contains zero dissolved oxygen. If a blue color does develop, proceed to Step 10.

10. Titrate the sample until the blue color begins to disappear. Record the number of mL of PAO used. If the blue color returns, this indicates that nitrate ions are present in the sample.

11. The amount of oxygen dissolved in the original solution will be equal to the total number of mL of PAO used in the titration provided significant interfering substances are not present.

mg/L DO = mL PAO

(DO—BOD)

OUTLINE OF PROCEDURE FOR PREPARING REAGENTS

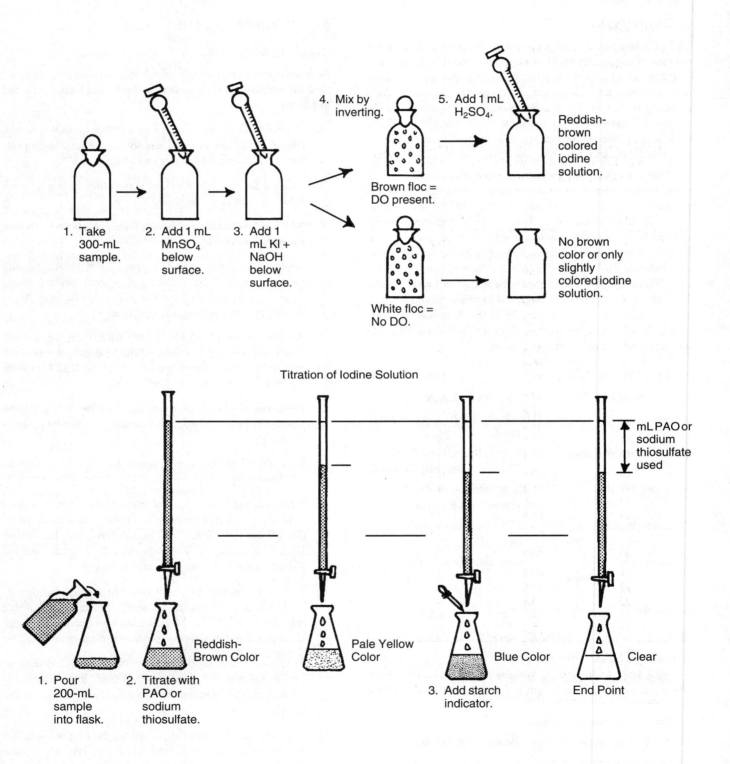

Brown floc = DO present.

White floc = No DO.

Reddish-brown colored iodine solution.

No brown color or only slightly colored iodine solution.

4. Mix by inverting.

5. Add 1 mL H_2SO_4.

1. Take 300-mL sample.

2. Add 1 mL $MnSO_4$ below surface.

3. Add 1 mL KI + NaOH below surface.

Titration of Iodine Solution

mL PAO or sodium thiosulfate used

1. Pour 200-mL sample into flask.

2. Titrate with PAO or sodium thiosulfate.

Reddish-Brown Color

Pale Yellow Color

3. Add starch indicator.

Blue Color

Clear

End Point

F. EXAMPLE

The DO titration of a 200-mL sample requires 5.0 mL of 0.025 N PAO. Therefore, the dissolved oxygen concentration in the sample is 5 mg/L.

G. CALCULATION

The dissolved oxygen saturation values are given in Table 14.5. Table 14.5 gives 100 percent DO saturation values for temperatures in °C and °F. Note that as the temperature of water increases, the DO saturation value (100% Saturation column) decreases.

TABLE 14.5   EFFECT OF TEMPERATURE ON OXYGEN SATURATION FOR A CHLORIDE CONCENTRATION OF ZERO mg/L

°C	°F	mg/L DO at 100% Saturation
0	32.0	14.6
1	33.8	14.2
2	35.6	13.8
3	37.4	13.5
4	39.2	13.1
5	41.0	12.8
6	42.8	12.5
7	44.6	12.2
8	46.4	11.9
9	48.2	11.6
10	50.0	11.3
11	51.8	11.1
12	53.6	10.8
13	55.4	10.6
14	57.2	10.4
15	60.0	10.2
16	61.8	10.0
17	63.6	9.7
18	65.4	9.5
19	67.2	9.4
20	68.0	9.2
21	69.8	9.0
22	71.6	8.8
23	73.4	8.7
24	75.2	8.5
25	77.0	8.4

Using Table 14.5, find the percent saturation of DO in the effluent of your secondary plant if the DO is 5.0 mg/L and the temperature is 20°C. At 20°C, 100 percent DO saturation is 9.2 mg/L (highlighted in Table 14.5).

$$\text{DO Saturation, \%} = \frac{\text{DO of Sample, mg/L} \times 100\%}{\text{DO at 100\% Saturation, mg/L}}$$

$$= \frac{5.0 \text{ mg/L}}{9.2 \text{ mg/L}} \times 100\%$$

$$= .54 \times 100\%$$

$$= 54\%$$

H. DO PROBE

1. DISCUSSION

Measuring the dissolved oxygen (DO) concentration with a probe and electronic readout meter is a satisfactory substitute for the Sodium Azide Modification of the Winkler Method under many circumstances. Using the DO probe is recommended when samples contain substances that interfere with the Modified Winkler procedure. These substances include sulfite, thiosulfate, polythionate, mercaptans, free chlorine or hypochlorite, organic substances readily hydrolyzed in alkaline solutions, free iodine, intense color or turbidity, and biological flocs. A record of the dissolved oxygen content of aeration tanks and receiving waters can be created using a probe. In determining the BOD of samples, a probe also may be used to determine the DO initially and again after the five-day incubation period of the BLANK[15] and sample dilutions.

2. PROCEDURE

Follow the manufacturer's instructions for using the DO probe.

3. CALIBRATION

To ensure that the DO probe reading provides the dissolved oxygen content of the sample, the probe must be calibrated as follows:

a. Take a sample that does not contain substances that interfere with either the probe reading or the Modified Winkler procedure.

b. Split the sample.

c. Measure the DO in one portion of the sample using the Modified Winkler procedure.

d. Compare this result with the DO probe reading on the other portion of the sample.

e. Adjust the probe reading to agree with the results from the Modified Winkler procedure.

15. *Blank.*   A bottle containing only dilution water or distilled water; the sample being tested is not added. Tests are frequently run on a sample and a blank and the differences are compared. The procedure helps to eliminate or reduce test result errors that could be caused when the dilution water or distilled water used is contaminated.

(DO—BOD)

When calibrating the probe in an aeration tank of the activated sludge process, do not attempt to measure the dissolved oxygen in the aerator and then adjust the probe. The biological flocs in the aerator will interfere with the Modified Winkler procedure, and the copper sulfate-sulfamic acid procedure is not sufficiently accurate to calibrate the probe. An aeration tank probe may be calibrated by splitting an effluent sample, measuring the DO by the Modified Winkler procedure, and comparing results with the probe readings. Always keep the membrane in the tip of the probe from drying out because the probe can lose its accuracy until reconditioned.

4. PRECAUTIONS

a. Periodically check the calibration of the probe.

b. Keep the membrane in the tip of the probe from drying out.

c. Check the probe for dissolved inorganic salts such as those found in seawater, which can influence probe readings.

d. Be aware that reactive compounds such as reactive gases and sulfur compounds can interfere with probe output.

e. Avoid placing the probe directly over a diffuser because you want to measure the dissolved oxygen in the water being treated, not the oxygen in the air supply to the aerator.

QUESTIONS

Please write your answers to the following questions and compare them with those on page 340.

14.5C   Calculate the percent dissolved oxygen saturation if the receiving water DO is 7.9 mg/L and the temperature is 10°C.

14.5D   How would you calibrate the DO probe in an aeration tank?

Biochemical Oxygen Demand (BOD)

A. DISCUSSION

The BOD test measures the amount of oxygen used by microorganisms as they consume the substrate (food) in wastewater when placed in a controlled temperature for five days. The BOD test is an estimate of the availability of food in the sample (food or organisms that take up oxygen) expressed in terms of oxygen use. The amount of organic material in wastewater can be estimated indirectly by measuring oxygen use with the BOD test. DO (dissolved oxygen) is measured and recorded at the beginning of the test. During the five-day period, microorganisms in the sample break down complex organic matter in the sample, using up oxygen in the process. After the five-day dark incubation period, the DO is again determined. The BOD is then calculated on the basis of the reduction of DO and the size of the sample. Results of the BOD test indicate the rate of oxidation and provide an indirect estimate of the availability to organisms of the waste.

Actual environmental conditions such as temperature, organism population, water movement, sunlight, and oxygen concentration cannot be accurately reproduced in the laboratory. Results obtained from this test must take these factors into account when relating BOD test results to receiving water oxygen demands.

Samples for the BOD test should be collected before chlorination because chlorine interferes with the organisms in the test. It is difficult to obtain accurate results with dechlorinated samples.

Samples are incubated for a standard period of five days because a fraction of the total BOD will be exerted during this period. The ultimate or total BOD is normally not run for plant control. A limitation of the BOD test is that the results are not available until five days after the sample is collected.

B. WHAT IS TESTED?

Sample	Common Range, mg/L
Influent	150 to 400
Primary Effluent	60 to 160
Secondary Effluent	5 to 30
Digester Supernatant	1,000 to 4,000+
Industrial Wastes	100 to 3,000+

C. APPARATUS

1. 300 mL BOD bottles with ground glass stoppers

2. Incubator, 20°C ± 1°C

3. Pipets, 10 mL graduated, $\frac{1}{32}$ inch to $\frac{1}{16}$ inch (0.8 mm to 1.6 mm) diameter tip

4. Buret and stand

5. Erlenmeyer flask, 500 mL

D. REAGENTS

Standardized solutions may be purchased from chemical suppliers.

1. Using distilled water. Water used for solutions and for preparation of the solution water must be of the highest quality. The distilled water must contain less than 0.1 mg/L copper and be free of chlorine, chloramines, caustic, organic material, and acids. Ordinary distilled water such as that used in a car battery is not appropriate for use in reagent solutions.

2. Using dilution water. Prepare the dilution water using standard methods. Add 1 mL each of phosphate buffer, magnesium sulfate, calcium chloride, and ferric chloride solutions for each liter of distilled water. Saturate with DO by shaking the solution in a partially filled bottle or by aerating with filtered air. Store at a temperature as close to 20°C as possible for at least 24 hours to allow the water to stabilize. This water should not show a drop in DO of more than 0.2 mg/L on incubation for five days.

Many plants do not prepare reagents. Small plants and plants that do not run many tests find it quicker and easier to purchase commercially prepared reagents. These reagents may be available in the desired strength or they may consist of dry *PILLOWS* [16] that are added to the sample, rather than the liquid reagent. Check with your chemical supplier for these reagents.

E. *PROCEDURE FOR TESTING UNCHLORINATED SAMPLES*

This test measures the oxygen used or depleted by a measured quantity of wastewater sample seeded into a reservoir of dilution water saturated with oxygen during a five-day period at 20°C. The result is compared to an unseeded or blank reservoir of dilution water by subtracting the difference and multiplying by a factor for dilution.

To prepare for the BOD test procedure, follow these steps:

1. Use 300-mL capacity BOD bottles with graduations and ground-glass stoppers. To clean the bottles, carefully rinse with tap water followed by distilled water. Number each BOD bottle and fitted stopper. Insert each numbered stopper into the BOD bottle with the same number. If the numbers do not match, errors could occur in the test results.

2. Prepare one or more dilutions of the sample to cover the estimated range of BOD values. From the estimated BOD, calculate the volume of raw sample to be added to the BOD bottle based on the fact that the most valid DO depletion is 4 mg/L. Therefore:

$$\begin{aligned} \text{mL of Sample Added} \atop \text{per 300 mL} &= \frac{(4 \text{ mg/L})(300 \text{ mL})}{\text{Estimated BOD, mg/L}} \\ &= \frac{1{,}200}{\text{Estimated BOD, mg/L}} \end{aligned}$$

EXAMPLES

a. Estimated BOD = 400 mg/L

$$\text{mL of Sample Added to BOD Bottle} = \frac{1{,}200}{400}$$
$$= 3 \text{ mL}$$

b. Estimated BOD = 200 mg/L: use 6 mL

100 mg/L: use 12 mL

20 mg/L: use 60 mL

When the BOD is unknown, select more than one sample size. For example, place several samples—1 mL, 3 mL, 6 mL, and 12 mL—into four BOD bottles.

For samples with very high BOD values, it may be difficult to accurately measure small volumes or to get a truly representative sample. In such a case, initial dilution should first be made on the sample. A dilution of 1:10 is convenient.

To perform the BOD test, follow these steps:

1. Fill two bottles completely with dilution water siphoned from its container. Insert the stoppers tightly so that no air is trapped beneath them. These bottles are numbered "1" and "2" in the Outline of Procedure for BOD on page 324.

2. Next, for each sample to be tested, carefully measure out the two portions of sample and place them into two new BOD bottles, numbered "3" and "4." Add dilution water until the bottles are completely filled. Insert the stoppers. Avoid entrapping air bubbles. Be sure that there are water seals on the stoppers. A water seal is obtained by water in the small indentation between the stopper and the rim of the bottle.

3. On bottles "2" and "4," immediately determine the initial dissolved oxygen.

4. Incubate the remaining dilution water blank and diluted sample in the dark at 20°C for five days. These are bottles "1" and "3."

5. At the end of exactly five days (±3 hours), test bottles "1"and "3" for their dissolved oxygen content by using the sodium azide modification of the Winkler method or a DO probe. At the end of five days, the oxygen content should be at least 1 mg/L. Also, a depletion of 2 mg/L or more is desirable. Bottles "1"and "2" (blanks) are only used to check the dilution water quality. Their difference should be less than 0.2 mg/L after five days if the quality is good and free of impurities. The difference in blank readings is not used as a blank correction, but merely as a check on the quality of dilution water. Differences of greater than 0.2 mg/L could possibly be due to contamination or dirty BOD bottles.

F. *PRECAUTIONS*

1. The temperature of the incubator must be 20°C. Any other temperature changes the rate of oxygen used.

2. The dilution water must be made according to the standard methods (as outlined on page 322, Section D, "Reagents") for the most favorable growth rate of the bacteria. This water must be free of copper, which is often present when copper stills are used by commercial dealers. Use all glass or stainless-steel stills or demineralized water.

3. The wastewater must also be free of toxic wastes such as hexavalent chromium.

4. If you use a cleaning agent to wash BOD bottles, thoroughly wash and rinse the sample bottles several times to remove all residue. Cleaning agents are toxic and if any residue remains in a BOD bottle, the BOD test could be ruined.

5. Unchlorinated wastewater normally contains an ample supply of seed bacteria, so seeding is usually not necessary.

6. Maintain a suitable environment for the organisms, since this is a bioassay (BUY-o-AS-say), that is, living organisms are used for the test; environmental conditions must be quite exact and must be considered when reviewing results.

---

16. *Pillows.* Plastic tubes shaped like pillows that contain exact amounts of chemicals or reagents. Cut open the pillow, pour the reagents into the sample being tested, mix thoroughly, and follow test procedures.

(DO—BOD)

OUTLINE OF PROCEDURE FOR BOD
(Unchlorinated Samples)

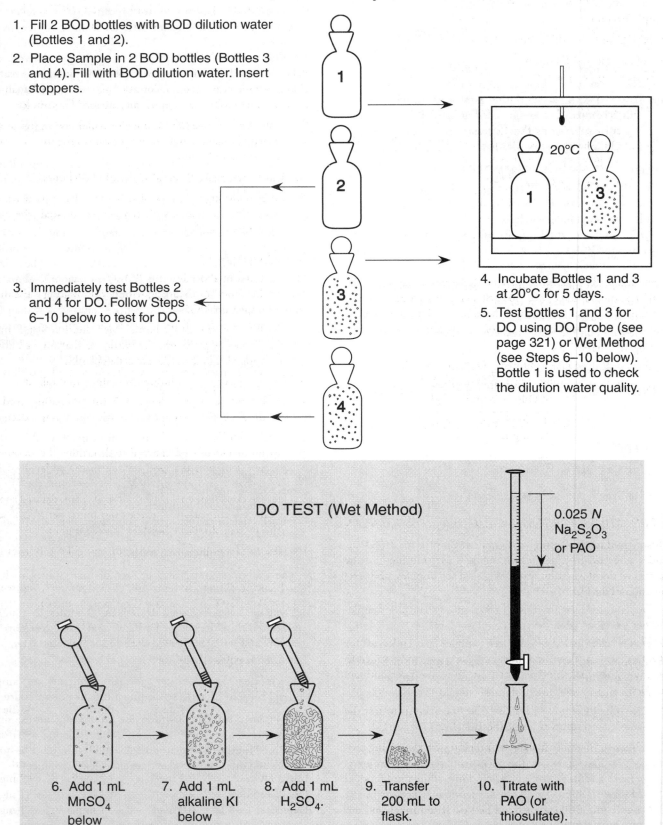

1. Fill 2 BOD bottles with BOD dilution water (Bottles 1 and 2).

2. Place Sample in 2 BOD bottles (Bottles 3 and 4). Fill with BOD dilution water. Insert stoppers.

3. Immediately test Bottles 2 and 4 for DO. Follow Steps 6–10 below to test for DO.

4. Incubate Bottles 1 and 3 at 20°C for 5 days.

5. Test Bottles 1 and 3 for DO using DO Probe (see page 321) or Wet Method (see Steps 6–10 below). Bottle 1 is used to check the dilution water quality.

20°C

DO TEST (Wet Method)

0.025 N $Na_2S_2O_3$ or PAO

6. Add 1 mL $MnSO_4$ below surface.

7. Add 1 mL alkaline KI below surface.

8. Add 1 mL H_2SO_4.

9. Transfer 200 mL to flask.

10. Titrate with PAO (or thiosulfate).

G. EXAMPLE

BOD Bottle Volume	= 300 mL
Sample Volume	= 15 mL
Initial DO of Diluted Sample	= 8.0 mg/L
DO of Sample and Dilution After 5-Day Incubation	= 4.0 mg/L

H. CALCULATIONS

$$\text{BOD, mg/L} = \begin{bmatrix} \text{Initial DO} \\ \text{of Diluted} \\ \text{Sample, mg/L} \end{bmatrix} \begin{matrix} \text{DO of Diluted} \\ - \text{ Sample After 5-day} \\ \text{Incubation, mg/L} \end{matrix} \begin{bmatrix} \text{BOD Bottle Vol, mL} \\ \overline{\text{Sample Volume, mL}} \end{bmatrix}$$

$$= (8.0 \text{ mg/L} - 4.0 \text{ mg/L}) \left[\frac{300 \text{ mL}}{15 \text{ mL}} \right]$$

$$= (4.0 \text{ mg/L})(20)$$

$$= 80 \text{ mg/L}$$

For acceptable results, the percent depletion of oxygen in the BOD test should range from 30 to 80 percent depletion. Also, if a residual DO of at least 1 mg/L remains or a DO uptake of at least 2 mg/L results after five days of incubation, the results also are considered reliable.

$$\% \text{ Depletion} = \frac{\left(\begin{matrix} \text{DO of Diluted} \\ \text{Sample, mg/L} \end{matrix} - \begin{matrix} \text{DO After 5} \\ \text{Days, mg/L} \end{matrix} \right)}{\text{DO of Diluted Sample, mgL}} \times 100\%$$

$$= \frac{(8.0 \text{ mg/L} - 4.0 \text{ mg/L})}{8.0 \text{ mg/L}} \times 100\%$$

$$= \frac{4}{8} \times 100\%$$

$$= 50\%$$

I. PROCEDURE FOR TESTING CHLORINATED SAMPLES

This procedure for testing chlorinated samples for BOD includes three major tasks: testing samples for chlorine and the process of dechlorination, preparing seed for samples, and testing samples for BOD.

Dechlorinating Samples

Whenever chlorinated wastewater samples are collected for BOD analysis, sufficient dechlorinating agent must be added to destroy the residual chlorine, if present. To test and dechlorinate samples:

1. Use the procedure for residual chlorine analysis to test the sample for residual chlorine (Section 14.51, page 314).

2. If there is no residual chlorine, proceed with the testing procedure that follows in this section. If residual chlorine is present, add sufficient 0.025 N sodium sulfite until residual chlorine is absent. If a chlorine residual is present in the sample, the BOD results will be low. If too much sodium sulfite is added, the BOD results will be high. Prepare 0.025 N sodium sulfite by dissolving 1.575 grams of anhydrous Na_2SO_3 in 1,000 mL of distilled water. NOTE: This solution is not stable and must be prepared daily.

Preparing Seed for Samples

When a sample contains very few microorganisms as a result of chlorination or extreme pH, for example, seed microorganisms must be added to the sample. To prepare seed for samples:

3. Collect about one liter of unchlorinated raw wastewater or primary effluent about 24 hours prior to the time you wish to start the BOD test.

4. Let the sample settle at 20°C for 24 to 36 hours.

5. Filter it through glass wool to remove large particles and grease clumps. Use this filtered sample for seed when testing samples for DO.

Testing Samples for DO

The "Outline of Procedure for BOD" on page 326 shows the steps in this procedure, starting with dechlorinating the sample. To test samples for DO:

6. Fill two 300-mL BOD bottles with dilution water. Insert the stoppers tightly into the bottles so that no air bubbles are trapped. These bottles are called blanks.

7. Set up one or more dilutions of samples in duplicate.

8. Add 1 mL seed from Steps 3–5 (in "Preparing Seed for Samples" above) to each BOD bottle containing a dechlorinated sample.

9. Set up blanks of seed material to determine the amount of oxygen depletion that will be due to the added seed material. Use 3, 6, and 9 mL of seed to determine the five-day oxygen depletion due to 1 mL of seed. Seed material should produce a correction of at least 0.6 mg/L per mL of seed. Also prepare duplicate bottles of these samples.

10. Test one set of duplicate bottles for initial DO (Bottles 1, 3, and 5).

11. Incubate dilution water blank, diluted samples, and seeded blanks at 20°C for five days (Bottles 2, 4, and 6).

12. At the end of five days, test the bottles for DO using a DO probe or the wet method (Steps 13–17).

13. Calculate five-day BOD:

EXAMPLE

	Bottle 2	Bottle 4	Bottle 6
Sample	blank	effluent	seed blank
Sample volume	—	10 mL	0
Seed volume	—	1 mL	3 mL
Initial DO	9.2 mg/L	9.2 mg/L	9.2 mg/L
DO after 5 days	9.2 mg/L	5.1 mg/L	7.1 mg/L
Depletion	0.0 mg/L	4.1 mg/L	2.1 mg/L

(DO—BOD)

OUTLINE OF PROCEDURE FOR BOD
(Chlorinated Samples)

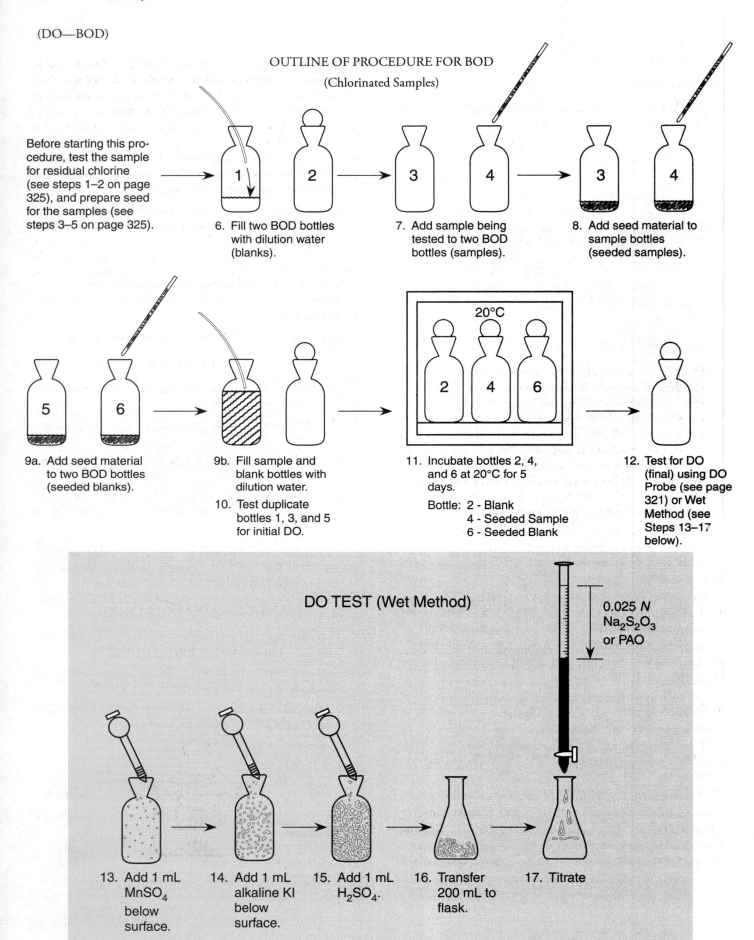

Before starting this procedure, test the sample for residual chlorine (see steps 1–2 on page 325), and prepare seed for the samples (see steps 3–5 on page 325).

6. Fill two BOD bottles with dilution water (blanks).

7. Add sample being tested to two BOD bottles (samples).

8. Add seed material to sample bottles (seeded samples).

9a. Add seed material to two BOD bottles (seeded blanks).

9b. Fill sample and blank bottles with dilution water.

10. Test duplicate bottles 1, 3, and 5 for initial DO.

11. Incubate bottles 2, 4, and 6 at 20°C for 5 days.

    Bottle: 2 - Blank
            4 - Seeded Sample
            6 - Seeded Blank

12. Test for DO (final) using DO Probe (see page 321) or Wet Method (see Steps 13–17 below).

DO TEST (Wet Method)

0.025 N
$Na_2S_2O_3$
or PAO

13. Add 1 mL $MnSO_4$ below surface.

14. Add 1 mL alkaline KI below surface.

15. Add 1 mL H_2SO_4.

16. Transfer 200 mL to flask.

17. Titrate

(DO—BOD)

CALCULATION

For seed correction:

$$\frac{\text{mg/L DO Depletion Caused}}{\text{by 1 mL of Seed}} = \frac{\text{5-day Depletion of Seed Sample}}{\text{mL of Seed}}$$

$$= \frac{9.2 \text{ mg/L} - 7.1 \text{ mg/L}}{3 \text{ mL}}$$

$$= 0.7 \text{ mg/L/mL}$$

5-day BOD:

$$\frac{\text{BOD,}}{\text{mg/L}} = \frac{[(\text{Initial DO} - \text{DO After 5 days}) - (\text{Seed Correction})] \times 300 \text{ mL}}{\text{mL of Sample Volume}}$$

$$= \frac{[(9.2 \text{ mg/L} - 5.1 \text{ mg/L}) - (0.7 \text{ mg/L/mL})(1 \text{ mL})] \times 300 \text{ mL}}{10 \text{ mL}}$$

$$= \frac{(4.1 \text{ mg/L} - 0.7 \text{ mg/L})300 \text{ mL}}{10 \text{ mL}}$$

$$= 102 \text{ mg/L}$$

J. PRECAUTIONS

1. On effluent samples where the DO is run on the sample and the blue color bounces back (returns) on the end point titration, this indicates nitrite interference and can cause the BOD test result to be higher than the actual result by as much as 10 to 15 percent. Consider this factor when interpreting your results. The end point also may waver because of decomposition of azide in an old reagent or resuspension of sample solids. To correct a wavering end point, prepare a new alkaline-azide solution or use more of the old solution because it may be decomposing (losing strength).

2. Use a minimum of two dilutions per sample. Use only analyses with oxygen depletions of greater than 2 mg/L and residual DOs of greater than 1 mg/L after five days of incubation at 20°C.

3. Samples should be well mixed before dilutions are made. Use a wide-tip pipet to make dilutions because it will not clog with suspended solids.

4. Wastewaters that have been partially nitrified may produce high BOD results. This increased oxygen demand results from the oxidation of ammonia to nitrate. The use of chemicals such as allythiourea or other commercially available nitrification inhibitors in the dilution water will inhibit the nitrifiers and alleviate this problem.

5. For effluent samples with a low DO and BOD, saturate the sample with DO by shaking it in a partially filled bottle.

K. PROCEDURE FOR INDUSTRIAL WASTE SAMPLES

Some industrial waste samples may require special seeding because of a low microbial population or because the wastes contain organic compounds that are not readily oxidized by domestic wastewater seed. To obtain the necessary specialized seed material (microorganisms) adapted or acclimated to the industrial organic compounds, collect a sample of adapted seed from the effluent of a biological treatment process (activated sludge aeration tank) treating the industrial waste. When this source of adapted seed is not available, develop the adapted seed in the laboratory by continuously aerating a large sample of water and feeding it daily with small portions of the particular waste, together with soil or settled domestic wastewater, until a satisfactory microbial population has developed.

Once a satisfactory adapted seed is available, follow the procedures in Section I, "Procedure for Testing Chlorinated Samples, Testing Samples for DO," on page 325. Start with Step 6, "Fill two 300-mL BOD bottles with dilution water." Follow the remaining steps, except use the industrial waste sample that you collected instead of the dechlorinated sample.

L. REFERENCE

See pages 4-136 and 5-2, *Standard Methods,* 21st Edition.

QUESTIONS

Please write your answers to the following questions and compare them with those on page 340.

14.5E How would you determine the amount of organic material in wastewater?

14.5F Why should samples for the BOD test be collected before chlorination?

14.5G How would you prepare dilutions to measure the BOD of cannery waste having an expected BOD of 2,000 mg/L?

14.5H What is the BOD of a wastewater sample if a 2-mL sample in a 300-mL BOD bottle had an initial DO of 7.5 mg/L and a final DO of 3.9 mg/L?

END OF LESSON 6 OF 7 LESSONS

on

LABORATORY PROCEDURES

Please answer the discussion and review questions next.

DISCUSSION AND REVIEW QUESTIONS

Chapter 14.   LABORATORY PROCEDURES

(Lesson 6 of 7 Lessons)

Please write your answers to the following questions to determine how well you understand the material in the lesson. The question numbering continues from Lesson 5.

22. Why is the effluent from a treatment plant chlorinated?

23. What is the formula for calculating the percent saturation of DO?

24. What is a blank, as defined in laboratory procedures?

25. What precautions should be exercised when using a DO probe?

26. What is a limitation of the BOD test?

27. What precautions should be taken when running a BOD test on an unchlorinated sample?

28. Calculate the BOD of a 5-mL sample if the initial DO of the diluted sample was 7.5 mg/L and the DO of the diluted sample was 3.0 mg/L after a five-day incubation period.

CHAPTER 14.   LABORATORY PROCEDURES

(Lesson 7 of 7 Lessons)

3. pH (Hydrogen Ion)

A. *DISCUSSION*

The intensity of the basic or acidic strength of water is expressed by its pH.

Mathematically, pH is the logarithm of the reciprocal of the hydrogen ion activity, or the negative logarithm of the hydrogen ion activity.

$$pH = Log \frac{1}{[H^+]} = -Log\ [H^+]$$

EXAMPLE

If a wastewater sample has a pH of 1, then the hydrogen ion activity $[H^+] = 10^{-1} = 0.1$.

If pH = 7, then $[H^+] = 10^{-7} = 0.0000001$.

pH Scale

0	Increasing acidity—7—Increasing alkalinity	14

$1\leftarrow2\leftarrow3\leftarrow4\leftarrow5\leftarrow6$ ⌇ $8\rightarrow9\rightarrow10\rightarrow11\rightarrow12\rightarrow13$

Neutral
6 through 8

In a solution, both hydrogen ions $[H^+]$ and the hydroxyl ions $[OH^-]$ are always present. At a pH of 7, the activity of both hydrogen and hydroxyl ions equals 10^{-7} moles per liter. When the pH is less than 7, the activity of hydrogen ions is greater than the hydroxyl ions. The hydroxyl ion activity is greater than the hydrogen ions in solutions with a pH greater than 7.

The pH test indicates whether a treatment process may continue to function properly at the pH measured. Each process in the plant has its own favorable range of pH, which must be checked routinely. Generally, a pH value from 6 to 8 is acceptable for best organism activity. Most wastewater contains many dissolved solids and buffers that tend to minimize pH changes.

B. *WHAT IS TESTED?*

Wastewater	Common Range
Influent or Raw Wastewater (domestic)	6.8 to 8.0
Raw Sludge (domestic)	5.6 to 7.0
Digester Recirculated Sludge or Supernatant	6.8 to 7.2
Plant Effluent Depending on Type of Treatment	6.8 to 8.0

C. *MINIMUM APPARATUS LIST*

1. pH meter
2. Glass electrode
3. Reference electrode
4. Magnetic stirrer (optional)
5. Beaker for sample

D. *REAGENTS*

1. Buffer tablets of various pH values
2. Distilled water

E. *PROCEDURE*

1. Due to the differences between the various makes and models of pH meters commercially available, specific instructions cannot be provided for the correct operation of all instruments. In each case, follow the manufacturer's instructions for preparing the electrodes and operating the instrument.

2. Standardize the instrument against a buffer solution with a pH approaching that of the sample.

3. Rinse electrodes thoroughly with distilled water after removal from buffer solution.

4. Place electrodes in sample and measure pH.

5. Remove electrodes from sample, rinse thoroughly with distilled water.

6. Immerse electrode ends in beaker of pH 7 buffer solution.

7. Shut off meter.

F. *PRECAUTIONS*

1. To avoid faulty instrument calibration, prepare fresh buffer solutions weekly from commercially available buffer tablets.

2. pH meter, buffer solution, and samples should all be at the same temperature (constant) because temperature variations will give erroneous results. Allow a few minutes for the probe to adjust to the buffers before calibrating a pH meter to ensure accurate pH readings.

3. Watch for erratic results arising from electrodes, faulty connections, or fouling of electrodes with interfering matter. Film may be removed from electrodes by placing isopropanol on a tissue or cotton swab to clean the probe.

4. If measuring pH in colored samples or samples with high solids content, or if taking measurements that need to be

(pH—Total Solids)

reported to the USEPA, use a pH electrode and meter instead of a colorimetric method or test papers. pH meters are capable of providing ± 0.1 pH accuracy in most applications. In contrast, colorimetric tests provide ± 0.1 pH accuracy only in a limited range, and pH papers provide even less accuracy.

QUESTION

Please write your answers to the following questions and compare them with those on page 341.

14.5I   What precautions should be exercised when using a pH meter?

4. Total Solids (Residue)

The term total solids (residue) refers to the matter suspended and dissolved in wastewater or water. The total solids (residue) test measures both suspended and dissolved solids in wastewater.

Total Solids

A. *DISCUSSION*

Total solids is the term applied to material left in a container after evaporation of a sample in an oven at 103° to 105°C.

B. *WHAT IS TESTED?*

Sample	Common Range, mg/L
Influent and Effluent Sludges	300 to 1,200

C. *APPARATUS REQUIRED*

1. Evaporating dishes (100 mL capacity)
2. Drying oven
3. Desiccator
4. Analytical balance
5. Muffle furnace

D. *PROCEDURE*

The procedure for determining total solids is discussed in this section and is outlined on page 331. Aluminum evaporating dishes do not require preparation.

Preparation of Evaporating Dish

1. Ignite a clean, porcelain evaporating dish at 550° ± 50°C for one hour in a muffle furnace.
2. Cool the dish in a desiccator.
3. Weigh, and record the weight. Store the dish in the desiccator until ready for use.

Sample Analysis

4. Shake the sample vigorously and pour a measured amount into the pre-weighed dish. This can be done by using a graduated cylinder to deliver the desired volume. Due to the difficulty in accurately measuring the volume of some sludge samples, an unknown quantity of sludge can be poured into the dish and then weighed to determine the initial sample quantity. This is possible because the specific gravity of sludge is only slightly higher than 1.0. Choose a sample volume that will yield a maximum residue of 25 to 250 milligrams. If necessary, add additional portions of sample to the dish.

5. Place the sample in an oven and evaporate to dryness. The oven temperature should be 98°C (2°C lower than boiling) to prevent boiling and spattering. Once the water has evaporated, raise the temperature to 103°C and maintain this temperature for at least one hour. Repeat drying cycle until a constant weight is obtained or until weight loss is less than 0.5 milligrams. The exact drying time is usually determined by experience, but at 103°C some sludge samples will need to be dried overnight.

6. Remove the dried sample from the oven. Place it in the desiccator and allow it to cool completely.

7. Weigh the dish plus total solids and save the solids if the volatile portion is to be determined.

E. *PRECAUTIONS*

1. Wastewater and sludge may contain infectious organisms and must be handled with care.
2. Use tongs or gloves when handling hot items.
3. Make sure the balance on the scale is properly leveled, zeroed, and maintained.
4. Do not weigh items when they are hot because heat causes errors by creating convection currents in the balance.

F. *EXAMPLE*

Results from an effluent sample were:

Weight of Clean Dish   = 80.1526 grams

Weight of Dish and Residue = 80.1732 grams

Sample Volume   = 100 mL

G. *CALCULATIONS*

1. Total Solids, mg/L $= \dfrac{(A - B)(1{,}000 \text{ mL/L})}{\text{mL of Sample}}$

where A = Weight of Dish and Residue in milligrams

and B = Weight of Dish in milligrams

2. From example,

$$\text{Total Solids,} \atop \text{mg/L} = \frac{(A - B)(1{,}000 \text{ mL/L})}{\text{mL of Sample}}$$

$$= \frac{(80{,}173.2 \text{ mg} - 80{,}152.6 \text{ mg})(1{,}000 \text{ mL/L})}{100 \text{ mL}}$$

$$= 206 \text{ mg/L}$$

H. *REFERENCE*

See page 2-54, *Standard Methods,* 21st Edition.

OUTLINE OF PROCEDURE FOR TOTAL SOLIDS

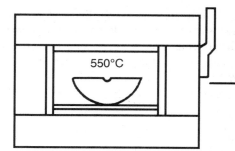

1. Ignite evaporating dish at 550°C.

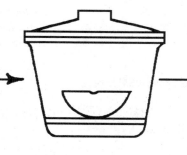

2. Cool in desiccator.

3. Weigh dish and store in desiccator.

4. Pour measured amount into a pre-weighed dish.

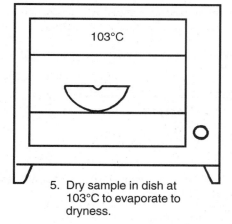

5. Dry sample in dish at 103°C to evaporate to dryness.

6. Cool dish in desiccator.

7. Weigh dish and total solids.

(Total Solids)

Total Dissolved (Filterable) Solids

A. DISCUSSION

Total dissolved solids (TDS) refers to material that passes through a standard glass-fiber filter disk and remains after evaporation at 180°C.

B. WHAT IS TESTED?

Sample	Common Range, mg/L
Influent and Effluent	150 to 600

C. APPARATUS REQUIRED

1. Glass-fiber filter discs (Reeve Angel Type 934A, 984H; or Gelman Type A/E)
2. Flask, suction 500 mL
3. Filter holder or Gooch crucible adapter
4. Gooch crucibles (25 mL if 2.2 cm filter used)
5. Evaporating dishes, 100 mL
6. Drying oven, at 180°C
7. Steam bath
8. Vacuum source
9. Desiccator
10. Analytical balance
11. Muffle furnace, at 550°C

D. PROCEDURE

Preparation of Dish

1. Ignite a clean, porcelain evaporating dish at 550 ± 50°C for one hour in muffle furnace.
2. Cool the dish in a desiccator.
3. Weigh the dish and record weight. Store the dish in the desiccator until ready to use.

Preparation of Glass-Fiber Filter Disk

4. Place the glass-fiber filter disk on the filter apparatus or insert into the bottom of a suitable Gooch crucible.
5. While the vacuum is applied, wash the filter disk with three successive 20-mL volumes of distilled water. Continue the suction to remove all traces of water from the disk and discard the washings.

Sample Analysis

6. Shake the sample vigorously and transfer 125 to 150 mL to a funnel or Gooch crucible using a 150-mL graduated cylinder.
7. Filter out suspended material from the sample through the glass-fiber filter and continue to apply vacuum for about three minutes after filtration is complete to remove as much water as possible.

8. Transfer 100 mL of the filtrate to the weighed evaporating dish and evaporate to dryness on a steam bath. Because excessive residue in the evaporating dish may form a water-entrapping crust, use a sample that yields no more than 200 mg of residue for this test.
9. Dry the evaporated sample for at least one hour at 180°C.
10. Cool the dish in a desiccator.
11. Weigh the dish and total solids. Repeat the drying cycle until a constant weight is obtained or until the weight loss is less than 0.5 mg.

E. EXAMPLE

Weighing results:

Empty Dish   = 47,002.8 mg

Dish + Residue = 47,045.3 mg

Sample Volume = 100 mL

F. CALCULATIONS

1. $$\text{Total Dissolved Solids, mg/L} = \frac{(A - B)(1{,}000 \text{ mL/L})}{\text{Sample Volume, mL}}$$

   where A = Weight of Dish and Residue in milligrams (mg)

   and B = Weight of Empty Dish in milligrams (mg)

2. From example:

$$\text{Total Dissolved Solids, mg/L} = \frac{(A - B)(1{,}000 \text{ mL/L})}{\text{Sample Volume, mL}}$$

$$= \frac{(47{,}045.3 \text{ mg} - 47{,}002.8 \text{ mg})(1{,}000 \text{ mL/L})}{100 \text{ mL}}$$

$$= 425 \text{ mg/L}$$

G. REFERENCE

See page 2-55, *Standard Methods,* 21th Edition.

Suspended Solids

See Section 14.43 for the procedure used to measure suspended solids.

QUESTION

Please write your answers to the following questions and compare them with those on page 341.

14.5J Determine the total solids (residue) and total dissolved (filterable) solids from the following lab test results.

Weight of Empty Dish	= 64,328.9 mg
Weight of Dish and Residue	= 64,381.2 mg
Weight of Dish and Dissolved Solids	= 64,351.2 mg
Sample Volume	= 100 mL

(Total Solids)

OUTLINE OF PROCEDURE FOR TOTAL SOLIDS

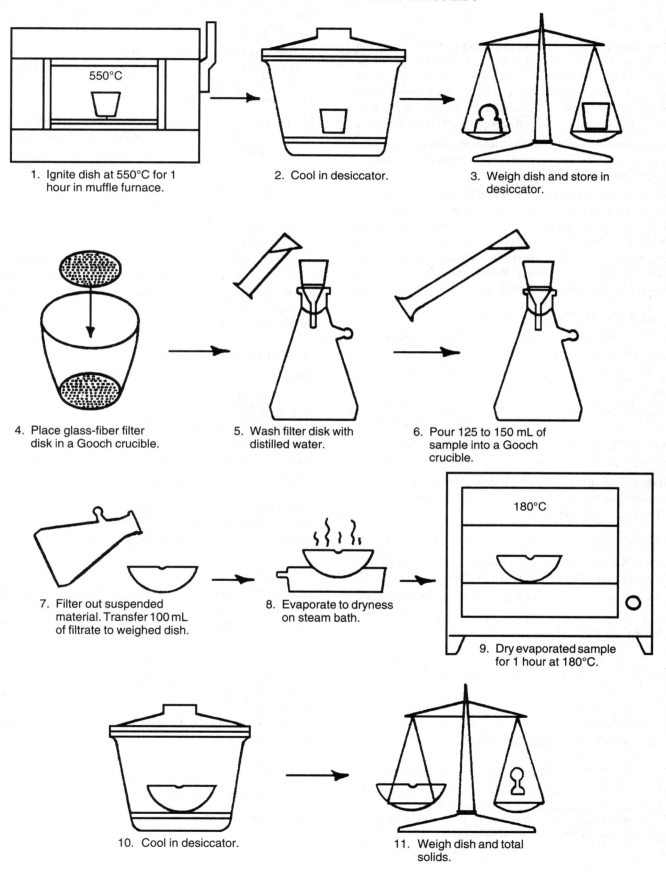

1. Ignite dish at 550°C for 1 hour in muffle furnace.

2. Cool in desiccator.

3. Weigh dish and store in desiccator.

4. Place glass-fiber filter disk in a Gooch crucible.

5. Wash filter disk with distilled water.

6. Pour 125 to 150 mL of sample into a Gooch crucible.

7. Filter out suspended material. Transfer 100 mL of filtrate to weighed dish.

8. Evaporate to dryness on steam bath.

9. Dry evaporated sample for 1 hour at 180°C.

10. Cool in desiccator.

11. Weigh dish and total solids.

(Temperature)

5. Temperature

A. DISCUSSION

Temperature is one of the most important factors affecting biological growth. Temperature measurements can be helpful in detecting changes in raw wastewater quality. For example, an influent temperature drop may be caused by large volumes of cold water from infiltration. An increase in temperature may be caused by industrial hot water discharges reaching your plant.

Temperature is one of the most frequently taken tests. One of the many uses is to calculate the percent saturation of dissolved oxygen in the DO test.

A temperature measurement should be taken where samples are collected for other tests. This test is always immediately performed on a grab sample because temperature changes so rapidly. The thermometer should remain immersed in the liquid while being read for accurate results. When removed from the liquid, the reading will change. Record temperature on a suitable work sheet, include the time, location, and sampler's name.

B. WHAT IS TESTED?

Sample	Common Range
Influent[17]	65°F to 85°F[18] (18°C to 29°C)
Effluent[17]	60°F to 95°F (16°C to 35°C) or higher from ponds
Receiving Water	60°F (16°C) to ambient temperature[19]
Digester (Recirculated Sludge before Heat Exchanger and Supernatant)	60°F to 100°F (16°C to 38°C)

C. APPARATUS

1. One National Bureau of Standards (NBS) certified thermometer for calibration of the other thermometers

2. One Fahrenheit mercury-filled, 1° subdivided thermometer

3. One Celsius (formerly called Centigrade) mercury-filled, 1° subdivided thermometer

4. One metal case to fit each thermometer

There are three types of thermometers and two scales.

Scales

1. Fahrenheit, marked °F

2. Celsius (formerly Centigrade), marked °C

Types of Thermometers

Thermometers are calibrated for either total immersion or partial immersion.

1. Total immersion. This type of thermometer must be totally immersed in the wastewater sample to give a correct reading. This reading will change most rapidly when removed from the liquid.

2. Partial immersion. This type of thermometer has a solid, etched circle around the stem of the thermometer below the point where the scale starts. It must be immersed in the sample to the depth of the circle to give a correct reading.

3. Dial. This type of thermometer has a dial that can be read easily while the thermometer is still immersed.

All thermometers should be calibrated (checked) against an accurate NBS-certified thermometer because some thermometers are substantially inaccurate when purchased (by as much as 6°F or 3°C). Some dial thermometers can be recalibrated (adjusted) to read the correct temperature of an NBS thermometer. To avoid breaking or damaging a glass thermometer, store it in a shielded metal case.

D. REAGENTS

None

E. PROCEDURES

Check your thermometer accuracy against an NBS-certified thermometer by measuring the temperature of a sample with both thermometers simultaneously. Use a large volume of sample, preferably at least a 2-pound coffee can or equivalent volume. The temperature will change less in a large volume than in a small volume. Collect the sample in a container and immediately measure and record the temperature. Do not touch the bottom or sides of the sample container with the thermometer.

F. EXAMPLE

To measure the influent temperature, obtain a sample in a large coffee can, immediately immerse the thermometer in the can, and record the temperature once the reading becomes constant.

G. CALCULATIONS

Normally, we measure and record temperatures using a thermometer with the proper scale. However, we could measure a temperature in °F and convert to °C, or measure in °C and convert to °F. The following formulas are used to convert temperatures from one scale to the other.

---

17. If dissolved oxygen (DO) measurements are performed on any samples, the temperature should be measured and recorded.
18. Depends on season, location, and temperature of water supply.
19. *Ambient* (AM-bee-ent) *Temperature.* Temperature of the surroundings.

1. Measure in °F, convert to °C.

$$°C = (°F - 32°F) \times \frac{5}{9}$$

2. Measure in °C, convert to °F.

$$°F = \left(C° \times \frac{9}{5}\right) + 32°F$$

EXAMPLE: The measured influent temperature was 77°F. Calculate the temperature in °C.

$$°C = (°F - 32°) \times \frac{5}{9}$$

$$= (77° - 32°) \times \frac{5}{9}$$

$$= 45° \times \frac{5}{9}$$

$$= 5° \times 5$$

$$= 25°C$$

QUESTIONS

Please write your answers to the following questions and compare them with those on page 341.

14.5K What could a change in influent temperature indicate?

14.5L Why should the thermometer remain immersed in the liquid while being read?

14.5M Why should thermometers be calibrated against an accurate National Bureau of Standards (NBS) certified thermometer?

6. Turbidity

A. DISCUSSION

The term *TURBIDITY*[20] describes the physical clarity of water or wastewater. Turbidity can be caused by the presence of suspended matter such as mud, finely divided organic and inorganic matter, and microscopic organisms such as algae.

The turbidity measurement is useful for plant effluent monitoring; however problems can be encountered when one instrument's reading is compared with those of another. Commercial turbidimeters come in many shapes and sizes. They each can read different turbidity values on the same sample even though they have been calibrated using the procedure given later in this section (see "Procedures"). The operator should simply be aware of this shortcoming.

The accepted method used to measure turbidity is called nephelometric. The nephelometric turbidimeter is designed to measure particle-reflected light at an angle of 90 degrees. The greater the intensity of the scattered light, the higher the turbidity.

The *TURBIDITY UNIT*[21] (TU) is an empirical quantity that is based on the amount of light that is scattered by particles of a polymer reference standard called formazin, which produces particles that scatter light in a reproducible manner. Formazin, the primary turbidity standard, is an aqueous suspension of an insoluble polymer formed by the condensation reaction between hydrazine sulfate and hexamethylenetetramine.

Secondary turbidity standards are suspensions of various materials formulated to match the primary formazin solutions. These secondary standards are generally used because of their convenience and the instability of dilute formazin primary standard solutions. Examples of secondary standards include those that are supplied by the turbidimeter manufacturer with the instrument. Periodic checks of these secondary standards against the primary formazin standard are a must, and provide assurance of measurement accuracy.

B. WHAT IS TESTED?

Sample	Common Range, NTU
Effluent	10 to 50

C. APPARATUS

1. Turbidimeter: The turbidimeter should consist of a nephelometer with a light source illuminating the samples and one or more photoelectric detectors with a readout device to indicate the intensity of scattered light. Turbidimeters used to test plant effluents should be approved by the US Environmental Protection Agency.

2. Sample tubes

---

20. *Turbidity* (ter-BID-it-tee).   The cloudy appearance of water caused by the presence of suspended and colloidal matter. In the waterworks field, a turbidity measurement is used to indicate the clarity of water. Technically, turbidity is an optical property of the water based on the amount of light reflected by suspended particles. Turbidity cannot be directly equated to suspended solids because white particles reflect more light than dark-colored particles and many small particles will reflect more light than an equivalent large particle.

21. *Turbidity* (ter-BID-it-tee) *Units (TU)*.   Turbidity units are a measure of the cloudiness of water. If measured by a nephelometric (deflected light) instrumental procedure, turbidity units are expressed in nephelometric turbidity units (NTU) or simply TU. Those turbidity units obtained by visual methods are expressed in Jackson turbidity units (JTU), which are a measure of the cloudiness of water; they are used to indicate the clarity of water. There is no real connection between NTUs and JTUs. The Jackson turbidimeter is a visual method and the nephelometer is an instrumental method based on deflected light.

(Turbidity)

D. REAGENTS

1. Turbidity-free water: Pass distilled water through a membrane filter having a pore size no greater than 100 microns. Discard the first 200 mL collected. If filtration does not reduce turbidity, use distilled water.

2. Stock Formazin turbidity suspension:[22]

   a. Solution I. Dissolve 1.000 gm hydrazine sulfate in distilled water and dilute to 100 mL in a volumetric flask.

   b. Solution II. Dissolve 10.00 gm hexamethylenetetramine in distilled water and dilute to 100 mL in a volumetric flask.

   c. In a 100-mL volumetric flask, add (using 5 mL volumetric pipets) 5.0 mL Solution I and 5.0 mL of Solution II. Mix and allow to stand 24 hours at 25°C. Then dilute to the mark and mix. The turbidity of this suspension is 400 NTU.

   d. Prepare solutions and suspensions monthly.

3. Standard turbidity suspensions. Dilute 10.00 mL stock turbidity suspension to 100 mL with turbidity-free water. Prepare weekly. The turbidity of this suspension is defined as 40 NTU.

4. Dilute turbidity standards. Dilute portions of the standard turbidity suspension with turbidity-free water as required. Prepare weekly.

E. PROCEDURE

1. Turbidimeter calibration. The manufacturer's operating instructions should be followed. Measure standards on the turbidimeter covering the range of interest. If the instrument is already calibrated in standard turbidity units, this procedure will check the accuracy of the calibration scales. At least one standard should be run in each instrument range to be used. Some instruments permit adjustments of sensitivity so that scale values will correspond to turbidities. Reliance on a manufacturer's solid scattering standard for setting overall instrument sensitivity for all ranges is not an acceptable practice unless the turbidimeter has been shown to be free of drift on all ranges. If a pre-calibrated scale is not supplied, then calibration curves should be prepared for each range of the instrument.

2. Turbidities of less than 40 units: Shake the sample to thoroughly disperse the solids. Wait until air bubbles disappear, then pour the sample into the turbidimeter tube. Read the turbidity directly from the instrument scale or from the appropriate calibration curve.

3. Turbidities exceeding 40 units: Dilute the sample with one or more volumes of turbidity-free water until the turbidity falls below 40 units. The turbidity of the original sample is then computed from the turbidity of the diluted sample and the dilution factor.

F. EXAMPLE

If 5 volumes of turbidity-free water were added to 1 volume of sample, and the diluted sample showed a turbidity of 20 units, then the turbidity of the original sample was 120 units.

G. CALCULATIONS

Sample Reading × Dilution = Actual Turbidity

Report results as follows:

NTU	Record to Nearest:
0.0 to 1.0	0.05
1 to 10	0.01
10 to 40	1
40 to 100	5
100 to 400	10
400 to 1,000	50
>1,000	100

H. NOTES AND PRECAUTIONS

1. A commercially available polymer standard that requires no preparation is also approved for use. This standard is identified as AMCO Clear, available from GFS Chemicals, Inc., PO Box 245, Powell, OH 43065.

2. Sample tubes must be kept scrupulously clean, both inside and out. Discard them when they become scratched or etched. Never handle the area where the light strikes the tube during the test.

3. Fill the tubes with samples and standards that have been agitated thoroughly, and allow sufficient time for bubbles to escape.

I. REFERENCE

See page 2-8, *Standard Methods,* 21st Edition.

QUESTION

Please write your answers to the following questions and compare them with those on page 341.

14.5N   What causes turbidity in wastewater?

14.6   ARITHMETIC ASSIGNMENT

Turn to the Appendix, "How to Solve Wastewater Treatment Plant Arithmetic Problems," at the back of this manual and read Section A.25, "Laboratory." Also, work the example problems and check the arithmetic using your calculator. You should be able to get the same answers.

END OF LESSON 7 OF 7 LESSONS

on

LABORATORY PROCEDURES

Please answer the discussion and review questions next.

---

22. Stock secondary standard turbidity suspensions that require no preparation are available from commercial suppliers.

DISCUSSION AND REVIEW QUESTIONS
Chapter 14.   LABORATORY PROCEDURES
(Lesson 7 of 7 Lessons)

Please write your answers to the following questions to determine how well you understand the material in the lesson. The question numbering continues from Lesson 6.

29. What does the total solids (residue) test measure?

30. What is the ambient temperature?

31. Convert a temperature reading of 50°F to degrees Celsius.

SUGGESTED ANSWERS
Chapter 14.   LABORATORY PROCEDURES

ANSWERS TO QUESTIONS IN LESSON 1

Answers to questions on page 278.

14.1A   Descriptions of laboratory equipment and their uses:

Item	Description	Use or Purpose
1. Beakers	Short, wide cylinders in sizes from 1 mL to 4,000 mL.	Mixing chemicals.
2. Graduated Cylinders	Long, narrow cylinders in sizes from 5 mL to 4,000 mL.	Measuring Volumes.
3. Pipets	Very thin tubes with a pointed tip in sizes from 0.1 mL to 100 mL.	Delivering accurate volumes
4 Burets	Long tubes with graduated walls and a stopcock in sizes from 10 to 1,000 mL.	Delivering and measuring accurate volumes used in titrations.

14.1B   Never heat graduated cylinders in an open flame because they will break.

14.1C   A bench sheet is used to record data and arrange data in an orderly manner. Bench sheets also are called laboratory worksheets.

ANSWERS TO QUESTIONS IN LESSON 2

Answers to questions on page 286.

14.2A   A bulb should always be used to pipet wastewater or polluted water to prevent serious illness or death.

14.2B   Immunizations are recommended to reduce the possibility of contracting diseases.

14.2C   True. You can add acid to water, but never water to acid—as you would add chemicals to the water in a swimming pool.

14.2D   Immediately wash area where acid spilled with water and neutralize the acid with sodium carbonate or bicarbonate.

ANSWERS TO QUESTIONS IN LESSON 3

Answers to questions on page 292.

14.3A   The greatest errors found in laboratory results are usually caused by improper sampling, poor preservation, or lack of enough mixing during compositing and testing.

14.3B   A representative sample must be collected or the test results will not have any significant meaning. To efficiently operate a wastewater treatment plant, the operator must rely on test results to indicate what is happening in the treatment process.

14.3C   A proportional composite sample may be prepared by collecting a sample every hour. The size of this sample is proportional to the rate of flow when the sample is collected. All of these proportional samples are mixed together to produce a proportional composite sample. If an equal volume of sample was collected each hour and mixed, this would simply be a composite sample.

ANSWERS TO QUESTIONS IN LESSON 4

Answers to questions on page 294.

14.4A   The clarity test indicates the quality of the effluent with respect to color, solids, and turbidity.

14.4B   When clarity is measured under different conditions the results will not be comparable from day to day. You would not be able to tell whether your plant performance is improving, staying the same, or deteriorating.

Answer to question on page 295.

14.4C   1. You would measure the amount of H_2S in the atmosphere to provide an indication of the rate of corrosion taking place in a sewer collection system.

   2. H_2S tests are run to determine the rate of H_2S increase. H_2S in wastewater produces a typical rotten egg odor. It is indicative of anaerobic decomposition of organics in wastewater, which occurs in the absence of oxygen. High levels of H_2S are toxic to your respiratory system and can create flammable and explosive conditions.

Answer to question on page 297.

14.4D   Estimate the volume of solids pumped to a digester in gallons per day (GPD) if the flow is 1 MGD, the influent settleable solids are 10 mL/L, and the effluent settleable solids are 0.4 mL/L for a primary clarifier.

$$\text{Sludge to Digester, GPD} = (\text{Total SS Removed, mL/L})(1{,}000\text{ mg/mL})(\text{Flow, MGD})$$

$$= (10\text{ mL/L} - 0.4\text{ mL/L})(1{,}000\text{ mg/mL})(1\text{ M gal/day})$$

$$= \frac{9.6\text{ mL}}{\text{M mg}} \times \frac{1{,}000\text{ mg}}{\text{mL}} \times \frac{1\text{ M gal}}{\text{day}}$$

$$= 9{,}600\text{ GPD}$$

This value may be reduced by 30 to 75 percent due to compaction of the sludge in the clarifier.

Answers to questions on page 302.

14.4E   Some suspended material in wastewater will not be removed because the specific gravity is very near that of water and the material is not light enough to float or heavy enough to settle.

14.4F   Solids calculations are shown in detail here to illustrate the computational approach and the units involved. After you understand this approach, using the laboratory worksheet is more convenient.

   1. Calculate total suspended solids, mg/L

   Volume of Sample, mL = 100 mL
   Weight of Dried Sample + Dish, gm = 19.3902 gm
   Weight of Dish (Tare Weight), gm = 19.3241 gm
   Dry Weight, gm = 0.0661 gm
   or = 66.1 mg

$$\text{Total Susp Solids, mg/L} = \frac{\text{Weight of Solids, mg} \times 1{,}000\text{ mL/L}}{\text{Volume of Sample, mL}}$$

$$= \frac{66.1\text{ mg} \times 1{,}000\text{ mL/L}}{100\text{ mL}}$$

$$= 661\text{ mg/L}$$

   2. Calculate volatile suspended solids, mg/L

   Weight of Dried Sample + Dish, gm = 19.3902 gm
   Weight of Ash + Dish, gm = 19.3469 gm
   Weight Volatile Solids, gm = 0.0433 gm
   or = 43.3 mg

$$\text{Vol Susp Solids, mg/L} = \frac{\text{Weight of Vol Sol, mg} \times 1{,}000\text{ mL/L}}{\text{Volume of Sample, mL}}$$

$$= \frac{(43.3\text{ mg})(1{,}000\text{ mL/L})}{100\text{ mL}}$$

$$= 433\text{ mg/L}$$

   3. Calculate volatile solids, %

$$\text{\% Volatile Solids} = \frac{\text{Weight Volatile, mg} \times 100\%}{\text{Weight Dry, mg}}$$

$$= \frac{43.3\text{ mg}}{66.1\text{ mg}} \times 100\%$$

$$= 65.5\%$$

   4. Calculate fixed suspended solids, mg/L

   Total Suspended Solids, % = 661 mg/L
   Volatile Suspended Solids, % = 433 mg/L
   Fixed Suspended Solids, % = 228 mg/L

   5. Calculate fixed solids, %

   Total Solids, % = 100.00%
   Volatile Solids, % = 65.50%
   Fixed Solids, % = 34.5%

or

$$\% \text{ Fixed} = \frac{\text{Fixed, mg}}{\text{Total, mg}} \times 100\%$$

$$= \frac{22.8 \text{ mg}}{66.1 \text{ mg}} \times 100\%$$

$$= 34.5\% \text{ (Check)}$$

14.4G Suspended solids data from a wastewater treatment plant are given below:

Influent Suspended Solids = 221 mg/L

Primary Effluent SS = 159 mg/L

Final Effluent SS = 33 mg/L

Compute the percent removal of suspended solids by the:

1. Primary clarifier,
2. Secondary process (removal between primary effluent and secondary effluent), and
3. Overall plant.

1. Calculate the percent removal of suspended solids by the primary clarifier.

In = Suspended solids entering the plant or unit (influent)

Out = Suspended solids leaving the plant or unit (primary effluent)

$$\% \text{ Removal} = \frac{(\text{In} - \text{Out})}{\text{In}} \times 100\%$$

$$= \frac{(221 \text{ mg/L} - 159 \text{ mg/L})}{221 \text{ mg/L}} \times 100\%$$

$$= \frac{62}{221} \times 100\%$$

$$= 28\% \text{ Removal by Primary Clarifier}$$

2. Calculate the percent removal by secondary process.

In = 159 mg/L SS in Primary Effluent

Out = 33 mg/L SS in Final Effluent

$$\% \text{ Removal} = \frac{(\text{In} - \text{Out})}{\text{In}} \times 100\%$$

$$= \frac{(159 \text{ mg/L} - 33 \text{ mg/L})}{159 \text{ mg/L}} \times 100\%$$

$$= \frac{126}{159} \times 100\%$$

$$= 79.2\% \text{ Removal from Primary Effluent to Final Effluent}$$

3. Calculate the percent removal by the plant overall or overall plant efficiency.

In = 221 mg/L SS in Plant Influent

Out = 33 mg/L SS in Plant Effluent

$$\% \text{ Removal} = \frac{(\text{In} - \text{Out})}{\text{In}} \times 100\%$$

$$= \frac{(221 \text{ mg/L} - 33 \text{ mg/L})}{221 \text{ mg/L}} \times 100\%$$

$$= \frac{188}{221} \times 100\%$$

$$= 85\% \text{ Overall Plant Removal}$$

14.4H Calculate the pounds of solids removed per day by each unit:

$$\begin{array}{l} \text{Amount} \\ \text{Removed,} \\ \text{lbs/day} \end{array} = \frac{\text{Conc Rem,}}{\text{mg/L}} \times \frac{\text{Flow,}}{\text{MGD}} \times 8.34 \text{ lbs/gal}$$

where MGD = million gallons per day

1. Influent, mg/L = 221 mg/L

   Primary Effluent, mg/L = 159 mg/L

   Primary Removal, mg/L = 62 mg/L

$$\begin{array}{l} \text{Amount Removed,} \\ \text{lbs/day (Primary)} \end{array} = (62 \text{ mg/L})(1.5 \text{ MGD})(8.34 \text{ lbs/gal})$$

$$= 775.6 \text{ lbs/day Removed by Primary}$$

2. Primary Effluent, mg/L = 159 mg/L

   Final Effluent, mg/L = 33 mg/L

   Secondary Removal, mg/L = 126 mg/L

$$\begin{array}{l} \text{Amount Removed,} \\ \text{lbs/day (Secondary)} \end{array} = (126 \text{ mg/L})(1.5 \text{ MGD})(8.34 \text{ lbs/gal})$$

$$= 1,576 \text{ lbs/day Removed by Secondary}$$

3. Influent, mg/L = 221 mg/L

   Final Effluent, mg/L = 33 mg/L

   Overall Removal, mg/L = 188 mg/L

$$\begin{array}{l} \text{Amount Removed,} \\ \text{lbs/day} \end{array} = (188 \text{ mg/L})(1.5 \text{ MGD})(8.34 \text{ lbs/gal})$$

$$= 2,352 \text{ lbs/day Removed by Plant}$$

$$= \frac{\text{Primary Rem,}}{\text{lbs/day}} + \frac{\text{Secondary Rem,}}{\text{lbs/day}}$$

or

$$= 776 + 1,576$$

$$= 2,352 \text{ (Check)}$$

Answers to questions on page 304.

14.4I Volatile solids are composed of organic compounds of either plant or animal origin.

14.4J Volatile solids in a treatment plant indicate the portion of waste material that may be treated by biological processes.

ANSWERS TO QUESTIONS IN LESSON 5

Answers to questions on page 309.

14.4K   Mixed liquor settleability tests are run to get an indication of what will happen to the mixed liquor in the final clarifier—the rate of sludge settling, turbidity, color, and volume of sludge at the end of 60 minutes.

14.4L   The SVI is the volume in mL occupied by one gram of mixed liquor suspended solids after 30 minutes of settling.

14.4M   The SVI test is used to indicate the condition of sludge and changes in sludge characteristics.

14.4N   Sludge Density Index (SDI) = 100/SVI.

Answer to question on page 309.

14.4O   The sludge age of 200,000 gallon aeration tank that has 2,000 mg/L mixed liquor suspended solids, a primary effluent of 115 mg/L SS, and an average flow of 1.8 MGD:

$$\text{Sludge Age, days} = \frac{\text{Vol of Aeration Tank, MG} \times \text{Susp Solids, mg/L}}{\text{Flow, MGD} \times \text{Primary Effluent, mg/L}}$$

$$= \frac{0.2\ \text{MG} \times 2,000\ \text{mg/L}}{1.8\ \text{MGD} \times 115\ \text{mg/L}}$$

$$= 1.93\ \text{days}$$

Answer to question on page 311.

14.4P   When the DO in the aeration tank is very low, the copper sulfate-sulfamic acid test procedure can give DO results higher than actual DO in the aerator. The results may be high because oxygen may enter the sample from the air when the sample is collected, when the copper sulfate-sulfamic acid inhibitor is added, while the solids are settling, or when the sample is transferred to a BOD bottle for the DO test.

Answer to question on page 312.

14.4Q   The advantages of the centrifuge method over the regular suspended solids test is that it is a quick and easy way to estimate suspended solids in the aerators. It also provides reliable results when the suspended solids concentration is below 1,500 mg/L.

Answer to question on page 313.

14.4R   In order to calculate the mean cell residence time (MCRT), measure the representative composite samples of mixed liquor suspended solids, effluent suspended solids, and waste sludge suspended solids. Influent flow and waste sludge flow measurements are also required.

ANSWERS TO QUESTIONS IN LESSON 6

Answers to questions on page 318.

14.5A   Plant effluents should be chlorinated for disinfection purposes to protect the bacteriological quality of the receiving waters.

14.5B   The iodometric titration method gives good results with samples containing wastewater, such as plant effluent or receiving waters. Amperometric titration gives the best results for wastewater samples, but the equipment is expensive.

Answers to questions on page 322.

14.5C   Calculate the percent dissolved oxygen saturation if the receiving water DO is 7.9 mg/L and the temperature is 10°C.

$$\text{DO Saturation, \%} = \frac{\text{DO of Sample, mg/L} \times 100\%}{\text{DO at 100\% Saturation, mg/L}}$$

$$= \frac{(7.9\ \text{mg/L})100\%}{11.3\ \text{mg/L}}$$

$$= 70\%$$

14.5D   To calibrate the DO probe in an aeration tank, a sample of the effluent can be collected and split. The DO of the effluent is measured by the Modified Winkler procedure, and the probe DO reading is compared with and adjusted to agree with the Winkler results.

Answers to questions on page 327.

14.5E   The amount of organic material in wastewater can be estimated indirectly by measuring oxygen use with the BOD test.

14.5F   Samples for the BOD test should be collected before chlorination because chlorine interferes with the organisms in the test. It is difficult to obtain accurate results with dechlorinated samples.

14.5G   To prepare dilutions for a cannery waste with an expected BOD of 2,000 mg/L, take 10 mL of sample and add 90 mL of dilution water to obtain a new sample with an estimated BOD of 200 mg/L (10 to 1 dilution):

$$\text{BOD Dilution, mL} = \frac{1,200}{\text{Estimated BOD, mg/L}}$$

$$= \frac{1,200}{200}$$

$$= 6\ \text{mL}$$

14.5H   What is the BOD of a sample of wastewater if a 2-mL sample in a 300-mL BOD bottle had an initial DO of 7.5 mg/L and a final DO of 3.9 mg/L?

$$\text{BOD, mg/L} = \begin{bmatrix} \text{Initial DO} \\ \text{of Diluted} \\ \text{Sample, mg/L} \end{bmatrix} \begin{bmatrix} \text{DO of Diluted} \\ -\text{ Sample After 5-day} \\ \text{Incubation, mg/L} \end{bmatrix} \begin{bmatrix} \text{BOD Bottle Vol, mL} \\ \text{Sample Volume, mL} \end{bmatrix}$$

$$= (7.5\ \text{mg/L} - 3.9\ \text{mg/L}) \begin{bmatrix} \dfrac{300\ \text{mL}}{2\ \text{mL}} \end{bmatrix}$$

$$= (3.6\ \text{mg/L})(150)$$

$$= 540\ \text{mg/L}$$

ANSWERS TO QUESTIONS IN LESSON 7

Answer to question on page 330.

14.5I Precautions to exercise when using a pH meter include:

1. Prepare fresh buffer solution weekly for calibration purposes.
2. pH meter, buffer solutions, and samples should all be at the same temperature.
3. Watch for erratic results arising from faulty connections or fouling of electrodes with interfering matter.

Answer to question on page 332.

14.5J Determine the total solids (residue) and total dissolved (filterable) solids from the following lab test results.

Known

Empty Dish, mg   = 64,328.9 mg

Dish and Residue, mg   = 64,381.2 mg

Dish and Dissolved Solids, mg = 64,351.2 mg

Sample Volume   = 100 mL

Unknown

1. Total Solids, mg/L
2. Total Dissolved Solids, mg/L

1. Calculate Total Solids (Residue)

$$\text{Total Solids, mg/L} = \frac{(A - B)(1{,}000 \text{ mL/L})}{\text{mL of Sample}}$$

$$= \frac{(64{,}381.2 \text{ mg} - 64{,}328.9 \text{ mg})(1{,}000 \text{ mL/L})}{100 \text{ mL}}$$

$$= 523 \text{ mg/L}$$

2. Calculate Total Dissolved Solids (Filterable)

$$\text{Total Dissolved Solids, mg/L} = \frac{(A - B)(1{,}000 \text{ mL/L})}{\text{mL of Sample}}$$

$$= \frac{(64{,}351.2 \text{ mg} - 64{,}328.9 \text{ mg})(1{,}000 \text{ mL/L})}{100 \text{ mL}}$$

$$= 223 \text{ mg/L}$$

Answers to questions on page 335.

14.5K Changes in influent temperature could indicate a new influent source. A drop in temperature may be caused by large volumes of cold water from infiltration. An increase in temperature also may be caused by industrial hot water discharges reaching your plant.

14.5L The thermometer should remain immersed in the liquid while being read for accurate results. When removed from the liquid, the reading will change.

14.5M All thermometers should be calibrated (checked) against an accurate NBS-certified thermometer because some thermometers are substantially inaccurate when purchased (by as much as 6°F or 3°C).

Answer to question on page 336.

14.5N Turbidity in wastewater is caused by the presence of suspended matter such as clay, silt, finely divided organic and inorganic matter, and microscopic organisms.

CHAPTER 15

MANAGEMENT

by

John Brady

and

Ken Kerri

Adapted from *UTILITY MANAGEMENT*

by Lorene Lindsay

TABLE OF CONTENTS
Chapter 15. MANAGEMENT

OBJECTIVES

Chapter 15.   MANAGEMENT

1. Identify the functions of a manager.

2. Describe the benefits of short-term, long-term, and emergency planning.

3. Define the following terms:

   a. Authority
   b. Responsibility
   c. Delegation
   d. Accountability
   e. Unity of command

4. Read and construct an organizational chart identifying lines of authority and responsibility.

5. Prepare a written or oral report on the utility's operations.

6. Communicate effectively within the organization, with media representatives, and with the community.

7. Describe the financial strength of your utility.

8. Calculate your utility's operating ratio, coverage ratio, and simple payback.

9. Prepare a contingency plan for emergencies.

10. Prepare a plan to strengthen the security of your utility.

11. Set up a safety program for your utility.

12. Develop and implement capital improvement plans for your utility.

13. List and describe the types of funding sources available to the utility.

14. Collect, organize, file, retrieve, use, and dispose of utility records.

WORDS
Chapter 15.   MANAGEMENT

ACCOUNTABILITY ACCOUNTABILITY

When a manager gives power/responsibility to an employee, the employee ensures that the manager is informed of results or events.

AIR GAP AIR GAP

An open, vertical drop, or vertical empty space, between a drinking (potable) water supply and potentially contaminated water. This gap prevents the contamination of drinking water by backsiphonage because there is no way potentially contaminated water can reach the drinking water supply.

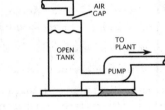

AUTHORITY AUTHORITY

The power and resources to do a specific job or to get that job done.

BACKFLOW BACKFLOW

A reverse flow condition, created by a difference in water pressures, that causes water to flow back into the distribution pipes of a potable water supply from any source or sources other than an intended source. Also see BACKSIPHONAGE.

BOND BOND

(1) A written promise to pay a specified sum of money (called the face value) at a fixed time in the future (called the date of maturity). A bond also carries interest at a fixed rate, payable periodically. The difference between a note and a bond is that a bond usually runs for a longer period of time and requires greater formality. Utility agencies use bonds as a means of obtaining large amounts of money for capital improvements.

(2) A warranty by an underwriting organization, such as an insurance company, guaranteeing honesty, performance, or payment by a contractor.

CALL DATE CALL DATE

First date a bond can be paid off.

CERTIFICATION EXAMINATION CERTIFICATION EXAMINATION

An examination administered by a state agency or professional association that operators take to indicate a level of professional competence. In the United States, certification of operators of water treatment plants, wastewater treatment plants, water distribution systems, and small water supply systems is mandatory. In many states, certification of wastewater collection system operators, industrial wastewater treatment plant operators, pretreatment facility inspectors, and small wastewater system operators is voluntary; however, current trends indicate that more states, provinces, and employers will require these operators to be certified in the future. Operator certification is mandatory in the United States for the Chief Operators of water treatment plants, water distribution systems, and wastewater treatment plants.

CERTIFIED OPERATOR CERTIFIED OPERATOR

A person who has the education and experience required to operate a specific class of treatment facility as indicated by possessing a certificate of professional competence given by a state agency or professional association.

CODE OF FEDERAL REGULATIONS (CFR)

CODE OF FEDERAL REGULATIONS (CFR)

A publication of the US government that contains all of the proposed and finalized federal regulations, including safety and environmental regulations.

CONFINED SPACE

CONFINED SPACE

Confined space means a space that:

(1) Is large enough and so configured that an employee can bodily enter and perform assigned work; and

(2) Has limited or restricted means for entry or exit (for example, manholes, tanks, vessels, silos, storage bins, hoppers, vaults, and pits are spaces that may have limited means of entry); and

(3) Is not designed for continuous employee occupancy.

Also see DANGEROUS AIR CONTAMINATION and OXYGEN DEFICIENCY.

CONFINED SPACE, PERMIT-REQUIRED (PERMIT SPACE)

CONFINED SPACE, PERMIT-REQUIRED (PERMIT SPACE)

A confined space that has one or more of the following characteristics:

(1) Contains or has a potential to contain a hazardous atmosphere,

(2) Contains a material that has the potential for engulfing an entrant,

(3) Has an internal configuration such that an entrant could be trapped or asphyxiated by inwardly converging walls or by a floor that slopes downward and tapers to a smaller cross section, or

(4) Contains any other recognized serious safety or health hazard.

COVERAGE RATIO

COVERAGE RATIO

The coverage ratio is a measure of the ability of the utility to pay the principal and interest on loans and bonds (this is known as debt service) in addition to any unexpected expenses.

DEBT SERVICE

DEBT SERVICE

The amount of money required annually to pay the (1) interest on outstanding debts, or (2) funds due on a maturing bonded debt or the redemption of bonds.

DELEGATION

DELEGATION

The act in which power is given to another person in the organization to accomplish a specific job.

MATERIAL SAFETY DATA SHEET (MSDS)

MATERIAL SAFETY DATA SHEET (MSDS)

A document that provides pertinent information and a profile of a particular hazardous substance or mixture. An MSDS is normally developed by the manufacturer or formulator of the hazardous substance or mixture. The MSDS is required to be made available to employees and operators or inspectors whenever there is the likelihood of the hazardous substance or mixture being introduced into the workplace. Some manufacturers are preparing MSDSs for products that are not considered to be hazardous to show that the product or substance is not hazardous.

OSHA (O-shuh)

OSHA

The Williams-Steiger Occupational Safety and Health Act of 1970 (OSHA) is a federal law designed to protect the health and safety of workers, including collection system and treatment plant operators. The Act regulates the design, construction, operation, and maintenance of industrial plants and wastewater collection systems and treatment plants. The Act does not apply directly to municipalities, *except* in those states that have approved plans and have asserted jurisdiction under Section 18 of the OSHA Act. *However, contract operators and private facilities do have to comply with OSHA requirements.* Wastewater treatment plants have come under stricter regulation in all phases of activity as a result of OSHA standards. OSHA also refers to the federal and state agencies that administer the OSHA regulations.

OPERATING RATIO

OPERATING RATIO

The operating ratio is a measure of the total revenues divided by the total operating expenses.

ORGANIZING

ORGANIZING

Deciding who does what work and delegating authority to the appropriate persons.

PLANNING

PLANNING

Management of utilities to build the resources and financial capability to provide for future needs.

PRESENT WORTH

PRESENT WORTH

The value of a long-term project expressed in today's dollars. Present worth is calculated by converting (discounting) all future benefits and costs over the life of the project to a single economic value at the start of the project. Calculating the present worth of alternative projects makes it possible to compare them and select the one with the largest positive (beneficial) present worth or minimum present cost.

RESPONSIBILITY

RESPONSIBILITY

Answering to those above in the chain of command to explain how and why you have used your authority.

TAILGATE SAFETY MEETING

TAILGATE SAFETY MEETING

Brief (10 to 20 minutes) safety meetings held every 7 to 10 working days. The term comes from the safety meetings regularly held by the construction industry around the tailgate of a truck.

CHAPTER 15. MANAGEMENT

(Lesson 1 of 2 Lessons)

15.0 NEED FOR MANAGEMENT

The management of a public or private utility, large or small, is a complex and challenging job. Communities are concerned about their drinking water and their wastewater as well as a demand for service to be provided without interruption. They are aware of past environmental disasters and they want to protect their communities, but they want this uninterrupted service and environmental protection with a minimum investment of money. In addition to the local community demands, the utility manager must also keep up with increasingly stringent regulations and monitoring from regulatory agencies. While meeting these external concerns, the manager faces the normal challenges from within the organization: personnel, resources, equipment, and preparing for the future. An important responsibility for managers is to communicate effectively with councils, boards, or commissions that control the administrative policies of the utility. For the successful manager, all of these responsibilities combine to create an exciting and rewarding job.

A brief quiz is given in Table 15.1 that asks some basic management questions. This quiz can be used as a guide to management areas that may need some attention in your utility. You should be able to answer yes to most of the questions; however, all utilities have areas that can be improved.

In the environmental field, as well as other fields, the workforce is changing. Most workplaces are becoming increasingly diverse as people from different genders, races, cultures, ethnic origins, and lifestyles find themselves working together. Workplace diversity may also encompass differences in age, personality, cognitive style, educational background, religious beliefs, work experience, physical ability, and more. Forward thinking managers recognize that employees from these diverse backgrounds bring individual talents and experiences to the job.

The utility manager's task includes providing adequate support services and opportunities for all operators, considering the demanding and often physically challenging nature of the work. This includes dealing effectively with issues such as training, communication, flexibility, and change. The successful manager creates a workplace that respects and includes differences, recognizing the unique contributions that individuals with many types of differences can make, and fostering a work environment that maximizes the potential of all employees.

Changes in the environmental workplace also are created by advances in technology. The environmental field has exploded with new technologies, such as video monitoring, robotics in the collection system, and computer-controlled treatment processes.

The utility manager must keep up with these changes and provide the leadership to keep everyone at the utility up to speed on new ways of doing things. In addition, the utility manager must provide a safer, cleaner work environment while constantly training and retraining employees to understand new technologies.

QUESTIONS

Please write your answers to the following questions and compare them with those on page 389.

15.0A What are the local community demands on a utility manager?

15.0B What two factors have contributed to changes in the environmental field and workplace?

15.1 FUNCTIONS OF A MANAGER

The functions of a utility manager are the same as for the chief executive officer (CEO) of any big company: planning, organizing, staffing, directing, and controlling. In many small communities, the utility manager may be the only one who has these responsibilities and the community depends on the manager to handle everything.

Planning (see Section 15.2) consists of determining the goals, policies, procedures, and other elements to achieve the goals and objectives of the agency. Planning requires the manager to collect and analyze data, consider alternatives, and then make decisions. Planning must be done before the other managing functions. Planning may be the most difficult in smaller communities,

TABLE 15.1   HOW WELL DOES YOUR SYSTEM MANAGE?

The following self-test is designed for small water or wastewater treatment facilities to provide a guide for identifying areas of concern and for improving small system management.

1. Is the treatment system budget separate from other accounts so that the true cost of the treatment system can be determined?

2. Are the funds adequate to cover operating costs, debt service, and future capital improvements?

3. Do operational personnel have input into the budgetary process?

4. Is there a monthly or quarterly review of the actual operating costs compared to the budgeted costs?

5. Does the user charge system adequately reflect the cost of treatment?

6. Are all users properly metered and does the unaccounted-for water not exceed 20 percent of the total flow?

7. Are plant discharges and monitoring tests representative of plant performance?

8. Are operational control decisions based on process control testing within the plant?

9. Are provisions made for continued training for plant personnel?

10. Are qualified personnel available to fill job vacancies and is job turnover relatively low?

11. Are the energy costs for the system not more than 20 to 30 percent of the total operating costs?

12. Is the ratio of corrective (reactive) maintenance to preventive (proactive) maintenance remaining stable and is it less than 1.0?

13. Are maintenance records available for review?

14. Is the spare parts inventory adequate to prevent long delays in equipment repairs?

15. Are old or outdated pieces of equipment replaced as necessary to prevent excessive equipment downtime, inefficient process performance, or unreliability?

16. Are technical resources and tools available for repairing, maintaining, and installing equipment?

17. Is the utility's equipment providing the expected design performance?

18. Are standby units for key equipment available to maintain process performance during breakdowns or during preventive maintenance activities?

19. Are the plant processes adequate to meet the demand for treatment?

20. Does the facility have an adequate emergency response plan including a wastewater bypass storage procedure or an alternate water source.

where the future may involve a decline in population instead of growth.

Organizing (see Section 15.3) means that the manager decides who does what work and delegates authority to the appropriate operators. The organizational function in some utilities may be fairly loose while some communities are very tightly controlled.

Staffing (see Section 15.4) is the recruiting of new operators and staff and determining if there are enough qualified operators and staff to fill available positions. The utility manager's staffing responsibilities include selecting and training employees, evaluating their performance, and providing opportunities for advancement for operators and staff in the agency.

Directing includes guiding, teaching, motivating, and supervising operators and utility staff members. Direction includes issuing orders and instructions so that activities at the facilities or in the field are performed safely and are properly completed.

Controlling involves taking the steps necessary to ensure that essential activities are performed so that objectives will be achieved as planned. Controlling means ensuring that progress is being made toward objectives and taking corrective action as necessary. The utility manager is directly involved in controlling the treatment process to ensure that water or wastewater is being properly treated and to make sure that the utility is meeting its short- and long-term goals.

15.2   PLANNING[1]

A very large portion of any manager's typical work day will be spent on activities that can be described as planning activities since nearly every area of a manager's responsibilities requires some type of planning.

Planning is one of the most important functions of utility management and one of the most difficult. Communities must have good, clean water and the management of water or wastewater utilities must include building the resources and financial capability to provide for future needs. The utility must plan for future growth, including commercial and industrial development, and be ready to provide the water and waste treatment systems that will be needed as the community grows. The most difficult problem for some small communities is recognizing and planning for a decline in population. The utility manager must develop reliable information to plan for growth or decline. Decisions must be made about goals, both short- and long-term. The manager must prepare plans for the next two years and the next 10 to 20 years. Remember that utility planning should include operational personnel, local officials (decision makers), and the public. Everyone must understand the importance of planning and be willing to contribute to the process.

Operation and maintenance of a utility also involves planning by the utility manager. A preventive maintenance program should be established to keep the system performing as intended and to protect the community's investment in water or wastewater facilities. (Section 15.9 describes the various types of maintenance and the benefits of establishing maintenance programs.)

The utility also must have an emergency response plan to deal with natural or human disasters. Without adequate planning, your utility will be facing system failures, the inability to meet compliance regulations, and inadequate service capacity to meet community needs. Plan today and avoid disaster tomorrow. (Section 15.10, "Emergency Response," describes the basic elements of an emergency operations plan.)

15.3   ORGANIZING[2]

A utility should have a written organizational chart and written policies. In some communities, the organizational chart and policies are part of the overall community plan. In either case, the utility manager and all plant personnel should have a copy of the organizational chart and written policies of the utility.

The purpose of the organizational chart is to show who reports to whom and to identify the lines of authority. The chart should show each person or job position in the organization with a direct line showing to whom each person reports in the organization. Remember, an employee can serve only one supervisor (unity of command) and each supervisor should ideally manage only six or seven employees. The organizational chart should include a job description for each of the positions on the chart. When the organizational chart is in place, employees know who is their immediate boss and confusion about job tasks is eliminated.

To understand organization and its role in management, we need to understand some other terms including authority, responsibility, delegation, and accountability. *AUTHORITY* means having the power and resources to do a specific job or to get that job done. Authority may be given to an employee due to their position in the organization (this is formal authority) or authority may be given to the employee informally by their co-workers when the employee has earned their respect. *RESPONSIBILITY* may be described as answering to those above in the chain of command to explain how and why you have used your authority. *DELEGATION* is the act in which power is given to another person in the organization to accomplish a specific job. Finally, when a manager gives power/responsibility to an employee, then the employee is *ACCOUNTABLE*[3] for the results.

Organization and effective delegation are very important to keep any utility operating efficiently. Delegation is uncomfortable for many managers because it requires giving up power and responsibility. Many managers believe that they can do the job better than others, they believe that other employees are not well trained or experienced, and they are afraid of mistakes. The utility manager retains some responsibility even after delegating to another employee and, therefore, the manager is often reluctant to delegate or may delegate the responsibility but not the authority to get the job done. For the utility manager, good organization means that employees are ready to accept responsibility and have the power and resources to make sure that the job gets done.

Employees should not be asked to accept responsibilities for job tasks that are beyond their level of authority in the organizational structure. For example, an operator or lead utility worker should not be asked to accept responsibility for additional lab testing unless qualified. The responsibility for additional lab testing must be delegated to the lab supervisor. Authority and responsibility must be delegated properly to be effective. When these three components—proper job assignments, authority, and responsibility—are all present, the supervisor has successfully delegated. The success of delegation is dependent upon all three components.

---

1. *Planning.*   Management of utilities to build the resources and financial capability to provide for future needs.
2. *Organizing.*   Deciding who does what work and delegating authority to the appropriate persons.
3. *Accountability.*   When a manager gives power/responsibility to an employee, the employee ensures that the manager is informed of results or events.

An important and often overlooked part of delegation is follow-up by the supervisor. A good manager will delegate and follow up on progress to make sure that the employee has the necessary resources to get the job done. Well-organized managers can delegate effectively and do not try to do all the work themselves, but are responsible for getting good results.

QUESTIONS

Please write your answers to the following questions and compare them with those on page 389.

15.2A   Who must be included in utility planning?

15.3A   What is the purpose of an organizational chart?

15.3B   Why is it sometimes difficult or uncomfortable for supervisors or managers to delegate?

15.3C   What is an important and often overlooked part of delegation?

NOTICE

The information provided in this section on staffing should **not** be viewed as **legal advice**. The purpose of this section is simply to identify and describe in general terms the major components of a utility manager's responsibilities in the area of staffing. Personnel administration is affected by many federal and state regulations. Legal requirements of legislation such as the Americans with Disabilities Act (ADA), Equal Employment Opportunity (EEO) Act, Family and Medical Leave Act (FMLA), and wages and hours laws are complex and beyond the scope of this manual. If your utility does not have established personnel policies and procedures, consider getting help from a labor law attorney to develop appropriate policies. At the very least, you should get help from a recruitment specialist to develop and document hiring procedures that meet the federal guidelines for Equal Employment Opportunity.

15.4   STAFFING

15.40   The Utility Manager's Responsibilities

The utility manager is also responsible for staffing, which includes hiring new employees, training employees, and evaluating job performance. The utility should have established procedures for job hiring that include requirements for advertising the position, application procedures, and the procedures for conducting interviews.

In the area of staffing, more than any other area of responsibility, a manager must be extremely cautious and consider the consequences before taking action. Personnel management practices are changing dramatically and continue to be redefined almost daily by the courts. A manager who violates an employee's or job applicant's rights can be held both personally and professionally liable in court. Throughout this section on staffing you will repeatedly find references to two terms: "job-related" and "documentation." These are key concepts in personnel management today. Any personnel action you take must be job-related, from the questions you ask during interviews to disciplinary actions or promotions. And, while almost no one wants more paperwork, documentation of personnel actions detailing what you did, when you did it, and why you did it (the reasons will be job-related, of course) is absolutely essential. There is no way to predict when you might be called upon to defend your actions in court. Good records not only serve to refresh your memory about past events but can also be used to demonstrate your pattern of lawful behavior over time.

15.41   How Many Employees Are Needed?

There is a common tendency for organizations to add personnel in response to changing conditions without first examining how the existing workforce might be reorganized to achieve greater efficiency and meet the new work demands. In water and wastewater utilities, aging of the system, changes in use, and expansion of the system often mean changes in the operation and maintenance tasks being performed. The manager of a utility should periodically review the agency's work requirements and staffing to ensure that the utility is operating as efficiently as possible. A good time to conduct such a review is during the annual budgeting process or when you are considering hiring a new employee because the workload seems to be greater than the current staff can adequately handle. Another option that a manager might consider is overtime. When you consider the lower overhead costs associated with overtime, it can be an inexpensive alternative for meeting temporary needs.

The staffing analysis procedure outlined in this section illustrates how to conduct a comprehensive analysis of the type needed for a complete reorganization of the agency. In practice, however, a complete reorganization may not be desirable or even possible. Frequent organizational changes can make employees anxious about their jobs and may interfere with their work performance. Some employees show strong resistance to any change in job responsibilities. Nonetheless, by thoroughly examining the functions and staffing of the utility on a periodic basis, the manager may spot trends (such as an increase in the amount of time spent maintaining certain equipment or portions of the system) or discover inefficiencies that could be corrected over a period of time.

The first step in analyzing the utility's staffing needs is to prepare a detailed list of all the tasks to be performed to operate and maintain the utility. Next, estimate the number of staff hours per year required to perform each task. Be sure to include the time required for supervision and training.

When you have completed the task analysis, prepare a list of the utility's current employees. Assign tasks to each employee based on the person's skills and abilities. To the extent possible, try to minimize the number of different work activities assigned to each person but also keep in mind the need to provide opportunities for career advancement. One full-time staff year equals 260 days including vacation and holiday time: (52 wk/yr)(5 days/wk) = 260 days/yr.

Consideration must be given to the amount and causes of employee overtime, and to identify possible ways to correct the problem. If excessive overtime and the related costs are due to lack of staff, this information can be used in a request for increased staffing.

You can expect to find that this ideal staffing arrangement does not exactly match up with your current employees' job assignments. Most likely, you will also find that the number of staff hours required does not exactly equal the number of staff hours available. Your responsibility as a manager is to create the best possible fit between the work to be done and the personnel/skills available to do it. In addition to shifting work assignments between employees, other options you might consider are contracting out some types of work, hiring part-time or seasonal staff, or setting up a second shift (to make fuller use of existing equipment). Of course, you may find that it is time to hire another full- or part-time operator.

15.42   Qualifications Profile

Hiring new employees requires careful planning before the personal interview process. In an effort to limit discriminatory hiring practices, the law and administrative policy have carefully defined the hiring methods and guidelines employers may use.

The selection method and examination process used to evaluate applicants must be limited to the applicant's knowledge, skills, and abilities to perform relevant job-related activities. In all but rare cases, factors such as age and level of education may not be used to screen candidates in place of performance testing. A description of the duties and qualifications for the job must be clearly defined in writing. The job description may be used to develop a qualifications profile. This qualifications profile clearly and precisely identifies the required job qualifications. All job qualifications must be relevant to the actual job duties that will be performed in that position. The following list of typical job qualifications may be used to help you develop your own qualifications profiles with advice from a recruitment specialist.

1. General Requirements:
   a. Knowledge of methods, tools, equipment, and materials used in water/wastewater utilities
   b. Knowledge of work hazards and applicable safety precautions
   c. Ability to establish and maintain effective working relations with employees and the general public
   d. Possession of a valid state driver's license for the class of equipment the employee is expected to drive

2. General Educational Development:
   a. Reasoning Ability: Apply common-sense understanding to carry out instructions furnished in oral, written, or diagrammatic form.
   b. Mathematical Ability: Use a pocket calculator to make arithmetic calculations relevant to the utility's operation and maintenance processes.
   c. Language Ability: Communicate with fellow employees and train subordinates in work methods. Fill out maintenance report forms.

3. Specific Vocational Preparation: Three years of experience in water/wastewater utility operation and maintenance.

4. Interests: May or may not be relevant to knowledge, skills, and ability; for example, an interest in activities concerned with objects and machines, ecology, or business management.

5. Temperament: Must adjust to a variety of tasks requiring frequent change and must routinely use established standards and procedures.

6. Physical Demands: Medium to heavy work involving lifting, climbing, kneeling, crouching, crawling, reaching, hearing, and seeing. Must be able to lift and carry _____ number of pounds for a distance of _____ feet.

7. Working Conditions: The work is outdoors and involves wet conditions, cramped and awkward spaces, noise, risks of bodily injury, and exposure to weather extremes.

For more information and details regarding staffing, we recommend reading Sections 5.3, "Applications and the Selection Process," and 5.5, "Employment Policies and Procedures," in *Utility Management* in this series of operator training manuals.

15.43   New Employee Orientation

During the first day of work, a new employee should be given all the information available in written and verbal form on the policies and practices of the utility including compensation, benefits, attendance expectations, alcohol and drug testing (if the utility does this), and employer/employee relations. Answer any questions from the new employee at this time and try to explain the overall structure of the utility as well as identify who can answer employee questions when they arise. Introduce the new employee to co-workers and tour the work area. Every utility should have a safety training session for all new employees and specific safety training for some job categories. Provide safety training (see Section 15.12, "Safety Program") for new employees on the first day of employment or as soon thereafter as possible. Establishing safe work practices is a very important function of management. The utility agency should have a checklist of all the topics or tasks that are a part of the orientation. Each item should be checked off as delivered and then signed by the employee.

QUESTIONS

Please write your answers to the following questions and compare them with those on page 389.

15.4A   What do staffing responsibilities include?

15.4B   What are two key personnel management concepts a manager should always keep in mind?

15.4C   List the steps involved in a staffing analysis.

15.4D   What is a qualifications profile?

15.4E   When should a new employee's safety training begin?

15.5   COMMUNICATION

Good communication is an essential part of good management skills. Both written and oral communication skills are needed to effectively organize and direct the operation of a treatment facility. Remember that communication is a two-part process; information must be given and it must be understood. Good listening skills are as important in communication as the information you need to communicate. As the manager of a wastewater utility, you will need to communicate with employees, with your governing body, and with the public. Your communication style will be slightly different with each of these groups but you should be able to adjust easily to your audience.

15.50   Oral Communication

Oral communication may be informal, such as talking with employees, or it may be formal, such as giving a technical presentation. In both cases, your words should be appropriate to the audience, for example, avoid technical jargon when talking with nontechnical audiences. As you talk, you should be observing your audience to be sure that what you are saying is getting across. If you are talking with an employee, it is a good idea to ask for feedback from the employee, especially if you are giving instructions. When the employee is talking, watch and listen carefully and clarify areas that seem unclear. Likewise, in a more formal presentation, watching your audience will give you feedback about how well your message is being received. Some tips for preparing a formal speech are given in Table 15.2.

15.51   Written Communication

Written communication is more demanding than oral communication and requires more careful preparation. Again, keep your audience in mind and use language that will be understood. Written communication requires more organization since you cannot clarify and explain ideas in response to your audience. Before you begin, you should have a clear idea of exactly what you wish to communicate, then keep your language as concise as possible. Extra words and phrases tend to confuse and clutter your message. Good writing skills develop slowly, but you should be able to find good writing classes in your community if you need help improving your skills. In addition, many publications and computer software programs are available that can assist you in writing the most commonly needed documents such as memos, letters, press releases, résumés, monitoring reports, and the annual report.

Before you can write a report you must first organize your thoughts. Ask yourself, what is the objective of this report? Am I trying to persuade someone of something? What information is important to communicate in this report? For whom is the report being written? How can I make it interesting? What does the reader want to learn from this report?

After you have answered the above questions, the next step is to prepare a general outline of how you intend to proceed with the preparation of the report. List not only key topics, but try to list all of the related topics. Then, arrange the key topics in sequence so there is a workable, smooth flow from one topic to the next. Do not attempt to make your outline perfect. It is just a guide. It should be flexible. As you write, you will find that you need to remove nonessential points and expand on more important points.

You might, for example, outline the following points in preparation for writing a report on a polymer testing program.

* A problem condition of high turbidity was discovered
* Polymer testing offered the best means of reducing turbidity
* Funds, equipment, and material were acquired
* Operators were trained
* Tests were conducted
* Results and conclusions were reached
* Corrective actions were planned and taken
* Conclusion, the tests did or did not produce the anticipated results or correct the problem

Once you are fairly sure you have included all the major topics you will want to discuss, go through the outline and write down facts you want to include on each topic. As you work through it, you may decide to move material from one topic to another. The outline will help you organize your ideas and facts.

When your outline is complete, you will have the essentials of your report. Now you need to tailor it to the audience that will

TABLE 15.2   TIPS FOR GIVING AN ORAL PRESENTATION

1. Arrive early. Give yourself plenty of time to become familiar with the room, practice using your audiovisuals, and make any necessary changes in room setup.

2. Be ready for mistakes. Number the pages in your presentation and your audiovisuals. Check the order of the pages before the meeting begins.

3. Pace yourself. Do not speak too quickly, speak slowly and carefully. Keep a careful eye on audience reaction to be sure that you are speaking at a pace that can be understood.

4. Project yourself. Speak loudly and look at the audience. Do not talk with your back to the audience. Check that those in the back can hear you.

5. Be natural. Try not to read your presentation. Practice ahead of time so that you can speak normally and keep eye contact with your audience.

6. Connect with the audience. Try to smile and make eye contact with the audience.

7. Involve the audience. Allow for audience questions and invite their comments.

8. Repeat audience questions. Always repeat the question so everyone can hear and to be sure that you hear the question correctly.

9. Know when to stop. Keep your remarks within the time allocated for your presentation and be aware that long, rambling speeches create a negative impression on the audience.

10. Use audiovisuals that can be heard and read. Audiovisuals should enhance and reinforce your words. Be sure that all members of the audience can hear, see, and read your audiovisuals. Normal typewritten text is not readable on overheads; use large type so everyone can see. Use no more than 5 to 7 key ideas per overhead.

11. Organize your presentation. Prepare an introduction, body of the speech, and conclusion. "Tell them what you are going to say, say it, and then tell them what you said." The presentation should have 3 to 5 main points presented in some logical order, for example, chronologically or from simple to complex.

be reading it. Take a few minutes to think about your audience. What information do they want? What aspects of the topics will they be most interested in reading? Each of the following groups may be interested in specific topics in the report. Consider these interests as you write.

1. Management

Management will have specific interests that relate to the cost effectiveness of the program. A report to management should include a summary that presents the essential information, procedures used, an analysis of the data (including trends), and conclusions. Be sure to include complete cost information. Did the benefits warrant the costs? As a result of the tests, can future expenses be reduced? Backup information and field data can be included in an appendix for those who want more information.

2. Other Utilities

Other utilities will be interested in costs but will also want more detailed information about how the program was performed. They will also be interested in the results and benefits of the program. Explain how the tests were done, the procedures, size of the crew, equipment used, source and availability of materials, and difficulties encountered and how they were overcome.

3. Citizen Groups

Citizens' interest will be more general. What is a polymer test and why is it needed? Is the polymer harmless? Will it injure fish or birds? How does the polymer test work? Who pays for the test? How much will it cost to implement results and will they be effective?

Your report may be written to include all of these groups. Adjust the outline to include the topics of interest to each group and identify the topics so readers can find the information most interesting to them. Keep the following information in mind as you write your report:

• Drafts. Good reports are not perfect the first time. Re-read and improve your report several times.

• Facts. Confine your writing to the facts and events that occurred. Include figures and statistics only when they make the report more effective. Include only the relevant facts. Large amounts of data should be put in the appendix. Do not clutter the report with unimportant data. Do not make subjective comments or offer personal opinions.

- Continuity. To be interesting and understood, a report must have continuity. It must make sense to the reader and be organized logically. In the report on the polymer testing, the report should be organized to show you had a problem, you had to find a way to identify where the problem existed, you did the testing, and you identified the problems and the corrective actions.

- Effective. To be effective, a report should achieve the objective for which it was written. In this example, we wanted to justify the costs for the program to management, help other utilities in conducting a similar program, and help citizens to understand what we were doing and why.

- Candid. A good report should be frank and straightforward. Keep the language appropriate for the audience. Do not try to impress your readers with technical terms they do not understand. Your purpose is to communicate information. Keep the information accurate and easy to understand.

The annual report is an important part of the management of the utility. It is one of the most involved writing projects that the utility must put together. The annual report should be a review of what and how the utility operated during the past year and it should also include the goals for the next year. In many small communities, the annual report may be presented orally to the city council rather than written. If this report is well written, it can be used to highlight accomplishments and provide support for future planning.

The first step to organizing the report is to make a list of three or four major accomplishments of the last year, then make a list of the top three goals for next year. These accomplishments and goals should be the focus of the report. The annual report should be a summary of the expenses, treatment services provided, and revenues generated over the last year. As you organize this information, keep those accomplishments in mind and let the data tell the story of how the utility accomplished last year's tasks. The data by itself may seem boring but as you organize the data it becomes a meaningful description of the year's accomplishments. Conclude with projections for next year. The facts and figures should tell the audience how you plan to accomplish your goals for the next year. The annual report may be simple or complex depending on your community needs. A sample table of contents for a medium-sized utility is given in Table 15.3. When you are finished, the annual report will be a valuable planning tool for the utility and can be used to build support for new projects.

QUESTIONS

Please write your answers to the following questions and compare them with those on page 389.

15.5A   What kinds of communication skills are needed by a manager?

15.5B   What are the most common written documents that a utility manager must write?

15.5C   What should be included in the annual report?

TABLE 15.3 SAMPLE TABLE OF CONTENTS FOR A UTILITY'S ANNUAL REPORT

TABLE OF CONTENTS

Executive Summary

Summary of the Treatment Processes Including Flows and Costs

Review of Goals and Objectives for the Year

Special Projects Completed

Professional Awards or Recognition for the Utility or Its Staff

General Operating Conditions Including Regulatory Requirements

Expectations for the Next Year—Goals and Objectives

Recommended Changes for the Utility in Organizing, Staffing, Equipment, of Resources Summary

System deficiencies including regulatory notifications and causes of problems

Appendixes: Operating Data

Budget

Information on Special Projects

15.6 CONDUCTING MEETINGS

As a utility manager, you will be asked to conduct meetings. These meetings may be with employees, your governing board, the public, or with other professionals in your field. Many new managers fail to prepare for these meetings and the meetings end up as a terrible waste of time. As a manager, you need to learn to conduct meetings in a way that is productive and guides the participants into an active role. The following steps should be taken to conduct a productive meeting.

Before the meeting

- Prepare an agenda and distribute it to all participants.
- Find an adequate meeting room.
- Set a beginning and ending time for the meeting.

During the meeting

- Start the meeting on time.
- Clearly state the purpose and objectives of the meeting.
- Involve all the participants.
- Do not let one or two individuals dominate the meeting.
- Keep the discussion on track and on time with the agenda.
- When the group makes a decision or reaches consensus, restate your understanding of the results.
- Make clear assignments for participants and review them with everyone during the meeting.
- Establish a time schedule for progress reports or completion of assignments.

After the meeting

- Send out minutes of the meeting.
- Send out reminders, when appropriate, about any assignments made for participants, and the next meeting time.

QUESTIONS

Please write your answers to the following questions and compare them with those on page 389.

15.6A   With whom may a utility manager be asked to conduct meetings?

15.6B   What should be done before a meeting?

15.7 PUBLIC RELATIONS

15.70 Establish Objectives

The first step in organizing an effective public relations campaign is to establish objectives. The only way to know whether your program is a success is to have a clear idea of what you expect to achieve—for example, better customer relations, greater water conservation, and enhanced organizational credibility. Each objective must be specific, achievable, and measurable. It is also important to know your audience and tailor various elements of your public relations effort to specific groups you wish to reach, such as community leaders, school children, or the average customer. Your objective may be the same in each case, but what you say and how you say it will depend upon your target audience.

15.71 Utility Operations

Good public relations begin at home. Any time you or a member of your utility comes in contact with the public, you will have an impact on the quality of your public image. Dedicated, service-oriented employees provide for better public relations than paid advertising or complicated public relations campaigns. For most people, contact with an agency employee establishes their first impression of the competence of the organization, and those initial opinions are difficult to change.

In addition to ensuring that employees are adequately trained to do their jobs and knowledgeable about the utility's operations, management has the responsibility to keep employees informed about the organization's plans, practices, and goals. Newsletters, bulletin boards, and regular, open communication between supervisors and subordinates will help build understanding and contribute to a team spirit.

Despite the old adage to the contrary, the customer is not always right. Management should try to instill among its employees the attitude that while the customer may be confused or unclear about the situation, everyone is entitled to courteous treatment and a factual explanation. Whenever possible, employees should phrase responses as positively, or neutrally, as possible, avoiding negative language. For example, "Your complaint" is better stated as "Your question." "You should have…" is likely to make the customer defensive, while "Will you please…"

is courteous and respectful. "You made a mistake" emphasizes the negative, "What we will do…" is a positive, problem-solving approach.

15.72 The Mass Media

We live in the age of communications, and one of the most effective and least expensive ways to reach people is through the mass media—radio, television, newspapers, and the Internet. Each medium has different needs and deadlines, and obtaining coverage for your issue or event is easier if you are aware of these constraints. Television must have strong visuals, for example. When scheduling a press conference, provide an interesting setting and be prepared to suggest good shots to the reporter. Radio's main advantage over television and newspapers is immediacy, so have a spokesperson available and prepared to give the interview over the telephone, if necessary. Newspapers give more thorough, in-depth coverage to stories than do the broadcast media, so be prepared to spend extra time with print reporters and provide written backup information and additional contacts.

It is not difficult to get press coverage for your event or press conference if a few simple guidelines are followed:

1. Demonstrate that your story is newsworthy, that it involves something unusual or interesting.

2. Make sure your story will fit the targeted format (television, radio, newspaper, or the Internet).

3. Provide a spokesperson who is interesting, articulate, and well prepared.

Whenever possible, all media contact should be restricted to the manager, a designated spokesperson, or, in some cases, an appointed attorney. An agency policy should clearly state that "employees must refer all mass media people to the proper spokesperson." If an employee wishes to contact the media, the employee should be accompanied by the manager.

15.73 Being Interviewed

Whether you are preparing for a scheduled interview or are simply contacted by the press on a breaking news story, here are some key hints to keep in mind when being interviewed.

1. Speak in personal terms, free of institutional jargon.

2. Do not argue or show anger if the reporter appears to be rude or overly aggressive.

3. If you do not know an answer, say so and offer to find out. Do not bluff.

4. If you say you will call back by a certain time, do so. Reporters face tight deadlines.

5. State your key points early in the interview, concisely and clearly. If the reporter wants more information, he or she will ask for it.

6. If a question contains language or concepts with which you disagree, do not repeat them, even to deny them.

7. Know your facts.

8. Never ask to see a story before it is printed or broadcast. Doing so indicates that you doubt the reporter's ability and professionalism.

15.74 Public Speaking

Direct contact with people in your community is another effective tool in promoting your utility. Though the audiences tend to be small, a personal, face-to-face presentation generally leaves a strong and long-lasting impact on the listener.

Depending upon the size of the organization, your utility may wish to establish a speaker's bureau and send a list of topics to service clubs in the area. Visits to high schools and college campuses can also be beneficial, and educators are often looking for new and interesting topics to supplement their curriculum.

Effective public speaking takes practice. It is important to be well prepared while retaining a personal, informal style. Find out how long your talk is expected to be, and do not exceed that time frame. Have a definite beginning, middle, and end to your presentation. Visual aids such as charts, slides, or models can assist in conveying your message. The use of humor and anecdotes can help to warm up the audience and build rapport between the speaker and the listener. Just be sure the humor is natural, not forced, and that the point of your story is accessible to the particular audience. Try to keep in mind that audiences only expect you to do your best. They are interested in learning about their utility and will appreciate that you are making a sincere effort to inform them about an important subject.

15.75 Telephone Contacts

First impressions are extremely important, and frequently a person's first contact with your utility is over the telephone. A person who answers the phone in a courteous, pleasant, and helpful manner goes a long way toward establishing a friendly, cooperative atmosphere. Be sure anyone answering telephone inquiries receives appropriate training and conveys a positive image for the utility.

Following a few simple guidelines will help to start your utility off on the right note with your customers:

1. *Answer calls promptly.* Your conversation will get off to a better start if the phone is answered by the third or fourth ring.

2. *Identify yourself.* This adds a personal note and lets the caller know whom he or she is talking to.

3. *Pay attention.* Do not conduct side conversations. Minimize distractions so you can give the caller your full attention, avoiding repetitions of names, addresses, and other pertinent facts.

4. *Minimize transfers.* Nobody likes to get the run-around. Few things are more frustrating to a caller than being transferred from office to office, repeating the situation, problem, or concern over and over again. Transfer only those calls that must be transferred, and make certain you are referring the caller to the right person. Then, explain why you are transferring the call. This lets the caller know you are referring him or her to a co-worker for a reason and reassures the customer that the problem or question will be dealt with. In some cases, it may be better to take a message and have someone return the call than to keep transferring the customer's call.

15.76 Customer Inquiries

No single set of rules can possibly apply to all types of customer questions or complaints about utility service. There are, however, basic principles to follow in responding to inquiries and concerns.

1. *Be prepared.* Your employees should be familiar enough with your utility's organization, services, and policies to either respond to the question or complaint or locate the person who can.

2. *Listen.* Ask the customer to describe the problem and listen carefully to the explanation. Take written notes of the facts and addresses.

3. *Do not argue.* Callers often express a great deal of pent-up frustration in their contacts with a utility. Give the caller your full attention. Once you have heard them out, most people will calm down and state their problems in more reasonable terms.

4. *Avoid jargon.* The average consumer lacks the technical knowledge to understand the complexities of water quality or wastewater treatment. Use plain, nontechnical language and avoid telling the consumer more than he or she needs to know.

5. *Summarize the problem.* Repeat your understanding of the situation back to the caller. This will assure the customer that you understand the problem and offer the opportunity to clear up any confusion or missed communication.

6. *Promise specific action.* Make an effort to give the customer an immediate, clear, and accurate answer to the problem. Be as specific as possible without promising something you cannot deliver.

In some cases, you may wish to have a representative of the utility visit the customer and observe the problem first hand. If the complaint involves water quality, take samples, if necessary, and report back to the customer to be sure the problem has been resolved.

Complaints can be a valuable asset in determining consumer acceptance and pinpointing problems. Customer calls are frequently your first indication that something may be wrong. Keep a record or log of phone contacts indicating who called, when, why, problem, and promises or action taken. Responding to complaints and inquiries promptly can save the utility money and staff resources, and minimize the number of customers who are inconvenienced. Still, education can greatly reduce complaints about utility services. Information brochures, utility bill inserts, and other educational tools help to inform customers and avoid future complaints.

15.77 Plant Tours

Tours of water and wastewater treatment plants can be an excellent way to inform the public about your utility's efforts. Political leaders, such as the City Council and members of the Board of Supervisors, should be invited and encouraged to tour the facilities, as should school groups and service clubs.

A brochure describing your utility's goals, accomplishments, operations, and processes can be a good supplement to the tour and should be handed out at the end of the visit. The more visually interesting the brochure is, the more likely that it will be read, and the use of color, photographs, graphics, or other design features is encouraged. If you have access to the necessary equipment, production of a videotape program about the utility can also add interest to the facility tour.

The tour itself should be conducted by an employee who is very familiar with plant operations and can answer the types of questions that are likely to arise. Consider including the following basics:

1. A description of the sources of water supply or wastewater (if appropriate)

2. History of the plant, the years of operation, modifications, and innovations over the years

3. Major plant design features, including plant capacity and safety features

4. Observation of the treatment processes

5. A visit to the laboratory

6. Anticipated improvements, expansions, and long-range plans for meeting future service needs

Plant tours can contribute to a utility's overall program to gain financing for capital improvements. If the City Council or other governing board has seen the treatment process first hand, it is more likely to understand the need for enhancement and support future funding.

As beneficial as plant tours may be for promoting public interest and confidence in the utility, security precautions should be carefully considered when planning for visitors to the facilities. For more information, see Section 15.11, "Homeland Defense."

15.8 FINANCIAL MANAGEMENT

Financial management for a utility should include providing financial stability for the utility, careful budgeting, and providing capital improvement funds for future utility expansion. These three areas must be examined on a routine basis to ensure the continued operation of the utility. They may be formally reviewed on an annual basis or more frequently when the utility is changing rapidly. The utility manager should understand what is required for each of the three areas and be able to develop record systems that keep the utility on track and financially prepared for the future. Also, the manager must identify areas of financial concern that are or will place the utility agency in peril. This information must then be presented to the administration for action.

15.80 Financial Stability

How do you measure financial stability for a utility? Two very simple calculations can be used to help you determine how healthy and stable the finances are for the utility. These two calculations are the OPERATING RATIO and the COVERAGE RATIO. The operating ratio is a measure of the total revenues divided by the total operating expenses. The coverage ratio is a measure of the ability of the utility to pay the principal and interest on loans and bonds (this is known as DEBT SERVICE[4]) in addition to any unexpected expenses. A utility that is in good financial shape will have an operating ratio and coverage ratio above 1.0. In fact, most bonds and loans require the utility to have a coverage ratio of at least 1.25. As state and federal funds for utility improvements have become much more difficult to obtain, these financial indicators have become more important for utilities. Being able to show and document the financial stability of the utility is an important part of getting funding for more capital improvements.

The operating ratio is perhaps the simplest measure of a utility's financial stability. In essence, the utility must be generating enough revenue to pay its operating expenses. The actual ratio is usually computed on a yearly basis, since many utilities may have monthly variations that do not reflect the overall performance.

---

4. *Debt Service.*   The amount of money required annually to pay the (1) interest on outstanding debts, or (2) funds due on a maturing bonded debt or the redemption of bonds.

The total revenue is calculated by adding up all revenue generated by user fees, hook-up charges, customer taxes or assessments, interest income, and special income. Next, determine the total operating expenses by adding up the expenses of the utility, including administrative costs, salaries, benefits, energy costs, chemicals, supplies, fuel, equipment costs, equipment replacement fund, principal and interest payments, and other miscellaneous expenses.

EXAMPLE 1

The total revenues for a utility are $1,686,000 and the operating expenses for the utility are $1,278,899. The debt service expenses are $560,000. What is the operating ratio? What is the coverage ratio?

Known		Unknown
Total Revenue, $	= $1,686,000	1. Operating Ratio
Operating Expenses, $	= $1,278,899	2. Coverage Ratio
Debt Service Expenses, $	= $560,000	

1. Calculate the operating ratio.

$$\text{Operating Ratio} = \frac{\text{Total Revenue, \$}}{\text{Operating Expenses, \$}}$$

$$= \frac{\$1,686,000}{\$1,278,899}$$

$$= 1.32$$

2. Calculate the coverage ratio.

 a. Calculate the nondebt expenses.

$$\text{Nondebt Expenses, \$} = \text{Operating Exp, \$} - \text{Debt Service Exp, \$}$$

$$= \$1,278,899 - \$560,000$$

$$= \$718,899$$

 b. Calculate the coverage ratio.

$$\frac{\text{Coverage}}{\text{Ratio}} = \frac{\text{Total Revenue, \$} - \text{Nondebt Expenses, \$}}{\text{Debt Service Expenses, \$}}$$

$$= \frac{\$1,686,000 - \$718,899}{\$560,000}$$

$$= 1.73$$

These calculations provide a good starting point for looking at the financial strength of the utility. Both of these calculations use the total revenue for the utility, which is an important component for any utility budgeting. As managers, we often focus on the expense side and forget to look carefully at the revenue side of utility management. The fees collected by the utility, including hook-up fees and user fees, must accurately reflect the cost of providing service. These fees must be reviewed annually and they must be increased as expenses rise to maintain financial stability. Some other areas to examine on the revenue side include how often and how well user fees are collected, the number of delinquent accounts, and the accuracy of meters (for drinking water utilities). Some small communities have found

they can cut their administrative costs significantly by switching to a quarterly billing cycle. The utility must have the support of the community to determine and collect user fees, and the utility must keep track of revenue generation as carefully as resource spending.

15.81   Budgeting

Budgeting for the utility is perhaps the most challenging task of the year for many managers. The list of needs usually is much larger than the possible revenue for the utility. The only way for the manager to prepare a good budget is to have good records from the year before. A system of recording or filing purchase orders or a requisition records system (see Section 15.134, "Procurement Records") must be in place to keep track of expenses and prevent spending money that is not in the budget.

To budget effectively, a manager needs to understand how the money was spent over the last year, the needs of the utility, and how the needs should be prioritized. The manager also must take into account cost increases that cannot be controlled while trying to minimize the expenses as much as possible. The following problem is an example of the types of decisions a manager must make to keep the budget in line while also improving service from the utility.

EXAMPLE 2

A wastewater pump that has been in operation for 25 years pumps a constant 600 GPM through 47 feet of dynamic head. The pump uses 6,071 kilowatt-hours of electricity per month, at a cost of $0.085 per kilowatt-hr. The old pump efficiency has dropped to 63 percent. Assuming a new pump that operates at 86 percent efficiency is available for $9,730.00, how long would it take to pay for replacing the old pump?

Known		Unknown
Electricity, kW-hr/mo	= 6,071 kW-hr/mo	New Pump
Electricity Cost, $/kW-hr	= $0.085/kW-hr	Payback
Old Pump Efficiency, %	= 63%	Time, yr
New Pump Efficiency, %	= 86%	
New Pump Cost, $	= $9,730	

1. Calculate old pump operating costs in dollars per month.

$$\begin{array}{l}\text{Old Pump} \\ \text{Operating} \\ \text{Costs, \$/mo}\end{array} = \left(\begin{array}{c}\text{Electricity,} \\ \text{kW-hr/mo}\end{array}\right)\left(\begin{array}{c}\text{Electricity Cost,} \\ \text{\$/kW-hr}\end{array}\right)$$

$$= (6,071 \text{ kW-hr/mo})(\$0.085/\text{kW-hr})$$

$$= \$516.04/\text{mo}$$

2. Calculate new pump operating electricity requirements.

$$\begin{array}{l}\text{New Pump} \\ \text{Electricity,} \\ \text{kW-hr/mo}\end{array} = (\text{Old Pump Elect, kW-hr/mo})\frac{(\text{Old Pump Eff, \%})}{(\text{New Pump Eff, \%})}$$

$$= (6,071 \text{ kW-hr/mo})\frac{(63\%)}{(86\%)}$$

$$= 4,447 \text{ kW-hr/mo}$$

3. Calculate new pump operating costs in dollars per month.

$$\text{New Pump Operating Costs, \$/mo} = \left(\begin{array}{c}\text{Electricity,}\\ \text{kW-hr/mo}\end{array}\right)\left(\begin{array}{c}\text{Electricity Cost,}\\ \text{\$/kW-hr}\end{array}\right)$$

$$= (4{,}447 \text{ kW-hr/mo})(\$0.085/\text{kW-hr})$$

$$= \$378.03/\text{mo}$$

4. Calculate annual cost savings of new pump.

$$\text{Cost Savings, \$/yr} = (\text{Old Costs, \$/mo} - \text{New Costs, \$/mo})(12 \text{ mo/yr})$$

$$= (\$516.04/\text{mo} - \$378.03/\text{mo})(12 \text{ mo/yr})$$

$$= \$1{,}656.12/\text{yr}$$

5. Calculate the new pump payback time in years.

$$\text{Payback Time, yr} = \frac{\text{Initial Cost, \$}}{\text{Savings, \$/yr}}$$

$$= \frac{\$9{,}730.00}{\$1{,}656.12/\text{yr}}$$

$$= 5.9 \text{ years}$$

In this example, a payback time of 5.9 years is acceptable and would probably justify the expense for a new pump. This calculation was a simple payback calculation, which did not take into account the maintenance on each pump, depreciation, and inflation. Many excellent references are available from EPA to help utility managers make more complex decisions about purchasing new equipment.

The annual report should be used to help develop the budget so that long-term planning will have its place in the budgeting process. The utility manager must track revenue generation and expenses with adequate records to budget effectively. The manager must also get input from other personnel in the utility as well as community leaders as the budgeting process proceeds. This input from others is invaluable to gain support for the budget and to keep the budget on track once adopted.

15.82   Equipment Repair/Replacement Fund

To adequately plan for the future, every utility must have a repair/ replacement fund. The purpose of this fund is to generate additional revenue to pay for the repair and replacement of capital equipment as the equipment wears out. To prepare adequately for this repair/replacement, the manager should make a list of all capital equipment (this is called an asset inventory) and estimate the replacement cost for each item. The expected life span of the equipment must be used to determine how much money should be collected over time. When a plant is new, the balance in the repair/replacement fund should be increasing each year. As the plant gets older, the funds will have to be used and the balance may get dangerously low as equipment breakdowns occur. Perhaps the hardest job for the utility manager is to maintain a positive balance in this account with the understanding that this account is not meant to generate a profit for the utility but rather to plan for future equipment needs. In wastewater facilities constructed with federal funds under the construction grant program, providing an adequate repair/replacement fund was one of the grant conditions, but if this repair/replacement fund has not been reviewed annually, it must be updated.

To set up a repair/replacement fund for your utility, you should first put together a list of the equipment required for each process in your utility. Once you have this list, you need to estimate the life expectancy of the equipment and the replacement cost. From this list you can predict the amount of money you should set aside each year so that when each piece of equipment wears out, you will have enough money to replace that piece of equipment. The EPA publications listed in Section 15.15, "Additional Reading," at the end of this chapter are excellent references for utility planning.

15.83   Capital Improvements and Funding in the Future

A capital improvements fund must be a part of the utility budget and included in the operating ratio. Your responsibility as the utility manager is to be sure that everyone, your governing body and the public, understands the capital improvement fund is not a profit for the utility but a replacement fund to keep the utility operating in the future.

Capital planning starts with a look at changes in the community. Where are the areas of growth in the community, where are the areas of decline, and what are the anticipated changes in industry within the community? After identifying the changing needs in the community, you should examine the existing utility structure. Identify your weak spots (in the collection or distribution system, or with in-plant processes). Make a list of the areas that will be experiencing growth, weak spots in the system, and anticipated new regulatory requirements. The list should include expected capital improvements that will need to be made over the next year, two years, five years, and ten years. You can use the information in your annual reports and other operational logs to help compile the list.

Once you have compiled this information, prioritize the list and make a timetable for improving each of the areas. Starting at the top of the priority list, estimate the costs for improvements and incorporate these costs into your capital improvement budget. The calculations you have made previously, including operating ratio, coverage ratio, and payback time, will all be useful in prioritizing and streamlining your list of needs. Another useful ratio is the corrective to preventive maintenance ratio.

You may find that some of your capital improvement needs could be met in more than one way. How do you decide which of several options is the most cost-effective? How do you compare fundamentally different solutions? For example, assume your community's population is growing rapidly and you will need to increase the capacity of your wastewater treatment facilities by the end of the next ten years. Your two existing wastewater treatment plants are both operating near 95 percent of design capacity. Possible solutions might include the following options, where portions of some of the options might be implemented immediately while other portions might be brought on line in five or ten years.

- Expand the existing wastewater treatment plants.

- Construct a new wastewater treatment plant to treat flows greater than the capacity of the two existing plants.

- Construct a new regional plant capable of handling all expected flows (shut down the two existing plants).

- Keep the newer of the two existing plants, shut down the older of the two plants, and construct one new, larger plant to treat flows from the older plant and the expected future flows.

Each of these possible solutions should be evaluated individually and then compared with the other alternatives. To compare alternative plans, you will need to calculate the present value (or *PRESENT WORTH*[5]) of each plan; that is, the costs and benefits of each plan in today's dollars. This is done by identifying all the costs and benefits of each alternative plan over the same time period or time horizon. Costs should include not only the initial purchase price or construction costs, but also financing costs over the life of the loans or bonds and all operation and maintenance costs. Benefits include all of the revenue that would be produced by this facility or equipment, including connection and user fees. With the help of an experienced accountant, apply standard inflation, depreciation, and other economic discount factors to calculate the present value of all the benefits and costs of each plan during the same planning period. This will give you the cost of each plan in the equivalent of today's dollars.

Remember to involve all of your local officials and the public in this capital improvement budget so they understand what will be needed.

Long-term capital improvements such as a new plant or a new treatment process are usually anticipated in your 10-year or 20-year projection. These long-term capital improvements usually require some additional financing. The basic ways for a utility to finance capital improvements are through general obligation bonds, revenue bonds, or loan funding programs.

General obligation bonds or *ad valorem* (based on value) taxes are assessed based on property taxes. These bonds usually have a lower interest rate and longer payback time, but the total bond limit is determined for the entire community. This means that the water or wastewater utility will have available only a portion of the total bond capacity of the community. These bonds are not often used for funding water and wastewater utility improvements today.

The second type of bond, the revenue bond, is a debt incurred by the community, often to finance utility improvements. This bond has no limit on the amount of funds available and the user charges provide repayment on the bond. To qualify for these bonds, the utility must show sound financial management and the ability to repay the bond. As the utility manager, you should be aware of the provisions of the bond. Be sure the bond has a call date, which is the first date when you can pay off the bond. The common practice is for a 20-year bond to have a 10-year call date and for a 15-year bond to have an 8-year call date. The bond will also have a call premium, which is the amount of extra funds needed to pay off the debt on the call date. You should try to get your bonds a call premium of no more than 102 percent par. This means that for a debt of $200,000 on the call date, the total payoff would be $204,000, which includes the extra two percent for the call premium. You will need to get help from a financial advisor to prepare for and issue the bonds. These advisors will help you negotiate the best bond structure for your community.

Special assessment bonds may be used to extend services into specific areas. The direct users pay the capital costs and the assessment is usually based on frontage or area of real estate. These special assessments carry a greater risk to investors but may be the best way to extend service to some areas.

The most common way to finance water and wastewater improvements in the past has been federal and state grant programs. The Block Grants from HUD are still available for some projects and Rural Utilities Service (RUS) loans may also be used as a funding source. In addition, state revolving fund (SRF) programs provide loans (but not direct grants) for improvements. The SRF program has already been implemented with wastewater improvements and the Safe Drinking Water Regulations include an SRF program for funding water treatment improvements. These SRF programs are very competitive and utilities must provide evidence of sound financial management to qualify for these loans. You should contact your state regulatory agency to find out more about the SRF program in your state.

---

5. *Present Worth.*   The value of a long-term project expressed in today's dollars. Present worth is calculated by converting (discounting) all future benefits and costs over the life of the project to a single economic value at the start of the project. Calculating the present worth of alternative projects makes it possible to compare them and select the one with the largest positive (beneficial) present worth or minimum present cost.

15.84 Financial Assistance

Many small utility systems need additional funds to repair and upgrade their systems. Potential funding sources include loans and grants from federal and state agencies, banks, foundations, and other sources. Some of the federal funding programs for small public utility systems include:

- Appalachian Regional Commission (ARC)

- Department of Housing and Urban Development (HUD) (provides Community Development Block Grants)

- Economic Development Administration (EDA)

- Indian Health Service (IHS)

- Rural Utilities Service (RUS) (formerly Farmer's Home Administration (FmHA) and Rural Development Administration (RDA))

For additional information, see "Financing Assistance Available for Small Public Water Systems," by Susan Campbell, Benjamin W. Lykins, Jr., and James A. Goodrich, *Journal American Water Works Association,* June 1993, pages 47–53.

Another valuable contact is the Environmental Financing Information Network (EFIN), which provides information on financing alternatives for state and local environmental programs and projects in the form of abstracts of publications, case studies, and contacts. EFIN is a component of the Center for Environmental Finance. For further information, contact US Environmental Protection Agency, Office of the Chief Financial Officer, Center for Environmental Finance, 1200 Pennsylvania Avenue, NW, Mail Code 2731A, Washington, DC 20460. Phone (202) 564-4994 and FAX (202) 565-2587.

Also, many states have one or more special financing mechanisms for small public utility systems. These funds may be in the form of grants, loans, bonds, or revolving loan funds. Contact your state for more information.

QUESTIONS

Please write your answers to the following questions and compare them with those on page 390.

15.8A List the three main areas of financial management for a utility.

15.8B How is a utility's operating ratio calculated?

15.8C Why is it important for a manager to consult with other utility personnel and with community leaders during the budgeting process?

15.8D How can long-term capital improvements be financed?

15.8E What is a revenue bond?

END OF LESSON 1 OF 2 LESSONS

on

MANAGEMENT

Please answer the discussion and review questions next.

DISCUSSION AND REVIEW QUESTIONS

Chapter 15.   MANAGEMENT

(Lesson 1 of 2 Lessons)

At the end of each lesson in this chapter, you will find discussion and review questions. Please write your answers to these questions to determine how well you understand the material in the lesson.

1. What are the different types of demands on a utility manager?

2. List the basic functions of a manager.

3. What can happen without adequate utility planning?

4. What is the purpose of an organizational chart?

5. Define the following terms:

   1. Authority

   2. Responsibility

   3. Delegation

   4. Accountability

6. When has a supervisor successfully delegated?

7. What information should be provided to a new employee during orientation?

8. With whom do managers need to communicate?

9. What information should be included in the utility's annual report?

10. List four steps that can be taken during a meeting to make sure it is a productive meeting.

11. What happens any time you or a member of your utility comes in contact with the public?

12. What attitude should management try to instill among its employees regarding the customer?

13. What is the value of customer complaints?

14. How do you measure financial stability for a utility?

15. How can a manager prepare a good budget?

CHAPTER 15.   MANAGEMENT

(Lesson 2 of 2 Lessons)

15.9   OPERATIONS AND MAINTENANCE

15.90   The Manager's Responsibilities

A utility manager's specific operation and maintenance (O&M) responsibilities vary depending on the size of the utility. At a small utility, the manager may oversee all utility operations while also serving as chief operator and supervising a small staff of operations and maintenance personnel. In larger utility agencies, the manager may have no direct, day-to-day responsibility for operations and maintenance but is ultimately responsible for efficient, cost-effective operation of the entire utility. Whether large or small, every utility needs an effective operations and maintenance program.

15.91   Purpose of O&M Programs

The purpose of an O&M program is to operate and maintain the system in accordance with design specifications. This involves a variety of operational procedures, maintenance procedures, and prerequisites such as training and certification, along with resource management and organizational management. The ability to effectively operate and maintain a water or wastewater utility so it performs as intended depends greatly on proper design (including selection of appropriate materials and equipment), construction and inspection, acceptance, and system start-up. Permanent system deficiencies that affect O&M of the system are frequently the result of these phases. O&M staff should be involved at the beginning of each project, including planning, design, construction, acceptance, and start-up. When a utility system is designed with future O&M considerations in mind, the result is a more effective O&M program in terms of O&M cost and performance.

Effective O&M programs are based on knowing what components make up the system, where they are located, and the condition of the components. With that information, proactive maintenance can be planned and scheduled, rehabilitation needs identified, and a long-term Capital Improvement Program (CIP) planned and budgeted. High-performing agencies have all developed performance measurements of their O&M program and track the information necessary to evaluate performance.

15.92   Types of Maintenance

Water or wastewater system maintenance can be either a proactive or a reactive activity. In general, maintenance activities can be classified as corrective maintenance, preventive maintenance, and predictive maintenance.

Corrective maintenance, including emergency maintenance, is reactive. For example, a piece of equipment or a system is allowed to operate until it fails, with little or no scheduled maintenance occurring prior to the failure. Only when the equipment or system fails is maintenance performed. Reliance on reactive maintenance will always result in poor system performance, especially as the system ages. Utility agencies taking a corrective maintenance approach are characterized by the following:

- The inability to plan and schedule work

- The inability to budget adequately

- Poor use of resources

- A high incidence of equipment and system failures

- High corrective or repair costs with related overtime pay

Emergency maintenance involves two types of emergencies: normal emergencies and extraordinary emergencies. Public utilities are faced with normal emergencies on a daily basis, whether it is a water main break or a blockage in a sewer. Normal emergencies can be reduced by an effective maintenance program. Extraordinary emergencies, such as high-intensity rainstorms, hurricanes, floods, and earthquakes, will always be unpredictable occurrences. However, the effects of extraordinary emergencies on the utility's performance can be minimized by implementation of a planned maintenance program and development of a comprehensive emergency response plan (see Section 15.10).

Preventive maintenance is proactive and is defined as a programmed, systematic approach to maintenance activities. This type of maintenance will always result in improved system performance except in the case where major chronic problems are the result of design or construction flaws that cannot be corrected by O&M activities. Proactive maintenance is performed on a periodic (preventive) basis or an as-needed (predictive) basis. Preventive maintenance can be scheduled on the basis of specific criteria such as equipment operating time since the last maintenance was performed, or passage of a certain amount of time (calendar period). For example, performing television inspection of a gravity sewer system on a five-year cycle (calendar period) would require that 20 percent of the system be televised each year. At the end of the five-year period, the cycle would start over again. Similarly, lubrication of motors is frequently based on running time.

The major elements of a good preventive maintenance program include the following:

1. Planning and scheduling

2. Records management

3. Spare parts management

4. Cost and budget control

5. Emergency repair procedures

6. Training program

Some benefits of taking a preventive maintenance approach are these:

1. Maintenance can be planned and scheduled

2. Work backlog can be identified

3. Adequate resources necessary to support the maintenance program can be budgeted

4. Capital improvement program (CIP) items can be identified and budgeted

5. Human and material resources can be used effectively

Predictive maintenance, which is also proactive, is a method of establishing baseline performance data, monitoring performance criteria over a period of time, and observing changes in performance so that failure can be predicted and maintenance can be performed on a planned, scheduled basis. Knowing the condition of the system makes it possible to plan and schedule maintenance as required and thus avoid unnecessary maintenance. An example of predictive maintenance in a collection system would be visually inspecting manholes or monitoring flows; when changes in flow conditions are observed, cleaning is scheduled accordingly.

In reality, every agency operates with corrective and emergency maintenance, preventive maintenance, and predictive maintenance methods. The goal, however, is to reduce the corrective and emergency maintenance efforts by performing preventive maintenance, which will minimize system failures that result in stoppages and overflows.

System performance is frequently a reliable indicator of how the system is operated and maintained. Agencies that rely primarily on corrective maintenance in operating and maintaining the system are never able to focus on preventive and predictive maintenance. With most of their resources directed at corrective maintenance activities, it is difficult to free up these resources to begin developing preventive maintenance programs. For an agency to develop an effective proactive maintenance program, it must add initial resources over and above those currently existing.

15.93 Benefits of Managing Maintenance

The goal of managing maintenance is to minimize investments of labor, materials, money, and equipment. In other words, we want to manage our human and material resources as effectively as possible, while delivering a high level of service to our customers. The benefits of an effective operation and maintenance program are as follows:

- Ensuring the availability of facilities and equipment as intended.

- Maintaining the reliability of the equipment and facilities as designed. Utility systems are required to operate 24 hours per day, 7 days per week, 365 days per year. Reliability is a critical component of the operation and maintenance program. If equipment and facilities are not reliable, then the ability of the system to perform as designed is impaired.

- Maintaining the value of the investment. Water and wastewater systems represent major capital investments for communities and are major capital assets of the community. If maintenance of the system is not managed, equipment and facilities will deteriorate through normal use and age. Maintaining the value of the capital asset is one of the utility manager's major responsibilities. Accomplishing this goal requires ongoing investment, both to maintain existing facilities and equipment and extend the life of the system, and to establish a comprehensive O&M program.

- Obtaining full use of the system throughout its design life.

- Collecting accurate information and data on which to base the operation and maintenance of the system and justify requests for the financial resources necessary to support it.

QUESTIONS

Please write your answers to the following questions and compare them with those on page 390.

15.9A   What is the purpose of an operation and maintenance (O&M) program?

15.9B   In general, maintenance activities can be classified as which three categories?

15.9C   List the major elements of a good preventive maintenance program.

15.94   Computer Control Systems

Computer control systems are used by operators and administrators to monitor and adjust the operation of equipment in their systems. One type of computer control system is the SCADA system. SCADA stands for Supervisory Control And Data Acquisition.

A SCADA system collects, stores, and analyzes information about all aspects of operation and maintenance; transmits alarm signals, when necessary; and allows fingertip control of alarms, equipment, and processes. SCADA provides the information that operators and their supervisors need to solve minor problems before they become major incidents. As the nerve center of a wastewater collection and treatment agency, the SCADA system allows operators to enhance the efficiency of their facilities by keeping them fully informed and fully in control.

Applications for SCADA systems include wastewater collection and pumping system monitoring, wastewater treatment plant control monitoring, combined sewer overflow (CSO) diversion monitoring, and other related applications. SCADA systems can vary from merely data collection and storage to total data analysis, interpretation, and process control.

A SCADA system might include liquid level, pressure, and flow sensors. The measured (sensed) information could be transmitted to a computer system, which stores, analyzes, and presents the information. The information may be read by an operator on dials or as digital readouts or analyzed and plotted by the computer as trend charts.

Most SCADA systems present a graphical picture of the overall system on a computer screen. In addition, detailed pictures of specific portions of the system can be examined by the operator following a request and instructions to the computer. The graphical displays on the computer screen can include current operating information. The operator can observe this information, analyze it for trends, or determine if it is within acceptable operating ranges, and then decide if any adjustments or changes are necessary.

SCADA systems are capable of analyzing data and providing operating, maintenance, regulatory, and annual reports. In some plants, operators rely on a SCADA system to help them prepare daily, weekly, and monthly maintenance schedules; monitor the spare parts inventory status; order additional spare parts, when necessary; print out work orders; and record completed work assignments. SCADA systems can also be used to enhance energy conservation programs and emergency response procedures. For more information on SCADA systems and typical wastewater collection system and treatment plant applications, see *Operation of Wastewater Treatment Plants,* Volume II, Chapter 20, "Treatment Plant Administration."

15.95   Cross-Connection Control

BACKFLOW[6] of contaminated water through cross-connections into community water systems is not just a theoretical problem. Contamination through cross-connections has consistently caused more waterborne disease outbreaks in the United States than any other reported factor. Cross-connections commonly occur in the following types of situations:

1. A sprinkler system using nonpotable water is connected to a potable water supply.

2. A potable water source used as a seal supply is connected to a pump delivering unapproved or nonpotable water.

3. A hose connected to the house is left in a swimming pool. When water is drawn from indoor taps, the hose sucks the pool water into the potable water supply.

4. A hose connected to the house is used to apply chemical fertilizers or pesticides. Without a vacuum breaker or other backflow-prevention device at the house connection, the chemicals will enter the potable water supply.

5. A hose connected to the house is used to flush a car radiator. If the hose is left in the radiator after it has been filled with antifreeze and if there is no backflow protection at the house connection, the antifreeze will enter the potable water supply.

Inspections have often disclosed numerous unprotected cross-connections between public water systems and other piped systems on consumers' premises, which might contain wastewater; stormwater; processed waters (containing a wide variety of chemicals); and untreated supplies from private wells, streams, and ocean waters. Therefore, an effective cross-connection control program is essential.

Inspect your plant to see if there are any cross-connections between your potable (drinking) water supply and unapproved water supplies such as pump seals, feed water to boilers, hose bibs below grade where they may be subject to flooding with wastewater or sludges, or any other location where wastewater could contaminate a domestic water supply. If any of these or other existing or potential cross-connections are found, be certain that your drinking water supply source is properly protected by the installation of an approved backflow prevention device.

Many treatment plants use an *AIR GAP*[7] system (Figure 15.1) to protect their drinking water supply. An air gap separation system provides a physical break between the wastewater treatment plant's municipal or fresh water supply (well) and the plant's treatment process systems. The physical separation of the two water systems effectively protects the potable water supply in case wastewater backs up from a treatment process.

Installation of air gap systems is controlled by health regulations. These systems should be inspected periodically by the local health department or the public water supply agency. Also, it is good practice to have your drinking water tested at least monthly for coliform group organisms because sometimes even the best of backflow prevention devices fail.

Never drink from outside water connections such as faucets and hoses. The hose you drink from may have been used to carry effluent or sludge.

You may find in your plant that it will be more economical to use bottled drinking water. If so, be sure to post conspicuous signs that your plant water is not potable at all outlets. This also applies to all hose bibs in the plant from which you may obtain water other than a potable source. This is a must in order to inform visitors or absent-minded or thirsty employees that the water from each marked location is not for drinking purposes.

---

6. *Backflow.*   A reverse flow condition, created by a difference in water pressures, that causes water to flow back into the distribution pipes of a potable water supply from any source or sources other than an intended source. Also see BACKSIPHONAGE.

7. *Air Gap.*   An open, vertical drop, or vertical empty space, between a drinking (potable) water supply and potentially contaminated water. This gap prevents the contamination of drinking water by backsiphonage because there is no way potentially contaminated water can reach the drinking water supply.

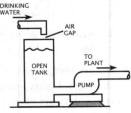

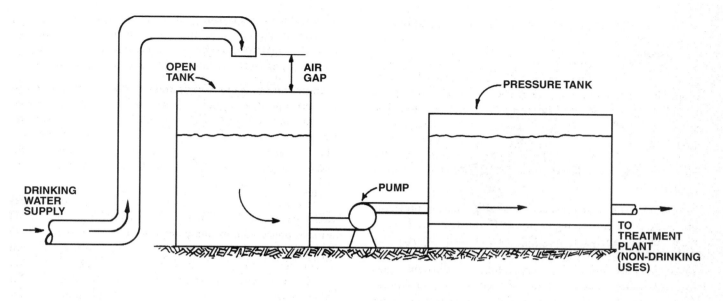

Fig. 15.1   Air gap device

15.96   Geographic Information System (GIS)

The geographic information system (GIS) is a computer program that combines mapping with detailed information about the physical structures within geographic areas. To create the database of information, entities within a mapped area, such as streets, manholes, collection system sewer pipe segments, and lift stations, are given "attributes." Attributes are simply the pieces of information about a particular feature or structure that are stored in a database. The attributes can be as basic as an address, lift station pumping capacity, or length of sewer pipe segment, or they may be as specific as pipe diameter, pipe material, and quadrant (coordinate) location. Attributes of a collection system pipe segment might include engineering information, maintenance information, and inspection information. Thus, an inventory of entities and their properties is created. The system allows the operator to periodically update the map entities and their corresponding attributes.

The power of a GIS is that information can be retrieved geographically. An operator can choose an area to look at by pointing to a specific place on the map or outlining (windowing) an area of the map. The system will display the requested section on the screen and show the attributes of the entities located on the map. A printed copy may also be requested. In most cases, CMMS (Computer-based Maintenance Management Systems) software has the ability to communicate with geographic information systems so that attribute information from the collection system can be copied into the GIS.

A GIS can generate work orders in the form of a map with the work to be performed outlined on the map. This minimizes paper work and gives the work crew precise information about where the work is to be performed. Completion of the work is recorded in the GIS to keep the work history for the area and entity up to date. Reports and other inquiries can be requested, as needed, for example, a listing of all collection system sewer pipe segments in a specific area could be generated for a report.

In many areas, GISs are being developed on an area-wide basis with many agencies, utilities, counties, cities, and state agencies participating. Usually, a county-wide base map is developed and then all participants provide attributes for their particular systems. For example, information on the sanitary sewer collection system might be one map layer, the second map layer might be the water distribution system, and the third layer might be the electric utility distribution system. In addition to sharing databases with CMMSs, GISs have the ability to operate smoothly with computer-aided design (CAD) systems.

QUESTIONS

Please write your answers to the following questions and compare them with those on page 390.

15.9D   What does SCADA stand for?

15.9E   What does a SCADA system do?

15.9F   What has caused more waterborne disease outbreaks in the United States than any other reported factor?

15.9G   What are some potential places where cross-connections could occur in a wastewater treatment plant?

15.9H   What makes a geographic information system (GIS) a potentially powerful tool for a wastewater utility operator?

15.10 EMERGENCY RESPONSE

Contingency planning is an essential facet of utility management and one that is often overlooked. Natural disasters such as floods, earthquakes, hurricanes, forest fires, avalanches, and blizzards are a more or less routine occurrence for some utilities. Although utilities in various locations will be vulnerable to somewhat different kinds of natural disasters, the effects of these disasters often will be similar. When such catastrophic emergencies occur, the utility must be prepared to minimize the effects of the event and have a plan for rapid recovery. Such preparation should be a specific obligation of every utility manager.

In addition to emergencies associated with natural disasters, all utilities suffer from common problems such as equipment breakdowns and leaking or broken pipes. During the past few years, there has also been an increasing amount of vandalism, civil disorder, toxic spills, and employee strikes, which have threatened to disrupt utility operations. In observing today's international tension and the potential for nuclear war or the effects of terrorist-induced chemical or biological warfare, water and wastewater utilities must seriously consider how to respond.

As a first step toward an effective contingency plan, each utility should make an assessment of its own vulnerability and then develop and implement a comprehensive plan of action. The following steps should be taken in assessing the vulnerability of a utility system:

1. Identify and describe the system components.

2. List assumed disaster characteristics.

3. Estimate disaster effects on system components.

4. Estimate customer demand for service following a potential disaster.

5. Identify key system components that would be primarily responsible for system failure.

If the assessment shows a system is unable to meet estimated requirements because of the failure of one or more critical components, the vulnerable elements have been identified. Repeating this procedure using several typical disasters will usually point out system weaknesses. Frequently, the same vulnerable element appears for a variety of assumed disaster events. Eliminate as many system weaknesses as possible. For the remaining weaknesses identified in the vulnerability assessment, estimate the probable extent of damage that would be caused by each of the emergency events most likely to occur in the area and develop and implement a comprehensive plan of action for responding to emergencies.

Although all system elements are important for the utility to function at peak efficiency, experience with disasters points out elements that are most subject to disruption. These elements are as follows:

1. The absence of trained personnel to make critical decisions and carry out orders

2. The loss of power to the utility's facilities

3. An inadequate amount of supplies and materials

4. Inadequate communication equipment

An emergency operations plan need not be too detailed, since all types of emergencies cannot be anticipated and a complex response program can be more confusing than helpful. Supervisory personnel must have a detailed description of their responsibilities during emergencies. They will need information, supplies, equipment, and the assistance of trained personnel. All these can be provided through a properly constructed emergency operations plan that is not extremely detailed.

The following outline can be used as the basis for developing an emergency operations plan:

1. Make a vulnerability assessment.

2. Inventory organizational personnel.

3. Provide for a recovery operation (plan).

4. Provide training programs for operators in carrying out the plan.

5. Coordinate with local and regional agencies such as the health, police, and fire departments to develop procedures for carrying out the plan.

6. Establish a communications procedure.

7. Provide protection for personnel, plant equipment, records, and maps.

By following these steps, an emergency plan can be developed and maintained even though changes in personnel may occur. Emergency simulation training sessions, including the use of standby power, equipment, and field test equipment will ensure that equipment and personnel are ready at times of emergency.

A list of phone numbers for operators to call in an emergency should be prepared and posted by a phone for emergency use. The list should include the following:

1. Plant supervisor

2. Director of public works or head of utility agency

3. Police

4. Fire

5. Doctor (2 or more)

6. Ambulance (2 or more)

7. Hospital (2 or more)

If appropriate for your utility, also include the following phone numbers on the emergency list:

8. Chlorine supplier and manufacturer

9. *CHEMTREC*, (800) 424-9300, for the hazardous chemical spills; sponsored by the Manufacturing Chemists Association

10. US Coast Guard's National Response Center, (800) 424-8802

11. Local and state poison control centers

12. Local hazardous materials spill response team

You should prepare a list for your plant *now*, if you have not already done so, and update the numbers annually.

QUESTIONS

Please write your answers to the following questions and compare them with those on page 390.

15.10A   What is the first step toward an effective contingency plan for emergencies?

15.10B   Why is too detailed an emergency operations plan not needed or even desirable?

15.10C   An emergency operations plan should include what specific information?

15.11 HOMELAND DEFENSE

World events in recent years have heightened concern in the United States over the security of the critical wastewater infrastructure. The nation's wastewater infrastructure, consisting of approximately 16,000 publicly owned wastewater treatment plants, 100,000 major pumping stations, 600,000 miles of sanitary sewers and another 200,000 miles of storm sewers, is one of America's most valuable resources, with treatment and collection systems valued at more than $2 trillion.

Taken together, the sanitary and storm sewers form an extensive network that runs near or beneath key buildings and roads and is physically close to many communication and transportation networks. Significant damage to the nation's wastewater facilities or collection systems could result in: loss of life; catastrophic environmental damage to rivers, lakes, and wetlands; contamination of drinking water supplies; long-term public health impacts; destruction of fish and shellfish production; and disruption to commerce, the economy, and our normal way of life.

Some actions should be taken at all times to reduce the possibility of a terrorist attack:

- Ensure that all visitors sign in and out of the facilities with a positive ID check.

- Reduce the number of visitors to a minimum.

- Discourage parking by the public near critical buildings to eliminate the chances of car bombs.

- Be cautious with suspicious packages that arrive.

- Be aware of the hazardous chemicals used and how to defend against spills.

- Keep emergency numbers posted near telephones and radios.

- Patrol the facilities frequently, looking for suspicious activity or behavior.

- Maintain, inspect, and use your personal protective equipment (PPE) (hard hats, respirators).

The following recommendations by the EPA[8] include many straightforward, common-sense actions a utility can take to increase security and reduce threats from terrorism.

Guarding Against Unplanned Physical Intrusion

- Lock all doors and set alarms at your office, pumping stations, treatment plants, and vaults, and make it a rule that doors are locked and alarms are set.

- Limit access to facilities and control access to pumping stations and chemical and fuel storage areas, giving close scrutiny to visitors and contractors.

- Post guards at treatment plants and post "Employees Only" signs in restricted areas.

- Secure hatches, metering vaults, manholes, and other access points to the sanitary collection system.

- Increase lighting in parking lots, treatment bays, and other areas with limited staffing.

- Control access to computer networks and control systems and change the passwords frequently.

- Do not leave keys in equipment or vehicles at any time.

Making Security a Priority for Employees

- Conduct background security checks on employees at hiring and periodically thereafter.

- Develop a security program with written plans and train employees frequently.

- Ensure all employees are aware of established procedures for communicating with law enforcement, public health, environmental protection, and emergency response organizations.

- Ensure that employees are fully aware of the importance of vigilance and the seriousness of breaches in security.

- Make note of unaccompanied strangers on the site and immediately notify designated security officers or local law enforcement agencies.

- If possible, consider varying the timing of operational procedures so that, to anyone watching for patterns, the pattern changes.

---

8. Adapted from "What Wastewater Utilities Can Do Now to Guard Against Terrorist and Security Threats," US Environmental Protection Agency, Office of Wastewater Management, October 2001.

- Upon the dismissal of an employee, change pass codes and make sure keys and access cards are returned.

- Provide customer service staff with training and checklists of how to handle a threat if it is called in.

Coordinating Actions for Effective Emergency Response

- Review existing emergency response plans and ensure that they are current and relevant.

- Make sure employees have the necessary training in emergency operating procedures.

- Develop clear procedures and chains of command for reporting and responding to threats and for coordinating with emergency management agencies, law enforcement personnel, environmental and public health officials, consumers, and the media. Practice the emergency procedures regularly.

- Ensure that key utility personnel (both on and off duty) have access to critical telephone numbers and contact information at all times. Keep the call list up to date.

- Develop close relationships with local law enforcement agencies and make sure they know where critical assets are located. Ask them to add your facilities to their routine rounds.

- Work with local industries to ensure that their pretreatment facilities are secure.

- Report to county or state health officials any illness among the employees that might be associated with wastewater contamination.

- Immediately report criminal threats, suspicious behavior, or attacks on wastewater utilities to law enforcement officials and the nearest field office of the Federal Bureau of Investigation.

Investing in Security and Infrastructure Improvements

- Assess the vulnerability of the collection system, major pumping stations, wastewater treatment plants, chemical and fuel storage areas, outfall pipes, and other key infrastructure elements.

- Assess the vulnerability of the stormwater collection system. Determine where large pipes run near or beneath government buildings, banks, commercial districts, industrial facilities, or are next to major communication and transportation networks. Move as quickly as possible with the most obvious and cost-effective physical improvements, such as perimeter fences, security lighting, and tamper-proofing manhole covers and valve boxes.

- Improve computer system and remote operational security.

- Use local citizen watches.

- Seek financing for more expensive and comprehensive system improvements.

Following 9/11, the Department of Homeland Security (DHS) established a 5-tiered Homeland Security Advisory System to provide a national framework for notification about the nature and degree of terrorist threats. The system identified five threat levels, which were color coded, beginning with green and increasing in severity through blue, yellow, orange, and red.

In a coordinating effort, the US Environmental Protection Agency (EPA) detailed the steps drinking water and wastewater utilities (water utilities) should consider implementing for each of the five threat levels to guard against terrorist and security threats. For each level, the measures focus on detection, preparedness, prevention, and protection. While the color-coded threat levels have been replaced, EPA's suggested protective measures are still valid. From the lowest risk of terrorist attack to the highest risk of terrorist attack, they are summarized as follows:

- Ongoing facility assessments; and the development, testing, and implementation of emergency plans

- Activating employee and public information plans; exercising communication channels with response teams and local agencies; and reviewing and exercising emergency plans

- Increasing surveillance of critical facilities; coordinating response plans with allied utilities and response teams and local agencies; and implementing emergency plans, as appropriate

- Limiting facility access to essential staff and contractors, and coordinating security efforts with local law enforcement officials and the armed forces, as appropriate

- The decision to close specific facilities and the redirection of staff resources to critical operations

In 2011, the color-coded Homeland Security Advisory System was replaced with the National Terrorism Advisory System (NTAS). This new alert system will more effectively communicate information about terrorist threats by providing timely, detailed information to the public, government agencies, first responders, airports and other transportation hubs, and the private sector.

Under the new system, the DHS will coordinate with other federal entities to issue formal, detailed alerts when the federal government receives information about a specific or credible terrorist threat. These alerts will include a clear statement that there is an "imminent threat" or "elevated threat." The alerts also will provide a concise summary of the potential threat, information about actions being taken to ensure public safety, and recommended steps that individuals and communities, businesses, and governments can take.

The NTAS alerts will be based on the nature of the threat: in some cases, alerts will be sent directly to law enforcement or affected areas of the private sector, while in others, alerts will be issued more broadly to the American people through both official and media channels—including a designated DHS webpage (www.dhs.gov/alerts), as well as social media channels including Facebook and Twitter @NTASAlerts. Additionally, NTAS will have a "sunset provision," meaning that individual threat alerts will be issued with a specified end date. Alerts may be extended if new information becomes available or if the threat evolves significantly. For more information on the National Terrorism Advisory System, visit www.dhs.gov/alerts.

To address the security of our nation's critical infrastructure, including its drinking water and wastewater systems, the US Environmental Protection Agency (EPA) has developed a series of Security Product Guides to assist treatment plant operators and utility managers in reducing risks from, and providing protection against, possible natural disasters and intentional terrorist attacks. These guides provide information on a variety of products:

- Physical security (such as walls, gates, and manhole locks to delay unauthorized entry into buildings or pipe systems)

- Electronic or cyber security (such as computer firewalls and remote monitoring systems that can report on outlying processes)

- Monitoring tools that can be used to identify anomalies in process streams or finished water that may represent potential threats

Individual products evaluated in these guides will be applicable to distribution systems, wastewater collection systems, pumping stations, treatment processes, main plant and remote sites, personnel entry, chemical delivery and storage, SCADA, and control systems for water and wastewater treatment systems. These EPA Security Product Guides are available online at http://cfpub.epa.gov/safewater/watersecurity/guide/.

QUESTION

Please write your answers to the following questions and compare them with those on page 391.

15.11A Under the National Terrorism Advisory System (NTAS), what actions will be taken by the Department of Homeland Security (DHS) in the event of a specific or credible terrorist threat?

15.12 SAFETY PROGRAM

The utility manager is responsible for the safety of the agency's personnel and the public exposed to the wastewater utility's operations. Therefore, the manager must develop and administer an effective safety program and must provide new employee safety training as well as ongoing training for all employees. The basic elements of a safety program include a safety policy statement, safety training and promotion, and accident investigation and reporting.

15.120 Policy Statement

A safety policy statement should be prepared by the top management of the utility. The purpose of the statement is to let employees know that the safety program has the full support of the agency and its management. The statement should achieve the following:

1. Define the goals and objectives of the program.

2. Identify the persons responsible for each element of the program.

3. Affirm management's intent to enforce safety regulations.

4. Describe the disciplinary actions that will be taken to enforce safe work practices.

Give a copy of the safety policy statement to every current employee and each new employee during orientation. Figure 15.2 is an example of a safety policy statement for a water supply utility.

SAFETY POLICY STATEMENT

It is the policy of the Las Vegas Valley Water District that every employee shall have a safe and healthy place to work. It is the District's responsibility; its greatest asset, the employees and their safety.

When a person enters the employ of the District, he or she has a right to expect to be provided a proper work environment, as well as proper equipment and tools, so that they will be able to devote their energies to the work without undue danger. Only under such circumstances can the association between employer and employee be mutually profitable and harmonious. It is the District's desire and intention to provide a safe workplace, safe equipment, proper materials, and to establish and insist on safe work methods and practices at all times. It is a basic responsibility of all District employees to make the SAFETY of human beings a matter for their daily and hourly concern. This responsibility must be accepted by everyone who works at the District, regardless of whether he or she functions in a management, supervisory, staff, or the operative capacity. Employees must use the SAFETY equipment provided; Rules of Conduct and SAFETY shall be observed; and, SAFETY equipment must not be destroyed or abused. Further, it is the policy of the Water District to be concerned with the safety of the general public. Accordingly, District employees have the responsibility of performing their duties in such a manner that the public's safety will not be jeopardized.

The joint cooperation of employees and management in the implementation and continuing observance of this policy will provide safe working conditions and relatively accident-free performance to the mutual benefit of all involved. The Water District considers the SAFETY of its personnel to be of primary importance, and asks each employee's full cooperation in making this policy effective.

Fig. 15.2 Safety policy statement
(Permission of Las Vegas Valley Water District)

15.121 Responsibilities

The following list of responsibilities for safety represents a typical list but may be incomplete if your agency is subject to stricter local, state, and/or federal regulations than what is shown here. Check with your safety professional.

Management's primary responsibility with regard to the safety program is to accomplish these tasks:

1. Formulate a written safety policy.

2. Provide a safe workplace.

3. Set achievable safety goals.

4. Provide adequate training.

5. Delegate authority to ensure that the program is properly implemented.

The manager is the key to any safety program. Implementation and enforcement of the program is the responsibility of the manager. The manager also has the responsibility to do the following:

1. Ensure that all employees are trained and periodically retrained in proper safe work practices.

2. Ensure that proper safety practices are implemented and continued as long as the policy is in effect.

3. Investigate all accidents and injuries to determine their cause; prepare and implement a written procedure for preventing reoccurrence of the accident.

4. Institute corrective measures where unsafe conditions or work methods exist.

5. Ensure that equipment, tools, and the work are maintained to comply with established safety standards.

6. Respond to and correct all identified unsafe conditions and activities.

The utility operators are the direct beneficiaries of a safety program. They share the responsibility to take these actions:

1. Observe prescribed work procedures with respect to personal safety and that of their co-workers.

2. Report any detected hazard to a manager immediately.

3. Report any accident, including a minor accident that causes minor injuries.

4. Report near-miss accidents so that hazards can be removed or procedures changed to avoid problems in the future.

5. Use all protective devices and safety equipment supplied to reduce the possibility of injury.

6. Refuse to perform unsafe orders.

15.122   First Aid

By definition, "first aid" means emergency treatment for injury or sudden illness, before regular medical treatment is available. Everyone in an organization should be able to give some degree of prompt treatment and attention to an injury.

First-aid training in the basic principles and practices of life-saving steps that can be taken in the early stages of an injury are available through the local Red Cross, Heart Association, local fire departments, and other organizations. Such training should be reinforced periodically so that the operator has a complete understanding of water safety, cardiopulmonary resuscitation (CPR), and other life-saving techniques. All operators need training in first aid, but it is especially important for those who regularly work with electrical equipment or must handle chlorine and other dangerous chemicals.

First aid has little to do with preventing accidents, but it has an important bearing upon the survival of the injured patient. A well-equipped first-aid chest or kit is essential for proper treatment. The kit should be inspected regularly by the safety officer to ensure that supplies are available when needed. First-aid kits should be prominently displayed throughout the treatment plant and in company vehicles. Special consideration must be given to the most hazardous areas of the plant such as shops, laboratories, and chemical handling facilities.

Regardless of size, each utility should establish standard operating procedures (SOPs) for first-aid treatment of injured personnel. All new operators should be instructed in the utility's first-aid program.

QUESTIONS

Please write your answers to the following questions and compare them with those on page 391.

15.12A   What are the utility manager's responsibilities with regard to safety?

15.12B   What should be included in a utility's policy statement on safety?

15.12C   What is first aid?

15.12D   First-aid training is most important for operators involved in what types of activities?

15.123   Hazard Communication Standard (HCS) and Worker Right-To-Know (RTK) Laws

Each year, thousands of new chemical compounds are produced for industrial, commercial, and household use. Frequently, the long-term effects of these chemicals are unknown. As a result, federal and state laws have been enacted to control all aspects of hazardous materials handling and use. These laws are more commonly known as Worker Right-To-Know (RTK) laws, which are enforced by OSHA. The law that has had the greatest impact is the *OCCUPATIONAL SAFETY AND HEALTH ACT OF 1970 (OSHA),*[9] Public Law 91-596, which took effect on December 29, 1970.

The intent of the OSHA safety regulations is to create a place of employment that is free from recognized hazards that could cause serious physical harm or death to an operator (or other employee). In many cases, the individual states have the authority under the OSHA standard to develop their own state RTK laws and most states have adopted their own laws. The Federal

---

9. *OSHA* (O-shuh).   The Williams-Steiger Occupational Safety and Health Act of 1970 (OSHA) is a federal law designed to protect the health and safety of workers, including collection system and treatment plant operators. The Act regulates the design, construction, operation, and maintenance of industrial plants and wastewater collection systems and treatment plants. The Act does not apply directly to municipalities, except in those states that have approved plans and have asserted jurisdiction under Section 18 of the OSHA Act. *However, contract operators and private facilities do have to comply with OSHA requirements.* Wastewater treatment plants have come under stricter regulation in all phases of activity as a result of OSHA standards. OSHA also refers to the federal and state agencies that administer the OSHA regulations.

OSHA Standard 29 *CFR*[10] 1910.1200—Hazard Communication forms the basis of most of these state RTK laws. Unfortunately, state laws vary significantly from state to state. The state laws that have been passed are at least as stringent as the federal standard and, in most cases, are even more stringent. State laws are also under continuous revision and, because a strong emphasis is being placed on hazardous materials and worker exposure, state laws can be expected to be amended in the future to apply to virtually everybody in the workplace. Managers should become familiar with both the state and federal OSHA regulations that apply to their organizations.

The basic elements of a hazard communication program are described in the following paragraphs.

1. Identify Hazardous Materials—While there are thousands of chemical compounds that could be considered hazardous, focus on the ones to which operators and other personnel in your utility are most likely to be exposed.

2. Obtain Chemical Information and Define Hazardous Conditions—Once the inventory of hazardous chemicals is complete, the next step is to obtain specific information on each of the chemicals. This information is generally incorporated into a standard format form called the *MATERIAL SAFETY DATA SHEET (MSDS).*[11] This information is commonly available from manufacturers. Many agencies request an MSDS when the purchase order is generated and will refuse to accept delivery of the shipment if the MSDS is not included. OSHA's Hazard Communication Standard (HCS) specifies information that must be included on MSDSs, but does not require that any particular format be followed in presenting this information (see 29 CFR 1910.1200 (g)). Figure 15.3 shows OSHA's standard MSDS form, containing all of the required information. In order to promote consistent presentation of information, however, OSHA now recommends that MSDSs follow the 16-section format established by the American National Standards Institute (ANSI) standard for preparation of MSDSs. OSHA is preparing a guidance document that will include instructions for composing individual sections of the MSDS. This guidance document will be posted on the agency's website in the near future. For more information, see http://www.osha.gov/dsg/hazcom/msdsformat.html.

The purpose of the MSDS is to have a readily available reference document that includes complete information on common names, safe exposure level, effects of exposure, symptoms of exposure, flammability rating, type of first-aid procedures, and other information about each hazardous substance. Operators should be trained to read and understand the MSDS forms. The forms themselves should be stored in a convenient location where they are readily available for reference.

3. Properly Label Hazards—Once the physical, chemical, and health hazards have been identified and listed, a labeling and training program must be implemented. To meet labeling requirements on hazardous materials, specialized labeling is available from a number of sources, including commercial label manufacturers. Exemptions to labeling requirements do exist, so consult your local safety regulatory agency for specific details.

4. Train Operators—The last element in the hazard communication program is training and making information available to utility personnel. A common-sense approach eliminates the confusing issue of which of the thousands of substances operators should be trained for, and concentrates on those that they will be exposed to or use in everyday maintenance routines.

The Hazard Communication Standard and the individual state requirements are obviously a very complex set of regulations. Remember, however, that the ultimate goal of these regulations is to provide additional operator protection. These standards and regulations, once the intent is understood, are relatively easy to implement.

15.124 Lockout/Tagout Procedures (OSHA 29 CFR 1910.147, "Control of Hazardous Energy")

Wastewater treatment plant operators are frequently required to work on mechanical, hydraulic, and electrical systems. All of these have the potential of causing serious injury unless some action is taken to prevent the accidental start-up of a system while maintenance work is being performed.

Many forms of energy are present in a typical wastewater collection and treatment system: electrical energy in a pump station, mechanical energy in a surge-relief valve, thermal energy from high temperatures, and energy associated with hydraulic pressure such as static head pressure in a pipeline. Typical equipment systems where these hazards exist are motor control centers, pumps, belts, and spring-loaded devices such as check valves. It is essential for the operator to take appropriate precautions to prevent the accidental discharge of energy from these systems because it could cause serious injury or death. Before performing any maintenance, be sure you thoroughly understand the systems you are working on and the precautions you need to take.

The following is a general lockout/tagout procedure that you can modify for your specific needs when performing maintenance where potential hazards exist.

---

10. *Code of Federal Regulations (CFR).* A publication of the US government that contains all of the proposed and finalized federal regulations, including safety and environmental regulations.

11. *Material Safety Data Sheet (MSDS).* A document that provides pertinent information and a profile of a particular hazardous substance or mixture. An MSDS is normally developed by the manufacturer or formulator of the hazardous substance or mixture. The MSDS is required to be made available to employees and operators or inspectors whenever there is the likelihood of the hazardous substance or mixture being introduced into the workplace. Some manufacturers are preparing MSDSs for products that are not considered to be hazardous to show that the product or substance is not hazardous.

Material Safety Data Sheet

May be used to comply with
OSHA's Hazard Communication Standard,
29 CFR 1910.1200 Standard must be
consulted for specific requirements.

U.S. Department of Labor

Occupational Safety and Health Administration
(Non-Mandatory Form)
Form Approved
OMB No. 1218-0072

IDENTITY *(As Used on Label and List)*

Note: Blank spaces are not permitted. If any item is not applicable, or no information is available, the space must be marked to indicate that.

Section I

Manufacturer's Name	Emergency Telephone Number
Address *(Number, Street, City, State, and ZIP Code)*	Telephone Number for Information
	Date Prepared
	Signature of Preparer *(optional)*

Section II — Hazardous Ingredients/Identity Information

Hazardous Components (Specific Chemical Identity; Common Name(s))	OSHA PEL	ACGIH TLV	Other Limits Recommended	% *(optional)*

Section III — Physical/Chemical Characteristics

Boiling Point		Specific Gravity (H_2O = 1)	
Vapor Pressure (mm Hg)		Melting Point	
Vapor Density (AIR = 1)		Evaporation Rate (Butyl Acetate = 1)	
Solubility in Water			
Appearance and Odor			

Section IV — Fire and Explosion Hazard Data

Flash Point (Method Used)	Flammable Limits	LEL	UEL
Extinguishing Media			
Special Fire Fighting Procedures			
Unusual Fire and Explosion Hazards			

(Reproduce locally)

OSHA 174, Sept. 1985

Fig. 15.3   OSHA's standard Material Safety Data Sheet

Section V — Reactivity Data

Stability	Unstable		Conditions to Avoid
	Stable		

Incompatibility *(Materials to Avoid)*

Hazardous Decomposition or Byproducts

Hazardous Polymerization	May Occur		Conditions to Avoid
	Will Not Occur		

Section VI — Health Hazard Data

Route(s) of Entry:	Inhalation?	Skin?	Ingestion?

Health Hazards *(Acute and Chronic)*

Carcinogenicity:	NTP?	IARC Monographs?	OSHA Regulated?

Signs and Symptoms of Exposure

Medical Conditions
Generally Aggravated by Exposure

Emergency and First Aid Procedures

Section VII — Precautions for Safe Handling and Use

Steps to Be Taken in Case Material is Released or Spilled

Waste Disposal Method

Precautions to Be Taken in Handling and Storing

Other Precautions

Section VIII — Control Measures

Respiratory Protection *(Specify Type)*

Ventilation	Local Exhaust	Special
	Mechanical *(General)*	Other

Protective Gloves	Eye Protection

Other Protective Clothing or Equipment

Work/Hygienic Practices

* U.S.G.P.O.:1986-491-529/45775

Fig. 15.3   OSHA's standard Material Safety Data Sheet (continued)

Whenever it is necessary to work on a piece of equipment or machinery, the following procedure must be adhered to:

1. The main power source must be locked out, in the OFF position, with a multiple lockout device and padlock. (*NOTE:* the pulling of fuses must not be considered a substitute for locking out.)

2. All pneumatic, hydraulic, and other fluid lines must be bled, drained, purged, or blanked off to prevent pressure or contents from causing movement and mechanisms under spring tension or compression must be blocked, clamped, or chained in position. Blocks may also be needed on some machinery or equipment to prevent gravitational movement.

3. Never place locks where the disconnect can be bypassed at other locations.

4. Multiple lockout devices must be used in all cases. Each employee who is engaged in working on machinery or equipment must install a lock on the multiple lockout device. Each lock must be individually keyed to prevent unauthorized removal of the lockout device. The operator using the padlock is the only person normally authorized to remove it (see step 9 below).

5. A tagging system must be used that will advise other employees of the work being performed.

6. *CAUTION:* It is mandatory that the disconnect or valve be tried to make sure it cannot be moved to ON. In addition, the machine controls themselves must be tried to make certain that the energy is OFF.

7. This procedure must be followed at the start of each shift or workday.

8. The immediate supervisor must check to determine that all requirements of this procedure have been complied with and must give the personnel involved the clearance to proceed with the required task.

9. In case of emergency, supervisory personnel may authorize removal of the lock after all precautions have been taken. An attempt must be made to contact the person who initially locked out the equipment (see step 5). A check shall be made to ensure that all personnel, tools, and other items will not be exposed to harm or injury before the lock is removed. The operator signing the tag must be notified of the removal of the lock at the start of the operator's next scheduled workday.

Your agency should develop a site-specific written lockout/tagout procedure and provide training to all of the operators who may use it. Each operator who must use lockout/tagout to safely do the job must be aware of applicable energy sources and the methods necessary for their effective isolation and control.

If you hire a contractor, you must inform each other of your respective lockout/tagout procedures. You must ensure that you and your employees or co-workers comply with the restrictions and prohibitions of the contractor's program.

Periodic inspections of your lockout/tagout program are also required (at least annually) to ensure that the requirements of the program are being followed. The inspection(s) must be done by someone other than the one(s) using the procedure.

15.125   Confined Space Entry Procedures

CONFINED SPACES[12,13] pose significant risks for a large number of workers, including many utility operators. OSHA has therefore defined very specific procedures to protect the health and safety of operators whose jobs require them to enter or work in a confined space. The regulations (which can be found in the Code of Federal Regulations at 29 CFR 1910.146) require conditions in the confined space to be tested and evaluated before anyone enters the space. If conditions exceed OSHA's limits for safe exposure, additional safety precautions must be taken and a confined space entry permit (Figure 15.4) must be approved by the appropriate authorities prior to anyone entering the space.

The managers of wastewater utilities may or may not be involved in the day-to-day details of enforcing the agency's confined space policy and procedures. However, every utility manager should be aware of the current OSHA requirements and should ensure that the utility's policies not only comply with current regulations, but that the agency's policies are vigorously enforced for the safety of all operators.

Managers should provide confined space entry training for all employees who will be working in or near confined spaces. It is a criminal offense to order an employee into a confined space without complying with all of the applicable regulations.

15.126   Reporting

Regardless of the size of the utility, recordkeeping is an important part of an effective safety program. All injuries should be reported, even if they are minor in nature, so as to establish a record in case the injury develops into a serious injury. It may be difficult at a later date to prove whether the accident occurred on or off the job. This information may determine who is responsible for the costs. The responsibility for reporting accidents affects several levels of personnel. First, of course, is the injured person. Next, it is the responsibility of the supervisor, and finally, it is the responsibility of management to review the causes of

---

12. *Confined Space.*   Confined space means a space that: (1) Is large enough and so configured that an employee can bodily enter and perform assigned work; and (2) Has limited or restricted means for entry or exit (for example, manholes, tanks, vessels, silos, storage bins, hoppers, vaults, and pits are spaces that may have limited means of entry); and (3) Is not designed for continuous employee occupancy. Also see DANGEROUS AIR CONTAMINATION and OXYGEN DEFICIENCY.

13. *Confined Space, Permit-Required (Permit Space).*   A confined space that has one or more of the following characteristics: (1) Contains or has a potential to contain a hazardous atmosphere, (2) Contains a material that has the potential for engulfing an entrant, (3) Has an internal configuration such that an entrant could be trapped or asphyxiated by inwardly converging walls or by a floor that slopes downward and tapers to a smaller cross section, or (4) Contains any other recognized serious safety or health hazard.

Date and Time Issued: _____ Date and Time Expires: _____ Job Site/Space I.D.: _____

Job Supervisor: _____ Equipment to be worked on: _____ Work to be performed: _____

Standby personnel: _____   _____   _____

1. Atmospheric Checks: Time _____ Oxygen _____ %   Toxic _____ ppm

   Explosive _____ % LEL   Carbon Monoxide _____ ppm

2. Tester's signature: _____

3. Source isolation: (No Entry)   N/A   Yes   No

   Pumps or lines blinded,
      disconnected, or blocked   ( )   ( )   ( )

4. Ventilation Modification: N/A   Yes   No

   Mechanical   ( )   ( )   ( )

   Natural ventilation only   ( )   ( )   ( )

5. Atmospheric check after isolation and ventilation: Time _____

   Oxygen _____ % > 19.5% < 23.5%   Toxic _____ ppm   < 10 ppm H_2S

   Explosive _____ % LEL   < 10%   Carbon Monoxide _____ ppm   < 35 ppm CO

Tester's signature: _____

6. Communication procedures: _____

7. Rescue procedures: _____

   _____

8. Entry, standby, and backup persons   Yes   No

   Successfully completed required training?   ( )   ( )

   Is training current?   ( )   ( )

9. Equipment: N/A   Yes   No

   Direct reading gas monitor tested   ( )   ( )   ( )

   Safety harnesses and lifelines for entry and standby persons   ( )   ( )   ( )

   Hoisting equipment   ( )   ( )   ( )

   Powered communications   ( )   ( )   ( )

   SCBAs for entry and standby persons   ( )   ( )   ( )

   Protective clothing   ( )   ( )   ( )

   All electric equipment listed for Class I, Division I,
      Groups A, B, C, and D, and nonsparking tools   ( )   ( )   ( )

10. Periodic atmospheric tests:

   Oxygen: ____% Time ____;   ____% Time ____;   ____% Time ____;   ____% Time ____;

   Explosive: ____% Time ____;   ____% Time ____;   ____% Time ____;   ____% Time ____;

   Toxic: ____ppm Time ____;   ____ppm Time ____;   ____ppm Time ____;   ____ppm Time ____;

   Carbon Monoxide: ____ppm Time ____;   ____ppm Time ____;   ____ppm Time ____;   ____ppm Time ____;

We have reviewed the work authorized by this permit and the information contained herein. Written instructions and safety procedures have been received and are understood. Entry cannot be approved if any brackets ( ) are marked in the "No" column. This permit is not valid unless all appropriate items are completed.

Permit Prepared By: (Supervisor) _____ Approved By: (Unit Supervisor) _____

Reviewed By: (CS Operations Personnel) _____

           (Entrant)   (Attendant)   (Entry Supervisor)

This permit to be kept at job site. Return job site copy to Safety Office following job completion.

Fig. 15.4   Confined space pre-entry checklist/entry permit

accidents and take steps to prevent such accidents from happening in the future.

Accident report forms may be very simple. However, they must record all details required by law and all data needed for statistical purposes. The forms shown in Figures 15.5 and 15.6 are examples for you to consider for use in your utility.

In addition to reports needed by the utility, other reports may be required by state or federal agencies. For example, vehicle accident reports must be submitted to local police departments. If a member of the public is injured, additional forms are needed because of possible subsequent claims for damages. If the accident is one of occupational injury, causing lost time, other reports may be required. Follow-up investigations to identify causes and responsibility may require the development of other specific types of record forms.

Emphasis on the prevention of future accidents cannot be overstressed. We must identify the causes of accidents and implement whatever measures are necessary to protect operators from becoming injured.

S SAFETY FIRST
A ACCIDENTS COST LIVES
F FASTER IS NOT ALWAYS BETTER
E EXPECT THE UNEXPECTED
T THINK BEFORE YOU ACT
Y YOU CAN MAKE THE DIFFERENCE

ACCIDENTS DON'T JUST HAPPEN...
THEY ARE CAUSED!

QUESTIONS

Please write your answers to the following questions and compare them with those on page 391.

15.12E   List the basic elements of a hazard communication program.

15.12F   What are a manager's responsibilities for ensuring the safety of operators entering or working in confined spaces?

15.12G   Why should a report be prepared whenever an injury occurs?

15.127   Training

If a safety program is to work well, management will have to accept responsibility for the following three components of training:

1. Safety education of all employees

2. Reinforced education in safety

3. Safety education in the use of tools and equipment

Or, to put it another way, the three most important controlling factors in safety are education, education, and more education.

Responsibility for overall training must be that of upper management. A program that will educate operators and then reinforce this education in safety must be planned systematically and promoted on a continuous basis. There are many avenues to achieving this goal.

The safety education program should start with the new operator. Even before employment, verify the operator's past record and qualifications and review the pre-employment physical examination. In the new operator's orientation, include instruction in the importance of safety at your utility or plant. Also, discuss the matter of proper reporting of accidents as well as the organization's policies and practices. Give new operators copies of all safety standard operating procedures (SOPs) and direct their attention to parts that directly involve them. Ask the safety officer to give a talk about utility policy, safety reports, and past accidents, and to orient the new operator toward the importance of safety to operators and to the organization.

The next consideration must be one of training the new operator in how to perform assigned work. Most supervisors think in terms of On-the-Job Training (OJT). However, OJT is not a good way of preventing accidents with an inexperienced operator. The idea is all right if the operator comes to the organization trained in how to perform the work, such as a treatment operator from another plant. Then you only need to explain your safety program and how your policies affect the new operator. For a new operator who is inexperienced in utility operation, the supervisor must give detailed consideration to the operator's welfare. In this instance, the training should include not only a safety talk, but the crew chief or supervisor must train the inexperienced operator in all aspects of treatment plant safety. This training includes instruction in the handling of chemicals, the dangers of electrical apparatus, fire hazards, and proper maintenance of equipment to prevent accidents. Special instructions will also be needed for specific work environments such as manholes, gases (chlorine and hydrogen sulfide (H_2S)), water safety, and any specific hazards that are unique to your facility. The new operator must be checked out on any equipment personnel may operate such as vehicles, forklifts, valve operators, and radios. All new operators should be required to participate in a safety orientation program during the first few days of their employment, and an overall training program in the first few months.

Date _____

Name of injured employee _____ Employee # _____ Area _____

Date of accident _____ Time _____ Employee's Occupation _____

Location of accident _____ Nature of injury _____

Name of doctor _____ Address _____

Name of hospital _____ Address _____

Witnesses (name & address) _____

PHYSICAL CAUSES

Indicate below by an "X" whether, in your opinion, the accident was caused by:

_____ Improper guarding   _____ No mechanical cause

_____ Defective substances or equipment   _____ Working methods

_____ Hazardous arrangement   _____ Lack of knowledge or skill

_____ Improper illumination   _____ Wrong attitude

_____ Improper dress or apparel   _____ Physical defect

_____ Not listed (describe briefly) _____

UNSAFE ACTS

Sometimes the injured person is not directly associated with the causes of an accident. Using an "X" to represent the injured worker and an "O" to represent any other person involved, indicate whether, in your opinion, the accident was caused by:

_____ Operating without authority   _____ Unsafe loading, placement & etc.

_____ Failure to secure or warn   _____ Took unsafe position

_____ Working at unsafe speed   _____ Worked on moving equipment

_____ Made safety device inoperative   _____ Teased, abused, distracted & etc.

_____ Unsafe equipment or hands instead of equipment   _____ Did not use safe clothing or personal protective equipment

_____ No unsafe act

_____ Not listed (describe briefly) _____

What job was the employee doing? _____

What specific action caused the accident? _____

What steps will be taken to prevent recurrence? _____

_____

_____

Date of Report _____ Immediate Supervisor _____

REVIEWING AUTHORITY

Comments: Comments:

_____   _____
Safety Officer   Date   Department Director   Date

Fig. 15.5   Supervisor's accident report

INJURED: COMPLETE THIS SECTION

Name _____ Age_____ Sex _____

Address _____

Title _____ Dept. Assigned _____

Place of Accident _____

Street or Intersection _____

Date _____ Hour_____ A.M. _____ P.M. _____

Type of Job You Were Doing When Injured

_____

Object Which Directly Injured You	Part of Body Injured

How Did Accident Happen? (Be specific and give details; use back of sheet if necessary.)

_____

First Aid Administered

_____

	Yes □	No □
Did You Report Accident or Exposure at Once? (Explain "No")	Yes □	No □
Did You Report Accident or Exposure to Supervisor? Give Name	Yes □	No □
Were There Witnesses to Accident or Exposure? Give Names	Yes □	No □
Did You See a Doctor? (If Yes, Give Name)	Yes □	No □
Are You Going to See a Doctor? (Give Name)	Yes □	No □

_____   _____
           Date   Signature

SUPERVISOR: COMPLETE THIS SECTION — (Return to Personnel as Soon as Possible)

	Yes □	No □
Was an Investigation of Unsafe Conditions and/or Unsafe Acts Made? If Yes, Please Submit Copy.	Yes □	No □
Was Injured Intoxicated or Behaving Inappropriately at Time of Accident? (Explain "Yes")	Yes □	No □

Date Disability   Last Day   Date Back
Commenced _____   Wages Earned _____   on Job _____

Date Report Completed _____ 20 ____   Signed By _____

Title _____

Distribution: Canary - Department Head, Pink - Supervisor, White - Personnel

Fig. 15.6 Accident report

The next step in safety education is reinforcement. Even if the operator is well trained, mistakes can occur; therefore, the education must be continual. Many organizations use the *TAILGATE SAFETY MEETING*[14] method as a means of maintaining the operator's interest in safety. The program should be conducted by the first line supervisor. Schedule the informal tailgate meeting for a suitable location, keep it short, avoid distractions, and be sure that everyone can hear. Hand out literature, if available. Tailgate talks should communicate to the operator specific considerations, new problems, and accident information. These topics should be published. One resource for such meetings can be those operators who have been involved in an accident. Although it is sometimes embarrassing to the injured, the victim is now the expert on how the accident occurred, what could have been done to prevent it, and how it felt to have the injury. Encourage all operators, new and old, to participate in tailgate safety sessions.

Use safety posters to reinforce safety training and to make operators aware of the location of dangerous areas or show the importance of good work habits. Such posters are available through the National Safety Council's catalog.[15]

Awards for good safety records are another means of keeping operators aware of the importance of safety. The awards could be given to individuals in recognition of a good safety record. Publicity about the awards may provide an incentive to the operators and demonstrates the organization's determination to maintain a good safety record. The awards may include safety lapel pins, certificates, or plaques showing number of years without an accident. Consider publishing a utility newsletter on safety tips or giving details concerning accidents that may be helpful to other operators in the organization. Awards may be given to the organization in recognition of its effort in preventing accidents or for its overall safety program. A suggestion program concerning safety will promote and reinforce the program and give recognition to the best suggestions. The goal of all these efforts is to reinforce concerns for the safety of all operators. If safety, as an idea, is present, then accidents can be prevented.

Education of the operator in the use of tools and equipment is necessary. As pointed out above, OJT is not the answer to a good safety record. A good safety record will be achieved only with good work habits and safe equipment. If the operator is trained in the proper use of equipment (hand tools or vehicles), the operator is less likely to misuse them. However, if the supervisor finds an operator misusing tools or equipment, then it is the supervisor's responsibility to reprimand the operator as a means of reinforcing utility policies. The careless operator who misuses equipment is a hazard to other operators. Careless operators will also be the cause of a poor safety record in the operator's division or department.

An important part of every job should be the consideration of its safety aspects by the supervisor. The supervisor should instruct the crew chief or operators about any dangers involved in job assignments. If a job is particularly dangerous, then the supervisor must bring that fact to everyone's attention and clarify utility policy in regard to unsafe acts and conditions.

If the operator is unsure of how to perform a job, then it is the operator's responsibility to ask for the training needed. Each operator must think, act, and promote safety if the organization is to achieve a good safety record. Training is the key to achieving this objective and training is everyone's responsibility—management, the supervisors, crew chiefs, and operators.

QUESTIONS

Please write your answers to the following questions and compare them with those on page 391.

15.12H   A new, inexperienced operator must receive instruction on what aspects of treatment plant safety?

15.12I   What should an operator do if unsure of how to perform a job?

15.128   Measuring

To be complete, a safety program must also include some means of identifying, measuring, and analyzing the effects of the program. The systematic classification of accidents, injuries, and lost time is the responsibility of the safety officer. This person should use an analytical method that would refer to types and classes of accidents. Reports should be prepared using statistics showing lost time, costs, type of injuries, and other data, based on a specific time interval.

Such data call attention to the effectiveness of the program and promote awareness of the types of accidents that are happening. Management can use this information to decide where the emphasis should be placed to avoid accidents. However, statistical data are of little value if a report is prepared and then set on the bookshelf or placed in a supervisor's desk drawer. The data must be distributed and read by all operating and maintenance personnel.

As an example, injuries can be classified as fractures, burns, bites, eye injuries, cuts, and bruises. Causes can be classified as related to heat, machinery, falls, handling objects, chemicals, unsafe acts, and miscellaneous. Cost can be classified as lost time, lost dollars, lost production, contaminated water, or any other means of showing the effects of the accidents.

---

14. *Tailgate Safety Meeting.*   Brief (10 to 20 minutes) safety meetings held every 7 to 10 working days. The term comes from the safety meetings regularly held by the construction industry around the tailgate of a truck.

15. Write or call your local safety council or National Safety Council, 1121 Spring Lake Drive, Itasca, IL 60143-3201.

Good analytical reporting will provide a great deal of detail without a lot of paper to read and comprehend. Keep the method of reporting simple and easy to understand by all operators, so they can identify with the causes and be aware of how to prevent the accident happening to themselves or other operators. Table 15.4 gives one method of summarizing the causes of various types of injuries.

TABLE 15.4 SUMMARY OF TYPES AND CAUSES OF INJURIES

Type of Injury	Primary Cause of Injury										
	Unsafe Act	Chemical	Falls	Handling Objects	Heat	Machinery	Falling Objects	Stepping	Striking	Miscellaneous	TOTAL
Fractures											
Sprains											
Eye Injuries											
Bites											
Cuts											
Bruises											
Burns											
Miscellaneous											

There are many other methods of compiling data. Table 15.4 could reflect cost in dollars or in work hours lost. Not all accidents mean time lost, but there can be other cost factors. The data analysis should also indicate if the accidents involve vehicles, company personnel, the public, company equipment, loss of chemical, or other factors. Results also should show direct and indirect costs to the agency, operator, and the public.

Once the statistical data have been compiled, someone must be responsible for reviewing it in order to take preventive actions. Frequently, such responsibility rests with the safety committee. In fact, safety committees may operate at several levels, for example management committee, working committee, or an accident review board. In any event, the committee must be active, be serious, and be fully supported by management.

Another means of determining the effectiveness of your safety program is by reviewing your agency's injury statistics and calculating injury rates during a given time period, typically a year. Injury rates are based on 200,000 operator work hours as a standardized conversion factor representing a workforce of 100 full-time workers who work 40 hours per week for 50 weeks per year (assuming two weeks for vacation and holidays). Typically, rates that measure total injuries (injury frequency rate) and lost work days (injury severity rate) are the most useful. The formulas for calculating these rates are:

$$\text{Injury Frequency Rate} = \left(\frac{\text{Number of Injuries}}{\text{Number of Hours Worked During the Period Covered}}\right) \times 200,000$$

and

$$\text{Injury Severity Rate} = \left(\frac{\text{Number of Lost Work Days}}{\text{Number of Hours Worked During the Period Covered}}\right) \times 200,000$$

These standard formulas were developed by the Department of Labor's Bureau of Labor Statistics. They allow "apples-to-apples" comparisons because they take into account operator variations such as headcount and individual work hours, as well as unequal time periods. You can use these injury rates to identify trends and spot safety needs or concerns. You can also compare your rates to those of similar agencies and established sources such as the Bureau of Labor Statistics or the National Safety Council.

The injury severity rate is considered the better indicator of safety performance because it measures the more serious injuries involving lost work days. The injury frequency rate is a good general indicator, but is limited because it gives equal weighting to both major and minor injuries.

Calculating injury frequency rates and injury severity rates involves a three-step process: (1) compile the number of hours worked during the period covered, (2) compile the number of injuries or lost work days, and (3) apply the formula as shown in the following examples.

EXAMPLE 3

A rural water company had a total of four operator injuries during the previous year. The total work hours were 71,856. Calculate the injury frequency rate.

Known	Unknown
Number of Injuries/yr = 4 injuries/yr	Injury Frequency Rate
Number of Hours Worked/yr = 71,856 hr/yr	

Calculate the injury frequency rate.

$$\text{Injury Frequency Rate} = \frac{(\text{Number of Injuries/yr})(200,000)}{\text{Number of Hours Worked/yr}}$$

$$= \frac{(4 \text{ injuries/yr})(200,000)}{71,856 \text{ hr/yr}}$$

$$= 11.1$$

This can be expressed as 11.1 injuries per 100 full-time employees (or 200,000 hours worked).

EXAMPLE 4

Of the four injuries suffered by the operators in Example 3, one was a disabling injury. Calculate the injury frequency rate for the disabling injuries.

Known	Unknown
Number of Disabling Injuries/yr = 1 injury/yr	Injury Frequency Rate (Disabling Injuries)
Number of Hours Worked/yr = 71,856 hr/yr	

Calculate the injury frequency rate for disabling injuries.

$$\text{Injury Frequency Rate} \atop \text{(Disabling Injuries)} = \frac{(\text{No. of Disabling Injuries/yr})(200,000)}{\text{No. of Hours Worked/yr}}$$

$$= \frac{(1 \text{ injury/yr})(200,000)}{71,856 \text{ hr/yr}}$$

$$= 2.8$$

This can be expressed as 2.8 disabling injuries per 100 full-time employees (or 200,000 hours worked).

EXAMPLE 5

The water company described in Examples 3 and 4 experienced five lost work days in one year due to injuries while the operators worked 71,856 hours. Calculate the injury severity rate.

Known	Unknown
Number of Lost Work Days/yr = 5 days/yr	Injury Severity Rate
Number of Hours Worked/yr = 71,856 hr/yr	

Calculate the injury severity rate.

$$\text{Injury Severity Rate} = \frac{(\text{Number of Lost Work Days/yr})(200,000)}{\text{Number of Hours Worked/yr}}$$

$$= \frac{(5 \text{ Days/yr})(200,000)}{71,856 \text{ hr/yr}}$$

$$= 13.9$$

This can be expressed as 13.9 lost work days per 100 full-time employees (or 200,000 hours worked).

Note that by using a one-year time interval in the previous injury rate calculations, the results may be used by the safety officer in preparing an annual report.

15.129 Human Factors

First, you may ask, what is a human factor? Well, it is not too often that a safety text considers human factors as part of the safety program. However, if these factors are understood and emphasis is given to their practical application, then many accidents can be prevented. Human factors engineering is the specialized study of technology relating to the design of operator-machine interfaces. That is to say, it examines ways in which machinery might be designed or altered to make it easier to use, safer, and more efficient for the operator. We hear a lot about making computers more user friendly, but human factors engineering is just as important to everyday operation of other machinery in the everyday plant.

Many accidents occur because the operator forgets the human factors. The ultimate responsibility for accidents due to human factors belongs to the management group. However, this does not relieve the operator of the responsibility to point out the human factors as they relate to safety. After all, it is the operator using the equipment who can best tell if it meets all the needs for an interrelationship between an operator and a machine.

The first step in the prevention of accidents takes place in the plant design. Even with excellent designs, accidents can and do happen. However, every step possible must be taken during design to ensure a maximum effort of providing a safe plant environment. Most often, the operator has little to do with design, and, therefore, needs to understand human factors engineering to be able to evaluate these factors as the plant is being operated. As newer plants become automated, this type of understanding may even be more important.

Other contributing human factors are the operator's mental and physical characteristics. The operator's decision-making abilities and general behavior (response time, sense of alarm, and perception of problems and danger) are all important factors. Ideally, tools and machines should function as intuitive extensions of the operator's natural senses and actions. Any factors disrupting this flow of action can cause an accident. Therefore, be on the lookout for such factors. When you find a system that cannot be operated in a smooth, logical sequence of steps, change it. You may prevent an accident. If the everyday behavior of an operator is inappropriate with regard to a specific job, reconsider the assignment to prevent an accident.

The human factor in safety is the responsibility of design engineers, supervisors, and operators. However, the operator who is doing the work will have a greater understanding of the operator-machine interface. For this reason, the operator is the appropriate person to evaluate the means of reducing the human factor's contribution to the cause of accidents, thereby improving the plant's safety record.

QUESTIONS

Please write your answers to the following questions and compare them with those on page 391.

15.12J   Statistical accident reports should contain what types of accident data?

15.12K   How can injuries be classified?

15.12L   How can causes of injuries be classified?

15.12M   How can costs of accidents be classified?

15.13   RECORDKEEPING

15.130   Purpose of Records

Accurate records are essential for effective utility management and to satisfy legal requirements. Records are also a valuable source of information. They can save time when trouble develops and provide proof that problems were identified and solved. Pertinent and complete records should be used as a basis for plant operation, interpreting the results of wastewater treatment, preparing preventive maintenance programs, and preparing budget requests. When accurately kept, records provide a sound basis for design of future changes or expansions of the treatment plant. If legal questions or problems occur in connection with the treatment processes or the operation of the plant, accurate and complete records will provide evidence of what actually occurred and what procedures were followed.

15.131   Types of Records

Many different types of records are required for effective management and operation of water supply, treatment, and distribution system facilities or wastewater collection, treatment, and disposal facilities. The following Sections, 15.132 through 15.136, describe some of the most important types of records that should be kept and Section 15.137 discusses how long records need to be kept.

15.132   Equipment and Maintenance Records

A good plant maintenance effort depends heavily on good recordkeeping. You will need to keep accurate records to monitor the operation and maintenance of each piece of plant equipment. Equipment control cards and maintenance work orders serve the following functions:

- Record important equipment data such as make, model, serial number, and date purchased.
- Record maintenance and repair work performed to date.
- Anticipate preventive maintenance needs.
- Schedule future maintenance work.

Whenever a piece of equipment is changed, repaired, or tested, the work performed should be recorded on an equipment history card of some type or in a computer maintenance program. Complete, up-to-date equipment records will enable the plant operators to evaluate the reliability of equipment and will provide the basis for a realistic preventive maintenance program.

15.133   Plant Operations Data

Plant operations logs can be as different as the treatment plants whose information they record. The differences in amount, nature, and format of data are so significant that any attempt to prepare a typical log would be very difficult. For detailed information and example recordkeeping forms for a water supply utility, see *Water Treatment Plant Operation,* Volume I, Chapter 10, "Plant Operation," Section 10.6, "Operating Records and Reports." For further information regarding the types of daily and monthly operating records required for a wastewater treatment plant, see *Operation of Wastewater Treatment Plants,* Volume II,

Chapter 19, "Records and Report Writing," Section 19.11, "Type of Records."

15.134   Procurement Records

Ordering repair parts and supplies usually is done when the on-hand quantity of a stocked part or chemical falls below the reorder point, a new item is added to stock, or an item has been requested that is not stocked. Most organizations require employees to submit a requisition (similar to the one shown in Figure 15.7) when they need to purchase equipment or supplies. When the requisition has been approved by the authorized person (a supervisor or purchasing agent, in most cases) the items are ordered using a form called a purchase order. A purchase order contains a number of important items, including (1) the date, (2) a complete description of each item and quantity needed, (3) prices, (4) the name of the vendor, and (5) a purchase order number.

A copy of the purchase order should be retained in a suspense file or on a clipboard until the ordered items arrive. This procedure helps keep track of the items that have been ordered but have not yet been received. The copy of the purchase order should be attached to the item delivery slip and then sent to an accounts payable procedure. This will ensure that items received will be paid for.

All supplies should be processed through the storeroom immediately upon arrival. When an item is received, it should be so recorded on an inventory card. The inventory card will keep track of the numbers of an item in stock, when last ordered, cost, and other information. Furthermore, by always logging in supplies immediately upon receipt, you are in a position to reject defective or damaged shipments and control shortages or errors in billing. Many utilities now use computer systems to keep track of orders and deliveries.

15.135   Inventory Records

An inventory consists of the supplies the treatment plant needs to keep on hand to operate the facility. These maintenance supplies may include repair parts, spare valves, electrical supplies, tools, and lubricants. The purpose of maintaining an inventory is to provide needed parts and supplies quickly, thereby reducing equipment downtime and work delays.

In deciding what supplies to stock, keep in mind the economics involved in buying and stocking an item as opposed to depending on outside availability to provide needed supplies. Is the item critical to continued plant or process operation? Should certain frequently used repair parts be kept on hand? Does the item have a shelf life?

Inventory costs can be held to a minimum by keeping on hand only those parts and supplies for which a definite need exists or that would take too long to obtain from an outside vendor. A definite need for an item is usually demonstrated by a history of regular use. Some items may be infrequently used but may be vital in the event of an emergency; these items should also be stocked. Take care to exclude any parts and supplies that may become obsolete, and do not stock parts for equipment scheduled for replacement.

P - 1

CITY OF SACRAMENTO

REQUISITION

1 BID NO.	2 PURCHASE ORDER NO.
DUE DATE	DATE
9 REQUISITION NO.	10 REFER QUESTIONS TO:
DATE	

3 DELIVER TO:	4 FUND GP.	5 FUND	6 ORGAN.	7 REV. COST CTR.	8 OBJECT

11 ITEM NO.	12 COMMODITY CODE NO.	13 DESCRIPTION	14 QUANTITY / UNIT	15 PRICE / UNIT	16 BRAND OFFERED (VENDORS: USE THESE COLS. ONLY)	17 LOT PRICE FOR EACH ITEM (VENDORS: USE THESE COLS. ONLY)

18 DEPT.

19 ORIG. BY

TEL. NO.

THIS REQUISITION WILL BE REPRODUCED TO CREATE A PURCHASE ORDER QUICKLY AND PROVIDE FASTER DELIVERY OF THESE ITEMS TO YOUR DEPARTMENT. PLEASE FOLLOW INSTRUCTIONS CAREFULLY.

SUB TOTAL

SALES TAX

INVOICE AMOUNT

20 EST. OF COST

22 SPECIAL REQUIREMENTS OR INSTRUCTIONS. PREVIOUS P. O. NO. OR BID NO., OR RECOMMENDED VENDORS

21 NUMBER OF REQUISITION ATTACHMENT SHEETS

23 CERTIFICATION IS HEREBY MADE THAT THE ABOVE IS A LEGAL CHARGE AGAINST THE APPROPRIATION INDICATED

DEPT. HEAD ________ BY ________

____
OFFICIAL TITLE

24

VENDOR ________ (A)

ADDRESS ________ (B)

CITY, STATE, AND ZIP ________ (C)

TERMS ________ TEL. NO. ________ (D)

DELIVERY ________ FOB DESTINATION ________ (E)

BY ________ TITLE ________

Fig. 15.7 Requisition/purchase order form

15.136 Personnel Records

Documentation of all aspects of personnel management provides an important measure of legal protection for the utility. If an employee files a lawsuit alleging discriminatory hiring practices or treatment, harassment, breach of contract, or other grievance, the utility's ability to defend its practices and procedures will depend almost entirely on complete and accurate records. Similarly, if an employee is injured on the job, written records can help establish whether the utility was responsible for the accident.

Each personnel action should be fully documented in writing and filed. Even verbal discussions of a supervisor with an employee about job performance should be summarized in writing upon completion of the conversation. Also, file copies of all written warnings and disciplinary actions.

An employee's personnel file should also contain a complete record of accomplishments, certificates earned, commendations received, and formal performance reviews.

As previously mentioned, personnel records often contain sensitive, confidential information; therefore, access to these records should be closely controlled.

15.137 Disposition of Plant Records

An important question is how long records should be kept. As a general rule, records should be kept for as long as they may be useful or as long as legally required. Some information will become useless after a short time, while other data may be valuable for many years. Data that might be used for future design or expansion should be kept indefinitely. Laboratory data will always be useful and should be kept indefinitely. Regulatory agencies may require you to keep certain water quality analyses (bacteriological test results) and customer complaint records on file for specified time periods (10 years for chemical analyses and bacteriological tests).

Even if old records are not consulted every day, this does not lessen their potential value. For orderly records handling and storage, set up a schedule to periodically review old records and to dispose of those records that are no longer needed. A decision can be made when a record is established regarding the time period for which it must be retained.

QUESTIONS

Please write your answers to the following questions and compare them with those on page 391.

15.13A　What are some of the benefits of keeping complete, up-to-date records?

15.13B　List the important items usually contained on a purchase order.

15.13C　What is the purpose of maintaining an inventory?

15.13D　As a general rule, how long should utility records be kept?

15.14　ACKNOWLEDGMENTS

During the writing of the material in this chapter, Lynne Scarpa, Phil Scott, Chris Smith, and Rich von Langen, all members of California Water Environment Association (CWEA), provided many excellent materials and suggestions for improvement. Their generous contributions are greatly appreciated.

15.15　ADDITIONAL READING

1. *Water Utility Management* (M5). Obtain from American Water Works Association (AWWA), Bookstore, 6666 West Quincy Avenue, Denver, CO 80235. Order No. 30005. ISBN 978-1-58321-361-2. Price to members, $78.50; non-members, $126.50; price includes cost of shipping and handling.

2. *A Water and Wastewater Manager's Guide for Staying Financially Healthy,* US Environmental Protection Agency. EPA No. 430-9-89-004. Obtain from National Technical Information Service (NTIS), 5301 Shawnee Road, Alexandria, VA 22312. Order No. PB90-114455. Price, $33.00, plus $6.00 shipping and handling.

3. *Wastewater Utility Recordkeeping, Reporting and Management Information Systems,* US Environmental Protection Agency. EPA No. 430-9-82-006. Obtain from National Technical Information Service (NTIS), 5301 Shawnee Road, Alexandria, VA 22312. Order No. PB83-109348. Price, $48.00, plus $6.00 shipping and handling.

4. *Supervision: Concepts and Practices of Management,* 12th Edition, Edwin Leonard, Jr. Obtain from Cengage Learning, Attn.: Order Fulfillment, PO Box 6904, Florence, KY 41022-6904. ISBN 978-1-111-96979-0. Price, $160.49, plus shipping and handling.

END OF LESSON 2 OF 2 LESSONS

on

MANAGEMENT

Please answer the discussion and review questions next.

DISCUSSION AND REVIEW QUESTIONS

Chapter 15.　MANAGEMENT

(Lesson 2 of 2 Lessons)

Please write your answers to the following questions to determine how well you understand the material in the lesson. The question numbering continues from Lesson 1.

16. What can happen when agencies rely primarily on corrective maintenance to keep the system running?

17. What does a SCADA system do?

18. How does an air gap separation system prevent contamination of the potable water supply at a wastewater treatment system?

19. What are the main advantages of a geographic information system's ability to generate work orders?

20. How would you assess the vulnerability of a utility system?

21. Prepare an outline that could be used as the basis for developing an emergency operations plan.

22. What is the intent of the OSHA safety regulations?

23. List four major types of records a utility must maintain.

24. Why is it important to document all aspects of personnel management?

SUGGESTED ANSWERS

Chapter 15.   MANAGEMENT

Answers to questions on page 351.

15.0A   Local community demands on a utility manager include protection from environmental disasters with a minimum investment of money.

15.0B   In the environmental field, as well as other fields, the workforce is changing. Most workplaces are becoming increasingly diverse as people from different genders, races, cultures, ethnic origins, and lifestyles find themselves working together. Changes in the environmental workplace also are created by advances in technology.

Answers to questions on page 353.

15.1A   The functions of a utility manager include planning, organizing, staffing, directing, and controlling.

15.1B   In small communities, the community depends on the utility manager to handle everything.

Answers to questions on page 354.

15.2A   Utility planning must include operational personnel, local officials (decision makers), and the public.

15.3A   The purpose of an organizational chart is to show who reports to whom and to identify the lines of authority.

15.3B   Delegation is uncomfortable for many managers because it requires giving up power and responsibility. Many managers believe that they can do the job better than others, they believe that other employees are not well trained or experienced, and they are afraid of mistakes. The utility manager retains some responsibility even after delegating to another employee and, therefore, the manager is often reluctant to delegate or may delegate the responsibility but not the authority to get the job done.

15.3C   An important and often overlooked part of delegation is follow-up by the supervisor.

Answers to questions on page 356.

15.4A   Staffing responsibilities include hiring new employees, training employees, and evaluating job performance.

15.4B   The two personnel management concepts a manager should always keep in mind are "job-related" and "documentation."

15.4C   The steps involved in a staffing analysis include:

1. List the tasks to be performed.
2. Estimate the number of staff hours per year required to perform each task.
3. List the utility's current employees.
4. Assign tasks based on each employee's skills and abilities.
5. Adjust the work assignments as necessary to achieve the best possible fit between the work to be done and the personnel/skills available to do it.

15.4D   A qualifications profile is a clear statement of the knowledge, skills, and abilities a person must possess to perform the essential job duties of a particular position.

15.4E   A new employee's safety training should begin on the first day of employment or as soon thereafter as possible.

Answers to questions on page 358.

15.5A   A manager needs both written and oral communication skills.

15.5B   The most common written documents that a utility manager must write include memos, business letters, press releases, monitoring reports, and the annual report.

15.5C   The annual report should be a review of what and how the utility operated during the past year and also the goals for the next year.

Answers to questions on page 359.

15.6A   A utility manager may be asked to conduct meetings with employees, the governing board, the public, and with other professionals in your field.

15.6B   Before a meeting, (1) prepare an agenda and distribute it to all participants, (2) find an adequate meeting room, and (3) set a beginning and ending time.

Answers to questions on page 361.

15.7A   The first step in organizing a public relations campaign is to establish objectives so you will have a clear idea of what you expect to achieve.

15.7B   Employees can be informed about the utility's plans, practices, and goals through newsletters, bulletin boards, and regular, open communication between supervisors and subordinates.

15.7C   Newspapers give more thorough, in-depth coverage to stories than do the broadcast media.

15.7D   Practice is the key to effective public speaking.

15.7E   Complaints can be a valuable asset in determining consumer acceptance and pinpointing problems. Customer calls are frequently the first indication that something may be wrong. Responding to complaints and inquiries promptly can save the utility money and staff resources, and minimize customer inconvenience.

Answers to questions on page 365.

15.8A   The three main areas of financial management for a utility include providing financial stability for the utility, careful budgeting, and providing capital improvement funds for future utility expansion.

15.8B   The operating ratio for a utility is calculated by dividing total revenues by total operating expenses.

15.8C   It is important for a manager to get input from other personnel in the utility as well as community leaders as the budgeting process proceeds in order to gain support for the budget and to keep the budget on track once adopted.

15.8D   The basic ways for a utility to finance capital improvements are through general obligation bonds, revenue bonds, or loan funding programs.

15.8E   A revenue bond is a debt incurred by the community, often to finance utility improvements. This bond has no limit on the amount of funds available and the user charges provide repayment on the bond.

Answers to questions on page 367.

15.9A   The purpose of an O&M program is to operate and maintain the system in accordance with design specifications. This involves a variety of operational procedures, maintenance procedures, and prerequisites such as training and certification, along with resource management and organizational management.

15.9B   In general, maintenance activities can be classified as corrective maintenance, preventive maintenance, and predictive maintenance.

15.9C   The major elements of a good preventive maintenance program include the following:

1. Planning and scheduling
2. Records management
3. Spare parts management
4. Cost and budget control
5. Emergency repair procedures
6. Training program

Answers to questions on page 369.

15.9D   SCADA stands for Supervisory Control And Data Acquisition.

15.9E   A SCADA system collects, stores, and analyzes information about all aspects of operation and maintenance; transmits alarm signals, when necessary; and allows fingertip control of alarms, equipment, and processes. SCADA provides the information that operators and their supervisors need to solve minor problems before they become major incidents.

15.9F   Contamination through cross-connections has consistently caused more waterborne disease outbreaks in the United States than any other reported factor.

15.9G   Potential places where cross-connections could occur in a wastewater treatment plant include pump seals, feed water to boilers, hose bibs below grade where they may be subject to flooding with wastewater or sludges, or any other location where wastewater could contaminate a domestic water supply.

15.9H   The power of a geographic information system (GIS) is that information can be retrieved geographically. An operator can choose an area to look at and the system will display the requested section on the screen and show the attributes of the entities located on the map. A printed copy may also be requested.

Answers to questions on page 371.

15.10A   The first step toward an effective contingency plan for emergencies is to make an assessment of vulnerability. Then, a comprehensive plan of action can be developed and implemented.

15.10B   A detailed emergency operations plan is not needed since all types of emergencies cannot be anticipated and a complex response program can be more confusing than helpful.

15.10C   An emergency operations plan should include:

1. A vulnerability assessment
2. An inventory of organizational personnel
3. Provisions for a recovery operation
4. Provisions for training programs for operators in carrying out the plan
5. Inclusion of coordination plans with health, police, and fire departments
6. Establishment of a communications procedure
7. Provisions for protection of personnel, plant equipment, records, and maps

Answer to question on page 373.

15.11A   Under the National Terrorism Advisory System (NTAS), the Department of Homeland Security (DHS) will coordinate with other federal entities to issue formal, detailed alerts when the federal government receives information about a specific or credible terrorist threat. These alerts will include a clear statement that there is an "imminent threat" or "elevated threat." The alerts also will provide a concise summary of the potential threat, information about actions being taken to ensure public safety, and recommended steps that individuals and communities, businesses, and governments can take.

Answers to questions on page 374.

15.12A   The utility manager is responsible for the safety of the agency's personnel and the public exposed to the wastewater utility's operations. Therefore, the manager must develop and administer an effective safety program and must provide new employee safety training as well as ongoing training for all employees.

15.12B   A safety policy statement should do the following:

1. Define the goals and objectives of the program.
2. Identify the persons responsible for each element of the program.
3. Affirm management's intent to enforce safety regulations.
4. Describe the disciplinary actions that will be taken to enforce safe work practices.

15.12C   First aid means emergency treatment for injury or sudden illness, before regular medical treatment is available.

15.12D   First-aid training is most important for operators who regularly work with electrical equipment and those who must handle chlorine and other dangerous chemicals.

Answer to question on page 380.

15.12E   The basic elements of a hazard communication program include the following:

1. Identify hazardous materials.
2. Obtain chemical information and define hazardous conditions.
3. Properly label hazards.
4. Train operators.

15.12F   A utility manager may or may not be involved in the day-to-day details of enforcing the agency's confined space policy and procedures. However, every utility manager should be aware of the current OSHA requirements and should ensure that the utility's policies not only comply with current regulations, but that the agency's policies are vigorously enforced for the safety of all operators.

15.12G   All injuries should be reported, even if they are minor in nature, so as to establish a record in case the injury develops into a serious injury. It may be difficult at a later date to prove whether the accident occurred on or off the job. This information may determine who is responsible for the costs.

Answer to questions on page 383.

15.12H   A new, inexperienced operator must receive instruction on all aspects of treatment plant safety. This training includes instruction in the handling of chemicals, the dangers of electrical apparatus, fire hazards, and proper maintenance of equipment to prevent accidents. Special instructions are required for specific work environments such as manholes, gases (chlorine and hydrogen sulfide (H_2S)), water safety, and any specific hazards that are unique to your facility. All new operators should be required to participate in a safety orientation program during the first few days of employment, and an overall training program in the first few months.

15.12I   If an operator is unsure of how to perform a job, then it is the operator's responsibility to ask for the training needed.

Answers to questions on page 385.

15.12J   Statistical accident reports should contain accident statistics showing lost time, costs, type of injuries, and other data, based on a specific time interval.

15.12K   Injuries can be classified as fractures, burns, bites, eye injuries, cuts, and bruises.

15.12L   Causes of injuries can be classified as related to heat, machinery, falls, handling objects, chemicals, unsafe acts, and miscellaneous.

15.12M   Costs of accidents can be classified as lost time, lost dollars, lost production, contaminated water, or any other means of showing the effects of the accidents.

Answers to questions on page 388.

15.13A   Keeping complete, up-to-date records contributes to more effective utility management, helps to satisfy legal requirements, provides valuable operations and maintenance information, and assists in preparing budget requests. Accurate records provide a sound basis for design of future changes or expansions of the treatment plant. If legal questions or problems occur in connection with the treatment processes or the operation of the plant, accurate and complete records will provide evidence of what actually occurred and what procedures were followed.

15.13B   A purchase order usually contains (1) the date, (2) a complete description of each item and quantity needed, (3) prices, (4) the name of the vendor, and (5) a purchase order number.

15.13C   The purpose of maintaining an inventory is to provide needed parts and supplies quickly, thereby reducing equipment downtime and work delays.

15.13D   As a general rule, utility records should be kept for as long as they may be useful or as long as legally required.

APPENDIX

SMALL WASTEWATER SYSTEM
OPERATION AND MAINTENANCE
(VOLUME II)

Comprehensive Review Questions and Suggested Answers

How to Solve Wastewater System Arithmetic Problems

Wastewater Abbreviations

Wastewater Words

Subject Index

COMPREHENSIVE REVIEW QUESTIONS

This section was prepared to help you review the material in this manual. You may wish to use it when preparing for civil service and certification examinations.

Since you have already completed this course, please *DO NOT SEND* your answers to California State University, Sacramento.

The questions are divided into four types:

1. True-False

2. Best Answer

3. Multiple Choice

4. Short Answer

To work this section, complete these steps:

1. Write your answer to each question.

2. Compare your answers to the suggested answers at the end of this section.

3. If you missed a question, reread the material in the manual.

True-False

1. Algae produce oxygen during the night hours and use oxygen during sunlight hours.

   1. True
   2. False

2. During the day, algae use carbon dioxide (raising the pH), while at night they produce carbon dioxide (lowering the pH).

   1. True
   2. False

3. Heavy chlorination at the pond recirculation point can assist in odor control, but it will probably interfere with treatment.

   1. True
   2. False

4. The calculation of pond BOD removal efficiency is figured in terms of percentage of BOD removed.

   1. True
   2. False

5. Algae are usually present in the effluent from ponds with continuous discharges.

   1. True
   2. False

6. The conversion of dissolved and suspended material to settled solids is the main objective of high-rate activated sludge processes.

   1. True
   2. False

7. A high-quality activated sludge plant effluent requires attention, understanding, and good plant operation.

   1. True
   2. False

8. Excessive return flows can cause a hydraulic overload on the clarifier.

   1. True
   2. False

9. Controlling wastes dumped into the collection system requires a pretreatment facility inspection program to ensure compliance.

   1. True
   2. False

10. The RBC media should move with the direction of flow to provide better treatment of the wastewater.

    1. True
    2. False

11. Damage to RBC equipment will be minimized if required maintenance is done as soon as the need is identified.

    1. True
    2. False

12. Slime growths should be washed off the media.

    1. True
    2. False

13. Chlorination is the most economical process for disinfection of potable water and treated wastewater.

    1. True
    2. False

14. Determining the most probable number (MPN) of coliform group organisms present is a test for individual pathogenic microorganisms.

    1. True
    2. False

15. All persons handling chlorine should be thoroughly aware of its hazardous properties.

    1. True
    2. False

16. Never adjust a hypochlorinator pump while it is off because doing so will damage the pump.

1. True
2. False

17. Free water surface (FWS) wetland treatment systems are recommended for individual home sites.

1. True
2. False

18. The wetland bed is built after the septic tank is installed.

1. True
2. False

19. Odors are regularly a nuisance in properly loaded wetlands.

1. True
2. False

20. More labor usually is required to operate and maintain fixed subsurface drip system components in comparison to spray systems.

1. True
2. False

21. All metric system units of length, volume, and weight (mass) use factors of 10 to express larger or smaller quantities of these units.

1. True
2. False

22. The larger the number in front of the N (normality), the more concentrated the solution.

1. True
2. False

23. A sample that has a low color intensity also will have a low percent transmittance, but a high absorbance.

1. True
2. False

24. Never drink from a beaker or other laboratory glassware.

1. True
2. False

25. Good housekeeping practices make significant contributions to an effective safety program in an active chemical storage room.

1. True
2. False

26. Always pour acid into water, never water into acid, to avoid a reaction (explosion) that could cause chemical burns.

1. True
2. False

27. An employee can serve only one supervisor (unity of command).

1. True
2. False

28. Hiring new employees requires careful planning after the interview process.

1. True
2. False

29. Direct contact with people in your community is an effective tool in promoting a small wastewater utility, even though the audiences tend to be small.

1. True
2. False

30. Supervisory control and data acquisition (SCADA) systems are capable of analyzing data and providing operating, maintenance, regulatory, and annual reports.

1. True
2. False

31. It is a criminal offense to order an employee into a confined space without complying with all of the applicable regulations.

1. True
2. False

Best Answer (Select only the closest or best answer.)

1. What is the purpose of a pond outlet baffle?

1. To distribute influent in the pond
2. To measure and record flows into the pond and possibly the effluent
3. To prevent scum and other surface debris from flowing to the next pond or into receiving waters
4. To remove coarse material from the pond influent

2. What do algae live on in the wastewater in ponds?

1. Carbon dioxide and other nutrients
2. Hardness and also iron and manganese
3. Microscopic organic microorganisms
4. Pathogenic microorganisms

3. Why should recirculation water be drawn from the surface of the source pond?

1. To avoid short-circuiting through the pond
2. To ensure the highest possible dissolved oxygen levels in the recirculated water
3. To keep grease balls out of the plant effluent
4. To prevent the development of scum mats

4. Who should the operator contact before attempting to apply any insecticide or pesticide?

1. The local health officer responsible for the well being of the community
2. The local official in charge of approving pesticide applications
3. The official responsible for EPA's pesticide control program
4. The production manager of the pesticide manufacturer

5. When should batch-operated ponds be allowed to discharge?

   1. Only after sufficient detention time has been provided for the settling cycle
   2. Only during the nonrecreational season when flows are high in the receiving waters
   3. Only when sufficient detention time has been provided to allow complete treatment
   4. Only when sufficient staff is available to monitor the discharge

6. Why should the time of day for collecting pond samples be varied occasionally when taking grab samples?

   1. So the operator can become familiar with the pond's characteristics at various times of the day
   2. So the operator can detect the impact of illegal or illicit discharges on the contents of the pond
   3. So the operator can gain information to train new operators
   4. So the operator can obtain information about the environment surrounding the pond

7. What does a gray pond color indicate?

   1. A declining pH and a lowered dissolved oxygen content
   2. A high pH and a satisfactory dissolved oxygen content
   3. A pond that is being overloaded or that is not working properly
   4. A pond that is undergoing a spring "turnover"

8. Why should an operator review the plans and specifications for a proposed pond?

   1. To be sure the office has sufficient space for the as-built plans (record drawings)
   2. To ensure the public is aware of potential problems
   3. To invite qualified contractors to bid on the construction
   4. To suggest improvements and changes before construction begins that will allow the operators and the ponds to do their jobs better

9. What are facultative bacteria?

   1. Bacteria that can use either dissolved molecular oxygen or oxygen obtained from food materials such as sulfate or nitrate ions
   2. Solids in the aeration tank that carry microorganisms that feed on wastewater
   3. The rate at which organisms use the oxygen in wastewater while stabilizing decomposable organic matter under aerobic conditions
   4. Very small organisms that can be seen only through a microscope

10. When critical factors in the aeration tank are under proper control, what will the organisms in the tank do?

    1. Convert soluble solids and agglomerate the fine particles into a floc mass
    2. Increase the level of new organisms consistent with the wasting activated sludge pumping rate
    3. Reproduce at rates below the demand created by the food supply
    4. Selectively eliminate nonproductive organisms

11. What is the purpose of air lifts?

    1. To catch rags and large debris and prevent them from entering the aeration tank
    2. To pump and move settled activated sludge from one compartment to another by the use of air
    3. To supply air to the plant
    4. To transfer oxygen from air to water for respiration by microorganisms

12. What is bulking?

    1. A condition produced by anaerobic bacteria and, if severe, produces hydrogen sulfide, turns black, and gives off foul odors
    2. Billowing solids that result when the settling tank sludge blanket becomes too deep and solids are swept up by the water and out over the effluent weirs
    3. Clouds of billowing sludge that occur throughout secondary clarifiers and sludge thickeners when the sludge does not settle properly
    4. Low solids in the return sludge that produce an almost clear liquid

13. Why is the proper type and grade of lubricating oil critical?

    1. If the oil is too thin or thick, it will interfere with the proper functioning of bearings and gears
    2. It is essential to maintain equipment warranties
    3. It is necessary to prevent leaks into wastewater being treated
    4. Spilled oil can create slippery surfaces

14. Why is the RBC process divided into multiple stages?

    1. To allow operator flexibility in controlling effluent quality
    2. To maximize the effectiveness of a given amount of media surface by separating competing biomass populations
    3. To optimize RBC performance when flows fluctuate
    4. To provide provisions for plant expansion when flows or population increases

15. Why might the bulkhead or baffle between stages one and two be removed during periods of severe RBC organic overload?

    1. To provide a greater amount of media surface area for the first stage of treatment
    2. To provide a greater dilution volume to reduce the overload concentration
    3. To provide more space for the overload organic solids to settle out
    4. To provide short circuiting and reduce the contact time of the overload with the media

16. What are saprophytic organisms?

    1. Destruction of all microorganisms
    2. Destruction or inactivation of pathogenic organisms
    3. Organisms living on dead or decaying organic matter
    4. Organisms thriving in the absence of dissolved oxygen

17. What is the ultimate measure of the effectiveness of the disinfection process?

    1. The bacteriological test result
    2. The clarity of the final effluent
    3. The health of the receiving water aquatic organisms
    4. The public health of the community

18. What should be an operator's first action when he/she experiences a mild case of chlorine exposure?

    1. Leave the contaminated area for a location with fresh air
    2. Notify other operators
    3. Repair the leak
    4. Request others to fix the leak

19. How can operators reduce the problems caused by insoluble calcium hypochlorite (HTH) plugging pumps, pipes, and fittings?

    1. By heating the water used to dissolve the insoluble calcium hypochlorite (HTH) before mixing it
    2. By installing a filter screen upstream from the pumps, pipes, and fittings to capture the insoluble solids
    3. By mixing the HTH powder in a separate mixing tank and allowing the insoluble solids to settle out before using the mixture
    4. By shifting (screening) the HTH before mixing it with water to capture the insoluble solids

20. How can operators prevent the formation of calcium carbonate coatings on hypochlorinators?

    1. By adding vinegar to the influent hypochlorite solution
    2. By feeding lime to overdose the calcium carbonate saturation index
    3. By generating hypochlorite on site
    4. By obtaining the dilution water from an ordinary home water softener

21. What problem do operators encounter when troubleshooting chlorine solution pumps?

    1. Discharge flow varies
    2. High discharge of chlorine solution
    3. High discharge pressures
    4. Low discharge of chlorine solution

22. Where is the primary treated wastewater (septic tank effluent) applied in a wetland treatment system?

    1. At the bottom, across the inlet end of the wetland bed
    2. At the bottom, across the outlet end of the wetland bed
    3. At the top, across the inlet end of the wetland bed
    4. At the top, across the outlet end of the wetland bed

23. Odorous compounds emitted from open water wetland areas are typically associated with anaerobic conditions that can be created by which factors?

    1. Excessive BOD and ammonia loadings
    2. Excessive coliforms and pathogens
    3. Excessive sulfide and heavy metal loadings
    4. Excessive sulfide and nitrite loadings

24. How is the physical control of a land disposal system achieved?

    1. By adjusting the valves or gates used to direct treated effluent to different disposal areas
    2. By constructing ditches, berms, and dikes to control flow
    3. By controlling access to the land disposal system
    4. By preventing flows from actually leaving the disposal site

25. Why are native grasses recommended in land treatment systems?

    1. Because they are preferred by local birds and animals
    2. Because they are readily available
    3. Because they can easily reseed the area
    4. Because they can thrive under local variables such as soils, moisture, and seasons

26. What is an aliquot?

    1. A method of determining the relative toxicity of a test sample of industrial wastes by using live test organisms
    2. An equally divided portion of a sample
    3. Free from the living germs of disease, fermentation, or putrefaction
    4. Temperature of the surroundings

27. What is a bioassay?

    1. A method of determining the relative toxicity of a test sample of industrial wastes by using live test organisms
    2. An equally divided portion of a sample
    3. Free from the living germs of disease, fermentation, or putrefaction
    4. Temperature of the surroundings

28. What is an indicator (chemical)?

    1. A bottle containing only dilution water or distilled water in which the sample being tested is not added
    2. A pure chemical substance that is used to make new products or is used in chemical tests to measure, detect, or examine other substances
    3. A solution or liquid whose chemical makeup neutralizes acids or bases without a great change in pH
    4. A substance that gives a visible change, usually of color, at a desired point in a chemical reaction, generally at a specified end point

29. What is the unit of measurement for volume in the metric system?

    1. Gram
    2. Liter
    3. Meter
    4. Second

30. What is the chemical symbol for sodium?

    1. Fe
    2. K
    3. Na
    4. So

31. How is a pipet used in the laboratory?

    1. To deliver accurate volume measurements
    2. To measure volume
    3. To mix chemicals
    4. To weigh chemicals

32. What does the utility manager do when controlling?

    1. Decides who does what work and delegates authority to the appropriate operators
    2. Guides, teaches, motivates, and supervises operators and utility staff members
    3. Recruits new operators and staff and determines if there are enough qualified operators and staff to fill available positions
    4. Takes the steps necessary to ensure that essential activities are performed so that objectives will be achieved as planned

33. When giving a talk or presentation, why is it important to watch the audience?

    1. To determine if there is space available for late arrivals to the presentation
    2. To ensure they are not discussing your presentation with others
    3. To observe whether they are appropriately dressed for your presentation
    4. To obtain feedback about how well your message is being received

34. Why is input from other personnel in the utility as well as community leaders invaluable when preparing a budget?

    1. To gain support for the budget and to keep the budget on track once adopted
    2. To generate additional revenue to pay for the repair and replacement of capital equipment as the equipment wears out
    3. To identify the lowest possible bidders and the one most likely to cooperate with the utility
    4. To minimize internal opposition from utility staff and to ensure cooperation during installation

35. When compiling data reflecting the costs of various workplace injuries, the results should show which costs to the agency, operator, and the public?

    1. Defensible and indefensible costs
    2. Direct and indirect costs
    3. Justified and unjustified costs
    4. Reported and unreported costs

36. Why should a copy of a purchase order be retained in a suspense file or on a clipboard until the ordered items arrive?

    1. To allow for prompt payment of an item when delivered
    2. To assist staff in determining where to look for the ordered items
    3. To ensure that all ordered items will be paid for
    4. To keep track of the items that were ordered but not yet received

Multiple Choice (Select all correct answers.)

1. What are the limitations of using a waste stabilization pond?

    1. Consistently producing a high quality effluent
    2. Limited process control capabilities
    3. Possibly contaminating groundwater, if not properly lined
    4. Possibly having high suspended solids levels in the effluent due to algae
    5. Treating wastes inconsistently, depending on climatic conditions

2. Bioflocculation is accelerated in a pond by which factors?

    1. High dissolved oxygen content
    2. Ice cover over the surface
    3. Increased fluctuation of pH levels
    4. Increased temperature
    5. Wave action

3. Which processes are used to remove algae from an effluent?

    1. Activated sludge
    2. Algae harvesting
    3. Dissolved air flotation
    4. Microscreening
    5. Slow sand filtration

4. What are probable causes of poor quality pond effluent?

    1. Excessive turbidity from scum mats
    2. Mixing/agitation equipment failure
    3. Organic overload
    4. Process chemicals dumped from illegal drug laboratories into rural area manholes
    5. Toxic material in the influent

5. Which pond lab tests are measured using composite samples?

    1. Biochemical oxygen demand (BOD)
    2. Coliforms
    3. Dissolved oxygen (DO)
    4. pH
    5. Suspended solids (SS)

6. Which factors are key considerations for duckweed systems?

    1. Allowing for a 30-day retention time, shallow pond depth, and frequent harvesting if nutrient removal is desired
    2. Having a plan in place for disposing of the harvested plants
    3. Having the capability to draw effluent from several levels and thereby avoid high algal concentrations near the water's surface during discharge
    4. Performing post-aeration of the effluent to meet dissolved oxygen (DO) requirements, when needed
    5. Using intermittent sand filters to polish effluent

7. The components required for an activated sludge wastewater treatment plant depend on which factors?

1. Makeup of the wastewater
2. Quantity of wastewater to be treated
3. Size of media available for the process
4. Water conservation practices of the community
5. Whether the wastewater has received any pretreatment before entering the activated sludge system

8. In a good activated sludge, the floc mass contains what types of organisms?

1. Bacteria
2. Fungi
3. Protozoa
4. Worms
5. Yeast

9. Which important items must be checked before backfilling the excavation around a package activated sludge tank?

1. Buried tanks are placed on a foundation of sand bedding material
2. Examining the tank's exterior corrosion protection material for any damage, and repairing it as needed
3. The anodes are properly installed and the electrical leads are well bonded to the tank and anodes
4. The space between the sides of the tank and the walls of the excavation are filled using Class A material
5. The tank is absolutely level in both directions

10. Which checks should be made daily in package activated sludge plants?

1. Brush weirs that are used to remove algae and captured materials
2. Check the aeration unit for proper operation and lubrication
3. Check the appearance of the aeration tank and final clarification compartment
4. Check the return sludge line for proper operation
5. Skim off grease and other floating material such as plastic and rubber goods

11. What are the probable causes of septic sludge in the activated sludge process?

1. Insufficient aeration
2. Return of sludge from aerobic digester
3. Return sludge rate that is too low
4. Shock organic load
5. Toxic waste load

12. Operators can develop an operating strategy to improve plant performance by answering which questions?

1. Do you avoid injuries by removing hazards and following safe procedures?
2. Do you visit your plant on a regular basis to observe process conditions, check equipment for proper operation, lubricate and maintain equipment, and clean process tanks and related equipment?
3. Does the return sludge flow rate produce a high concentration of solids with the minimum amount of water being returned to the aerator?
4. If an increase or decrease in organisms occurs, is the oxygen level adjusted accordingly to maintain proper solids settling and the production of a clear final effluent?
5. Is adequate sludge wasting being practiced to properly maintain favorable food for microorganism balance throughout the system?

13. What general observations of the oxidation ditch plant are important to help operators determine how the plant is operating?

1. Clarity of the effluent from the secondary clarifier discharged over the weirs
2. Clarity of the water's surface in the settling tank
3. Color of the mixed liquor in the ditch
4. Odor of the plant
5. Oxidation ditch surface is free of foam buildup

14. Which items should an operator verify are in place when reviewing plans and specifications for oxidation ditches?

1. In all climates, standby or auxiliary power is provided to operate critical equipment
2. In all climates, the discharge pipe is located where flood water will not flow back into the plant if there is a power outage or pump failure
3. In cold climates, a subsurface disposal pipeline and percolation field is provided for effluent disposal during winter months
4. In cold climates, all equipment requiring normal maintenance is housed or heated
5. In cold climates, the rotor assembly is provided with a lightweight cover to prevent motor icing

15. What are possible RBC effluent disposal methods?

1. Groundwater recharge
2. Industrial reuse
3. Irrigation
4. Surface waters (rivers, lakes, oceans)
5. Wetlands

16. Abnormal RBC operating conditions may develop as a result of which circumstances?

    1. Average wastewater temperatures
    2. High or low flows
    3. High or low solids loading
    4. Normal pH levels
    5. Power outages

17. Hypochlorite is available to disinfect wastewater effluents in which forms?

    1. Gas
    2. Granule
    3. Liquid
    4. Pellet
    5. Tablet

18. Which critical elements influence the effectiveness of disinfection?

    1. Dose rate and type of chemical
    2. Numbers and types of organisms
    3. pH
    4. Solids used to contain chlorine during disinfection
    5. Temperature

19. The feed rate of chlorine from a tablet hypochlorinator into the plant effluent is controlled by which methods?

    1. A peristaltic chemical feeder pump
    2. A V-notch weir that controls the flow and depth of wastewater through the tablet feeder housing
    3. A vibrating mechanism on the dry chemical feeder
    4. The most probable number (MPN) of the coliforms in the plant effluent
    5. The number of tubes that are in service to hold the tablets

20. What are the advantages of generating chlorine/hypochlorite on site?

    1. Fresh hypochlorite solution does not lose its strength when the disinfection system is not in use
    2. Hypochlorite solution can be generated as needed
    3. Hypochlorite spills are not hazardous
    4. On-site generation is safer for operators because they do not have to handle and transfer prepared liquid hypochlorite solutions
    5. Sufficient hypochlorite is readily available at all times

21. What are the advantages of a horizontal subsurface flow (HSSF) treatment wetland?

    1. Developing anaerobic conditions that release obnoxious odors
    2. Lowering the initial construction costs, if land is relatively cheap and available
    3. Promoting root binding, leading to short-circuiting and channeling
    4. Reducing operation and maintenance costs, compared to similar costs at a conventional mechanical plant
    5. Reducing the level of operational skill needed to maintain the system

22. What are the most critical items in which operator intervention is necessary when operating and maintaining a constructed wetland?

    1. Adjustment of water levels
    2. Cultivating the maturation of wetlands
    3. Ensuring health and safety
    4. Maintenance of berms and dikes
    5. Mechanical equipment maintenance

23. Which tasks do operators perform to maintain the desired plant communities in wetlands?

    1. Dividing and replanting decorative flowering species, such as irises, to enhance the system's appearance
    2. Encouraging deep root growth by slowly lowering the water level over several weeks during the dormant period
    3. Maintaining a three-inch layer of mulch on top of the media
    4. Removing weeds, trees, and shrubs from the beds, along with unwanted growth adjacent to the wetlands
    5. Using native, noninvasive vegetation species

24. What problems can be discovered when troubleshooting a land disposal system?

    1. An irrigated crop that is dead
    2. Deterioration of the lateral aluminum distribution piping
    3. No flow coming from some sprinkler nozzles
    4. Wastewater running off an irrigated area
    5. Water ponding in an irrigated area where ponding normally has not been observed

25. Which effluent levels and concentrations are a concern for crop growing operations?

    1. Boron
    2. Chloride
    3. Coliforms
    4. pH
    5. Total dissolved solids (TDS)

26. Which items should operators consider when reviewing the plans and specifications for a land treatment system?

    1. Being sure a sufficient buffer area is provided around spray areas to prevent mist from drifting onto nearby homes and yards
    2. Ensuring that an alternate place to pump or dispose of effluent is available in case of system failure
    3. Requiring pretreatment or that a sand trap is installed ahead of pumps if sand is a problem
    4. Verifying that screens are installed on the inlet side of the irrigation pumps to prevent the spray nozzles from becoming plugged
    5. Verifying the installation of specifically designed effluent spray irrigation heads that will not plug

27. In the field of wastewater analysis, measuring the intensity of color in a sample at a particular wavelength helps determine which component levels?

    1. BOD level
    2. Nitrate level
    3. Nitrite level
    4. Phosphorus level
    5. Turbidity level

28. Chemical indicators are used in which types of water testing?

    1. Gravimetric analysis
    2. QA/QC
    3. Qualitative analysis
    4. Quantitative colorimetric analysis
    5. Quantitative titration

29. Why must a laboratory develop and implement a written hazard communication program? To ensure that personnel dealing with chemicals are

    1. Allowed to develop and implement a written chemical hygiene plan
    2. Appropriately informed of any health and safety considerations
    3. Never exposed to toxic or hazardous levels of chemicals
    4. Provided with a description of how the OSHA criteria for labels and other forms of warnings will be met
    5. Trained in methods to safely handle and store chemicals

30. Acids are extremely corrosive to which items?

    1. Clothing
    2. Concrete
    3. Glass
    4. Human tissue
    5. Metal

31. Which items are toxic materials?

    1. Ammonia
    2. Carbon tetrachloride
    3. Chlorine
    4. Cyanide
    5. Hydrogen sulfide

32. Which tasks of a utility manager are considered planning functions?

    1. Collecting and analyzing data, considering alternatives, and then making decisions
    2. Deciding who does what work
    3. Determining the goals, policies, procedures, and other elements to achieve the goals and objectives of the agency
    4. Guiding, teaching, motivating, and supervising operators and utility staff members
    5. Taking the steps necessary to ensure that essential activities are performed so that objectives will be achieved as planned

33. During the first day of work, a new employee should be given all the information available on which topics?

    1. Alcohol and drug testing
    2. Attendance expectations
    3. Compensation and benefits
    4. Employer/employee relations
    5. Policies and practices of the utility

34. To obtain press coverage for an agency's event or press conference, which guidelines should be followed?

    1. All employees should be encouraged to talk to the mass media people
    2. Demonstrate that your story is newsworthy, that it involves something unusual or interesting
    3. Make sure your story will fit the targeted format (television, radio, newspaper, or Internet)
    4. Provide a spokesperson who is interesting, articulate, and well prepared
    5. Require all media to obtain agency approval of any media press releases involving the agency

35. What does a manager need to predict the amount of money to set aside each year for a repair/replacement fund?

    1. A list of the equipment required for each process in the utility
    2. A minimum and maximum target level of funding
    3. An estimate of the life expectancy of the equipment required for each process in the utility
    4. Expected or projected new technology available when equipment wears out
    5. The replacement cost of the equipment required for each process in the utility

36. Unprotected cross-connections between public water systems and other piped systems on consumers' premises might contain which substances?

    1. Processed waters
    2. Stormwater
    3. Untreated supplies from ocean waters
    4. Untreated supplies from private wells
    5. Wastewater

37. Which steps should be taken when assessing the vulnerability of a utility system?

    1. Estimate customer demand for service following a potential disaster
    2. Estimate disaster effects on system components
    3. Identify and describe the system components
    4. Identify key system components that would be primarily responsible for system failure
    5. List assumed disaster characteristics

38. What is the utility manager's responsibility with regard to implementation and enforcement of the safety program? The manager has the responsibility to

    1. Ensure that all employees are trained and periodically retrained in proper safe work practices
    2. Ensure that equipment, tools, and the work area are maintained to comply with established safety standards
    3. Institute corrective measures where unsafe conditions or work methods exist
    4. Investigate all accidents and injuries to determine who to blame
    5. Respond to and correct all identified unsafe conditions and activities

39. What are the basic elements of a hazard communication program?

    1. Identify hazardous materials
    2. Identify treatment processes to detoxify and dispose of wastes
    3. Obtain chemical information and define hazardous conditions
    4. Properly label hazards
    5. Train operators

40. The causes of accidents can be classified as related to which items?

    1. Chemicals
    2. Heat
    3. Machinery
    4. Prevention
    5. Unsafe acts

41. What information should be in an employee's personnel file?

    1. Certificates earned
    2. Commendations received
    3. Formal performance reviews
    4. Record of accomplishments
    5. Religious affiliations

Short Answer

1. Where is the chlorine contact basin located in a typical plant with ponds only?

2. What is photosynthesis?

3. Why should water be introduced into a new pond before any wastewater?

4. Why are chlorine compounds or chlorine solution not the best chemicals for odor control in a pond?

5. Why is it desirable for a pond to be isolated from neighbors?

6. How can scum be prevented from leaving a pond?

7. Why should activated sludge in the final clarifier be returned to the aeration tank as soon as possible?

8. How would you operate the aeration device in a package plant?

9. What items can cause problems for the operator of a package activated sludge plant?

10. During plant start-up, what does a dark gray color in the MLSS indicate and what would you do if this condition persists for more than several days?

11. How can an operator determine if the sludge wasting rate should be increased or decreased?

12. List the tasks that should be included when performing a regular inspection of rotating biological contactor equipment.

13. How do the slime growths (biomass) on the plastic media look under (a) normal conditions and (b) abnormal conditions?

14. Why must wastewater be disinfected?

15. What constituents in wastewater are mainly responsible for the chlorine demand?

16. Where might chlorine be applied in the treatment of wastewater?

17. Why must the chlorine solution be well mixed with the wastewater?

18. What precautions would you take before starting any chlorination system?

19. How can carbonate scale on hypochlorinators be removed?

20. How is emergent vegetation supported in a free water surface (FWS) wetland?

21. What basic tasks must an operator perform when operating and maintaining a treatment wetland?

22. How does land treatment work?

23. How long should the pumps be run while irrigating a plot of land?

24. Why must chemicals be properly labeled?

25. What precautions should personnel take to protect themselves from diseases when working in a wastewater treatment plant?

26. Where would you obtain a representative sample?

27. Why does the actual volume of solids pumped from a clarifier not agree exactly with calculations based on the settleable solids test?

28. What does sludge age measure?

29. Why is the effluent from a treatment plant chlorinated?

30. What is a limitation of the BOD test?

31. What does the total solids (residue) test measure?

32. What are the different types of demands on a utility manager?

33. When has a supervisor successfully delegated?

34. What happens any time you or a member of your utility comes in contact with the public?

35. What is the value of customer complaints?

36. What does a SCADA system do?

SUGGESTED ANSWERS
TO
COMPREHENSIVE REVIEW QUESTIONS

True-False

1. False — Algae produce oxygen during sunlight (not night) hours and use oxygen during night (not sunlight) hours.

2. True — During the day, algae use carbon dioxide (raising the pH), while at night they produce carbon dioxide (lowering the pH).

3. True — Heavy chlorination at the pond recirculation point can assist in odor control, but it will probably interfere with treatment.

4. True — The calculation of pond BOD removal efficiency is figured in terms of percentage of BOD removed.

5. True — Algae are usually present in the effluent from ponds with continuous discharges.

6. True — The conversion of dissolved and suspended material to settled solids is the main objective of high-rate activated sludge processes.

7. True — A high-quality activated sludge plant effluent requires attention, understanding, and good plant operation.

8. True — Excessive return flows can cause a hydraulic overload on the clarifier.

9. True — Controlling wastes dumped into the collection system requires a pretreatment facility inspection program to ensure compliance.

10. False — The RBC media should move against (not with) the direction of flow to provide better treatment of the wastewater.

11. True — Damage to RBC equipment will be minimized if required maintenance is done as soon as the need is identified.

12. False — Slime growths should not be washed off the media.

13. True — Chlorination is the most economical process for disinfection of potable water and treated wastewater.

14. False — Determining the most probable number (MPN) of coliform group organisms present is not a test for individual pathogenic microorganisms.

15. True — All persons handling chlorine should be thoroughly aware of its hazardous properties.

16. True — Never adjust a hypochlorinator pump while it is off because doing so will damage the pump.

17. False — Free water surface (FWS) wetland treatment systems are not recommended for individual home sites because of associated problems such as vector and disease control.

18. False — The wetland bed is built at the same time (not after) the septic tank is installed.

19. False — Odors are seldom (not regularly) a nuisance in properly loaded wetlands.

20. False — Less (not more) labor usually is required to operate and maintain fixed subsurface drip system components in comparison to spray systems.

21. True — All metric system units of length, volume, and weight (mass) use factors of 10 to express larger or smaller quantities of these units.

22. True — The larger the number in front of the N (normality), the more concentrated the solution.

23. False — A sample that has a low color intensity will have a high (not low) percent transmittance but a low (not high) absorbance.

24. True — Never drink from a beaker or other laboratory glassware.

25. True — Good housekeeping practices make significant contributions to an effective safety program in an active chemical storage room.

26. True — Always pour acid into water, never water into acid, to avoid a reaction (explosion) that could cause chemical burns.

27. True — An employee can serve only one supervisor (unity of command).

28. False — Hiring new employees requires careful planning before (not after) the interview process.

29. True — Direct contact with people in your community is an effective tool in promoting a small wastewater utility, even though the audiences tend to be small.

30. True — Supervisory control and data acquisition (SCADA) systems are capable of analyzing data and providing operating, maintenance, regulatory, and annual reports.

31. True — It is a criminal offense to order an employee into a confined space without complying with all of the applicable regulations.

Best Answer

1. 3 The purpose of a pond outlet baffle is to prevent scum and other surface debris from flowing to the next pond or into receiving waters.

2. 1 Algae live on carbon dioxide and other nutrients in wastewater.

3. 2 Recirculation water should be drawn from the surface of the source pond to ensure the highest possible dissolved oxygen levels in the recirculated water.

4. 2 Before attempting to apply any insecticide or pesticide, the operator should contact the local official in charge of approving pesticide applications.

5. 2 Batch-operated ponds should be allowed to discharge only during the nonrecreational season when flows are high in the receiving waters.

6. 1 The time of day for collecting pond samples should be varied occasionally when taking grab samples so the operator can become familiar with the pond's characteristics at various times of day.

7. 3 A gray pond color indicates that a pond that is being overloaded or that it is not working properly.

8. 4 An operator should review the plans and specifications for a proposed pond to suggest improvements and changes before construction begins that will allow the operators and the ponds to do their jobs better.

9. 1 Facultative bacteria are bacteria that can use either dissolved molecular oxygen or oxygen obtained from food materials such as sulfate or nitrate ions.

10. 1 When critical factors in the aeration tank are under proper control, the organisms in the tank will convert soluble solids and agglomerate the fine particles into a floc mass.

11. 2 The purpose of air lifts is to pump and move settled activated sludge from one compartment to another using air.

12. 3 Bulking occurs when clouds of billowing sludge that occur throughout secondary clarifiers and sludge thickeners when the sludge does not settle properly.

13. 2 The proper type and grade of lubricating oil is critical because if the oil is too thin or thick, it will interfere with the proper functioning of bearings and gears.

14. 2 The RBC process is divided into multiple stages to maximize the effectiveness of a given amount of media surface by separating competing biomass populations.

15. 1 The bulkhead or baffle between stages one and two may be removed during periods of severe RBC organic overload to provide a greater amount of media surface area for the first stage of treatment.

16. 3 Saprophytic organisms are organisms living on dead or decaying organic matter.

17. 1 The ultimate measure of the effectiveness of the disinfection process is the bacteriological test result.

18. 1 Whenever an operator experiences a mild case of chlorine exposure, his/her first action should be to leave the contaminated area for a location with fresh air.

19. 3 Operators can reduce the problems caused by insoluble calcium hypochlorite (HTH) plugging pumps, pipes, and fittings by mixing the HTH powder in a separate mixing tank and allowing the insoluble solids to settle out before using the mixture.

20. 4 Operators can prevent the formation of calcium carbonate coatings on hypochlorinators by obtaining the dilution water from an ordinary home water softener.

21. 4 A problem operators encounter when troubleshooting chlorine solution pumps is a low discharge of chlorine solution.

22. 1 The primary treated wastewater (septic tank effluent) is applied in a wetland treatment system at the bottom, across the inlet end of the wetland bed.

23. 1 Odorous compounds emitted from open water wetland areas are typically associated with anaerobic conditions that can be created by excessive BOD and ammonia loadings.

24. 1 The physical control of a land disposal system is achieved by adjusting the valves or gates used to direct treated effluent to different disposal areas.

25. 4 Native grasses are recommended in land treatment systems because they can thrive under local variables such as soils, moisture, and seasons.

26. 2 An aliquot is an equally divided portion of a sample.

27. 1 Bioassay is a method of determining the relative toxicity of a test sample of industrial wastes by using live test organisms.

28. 4 An indicator (chemical) is a substance that gives a visible change, usually of color, at a desired point in a chemical reaction, generally at a specified end point.

29. 2 The unit of measurement for volume in the metric system is the liter.

30. 3 The chemical symbol for sodium is Na.

31. 1 A pipet is used to deliver accurate volume measurements in the laboratory.

32. 4 When controlling, the utility manager takes the steps necessary to ensure that essential activities are performed so that objectives will be achieved as planned.

33. 4 When giving a talk or presentation, it is important to watch the audience to obtain feedback about how well your message is being received.

34. 1 Input from other personnel in the utility as well as community leaders is invaluable when preparing a budget in order to gain support for the budget and to keep the budget on track once adopted.

35. 2 When compiling data reflecting the costs of various workplace injuries, the results should show direct and indirect costs to the agency, operator, and the public.

36. 4   A copy of a purchase order should be retained in a suspense file or on a clipboard until the ordered items arrive to keep track of the items that were ordered but not yet received.

Multiple Choice

1. 2, 3, 4, 5   The limitations of using a waste stabilization pond include: (1) having limited process control capabilities; (2) possibly contaminating groundwater, if not properly lined; (3) possibly having high suspended solids levels in the effluent due to algae; (4) treating wastes inconsistently, depending on climatic conditions.

2. 1, 4, 5   Bioflocculation is accelerated in a pond by high dissolved oxygen content, by increased temperature, and by wave action.

3. 2, 3, 4, 5   The processes used to remove algae from an effluent include algae harvesting, dissolved air flotation, microscreening, and slow sand filtration.

4. 1, 2, 3, 4, 5   The probable causes of poor quality pond effluent include excessive turbidity from scum mats, mixing/agitation equipment failure, organic overload, process chemicals dumped from illegal drug laboratories into rural area manholes, and toxic material in the influent.

5. 1, 5   Pond lab tests that are measured using composite samples include biochemical oxygen demand (BOD) and suspended solids (SS).

6. 1, 2, 3, 4   Key considerations for duckweed systems include: (1) allowing for a 30-day retention time, shallow pond depth, and frequent harvesting if nutrient removal is desired; (2) having a plan in place for disposing of the harvested plants; (3) having the capability to draw effluent from several levels and thereby avoid algal concentrations near the water's surface during discharge; and (4) performing post-aeration of the effluent to meet dissolved oxygen (DO) requirements, when needed.

7. 1, 2, 5   The components required for an activated sludge wastewater treatment plant depend on the makeup of the wastewater, on the quantity of the wastewater to be treated, and on whether the wastewater has received any pretreatment before entering the activated sludge system.

8. 1, 2, 3, 4, 5   In a good activated sludge, the floc mass contains bacteria, fungi, protozoa, worms, and yeast.

9. 1, 2, 3, 4, 5   Important items that must be checked before backfilling the excavation around a package activated sludge tank include: (1) verifying that the buried tanks are placed on a foundation of sand bedding material, (2) examining the tank's exterior corrosion protection material for any damage and repairing it as needed, (3) verifying that the anodes are properly installed and that the electrical leads are well bonded to the tank and anodes, (4) verifying that the space between the sides of the tank and the walls of the excavation are filled using Class A material, and (5) verifying that the tank is absolutely level in both directions.

10. 1, 2, 3, 4, 5   Checks that should be made daily in package activated sludge plants include: (1) brushing weirs that are used to remove algae and captured materials, (2) checking the aeration unit for proper operation and lubrication, (3) checking the appearance of the aeration tank and final clarification compartment, (4) checking the return sludge line for proper operation, and (5) skimming off grease and other floating material such as plastic and rubber goods.

11. 1, 3, 4, 5   The probable causes of septic sludge in the activated sludge process include insufficient aeration, a return sludge rate that is too low, a shock organic load, and a toxic waste load.

12. 1, 2, 3, 4, 5   Operators can develop an operating strategy to improve plant performance by answering questions such as: (1) Do you avoid injuries by removing hazards and following safe procedures? (2) Do you visit your plant on a regular basis to observe process conditions, check equipment for proper operation, lubricate and maintain equipment, and clean process tanks and related equipment? (3) Does the return sludge flow rate produce a high concentration of solids with the minimum amount of water being returned to the aerator? (4) If an increase or decrease in organisms occurs, is the oxygen level adjusted accordingly to maintain proper solids settling and the production of a clear final effluent? and (5) Is adequate sludge wasting being practiced to properly maintain favorable food for microorganism balance throughout the system?

13. 1, 2, 3, 4, 5   General observations that are important to help operators determine how the plant is operating include: (1) observing the clarity of the effluent from the secondary clarifier discharged over the weirs, (2) observing the clarity of the water's surface in the settling tank, (3) observing the color of the mixed liquor in the ditch, (4) paying attention to the odor of the plant, and (5) observing if the oxidation ditch surface is free of foam buildup.

14. 1, 2, 3, 4, 5   When reviewing plans and specifications for oxidation ditches in all climates, operators should verify that: (1) standby or auxiliary power is provided to operate critical equipment, (2) the discharge pipe is located where flood water will not flow back into the plant if there is a power outage or pump failure, (3) a subsurface disposal pipeline and percolation field is provided for effluent disposal during winter months, (4) all equipment requiring normal maintenance is housed or heated, and (5) the rotor assembly is provided with a lightweight cover to prevent motor icing during winter months.

15. 1, 2, 3, 4, 5   Possible RBC effluent disposal methods include groundwater recharge, industrial reuse, irrigation, surface waters (rivers, lakes, oceans), and wetlands.

16. 2, 3, 5   Abnormal RBC operating conditions may develop as a result of high or low flows, high or low solids loading, and power outages.

17. 2, 3, 4, 5   Hypochlorite is available to disinfect wastewater effluents in granule form, liquid form, pellet form, and tablet form.

18. 1, 2, 3, 5   Critical elements that influence the effectiveness of disinfection include the dose rate and type of chemical, numbers and types of organisms, pH, and temperature.

19. 2, 5   The feed rate of chlorine from a tablet hypochlorinator is controlled by a V-notch weir that controls the flow and depth of wastewater through the tablet feeder housing and by the number of tubes that are in service to hold the tablets.

20. 1, 2, 4   The advantages of generating chlorine/hypochlorite on site include: (1) having fresh hypochlorite solution that does not lose its strength when the disinfection system is not in use, (2) being able to generate hypochlorite solution as needed, and (3) on-site generation being safer for operators because they do not have to handle and transfer prepared liquid hypochlorite solutions.

21. 2, 4, 5   The advantages of a horizontal subsurface flow (HSSF) treatment wetland include: (1) lowering the initial construction costs, if land is relatively cheap and available; (2) reducing operation and maintenance costs, compared to similar costs at a conventional mechanical plant; and (3) reducing the level of operational skill needed to maintain the system.

22. 1, 2, 3, 4, 5   The most critical items in which operator intervention is necessary when operating and maintaining a constructed wetland include the adjustment of water levels, cultivating the maturation of wetlands, ensuring health and safety, the maintenance of berms and dikes, and mechanical equipment maintenance.

23. 1, 2, 3, 4, 5   Tasks that operators perform to maintain the desired plant communities in wetlands include: (1) dividing and replanting decorative flowering species, such as irises, to enhance the system's appearance; (2) encouraging deep root growth by slowly lowering the water level over several weeks during the dormant period; (3) maintaining a three-inch layer of mulch on top of the media; (4) removing weeds, trees, and shrubs from the beds, along with unwanted growth adjacent to the wetlands; and (5) using native, noninvasive vegetation species.

24. 1, 2, 3, 4, 5   Problems that can be discovered when troubleshooting a land disposal system include an irrigated crop that is dead, the deterioration of the lateral aluminum distribution piping, discovering no flow coming from some sprinkler nozzles, wastewater running off an irrigated area, and water ponding in an irrigated area where ponding normally has not been observed.

25. 1, 2, 4, 5   The effluent levels and concentrations that are a concern for crop growing operations include boron, chloride, pH, and total dissolved solids (TDS).

26. 1, 2, 3, 4, 5   The items operators should consider when reviewing plans and specifications for a land treatment system include: (1) being sure a sufficient buffer area is provided around spray areas to prevent mist from drifting onto nearby homes and yards, (2) ensuring that an alternate place to pump or dispose of effluent is available in case of system failure, (3) requiring pretreatment or that a sand trap is installed ahead of pumps if sand is a problem, (4) verifying that screens are installed on the inlet side of the irrigation pumps to prevent the spray nozzles from becoming plugged, and (5) verifying the installation of specifically designed effluent spray irrigation heads that will not plug.

27. 2, 3, 4   In the field of wastewater analysis, measuring the intensity of color in a sample at a particular wavelength helps determine the nitrate, nitrite, and phosphorus levels.

28. 3, 4, 5   Types of water testing that use chemical indicators include qualitative analysis, quantitative colorimetric analysis, and quantitative titration.

29. 2, 4, 5   A laboratory must develop and implement a written hazard communication to ensure that personnel dealing with chemicals are: (1) appropriately informed of any health and safety considerations, (2) provided with a description of how the OSHA criteria for labels and other forms of warnings will be met, and (3) trained in methods to safely handle and store chemicals.

30. 1, 2, 4, 5   Acids are extremely corrosive to clothing, concrete, human tissue, and metals.

31. 1, 2, 3, 4, 5   Toxic materials include ammonia, carbon tetrachloride, chlorine, cyanide, and hydrogen sulfide.

32. 1, 3   Planning functions of a utility manager include collecting and analyzing data, considering alternatives, and then making decisions; and determining the goals, policies, procedures, and other elements to achieve the goals and objectives of the agency.

33. 1, 2, 3, 4, 5   During the first day of work, a new employee should be given all the information available on alcohol and drug testing, attendance expectations, compensation and benefits, employer/employee relations, and policies and practices of the utility.

34. 2, 3, 4   To obtain press coverage for an agency's event or press conference: (1) demonstrate that your story is newsworthy, that it involves something unusual or interesting; (2) make sure your story will fit the targeted format (television, radio, newspaper, or Internet); and (3) provide a spokesperson who is interesting, articulate, and well prepared.

35. 1, 3, 5   To predict the amount of money to set aside each year for a repair/replacement fund, a manager needs a list of the equipment required for each process in the utility, an estimate of the life expectancy of the equipment required for each process in the utility, and the replacement cost of the equipment required for each process in the utility.

36. 1, 2, 3, 4, 5   Unprotected cross-connections between public water systems and other piped systems on consumers' premises might contain processed waters, stormwater, untreated supplies from ocean waters, untreated supplies from private wells, and wastewater.

37. 1, 2, 3, 4, 5   Steps that should be taken when assessing the vulnerability of a utility system include: (1) estimating customer demand for service following a potential disaster, (2) estimating disaster effects on system components, (3) identifying and describing the system components, (4) identifying key system components that would be primarily responsible for system failure, and (5) listing assumed disaster characteristics.

38. 1, 2, 3, 5   With regard to implementation and enforcement of the safety program, the utility manager has the responsibility to: (1) ensure that all employees are trained and periodically retrained in proper safe work practices, (2) ensure that equipment, tools, and the work area are maintained to comply with established safety standards, (3) institute corrective measures where unsafe conditions or work methods exist, and (4) respond to and correct all identified unsafe conditions and activities.

39. 1, 3, 4, 5   The basic elements of a hazard communication program include identifying hazardous materials, obtaining chemical information and defining hazardous conditions, properly labeling hazards, and training operators.

40. 1, 2, 3, 5   The causes of accidents can be classified as related to chemicals, heat, machinery, and unsafe acts.

41. 1, 2, 3, 4   An employee's personnel file should include certificates earned, commendations received, formal performance reviews, and a record of accomplishments.

Short Answer

1. In a typical plant with ponds only, the chlorine contact basin is located after the ponds.

2. Photosynthesis is a process in which organisms, with the aid of chlorophyll, convert carbon dioxide and inorganic substances into oxygen and additional plant material, using sunlight for energy. All green plants grow by this process.

3. The water should be added to the pond in advance of the wastewater to prevent odor development from waste solids exposed to the atmosphere.

4. Chlorine compounds or chlorine solution are not the best chemicals for odor control in a pond because they may interfere with biological stabilization of the wastes.

5. Ponds should be isolated to prevent associated nuisances such as odors and insects from disturbing residential, commercial, and recreational neighbors, as well as to prevent possible traffic hazards caused by insects.

6. Scum can be prevented from leaving a pond by using a floating baffle.

7. Activated sludge organisms in the final clarifier are in a deteriorating condition due to a lack of oxygen and food and should be returned to the aeration tank as quickly as possible.

8. The aeration device is operated by adjusting the air rates until the solids settle properly. If the water in the aeration tank is murky or cloudy and has a rotten egg odor (H_2S), then not enough air is being supplied. To resolve this problem, increase the air supply.

9. Package activated sludge plant problems may be caused by solids in the effluent, odors, and foaming/frothing. These problems could be caused by too little or too much aeration, too few or too many solids in the aeration tank, improper return sludge rate, improper sludge wasting or disposal of waste activated sludge, and abnormal influent conditions such as excessive flows or solids or toxic wastes.

10. A dark gray color in the mixed liquor during start-up indicates a lack of bacterial buildup in the MLSS. If this condition persists for more than several days, inspect the return sludge system to see that it is operating properly.

11. The appearance of the oxidation ditch surface can be a helpful indication of whether the sludge wasting rate should be increased or decreased. If the foam on the surface is white and crisp, reduce the wasting rate. A thick, dark foam indicates that the wasting rate should be increased.

12. Daily equipment inspection includes:

    1. Touch the outer housing of the shaft bearing to determine if it is running hot. Use a pyrometer or thermometer if the temperature is too hot for your hand.
    2. Listen for unusual noises in motor bearings. Locate the causes of any unusual noises and correct them.
    3. Touch the motors to determine if they are running hot. Use a pyrometer or thermometer, if necessary. If hot, determine the cause and correct.
    4. Look around the drive train and shaft bearing for oil spills. If oil is visible, check oil levels in the speed reducers and chain drive system. Also, look for possible damaged or worn out gaskets or seals.
    5. Inspect the chain drive for alignment and tightness.
    6. Inspect the belts for proper tension.
    7. Be sure all guards located over moving parts and equipment are in place and properly installed.
    8. Clean up any spills, messes, or debris.

13. Under normal conditions, the slime growths (biomass) on the plastic media appear shaggy, uniform in coverage, brown-to-gray in color, and without algae. Biomass thickness is not excessive. A black slime or a white slime growth indicates abnormal conditions and that something in the influent is disrupting normal slime growth.

14. Wastewaters must be disinfected because disease-producing microorganisms are potentially present in all wastewaters. These microorganisms must be removed or killed before treated wastewater can be discharged.

15. Chlorine demand in wastewater is caused by the waste constituents (reducing compounds) that react with chlorine. These constituents include hydrogen sulfide, ammonia, ferrous iron, nitrogenous compounds, and organic matter.

16. Chlorine might be applied up-sewer, before the plant (prechlorination), in the plant, or after the other treatment processes (postchlorination) to treat wastewater.

17. Thorough mixing of the chlorine solution with wastewater is necessary for effective disinfection and consistent results.

18. Before starting any chlorination system, read the manufacturer's O&M manual for installation, start-up, operation, maintenance, and troubleshooting procedures.

19. The following procedure can be used to remove carbonate scale from hypochlorinators.

    1. Take a one-quart glass jar and fill it half full of tap water.
    2. Place one fluid ounce of 30 to 37 percent hydrochloric acid in the jar, or use white vinegar.
    3. Fill the remainder of the jar with tap water.
    4. Place the suction hose of the hypochlorinator in the jar and pump the entire contents of the jar through the system.
    5. Return the suction hose to the hypochlorite solution tank and resume normal operation.

20. Soil, sand, gravel, rock, or other suitable material is used to support the emergent vegetation and the water is maintained at a relatively shallow depth flowing over the surface.

21. Basic operation and maintenance tasks involve simple procedures similar to the requirements for operation of a facultative lagoon. The operator must be observant, take appropriate actions when problems develop, and conduct required monitoring and operational monitoring as necessary.

22. In land treatment systems, effluent is pretreated and applied to land by conventional irrigation methods. Treatment is provided by natural processes as the effluent flows through the soil and plants. Part of the wastewater returns to the hydrologic cycle by evapotranspiration and part of it returns by surface runoff or percolation through the soil to groundwater.

23. While irrigating a plot of land, run the pumps for the length of time determined in the start-up procedure or turn the pumps off earlier if water begins to pond on the fields.

24. Chemicals must be properly labeled to avoid poor results and safety hazards.

25. To protect themselves from diseases when working in a wastewater treatment plant, personnel should be aware of personal hygiene. You should wash your hands thoroughly with soap and water after contact with wastewater, and again before eating or smoking. Work clothes should not be worn home. Inoculations against certain diseases will provide immunization against them. Do not pipet anything by mouth, and never drink from a beaker or other laboratory glassware.

26. Representative samples should be collected from the material, water, or wastestream where it is well mixed.

27. The actual volume of solids pumped may disagree with the calculated volume from the settleable solids test due to compacting of the sludge in the clarifier, the sludge may settle differently in Imhoff cones than in the clarifier, and the clarifier may not capture all of the solids that should settle.

28. Sludge age is a measure of the length of time a pound of suspended solids is retained under aeration in the activated sludge process.

29. The effluent from a treatment plant is chlorinated for disinfection purposes. In many cases the chlorine residual must be removed before final discharge to a receiving stream to protect fish and other aquatic life.

30. A limitation of the BOD test is that the results are not available until five days after the sample is collected.

31. The total solids (residue) test measures both suspended and dissolved solids in wastewater.

32. The different types of demands on a utility manager include demands from the community, regulatory agencies, and from within the utility.

33. A supervisor has successfully delegated when proper job assignments, authority, and responsibility are all present.

34. Any time you or a member of your utility comes in contact with the public, you will have an impact on the quality of your public image.

35. Complaints can be a valuable asset in determining consumer acceptance and pinpointing problems. Customer calls are frequently the first indication that something may be wrong. Responding to complaints and inquiries promptly can save the utility money and staff resources.

36. A SCADA system collects, stores, and analyzes information about all aspects of operation and maintenance; transmits alarm signals, when necessary; and allows fingertip control of alarms, equipment, and processes.

APPENDIX

HOW TO SOLVE SMALL WASTEWATER SYSTEM ARITHMETIC PROBLEMS

(VOLUME II)

by

Ken Kerri

TABLE OF CONTENTS

HOW TO SOLVE SMALL WASTEWATER SYSTEM ARITHMETIC PROBLEMS

HOW TO SOLVE SMALL WASTEWATER SYSTEM ARITHMETIC PROBLEMS (VOLUME II)

A.0 BASIC CONVERSION FACTORS (ENGLISH SYSTEM)

UNITS

1,000,000	= 1 Million	1,000,000/1 Million

LENGTH

12 in	= 1 ft	12 in/ft
3 ft	= 1 yd	3 ft/yd
5,280 ft	= 1 mi	5,280 ft/mi

AREA

144 sq in	= 1 sq ft	144 sq in/sq ft
43,560 sq ft	= 1 acre	43,560 sq ft/ac

VOLUME

7.48 gal	= 1 cu ft	7.48 gal/cu ft
1,000 mL	= 1 liter	1,000 mL/L
3.785 L	= 1 gal	3.785 L/gal
231 cu in	= 1 gal	231 cu in/gal

WEIGHT

1,000 mg	= 1 gm	1,000 mg/gm
1,000 gm	= 1 kg	1,000 gm/kg
454 gm	= 1 lb	454 gm/lb
2.2 lbs	= 1 kg	2.2 lbs/kg

POWER

0.746 kW	= 1 HP	0.746 kW/HP

DENSITY

8.34 lbs	= 1 gal	8.34 lbs/gal
62.4 lbs	= 1 cu ft	62.4 lbs/cu ft

DOSAGE

17.1 mg/L	= 1 grain/gal	17.1 mg/L/gpg
64.7 mg	= 1 grain	64.7 mg/grain

PRESSURE

2.31 ft water	= 1 psi	2.31 ft water/psi
0.433 psi	= 1 ft water	0.433 psi/ft water
1.133 ft water	= 1 in mercury	1.133 ft water/in mercury

FLOW

694 GPM	= 1 MGD	694 GPM/MGD
1.55 CFS	= 1 MGD	1.55 CFS/MGD

TIME

60 sec	= 1 min	60 sec/min
60 min	= 1 hr	60 min/hr
24 hr	= 1 day	24 hr/day*

*This may be written either as 24 hr/day or 1 day/24 hours depending on which units we wish to convert to obtain our desired results.

A.1 BASIC FORMULAS

WASTEWATER STABILIZATION PONDS

1. $\text{BOD Removal, \%} = \dfrac{(\text{In} - \text{Out})}{\text{In}} \times 100\%$

2. $\text{Detention Time, days} = \dfrac{\text{Pond Volume, ac-ft}}{\text{Flow Rate, ac-ft/day}}$

   where

   $\text{Pond Area, acres} = \dfrac{(\text{Average Width, ft})(\text{Average Length, ft})}{43,560 \text{ sq ft/acre}}$

   $\text{Pond Volume, ac-ft} = (\text{Average Area, ac})(\text{Depth, ft})$

   $\text{Flow Rate, ac-ft/day} = \dfrac{\text{Flow, gal/day}}{(7.48 \text{ gal/cu ft})(43,560 \text{ sq ft/acre})}$

3. $\text{Population Loading, persons/ac} = \dfrac{(\text{Pop Served, persons})(43,560 \text{ sq ft/ac})}{\text{Pond Surface Area, sq ft}}$

4. $\text{Hydraulic Loading, inches/day} = \dfrac{\text{Depth of Pond, inches}}{\text{Detention Time, days}}$

5. $\text{Organic (BOD) Loading, lbs/day/acre} = \dfrac{(\text{Flow, MGD})(\text{BOD, mg/L})(8.34 \text{ lbs/gal})}{\text{Area, acres}}$

ACTIVATED SLUDGE (OXIDATION DITCHES)

6. $\dfrac{\text{BOD Loading,}}{\text{lbs/day}}$ = (Flow, MGD)(BOD, mg/L)(8.34 lbs/gal)

7. $\dfrac{\text{BOD Loading,}}{\text{lbs/day/1,000 cu ft}}$ = $\dfrac{\text{BOD, lbs/day}}{\text{Ditch Volume, 1,000 cu ft}}$

8. $\dfrac{\text{Food}}{\text{Microorganism}}$, lbs BOD/day/lb MLVSS = $\dfrac{\text{BOD, lbs/day}}{\text{MLVSS, lbs}}$

where

MLVSS, lbs = (Vol, M Gal)(MLSS, mg/L)(8.34 lbs/gal)

9. Sludge Age, days = $\dfrac{\text{Aeration Solids, lbs}}{\text{Solids Added, lbs/day}}$

where

$\dfrac{\text{Aeration}}{\text{Solids, lbs}}$ = (Vol, M Gal)(MLSS, mg/L)(8.34 lbs/gal)

and

$\dfrac{\text{Solids Added,}}{\text{lbs/day}}$ = (Flow, MGD)(Infl SS, mg/L)(8.34 lbs/gal)

10. Detention Time, hr = $\dfrac{\text{(Ditch Volume, M Gal)(24 hr/day)}}{\text{Flow, MGD}}$

ROTATING BIOLOGICAL CONTACTORS (RBCs)

11. Hydraulic Loading, GPD/sq ft = $\dfrac{\text{Flow, gal/day}}{\text{Surface Area, sq ft}}$

12. $\dfrac{\text{Organic (BOD) Loading,}}{\text{lbs/day/1,000 sq ft}}$ = $\dfrac{\text{Soluble BOD Applied, lbs/day}}{\text{Surface Area of Media (in 1,000 sq ft)}}$

where

$\dfrac{\text{Soluble BOD,}}{\text{mg/L}}$ = $\dfrac{\text{Total BOD,}}{\text{mg/L}} - \left[(K)\left(\dfrac{\text{Suspended Solids,}}{\text{mg/L}}\right)\right]$

NOTE: The K value of a wastewater is the ratio of suspended BOD to suspended solids; it is obtained by dividing suspended BOD by suspended solids. By calculating K values for many samples, we obtain a range of K for most domestic wastewaters of from 0.5 to 0.7. K is useful in estimating the soluble BOD when only the total BOD and suspended solids are known.

and

$\dfrac{\text{Soluble BOD,}}{\text{lbs/day}}$ = (Flow, MGD)$\left(\dfrac{\text{Soluble BOD,}}{\text{mg/L}}\right)$(8.34 lbs/gal)

DISINFECTION AND CHLORINATION

13. $\dfrac{\text{Chlorinated Lime Feed Rate,}}{\text{lbs/day}}$ = $\dfrac{\text{Chlorine Require, lbs/day}}{\text{Portion of Chlorine in lb of Hypochlorite, lbs}}$

14. $\dfrac{\text{Chlorine Feed or Dose, mg/L}}$ = $\dfrac{\text{Chlorine Feed, lbs/day}}{\text{(Flow, MGD)(8.34 lbs/gal)}}$

15. $\dfrac{\text{Chlorine Demand,}}{\text{mg/L}}$ = $\dfrac{\text{Chlorine Dose,}}{\text{mg/L}} - \dfrac{\text{Chlorine Residual,}}{\text{mg/L}}$

16. $\dfrac{\text{Chlorine Dose,}}{\text{mg/L}}$ = $\dfrac{\text{Chlorine Demand,}}{\text{mg/L}} + \dfrac{\text{Chlorine Residual,}}{\text{mg/L}}$

17. $\dfrac{\text{Hypochlorinator}}{\text{Setting, lbs/day}}$ = (Flow, MGD)(Dose, mg/L)(8.34 lbs/gal)

LAND TREATMENT

18. Area, acres = $\dfrac{\text{(Length, ft)(Width, ft)}}{\text{43,560 sq ft/acre}}$

19. Hydraulic Loading, MGD/ac = $\dfrac{\text{Flow, MGD}}{\text{Area, acres}}$

20. $\dfrac{\text{Hydraulic Loading,}}{\text{inches/day}}$ = $\dfrac{\text{(Flow, MGD)(1,000,000/M)(12 in/ft)}}{\text{(Length, ft)(Width, ft)(7.48 gal/cu ct)}}$

LABORATORY

21. Temperature, °C = (°F − 32°F) $\times \dfrac{5}{9}$

22. Temperature, °F = $\left(\text{C}° \times \dfrac{9}{5}\right)$ + 32°F

MANAGEMENT

23. Operating Ratio = $\dfrac{\text{Total Revenue, \$}}{\text{Operating Expenses, \$}}$

24. $\dfrac{\text{Coverage}}{\text{Ratio}}$ = $\dfrac{\text{Total Revenue, \$ − Nondebt Expenses, \$}}{\text{Debt Service Expenses, \$}}$

A.2 TYPICAL WASTEWATER SYSTEM PROBLEMS (ENGLISH SYSTEM)

A.20 Wastewater Stabilization Ponds

EXAMPLE 1

The influent BOD to a series of ponds is 240 mg/L, and the effluent BOD is 50 mg/L. What is the pond efficiency in removing BOD?

Known	Unknown
BOD In, mg/L = 240 mg/L	BOD Removal, %
BOD Out, mg/L = 50 mg/L	

Calculate the pond efficiency in removing BOD.

BOD Removal, % = $\dfrac{\text{(In − Out)}}{\text{In}} \times 100\%$

= $\dfrac{\text{(240 mg/L − 50 mg/L)}}{\text{240 mg/L}} \times 100\%$

= 79%

EXAMPLES 2, 3, 4, and 5

To calculate the different loadings on a pond, the information listed in the known column must be available.

Known

Average Depth, ft	= 4 ft
Average Width, ft	= 100 ft
Average Length, ft	= 150 ft
Flow, GPD	= 30,000 GPD
Flow, MGD	= 0.030 MGD
BOD, mg/L	= 160 mg/L
Population Served, persons	= 320 persons
One Acre	= 43,560 sq ft

Unknown

2. Detention Time, days
3. Population Loading, persons/acre
4. Hydraulic Loading, inches/day
5. Organic (BOD) Loading, lbs/day/acre

2. Calculate the detention time in days.

   a. Estimate the average pond area in acres.

   $$\text{Avg Pond Area, ac} = \frac{(\text{Avg Width, ft})(\text{Avg Length, ft})}{43,560 \text{ sq ft/ac}}$$

   $$= \frac{(100 \text{ ft})(150 \text{ ft})}{43,560 \text{ sq ft/ac}}$$

   $$= 0.344 \text{ acre}$$

   b. Determine the pond volume in acre-feet.

   $$\text{Pond Vol, ac-ft} = (\text{Avg Area, ac})(\text{Avg Depth, ft})$$

   $$= (0.344 \text{ ac})(4 \text{ ft})$$

   $$= 1.38 \text{ ac-ft}$$

   c. Convert flow from gallons per day to acre-feet per day.

   $$\text{Flow, ac-ft/day} = \frac{\text{Flow, GPD}}{(7.48 \text{ gal/cu ft})(43,560 \text{ sq ft/ac})}$$

   $$= \frac{30,000 \text{ gal/day}}{(7.48 \text{ gal/cu ft})(43,560 \text{ sq ft/ac})}$$

   $$= 0.092 \text{ ac-ft/day}$$

   d. Calculate the detention time in days.

   $$\text{Detention Time, days} = \frac{\text{Pond Volume, ac-ft}}{\text{Flow, ac-ft/day}}$$

   $$= \frac{1.38 \text{ ac-ft}}{0.092 \text{ ac-ft/day}}$$

   $$= 15 \text{ days}$$

3. Calculate the population loading in persons per acre.

   $$\text{Population Loading, persons/acre} = \frac{\text{Population Served, persons}}{\text{Avg Pond Area, acres*}}$$

   $$= \frac{320 \text{ persons}}{0.344 \text{ acre}}$$

   $$= 930 \text{ persons/acre}$$

NOTE: We used average width and length rather than surface width and length to calculate average pond area to simplify our calculations. Loading calculated on a population-served basis is normally expressed as

$$\text{Population Loading, persons/ac} = \frac{(\text{Pop Served, persons})(43,560 \text{ sq ft/ac})}{\text{Pond Surface Area, sq ft}}$$

4. Calculate the hydraulic loading in inches per day.

   $$\text{Hydraulic Loading, inches/day} = \frac{(\text{Depth of Pond, ft})(12 \text{ in/ft})}{\text{Detention Time, days}}$$

   $$= \frac{(4 \text{ ft})(12 \text{ in/ft})}{15 \text{ days}}$$

   $$= 3.2 \text{ inches/day}$$

5. Calculate the organic (BOD) loading in pounds per day per acre.

   $$\text{Organic (BOD) Loading, lbs/day/ac} = \frac{(\text{Flow, MGD})(\text{BOD, mg/L})(8.34 \text{ lbs/gal})}{\text{Pond Area, acres}}$$

   $$= \frac{(0.030 \text{ MGD})(160 \text{ mg/L})(8.34 \text{ lbs/gal})}{0.344 \text{ acres}}$$

   $$= 116 \text{ lbs BOD/day/acre}$$

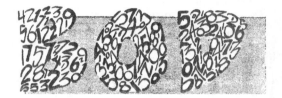

A.21   Activated Sludge (Oxidation Ditches)

EXAMPLES 6, 7, 8, 9, and 10

To calculate the different loadings on an oxidation ditch, the information listed in the known column must be available.

Known

Ditch Volume, cu ft	= 20,000 cu ft
	= 20.0 • 1,000 cu ft
Ditch Volume, MG	= 0.15 M Gal
Flow, MGD	= 0.2 MGD
BOD, mg/L	= 150 mg/L
Infl SS, mg/L	= 170 mg/L
MLSS, mg/L	= 3,500 mg/L
Volatile Matter in MLSS	= 70% or 0.70

Unknown

6. BOD Loading, lbs/day
7. BOD Loading, lbs/day/1,000 cu ft
8. $\dfrac{\text{Food}}{\text{Microorganism}}$, lbs BOD/day/lb MLVSS
9. Sludge Age, days
10. Detention Time, hours

6. Calculate the BOD loading in pounds per day.

   $$\text{BOD Loading, lbs/day} = (\text{Flow, MGD})(\text{BOD, mg/L})(8.34 \text{ lbs/gal})$$

   $$= (0.2 \text{ MGD})(150 \text{ mg/L})(8.34 \text{ lbs/gal})$$

   $$= 250 \text{ lbs BOD/day}$$

7. Calculate the BOD loading in pounds per day per 1,000 cubic feet.

$$\text{BOD Loading, lbs/day/} = \frac{\text{BOD, lbs/day}}{\text{Ditch Volume, 1,000 cu ft}}$$

$$= \frac{250 \text{ lbs BOD/day}}{20.0 \bullet 1,000 \text{ cu ft}}$$

$$= 12.5 \text{ lbs BOD/day/1,000 cu ft}$$

8. Calculate the food/microorganism ratio in pounds BOD per day per pound MLVSS.

a. Determine MLVSS in pounds.

$$\text{MLVSS, lbs} = (\text{Vol, MG})(\text{MLSS, mg/L})(\text{VM})(8.34 \text{ lbs/gal})$$

$$= (0.15 \text{ MG})(3,500 \text{ mg/L})(0.70)(8.34 \text{ lbs/gal})$$

$$= 3,065 \text{ lbs MLVSS}$$

b. Calculate the food/microorganism ratio in pounds BOD per day per pound MLVSS.

$$\text{F/M, lbs BOD/} = \frac{\text{BOD, lbs/day}}{\text{MLVSS, lbs}}$$

$$= \frac{250 \text{ lbs BOD/day}}{3,065 \text{ lbs MLVSS}}$$

$$= 0.08 \text{ lb BOD/day/lb MLVSS}$$

9. Calculate the sludge age in days.

a. Calculate the pounds of solids under aeration.

$$\text{Aeration Solids, lbs} = (\text{Ditch Vol, MG})(\text{MLSS, mg/L})(8.34 \text{ lbs/gal})$$

$$= (0.15 \text{ MG})(3,500 \text{ mg/L})(8.34 \text{ lbs/gal})$$

$$= 4,379 \text{ lbs}$$

b. Calculate the solids fed to the ditch in pounds per day.

$$\text{Solids Added, lbs/day} = (\text{Flow, MGD})(\text{Infl SS, mg/L})(8.34 \text{ lbs/gal})$$

$$= (0.2 \text{ MGD})(170 \text{ mg/L})(8.34 \text{ lbs/gal})$$

$$= 284 \text{ lbs/day}$$

c. Calculate the sludge age in days.

$$\text{Sludge Age, days} = \frac{\text{Aeration Solids, lbs}}{\text{Solids Added, lbs/day}}$$

$$= \frac{4,379 \text{ lbs}}{284 \text{ lbs/day}}$$

$$= 15.4 \text{ days}$$

10. Calculate the detention time in hours.

$$\text{Detention Time, hours} = \frac{(\text{Ditch Volume, MG})(24 \text{ hr/day})}{\text{Flow, MGD}}$$

$$= \frac{(0.15 \text{ MG})(24 \text{ hr/day})}{0.2 \text{ MGD}}$$

$$= 18 \text{ hours}$$

A.22   Rotating Biological Contactors (RBCs)

EXAMPLE 11

A rotating biological contactor treats a flow of 1.5 MGD. The surface area of the media is 500,000 square feet (provided by manufacturer). What is the hydraulic loading in GPD/sq ft?

Known		Unknown
Flow, MGD	= 1.5 MGD	Hydraulic Loading, GPD/sq ft
Surface Area, sq ft	= 500,000 sq ft	

Calculate the hydraulic loading in gallons per day of wastewater per square foot of media surface.

$$\text{Hydraulic Loading, GPD/sq ft} = \frac{\text{Flow, gal/day}}{\text{Surface Area, sq ft}}$$

$$= \frac{1,500,000 \text{ gal/day}}{500,000 \text{ sq ft}}$$

$$= 3.0 \text{ GPD/sq ft}$$

EXAMPLE 12

The rotating biological contactor in Example 11 treats an industrial influent with a total BOD of 220 mg/L and suspended solids of 300 mg/L. Assume a K value of 0.5 to calculate the soluble BOD. What is the organic (BOD) loading in pounds of soluble BOD per day per 1,000 square feet of media surface?

Known		Unknown
Flow, MGD	= 1.5 MGD	Organic (BOD) Loading, lbs/day/ 1,000 sq ft
Surface Area, sq ft	= 500,000 sq ft	
Total BOD, mg/L	= 220 mg/L	
SS, mg/L	= 300 mg/L	
K	= 0.5	

a. Estimate the soluble BOD treated by the rotating biological contactor in milligrams per liter.

$$\text{Soluble BOD, mg/L} = \text{Total BOD, mg/L} - (\text{K} \times \text{Suspended Solids, mg/L})$$

$$= 220 \text{ mg/L} - (0.5 \times 300 \text{ mg/L})$$

$$= 220 \text{ mg/L} - 150 \text{ mg/L}$$

$$= 70 \text{ mg/L}$$

b. Determine the soluble BOD applied to the rotating biological contactor in pounds of soluble BOD per day.

$$\text{Soluble BOD, lbs/day} = (\text{Flow, MGD})(\text{Soluble BOD, mg/L})(8.34 \text{ lbs/gal})$$

$$= (1.5 \text{ MGD})(70 \text{ mg/L})(8.34 \text{ lbs/gal})$$

$$= 876 \text{ lbs soluble BOD/day}$$

c. Calculate the organic (BOD) loading in pounds of soluble BOD per day per 1,000 square feet of media surface.

$$\text{Organic (BOD) Loading, lbs/day/ 1,000 sq ft} = \frac{\text{Soluble BOD, lbs/day}}{\text{Surface Area of Media (in 1,000 sq ft)}}$$

$$= \frac{876 \text{ lbs Soluble BOD/day}}{500 \bullet 1,000 \text{ sq ft}}$$

$$= 1.8 \text{ lbs BOD/day/1,000 sq ft}$$

A.23   Disinfection and Chlorination

EXAMPLE 13

A wastewater requires a chlorine feed rate of 12 pounds chlorine per day. How many pounds of chlorinated lime will be required per day to provide the needed chlorine? Assume each pound of chlorinated lime (hypochlorite) contains 0.34 pound of available chlorine.

Known		Unknown
Chlorine Required, lbs/day	= 12 lbs/day	Chlorinated Lime Feed Rate, lbs/day
Portion of Chlorine in lb of Hypochlorite, lbs	= 0.34 lb	

Calculate pounds of chlorinated lime required per day.

$$\text{Chlorinated Lime Feed Rate, lbs/day} = \frac{\text{Chlorine Required, lbs/day}}{\text{Portion of Chlorine in lb of Hypochlorite, lbs}}$$

$$= \frac{12 \text{ lbs/day}}{0.34 \text{ lb}}$$

$$= 35.3 \text{ lbs chlorinated lime/day}$$

EXAMPLES 14 and 15

A hypochlorinator is set to feed a dose of 15 pounds of chlorine per 24 hours. The wastewater flow is 0.4 MGD and the chlorine, as measured by the chlorine residual test after 30 minutes of contact time, is 0.4 mg/L. Find the chlorine dosage and the chlorine demand in mg/L.

Known		Unknown
Chlorine Feed, lbs/day	= 15 lbs/day	14. Chlorine Dose, mg/L
Flow, MGD	= 0.4 MGD	15. Chlorine Demand, mg/L
Chlorine Residual, mg/L	= 0.4 mg/L	

14. Calculate the chlorine dose in milligrams per liter.

$$\text{Chlorine Feed or Dose, mg/L} = \frac{\text{Chlorine Feed, lbs/day}}{(\text{Flow, MGD})(8.34 \text{ lbs/gal})}$$

$$= \frac{15 \text{ lbs/day}}{(0.4 \text{ MGD})(8.34 \text{ lbs/gal})}$$

$$= 4.5 \text{ mg/L}$$

15. Determine the chlorine demand in milligrams per liter.

$$\text{Chlorine Demand, mg/L} = \text{Chlorine Dose, mg/L} - \text{Chlorine Residual, mg/L}$$

$$= 4.5 \text{ mg/L} - 0.4 \text{ mg/L}$$

$$= 4.1 \text{ mg/L}$$

EXAMPLES 16 and 17

Calculate the chlorine dose in milligrams per liter and the hypochlorinator setting in pounds per 24 hours to treat a wastewater with a chlorine demand of 10 mg/L, when a chlorine residual of 2 mg/L is desired. The flow is 0.4 MGD.

Known		Unknown
Chlorine Demand, mg/L = 10 mg/L		16. Chlorine Dose, mg/L
Chlorine Residual, mg/L = 2 mg/L		17. Hypochlorinator Setting, lbs/day
Flow, MGD = 0.4 MGD		

16. Calculate the chlorine dose in milligrams per liter.

$$\text{Chlorine Dose, mg/L} = \text{Chlorine Demand, mg/L} + \text{Chlorine Residual, mg/L}$$

$$= 10 \text{ mg/L} + 2 \text{ mg/L}$$

$$= 12 \text{ mg/L}$$

17. Calculate the hypochlorinator setting in pounds per day.

$$\text{Hypochlorinator Setting, lbs/day} = (\text{Flow, MGD})(\text{Dose, mg/L})(8.34 \text{ lbs/gal})$$

$$= (0.4 \text{ MGD}(12 \text{ mg/L})(8.34 \text{ lbs/gal})$$

$$= 40 \text{ lbs/day}$$

$$= 40 \text{ lbs/24 hours}$$

A.24   Land Treatment

EXAMPLES 18, 19, and 20

A plot of land 1,600 feet long by 600 feet wide is used for a land disposal system. If 0.5 million gallons of water is applied to the land during a 24-hour period, calculate the land area in acres and the hydraulic loading in MGD per acre and also in inches per day.

Known		Unknown
Length, ft	= 1,600 ft	18. Area, acres
Width, ft	= 600 ft	19. Hydraulic Loading, MGD/ac
Flow, MGD	= 0.5 MGD	20. Hydraulic Loading, in/day

18. Determine the surface area in acres.

$$\text{Area, acres} = \frac{(\text{Length, ft})(\text{Width, ft})}{43,560 \text{ sq ft/acre}}$$

$$= \frac{(1,600 \text{ ft})(600 \text{ ft})}{43,560 \text{ sq ft/acre}}$$

$$= 22.0 \text{ acres}$$

19. Determine the hydraulic loading in million gallons per day per acre.

$$\text{Hydraulic Loading, MGD/ac} = \frac{\text{Flow, MGD}}{\text{Area, acres}}$$

$$= \frac{0.5 \text{ MGD}}{22.0 \text{ acres}}$$

$$= 0.023 \text{ MGD/acre}$$

20. Determine the hydraulic loading in inches per day.

$$\text{Hydraulic Loading, inches/day} = \frac{(\text{Flow, MGD})(1,000,000/\text{M})(12 \text{ in/ft})}{(\text{Length, ft})(\text{Width, ft})(7.48 \text{ gal/cu ft})}$$

$$= \frac{(0.5 \text{ MGD})(1,000,000/\text{M})(12 \text{ in/ft})}{(1,600 \text{ ft})(600 \text{ ft})(7.48 \text{ gal/cu ft})}$$

$$= 0.84 \text{ inch/day}$$

A.25   Laboratory

EXAMPLE 21

Convert a temperature of 59°F to degrees Celsius.

Known	Unknown
Temp, °F = 59°F	Temp, °C

Change 59°F to degrees Celsius.

$$\text{Temperature, °C} = (\text{°F} - 32°) \times \frac{5}{9}$$

$$= (59° - 32°) \times \frac{5}{9}$$

$$= 27° \times \frac{5}{9}$$

$$= 3° \times 5$$

$$= 15°C$$

EXAMPLE 22

Convert a temperature of 5°C to degrees Fahrenheit.

Known	Unknown
Temp, °C = 5°C	Temp, °F

Change 5°C to degrees Fahrenheit.

$$\text{Temperature, °F} = \left(\text{°C} \times \frac{9}{5}\right) + 32°$$

$$= \left(5° \times \frac{9}{5}\right) + 32°$$

$$= \frac{45°}{5} + 32°$$

$$= 9° + 32°$$

$$= 41°F$$

A.26   Management

EXAMPLES 23 and 24

The total revenues for a utility are $1,471,000 and the operating expenses for the utility are $1,108,000. The debt service expenses are $480,000. What is the operating ratio? What is the coverage ratio?

Known		Unknown
Total Revenue, $ = $1,471,000		23. Operating Ratio
Operating Expenses, $ = $1,108,000		24. Coverage Ratio
Debt Service Expenses, $ = $480,000		

23. Calculate the operating ratio.

$$\text{Operating Ratio} = \frac{\text{Total Revenue, \$}}{\text{Operating Expenses, \$}}$$

$$= \frac{\$1,471,000}{\$1,108,000}$$

$$= 1.33$$

24. Calculate the coverage ratio.

a. Calculate the nondebt expenses.

$$\text{Nondebt Expenses, \$} = \text{Operating Expenses, \$} - \text{Debt Service Expenses, \$}$$

$$= \$1,108,000 - \$480,000$$

$$= \$628,000$$

b. Calculate the coverage ratio.

$$\text{Coverage Ratio} = \frac{\text{Total Revenue, \$} - \text{Nondebt Expenses, \$}}{\text{Debt Service Expenses, \$}}$$

$$= \frac{\$1,471,000 - \$628,000}{\$480,000}$$

$$= 1.76$$

A.3   BASIC CONVERSION FACTORS (METRIC SYSTEM)

LENGTH

100 cm	= 1 m	100 cm/m
3.281 ft	= 1 m	3.281 ft/m

AREA

2.4711 ac	= 1 ha*	2.4711 ac/ha
10,000 sq m	= 1 ha	10,000 sq m/ha

VOLUME

1,000 mL	= 1 liter	1,000 mL/L
1,000 L	= 1 cu m	1,000 L/cu m
3.785 L	= 1 gal	3.785 L/gal

WEIGHT

1,000 mg	= 1 gm	1,000 mg/gm
1,000 gm	= 1 kg	1,000 gm/kg

DENSITY

1 kg	= 1 liter	1 kg/L

PRESSURE

10.015 m	= 1 kg/sq cm	10.015 m/kg/sq cm
1 Pascal	= 1 N/sq m	1 Pa/N/sq m
1 psi	= 6,895 Pa	1 psi/6,895 Pa

FLOW

3,785 cu m/day	= 1 MGD	3,785 cu m/day/MGD
3.785 ML/day	= 1 MGD	3.785 ML/day/MGD

*hectare

A.4   TYPICAL WASTEWATER SYSTEM PROBLEMS (METRIC SYSTEM)

A.40   Wastewater Stabilization Ponds

EXAMPLE 1

The influent BOD to a series of ponds is 180 mg/L, and the effluent BOD is 40 mg/L. What is the pond efficiency in removing BOD?

Known		Unknown
BOD In, mg/L	= 180 mg/L	BOD Removal, %
BOD Out, mg/L	= 40 mg/L	

Calculate the pond efficiency in removing BOD.

$$\text{BOD Removal, \%} = \frac{(\text{In} - \text{Out})}{\text{In}} \times 100\%$$

$$= \frac{(180 \text{ mg/L} - 40 \text{ mg/L})}{180 \text{ mg/L}} \times 100\%$$

$$= 78\%$$

EXAMPLES 2, 3, 4, and 5

To calculate the different loadings on a pond, the information listed in the known column must be available.

Known

Average Depth, m	= 1.4 m
Average Width, m	= 30 m
Average Length, m	= 45 m
Flow, cu m/day	= 150 cu m/day
BOD, mg/L	= 140 mg/L
Population Served, persons	= 320 persons

Unknown

2. Detention Time, days
3. Population Loading, persons/sq m
4. Hydraulic Loading, cm/day
5. Organic (BOD) Loading, gm/day/sq m

2. Calculate the detention time in days.

   a. Estimate the average pond area in square meters.

   $$\text{Avg Pond Area, sq m} = (\text{Avg Width, m})(\text{Avg Length, m})$$

   $$= (30 \text{ m})(45 \text{ m})$$

   $$= 1,350 \text{ sq m}$$

   b. Determine the pond volume in cubic meters.

   $$\text{Pond Vol, cu m} = (\text{Avg Area, sq m})(\text{Avg Depth, m})$$

   $$= (1,350 \text{ sq m})(1.4 \text{ m})$$

   $$= 1,890 \text{ cu m}$$

   c. Calculate the detention time in days.

   $$\text{Detention Time, days} = \frac{\text{Pond Volume, cu m}}{\text{Flow, cu m/day}}$$

   $$= \frac{1,890 \text{ cu m}}{150 \text{ cu m/day}}$$

   $$= 12.6 \text{ days}$$

3. Calculate the population loading in persons per square meters.

   $$\text{Population Loading, persons/sq m} = \frac{\text{Population Served, persons}}{\text{Avg Pond Area, sq m*}}$$

   $$= \frac{320 \text{ persons}}{1,350 \text{ sq m}}$$

   $$= 0.237 \text{ person/sq m}$$

NOTE: We used average width and length rather than surface width and length to calculate average pond area to simplify our calculations. Loading calculated on a population-served basis is normally expressed as

$$\text{Population Loading, persons/sq m} = \frac{\text{Pop Served, persons}}{\text{Pond Surface Area, sq m}}$$

4. Calculate the hydraulic loading in centimeters per day.

$$\text{Hydraulic Loading, cm/day} = \frac{(\text{Depth of Pond, m})(100 \text{ cm/m})}{\text{Detention Time, days}}$$

$$= \frac{(1.4 \text{ m})(100 \text{ cm/m})}{12.6 \text{ days}}$$

$$= 11.1 \text{ cm/days}$$

5. Calculate the organic (BOD) loading in grams per day per square meter.

$$\text{Organic (BOD) Loading, gm/day/sq m} = \frac{\left(\text{Flow, cu m/day}\right)\left(\text{BOD, mg/L}\right)\left(\frac{1{,}000 \text{ L}}{1 \text{ cu m}}\right)\left(\frac{1 \text{ gm}}{1{,}000 \text{ mg}}\right)}{\text{Pond Area, sq m}}$$

$$= \frac{(150 \text{ cu m/day})(140 \text{ mg/L})\left(\frac{1{,}000 \text{ L}}{1 \text{ cu m}}\right)\left(\frac{1 \text{ gm}}{1{,}000 \text{ mg}}\right)}{1{,}350 \text{ sq m}}$$

$$= 15.6 \text{ gm BOD/day/sq m}$$

A.41 Activated Sludge (Oxidation Ditches)

EXAMPLES 6, 7, 8, 9, and 10

To calculate the different loadings on an oxidation ditch, the information listed in the known column must be available.

Known

Ditch Volume, cu m	= 700 cu m
	= 0.70 • 1,000 cu m
Flow, cu m/day	= 800 cu m/day
BOD, mg/L	= 180 mg/L
Infl SS, mg/L	= 210 mg/L
MLSS, mg/L	= 3,800 mg/L
Volatile Matter in MLSS	= 70% or 0.70

Unknown

6. BOD Loading, kg/day
7. BOD Loading, kg/day/1,000 cu m
8. $\dfrac{\text{Food}}{\text{Microorganism}}$, kg BOD/day/kg MLVSS
9. Sludge Age, days
10. Detention Time, hours

6. Calculate the BOD loading in kilograms per day.

$$\text{BOD Loading, kg/day} = (\text{Flow, cu m/day})(\text{BOD, mg/L})\left(\frac{1 \text{ kg}}{1{,}000{,}000 \text{ mg}}\right)\left(\frac{1{,}000 \text{ L}}{1 \text{ cu m}}\right)$$

$$= (800 \text{ cu m/day})(180 \text{ mg/L})\left(\frac{1 \text{ kg}}{1{,}000{,}000 \text{ mg}}\right)\left(\frac{1{,}000 \text{ L}}{1 \text{ cu m}}\right)$$

$$= 144 \text{ kg BOD/day}$$

7. Calculate the BOD loading in kilograms per day per 1,000 cubic meters.

$$\text{BOD Loading, kg/day/1,000 cu m} = \frac{\text{BOD, kg/day}}{\text{Ditch Volume, 1,000 cu m}}$$

$$= \frac{144 \text{ kg BOD/day}}{0.7 \bullet 1{,}000 \text{ cu m}}$$

$$= 206 \text{ kg BOD/day/1,000 cu m}$$

8. Calculate the food/microorganism ratio in kilograms BOD per day per kilogram MLVSS.

   a. Determine MLVSS in kilograms.

$$\text{MLVSS, kg} = (\text{Vol, cu m})(\text{MLSS, mg/L})(\text{VM})\left(\frac{1 \text{ kg}}{1{,}000{,}000 \text{ mg}}\right)\left(\frac{1{,}000 \text{ L}}{1 \text{ cu m}}\right)$$

$$= (700 \text{ cu m})(3{,}800 \text{ mg/L})(0.70)\left(\frac{1 \text{ kg}}{1{,}000{,}000 \text{ mg}}\right)\left(\frac{1{,}000 \text{ L}}{1 \text{ cu m}}\right)$$

$$= 1{,}862 \text{ kg MLVSS}$$

   b. Calculate the food/microorganism ratio in kilograms BOD per day per kilogram MLVSS.

$$\text{F/M, kg BOD/day/kg MLVSS} = \frac{\text{BOD, kg/day}}{\text{MLVSS, kg}}$$

$$= \frac{144 \text{ kg BOD/day}}{1{,}862 \text{ kg MLVSS}}$$

$$= 0.08 \text{ kg BOD/day/kg MLVSS}$$

9. Calculate the sludge age in days.

   a. Calculate the kilograms of solids under aeration.

$$\text{Aeration Solids, kg} = (\text{Ditch Vol, cu m})(\text{MLSS, mg/L})\left(\frac{1 \text{ kg}}{1{,}000{,}000 \text{ mg}}\right)\left(\frac{1{,}000 \text{ L}}{1 \text{ cu m}}\right)$$

$$= (700 \text{ cu m})(3{,}800 \text{ mg/L})\left(\frac{1 \text{ kg}}{1{,}000{,}000 \text{ mg}}\right)\left(\frac{1{,}000 \text{ L}}{1 \text{ cu m}}\right)$$

$$= 2{,}660 \text{ kg}$$

   b. Calculate the solids fed to the ditch in kilograms per day.

$$\text{Solids Added, kg/day} = (\text{Flow, cu m/day})(\text{Infl SS, mg/L})\left(\frac{1 \text{ kg}}{1{,}000{,}000 \text{ mg}}\right)\left(\frac{1{,}000 \text{ L}}{1 \text{ cu m}}\right)$$

$$= (800 \text{ cu m/day})(210 \text{ mg/L})\left(\frac{1 \text{ kg}}{1{,}000{,}000 \text{ mg}}\right)\left(\frac{1{,}000 \text{ L}}{1 \text{ cu m}}\right)$$

$$= 168 \text{ kg/day}$$

   c. Calculate the sludge age in days.

$$\text{Sludge Age, days} = \frac{\text{Aeration Solids, kg}}{\text{Solids Added, kg/day}}$$

$$= \frac{2{,}660 \text{ kg}}{168 \text{ kg/day}}$$

$$= 15.8 \text{ days}$$

10. Calculate the detention time in hours.

$$\text{Detention Time, hours} = \frac{(\text{Ditch Volume, cu m})(24 \text{ hr/day})}{\text{Flow, cu m/day}}$$

$$= \frac{(700 \text{ cu m})(24 \text{ hr/day})}{800 \text{ cu m/day}}$$

$$= 21 \text{ hours}$$

A.42   Rotating Biological Contactors (RBCs)

EXAMPLE 11

A rotating biological contactor treats a flow of 5,000 cubic meters per day. The surface area of the media is 40,000 square meters (provided by manufacturer). What is the hydraulic loading in cu m/day/sq m?

Known		Unknown
Flow, cu m/day	= 5,000 cu m/day	Hydraulic Loading,
Surface Area, sq m	= 40,000 sq m	cu m/day/sq m

Calculate the hydraulic loading in cubic meters per day of wastewater per square meter of media surface.

$$\text{Hydraulic Loading, cu m/day/sq m} = \frac{\text{Flow, cu m/day}}{\text{Surface Area, sq m}}$$

$$= \frac{5,000 \text{ cu m/day}}{40,000 \text{ sq m}}$$

$$= 0.125 \text{ cu m/day/sq m}$$

EXAMPLE 12

The rotating biological contactor in Example 11 treats an industrial influent with a total BOD of 220 mg/L and suspended solids of 300 mg/L. Assume a K value of 0.5 to calculate the soluble BOD. What is the organic (BOD) loading in grams of soluble BOD per day per square meter of media surface?

Known		Unknown
Flow, cu m/day	= 5,000 cu m/day	Organic (BOD)
Surface Area, sq m	= 40,000 sq m	Loading,
Total BOD, mg/L	= 220 mg/L	gm/day/sq m
SS, mg/L	= 300 mg/L	
K	= 0.5	

a. Estimate the soluble BOD treated by the rotating biological contactor in milligrams per liter.

$$\text{Soluble BOD, mg/L} = \text{Total BOD, mg/L} - (\text{K} \times \text{Suspended Solids, mg/L})$$

$$= 220 \text{ mg/L} - (0.5 \times 300 \text{ mg/L})$$

$$= 220 \text{ mg/L} - 150 \text{ mg/L}$$

$$= 70 \text{ mg/L}$$

b. Determine the soluble BOD applied to the rotating biological contactor in grams of soluble BOD per day.

$$\text{Soluble BOD, gm/day} = \left(\begin{matrix}\text{Flow,}\\\text{cu m/day}\end{matrix}\right)\left(\begin{matrix}\text{Soluble BOD,}\\\text{mg/L}\end{matrix}\right)\left(\frac{1,000 \text{ kg}}{\text{cu m}}\right)\left(\frac{1,000 \text{ gm}}{\text{kg}}\right)$$

$$= \left(\frac{5,000 \text{ cu m}}{\text{day}}\right)\left(\frac{70 \text{ mg}}{1,000,000 \text{ mg}}\right)\left(\frac{1,000 \text{ kg}}{\text{cu m}}\right)\left(\frac{1,000 \text{ gm}}{\text{kg}}\right)$$

$$= 350,000 \text{ gm soluble BOD/day}$$

c. Calculate the organic (BOD) loading in grams of soluble BOD per day per square meter of media surface.

$$\text{Organic (BOD) Loading, gm/day/sq m} = \frac{\text{Soluble BOD, gm/day}}{\text{Surface Area of Media, sq m}}$$

$$= \frac{350,000 \text{ gm Soluble BOD/day}}{40,000 \text{ sq m}}$$

$$= 8.75 \text{ gm BOD/day/sq m}$$

A.43   Disinfection and Chlorination

EXAMPLE 13

A wastewater requires a chlorine feed rate of 6 kilograms chlorine per day. How many kilograms of chlorinated lime will be required per day to provide the needed chlorine? Assume each kilogram of chlorinated lime (hypochlorite) contains 0.34 kilogram of available chlorine.

Known		Unknown
Chlorine Required, kg/day	= 6 kg/day	Chlorinated Lime
Portion of Chlorine in kg of Hypochlorite, kg	= 0.34 kg	Feed Rate, kg/day

Calculate kilograms of chlorinated lime required per day.

$$\text{Chlorinated Lime Feed Rate, kg/day} = \frac{\text{Chlorine Required, kg/day}}{\text{Portion of Chlorine in kg of Hypochlorite, kg}}$$

$$= \frac{6 \text{ kg/day}}{0.34 \text{ kg}}$$

$$= 17.8 \text{ kg chlorinated lime/day}$$

EXAMPLES 14 and 15

A hypochlorinator is set to feed a dose of 15 kilograms of chlorine per 24 hours. The wastewater flow is 1,400 cu m/day and the chlorine, as measured by the chlorine residual test after 30 minutes of contact time, is 0.6 mg/L. Find the chlorine dosage and the chlorine demand in mg/L.

Known		Unknown
Chlorine Feed, kg/day	= 15 kg/day	14. Chlorine Dose, mg/L
Flow, cu m/day	= 1,400 cu m/day	15. Chlorine Demand, mg/L
Chlorine Residual, mg/L	= 0.6 mg/L	

14. Calculate the chlorine dose in milligrams per liter.

$$\text{Chlorine Feed or Dose, mg/L} = \frac{(\text{Chlorine Feed, kg/day})(1,000 \text{ gm/kg})(1,000 \text{ mg/gm})}{(\text{Flow, cu m/day})(1,000 \text{ L/cu m})}$$

$$= \frac{(15 \text{ kg/day})(1,000 \text{ gm/kg})(1,000 \text{ mg/gm})}{(1,400 \text{ cu m/day})(1,000 \text{ L/cu m})}$$

$$= 10.7 \text{ mg/L}$$

15. Determine the chlorine demand in milligrams per liter.

$$\text{Chlorine Demand, mg/L} = \text{Chlorine Dose, mg/L} - \text{Chlorine Residual, mg/L}$$

$$= 10.7 \text{ mg/L} - 0.6 \text{ mg/L}$$

$$= 10.1 \text{ mg/L}$$

EXAMPLES 16 and 17

Calculate the chlorine dose in milligrams per liter and the hypochlorinator setting in kilograms per 24 hours to treat a wastewater with a chlorine demand of 12 mg/L, when a chlorine residual of 1.5 mg/L is desired. The flow is 1,100 cu m per day.

Known		Unknown
Chlorine Demand, mg/L	= 12 mg/L	16. Chlorine Dose, mg/L
Chlorine Residual, mg/L	= 1.5 mg/L	17. Hypochlorinator Setting, kg/day
Flow, cu m/day	= 1,100 cu m/day	

16. Calculate the chlorine dose in milligrams per liter.

$$\text{Chlorine Dose, mg/L} = \text{Chlorine Demand, mg/L} + \text{Chlorine Residual, mg/L}$$

$$= 12 \text{ mg/L} + 1.5 \text{ mg/L}$$

$$= 13.5 \text{ mg/L}$$

17. Calculate the hypochlorinator setting in kilograms per day.

$$\text{Hypochlorinator Setting, kg/day} = \left(\text{Flow, cu m/day}\right)\left(\text{Dose, mg/L}\right)\left(\frac{1 \text{ kg}}{1,000,000 \text{ mg}}\right)\left(\frac{1,000 \text{ L}}{1 \text{ cu m}}\right)$$

$$= (1,100 \text{ cu m/day})(13.5 \text{ mg/L})\left(\frac{1 \text{ kg}}{1,000,000 \text{ mg}}\right)\left(\frac{1,000 \text{ L}}{1 \text{ cu m}}\right)$$

$$= 14.9 \text{ kg/day}$$

$$= 14.9 \text{ kg/24 hours}$$

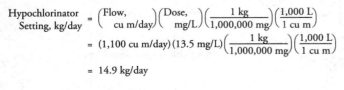

A.44 Land Treatment

EXAMPLES 18, 19, and 20

A plot of land 500 meters long by 200 meters wide is used for a land disposal system. If 2,000 cubic meters of water is applied to the land during a 24-hour period, calculate the land area in hectares and the hydraulic loading in cubic meters per day per hectare and also in centimeters per day.

Known		Unknown
Length, m	= 500 m	18. Area, hectares
Width, m	= 200 m	19. Hydraulic Loading, cu m/day/ha
Flow, cu m/day	= 2,000 cu m/day	20. Hydraulic Loading, cm/day

18. Determine the surface area in hectares.

$$\text{Area, hectares} = \frac{(\text{Length, m})(\text{Width, m})}{10,000 \text{ sq m/ha}}$$

$$= \frac{(500 \text{ m})(200 \text{ m})}{10,000 \text{ sq m/ha}}$$

$$= 10 \text{ hectares}$$

19. Determine the hydraulic loading in cubic meters per day per hectare.

$$\text{Hydraulic Loading, cu m/day/ha} = \frac{\text{Flow, cu m/day}}{\text{Area, ha}}$$

$$= \frac{2,000 \text{ cu m/day}}{10 \text{ ha}}$$

$$= 200 \text{ cu m/day/ha}$$

20. Determine the hydraulic loading in centimeters per day.

$$\text{Hydraulic Loading, cm/day} = \frac{(\text{Flow, cu m/day})(100 \text{ cm/m})}{(\text{Length, m})(\text{Width, m})}$$

$$= \frac{(2,000 \text{ cu m/day})(100 \text{ cm/m})}{(500 \text{ m})(200 \text{ m})}$$

$$= 2 \text{ cm/day}$$

A.45   Laboratory

EXAMPLE 21

Convert a temperature of 77°F to degrees Celsius.

Known	Unknown
Temp, °F = 77°F	Temp, °C

Change 77°F to degrees Celsius.

$$\text{Temperature, °C} = (\text{°F} - 32°) \times \frac{5}{9}$$

$$= (77° - 32°) \times \frac{5}{9}$$

$$= 45° \times \frac{5}{9}$$

$$= 5° \times 5$$

$$= 25°C$$

EXAMPLE 22

Convert a temperature of 15°C to degrees Fahrenheit.

Known	Unknown
Temp, °C = 15°C	Temp, °F

Change 15°C to degrees Fahrenheit.

$$\text{Temperature, °F} = \left(\text{°C} \times \frac{9}{5}\right) + 32°$$

$$= \left(15° \times \frac{9}{5}\right) + 32°$$

$$= \frac{135°}{5} + 32°$$

$$= 27° + 32°$$

$$= 59°F$$

A.46   Management

EXAMPLES 23 and 24

The total revenues for a utility are $1,283,000 and the operating expenses for the utility are $978,000. The debt service expenses are $320,000. What is the operating ratio? What is the coverage ratio?

Known		Unknown
Total Revenue, $	= $1,283,000	23. Operating Ratio
Operating Expenses, $	= $978,000	24. Coverage Ratio
Debt Service Expenses, $	= $320,000	

23. Calculate the operating ratio.

$$\text{Operating Ratio} = \frac{\text{Total Revenue, \$}}{\text{Operating Expenses, \$}}$$

$$= \frac{\$1,283,000}{\$978,000}$$

$$= 1.31$$

24. Calculate the coverage ratio.

a. Calculate the nondebt expenses.

$$\text{Nondebt Expenses, \$} = \text{Operating Expenses, \$} - \text{Debt Service Expenses, \$}$$

$$= \$978,000 - \$320,000$$

$$= \$658,000$$

b. Calculate the coverage ratio.

$$\text{Coverage Ratio} = \frac{\text{Total Revenue, \$} - \text{Nondebt Expenses, \$}}{\text{Debt Service Expenses, \$}}$$

$$= \frac{\$1,283,000 - \$658,000}{\$320,000}$$

$$= 1.95$$

ABBREVIATIONS

°C	degrees Celsius	km	kilometers
°F	degrees Fahrenheit	kN	kilonewtons
µ	micron	kPa	kilopascals
µg	microgram	kW	kilowatts
µm	micrometer	kWh	kilowatt-hours
ac	acres	L	liters
ac-ft	acre-feet	lb	pounds
amp	amperes	lbs/sq in	pounds per square inch
atm	atmosphere	M	mega
CFM	cubic feet per minute	M	million
CFS	cubic feet per second	*M*	molar (or molarity)
Ci	curie	m	meters
cm	centimeters	mA	milliampere
cu ft	cubic feet	meq	milliequivalent
cu in	cubic inches	mg	milligrams
cu m	cubic meters	MGD	million gallons per day
cu yd	cubic yards	mg/L	milligrams per liter
D	dalton	min	minutes
dB	decibel	mL	milliliters
ft	feet or foot	mm	millimeters
ft-lb/min	foot-pounds per minute	N	newton
g	gravity	*N*	normal (or normality)
gal	gallons	nm	nanometer
gal/day	gallons per day	ohm	ohm
GFD	gallons of flux per square foot per day	Pa	pascal
gm	grams	pCi	picocurie
GPCD	gallons per capita per day	pCi/L	picocuries per liter
GPD	gallons per day	ppb	parts per billion
gpg	grains per gallon	ppm	parts per million
GPM	gallons per minute	psf	pounds per square foot
GPY	gallons per year	psi	pounds per square inch
gr	grains	psig	pounds per square inch gauge
ha	hectares	RPM	revolutions per minute
HP	horsepower	SCFM	standard cubic feet per minute
hr	hours	sec	seconds
Hz	hertz	*SI*	*Le Système International d'Unités*
in	inches	sq ft	square feet
J	joules	sq in	square inches
k	kilos	W	watt
kg	kilograms	yd	yards

WASTEWATER WORDS

A Summary of the Words Defined

in

SMALL WASTEWATER SYSTEM OPERATION AND MAINTENANCE

PROJECT PRONUNCIATION KEY

by Warren L. Prentice

The Project Pronunciation Key is designed to aid you in the pronunciation of new words. While this key is based primarily on familiar sounds, it does not attempt to follow any particular pronunciation guide. This key is designed solely to aid operators in this program.

You may find it helpful to refer to other available sources for pronunciation help. Each current standard dictionary contains a guide to its own pronunciation key. Each key will be different from each other and from this key. Examples of the difference between the key used in this program and the *WEBSTER'S NEW WORLD COLLEGE DICTIONARY*[1] key are shown below.

In using this key, you should accent (say louder) the syllable that appears in capital letters. The following chart is presented to give examples of how to pronounce words using the project key.

| | SYLLABLE | | | | |
WORD	1st	2nd	3rd	4th	5th
aerobic	air	O	bick		
bacteria	back	TEER	e	uh	
contamination	kun	TAM	uh	NAY	shun

The first word, *AEROBIC*, has its second syllable accented. The second word, *BACTERIA*, has its second syllable accented. The third word, *CONTAMINATION*, has its second and fourth syllables accented.

Term	Project Key	Webster's Key
aerobic	air-O-bick	er ō´ bik
bacteria	back-TEER-e-uh	bak tir´ ē ə
contamination	kun-TAM-uh-NAY-shun	kən tam ə nā´ shən

---

[1] The *WEBSTER'S NEW WORLD COLLEGE DICTIONARY*, Fourth Edition, 1999, was chosen rather than an unabridged dictionary because of its availability to the operator. Other editions may be slightly different.

WORDS

>GREATER THAN >GREATER THAN

DO >5 mg/L would be read as DO GREATER THAN 5 mg/L.

<LESS THAN <LESS THAN

DO <5 mg/L would be read as DO LESS THAN 5 mg/L.

A

ACEOPS ACEOPS

See ALLIANCE OF CERTIFIED OPERATORS, LABORATORY ANALYSTS, INSPECTORS, AND SPECIALISTS (ACEOPS).

ABSORPTION (ab-SORP-shun) ABSORPTION

The taking in or soaking up of one substance into the body of another by molecular or chemical action (as tree roots absorb dissolved nutrients in the soil).

ACCOUNTABILITY ACCOUNTABILITY

When a manager gives power/responsibility to an employee, the employee ensures that the manager is informed of results or events.

ACTIVATED SLUDGE ACTIVATED SLUDGE

Sludge particles produced in raw or settled wastewater (primary effluent) by the growth of organisms (including zoogleal bacteria) in aeration tanks in the presence of dissolved oxygen. The term activated comes from the fact that the particles are teeming with bacteria, fungi, and protozoa. Activated sludge is different from primary sludge in that the sludge particles contain many living organisms that can feed on the incoming wastewater.

ACTIVATED SLUDGE PROCESS ACTIVATED SLUDGE PROCESS

A biological wastewater treatment process that speeds up the decomposition of wastes in the wastewater being treated. Activated sludge is added to wastewater and the mixture (mixed liquor) is aerated and agitated. After some time in the aeration tank, the activated sludge is allowed to settle out by sedimentation and is disposed of (wasted) or reused (returned to the aeration tank) as needed. The remaining wastewater then undergoes more treatment.

ACUTE HEALTH EFFECT ACUTE HEALTH EFFECT

An adverse effect on a human or animal body, with symptoms developing rapidly.

ADSORPTION (add-SORP-shun) ADSORPTION

The gathering of a gas, liquid, or dissolved substance on the surface or interface zone of another material.

ADVANCED WASTE TREATMENT ADVANCED WASTE TREATMENT

Any process of water renovation that upgrades treated wastewater to meet specific reuse requirements. May include general cleanup of water or removal of specific parts of wastes insufficiently removed by conventional treatment processes. Typical processes include chemical treatment and pressure filtration. Also called tertiary treatment.

AERATION (air-A-shun) LIQUOR AERATION LIQUOR

Mixed liquor. The contents of the aeration tank, including living organisms and material carried into the tank by either untreated wastewater or primary effluent.

AERATION (air-A-shun) TANK   AERATION TANK

The tank where raw or settled wastewater is mixed with return sludge and aerated. The same as aeration bay, aerator, or reactor.

AEROBES   AEROBES

Bacteria that must have dissolved oxygen (DO) to survive. Aerobes are aerobic bacteria.

AEROBIC (air-O-bick)   AEROBIC

A condition in which atmospheric or dissolved oxygen is present in the aquatic (water) environment.

AEROBIC BACTERIA (air-O-bick back-TEER-e-uh)   AEROBIC BACTERIA

Bacteria that will live and reproduce only in an environment containing oxygen that is available for their respiration (breathing), namely atmospheric oxygen or oxygen dissolved in water. Oxygen combined chemically, such as in water molecules (H_2O), cannot be used for respiration by aerobic bacteria.

AEROBIC (air-O-bick) DIGESTION   AEROBIC DIGESTION

The breakdown of wastes by microorganisms in the presence of dissolved oxygen. This digestion process may be used to treat only waste activated sludge, or trickling filter sludge and primary (raw) sludge, or waste sludge from activated sludge treatment plants designed without primary settling. The sludge to be treated is placed in a large aerated tank where aerobic microorganisms decompose the organic matter in the sludge. This is an extension of the activated sludge process.

AESTHETIC (es-THET-ick)   AESTHETIC

Attractive or appealing.

AGGLOMERATION (uh-glom-er-A-shun)   AGGLOMERATION

The growing or coming together of small scattered particles into larger flocs or particles, which settle rapidly. Also see FLOC.

AIR BINDING   AIR BINDING

The clogging of a filter, pipe, or pump due to the presence of air released from water. Air entering the filter media is harmful to both the filtration and backwash processes. Air can prevent the passage of water during the filtration process and can cause the loss of filter media during the backwash process.

AIR GAP   AIR GAP

An open, vertical drop, or vertical empty space, between a drinking (potable) water supply and potentially contaminated water. This gap prevents the contamination of drinking water by backsiphonage because there is no way potentially contaminated water can reach the drinking water supply.

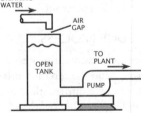

AIR LIFT PUMP   AIR LIFT PUMP

A special type of pump consisting of a vertical riser pipe submerged in the wastewater or sludge to be pumped. Compressed air is injected into a tail piece at the bottom of the pipe. Fine air bubbles mix with the wastewater or sludge to form a mixture lighter than the surrounding water, which causes the mixture to rise in the discharge pipe to the outlet.

AIR RELEASE   AIR RELEASE

A type of valve used to allow air caught in high spots in pipes to escape.

ALGAE (AL-jee)   ALGAE

Microscopic plants containing chlorophyll that live floating or suspended in water. They also may be attached to structures, rocks, or other submerged surfaces. Excess algal growths can impart tastes and odors to potable water. Algae produce oxygen during sunlight hours and use oxygen during the night hours. Their biological activities appreciably affect the pH, alkalinity, and dissolved oxygen of the water.

ALGAL (AL-gull) BLOOM

Sudden, massive growths of microscopic and macroscopic plant life, such as green or blue-green algae, which can, under the proper conditions, develop in lakes, reservoirs, and ponds.

ALIQUOT (AL-uh-kwot)

Representative portion of a sample. Often, an equally divided portion of a sample.

ALLIANCE OF CERTIFIED OPERATORS,
   LABORATORY ANALYSTS, INSPECTORS,
   AND SPECIALISTS (ACEOPS)

A professional organization for operators, laboratory analysts, inspectors, and specialists dedicated to improving professionalism; expanding training, certification, and job opportunities; increasing information exchange; and advocating the importance of certified operators, lab analysts, inspectors, and specialists. For information on membership, contact ACEOPS, 3130 Pierce Street, Suite 100, Sioux City, IA 51104-3942, phone (712) 258-3464, or email: Info@aceops.org.

AMBIENT (AM-bee-ent) TEMPERATURE

Temperature of the surroundings.

AMPERAGE (AM-purr-age)

The strength of an electric current measured in amperes. The amount of electric current flow, similar to the flow of water in gallons per minute.

AMPERE (AM-peer)

The unit used to measure current strength. The current produced by an electromotive force of one volt acting through a resistance of one ohm.

AMPEROMETRIC (am-purr-o-MET-rick)

A method of measurement that records electric current flowing or generated, rather than recording voltage. Amperometric titration is a means of measuring concentrations of certain substances in water.

AMPLITUDE

The maximum strength of an alternating current during its cycle, as distinguished from the mean or effective strength.

ANAEROBES

Bacteria that do not need dissolved oxygen (DO) to survive.

ANAEROBIC (AN-air-O-bick)

A condition in which atmospheric or dissolved oxygen (DO) is *NOT* present in the aquatic (water) environment.

ANAEROBIC BACTERIA (AN-air-O-bick back-TEER-e-uh)

Bacteria that live and reproduce in an environment containing no free or dissolved oxygen. Anaerobic bacteria obtain their oxygen supply by breaking down chemical compounds that contain oxygen, such as sulfate (SO_4^{2-}).

ANAEROBIC (AN-air-O-bick) DECOMPOSITION

The decay or breaking down of organic material in an environment containing no free or dissolved oxygen.

ANAEROBIC (AN-air-O-bick) DIGESTER

A wastewater solids treatment device in which the solids and water (about 5 percent solids, 95 percent water) are placed in a large tank where bacteria decompose the solids in the absence of dissolved oxygen.

ANODE (AN-ode)

The positive pole or electrode of an electrolytic system, such as a battery. The anode attracts negatively charged particles or ions (anions).

ALGAL BLOOM

ALIQUOT

ALLIANCE OF CERTIFIED OPERATORS,
LABORATORY ANALYSTS, INSPECTORS,
AND SPECIALISTS (ACEOPS)

AMBIENT TEMPERATURE

AMPERAGE

AMPERE

AMPEROMETRIC

AMPLITUDE

ANAEROBES

ANAEROBIC

ANAEROBIC BACTERIA

ANAEROBIC DECOMPOSITION

ANAEROBIC DIGESTER

ANODE

ANOXIC (an-OX-ick)   ANOXIC

A condition in which the aquatic (water) environment does not contain dissolved oxygen (DO), which is called an oxygen deficient condition. Generally refers to an environment in which chemically bound oxygen, such as in nitrate, is present. The term is similar to ANAEROBIC.

APPURTENANCE (uh-PURR-ten-nans)   APPURTENANCE

Machinery, appliances, structures, and other parts of the main structure necessary to allow it to operate as intended, but not considered part of the main structure.

AQUIFER (ACK-wi-fer)   AQUIFER

A natural, underground layer of porous, water-bearing materials (sand, gravel) usually capable of yielding a large amount or supply of water.

ARTIFICIAL GROUNDWATER TABLE   ARTIFICIAL GROUNDWATER TABLE

A groundwater table that is changed by artificial means. Examples of activities that artificially raise the level of a groundwater table include agricultural irrigation, dams, and excessive sewer line exfiltration. A groundwater table can be artificially lowered by sewer line infiltration, water wells, and similar drainage methods.

ASEPTIC (a-SEP-tick)   ASEPTIC

Free from the living germs of disease, fermentation, or putrefaction. Sterile.

ASHING   ASHING

Formation of an activated sludge floc in a clarifier effluent that is well oxidized and floats on the water surface (has the appearance of gray ash).

ASSET MANAGEMENT   ASSET MANAGEMENT

The process of maintaining the functionality and value of a utility's assets through repair, rehabilitation, and replacement. Examples of utility assets include buildings, tools, equipment, pipes, and machinery used to operate a water or wastewater system. The primary goal of asset management is to provide safe, reliable, and cost-effective service to a community over the useful life of a utility's assets.

AUGER (AW-grr)   AUGER

A sharp tool used to go through and break up or remove various materials that become lodged in sewers.

AUTHORITY   AUTHORITY

The power and resources to do a specific job or to get that job done.

AVAILABLE EXPANSION   AVAILABLE EXPANSION

The vertical distance from the sand surface to the underside of a trough in a sand filter. This distance is also called freeboard.

AVAILABLE MOISTURE CONTENT   AVAILABLE MOISTURE CONTENT

The quantity of water present that is available for use by vegetation.

B

BOD (pronounce as separate letters)   BOD

Biochemical Oxygen Demand. The rate at which organisms use the oxygen in water or wastewater while stabilizing decomposable organic matter under aerobic conditions. In decomposition, organic matter serves as food for the bacteria and energy results from its oxidation. BOD measurements are used as a surrogate measure of the organic strength of wastes in water.

BOD₅ BOD₅

BOD_5 refers to the five-day biochemical oxygen demand. The total amount of oxygen used by microorganisms decomposing organic matter increases each day until the ultimate BOD is reached, usually in 50 to 70 days. BOD usually refers to the five-day BOD or BOD_5.

BACKFLOW BACKFLOW

A reverse flow condition, created by a difference in water pressures, that causes water to flow back into the distribution pipes of a potable water supply from any source or sources other than an intended source. Also see BACKSIPHONAGE.

BACKUP BACKUP

An overflow or accumulation of water caused by clogging or by a stoppage.

BACTERIA (back-TEER-e-uh) BACTERIA

Bacteria are living organisms, microscopic in size, that usually consist of a single cell. Most bacteria use organic matter for their food and produce waste products as a result of their life processes.

BACTERIAL (back-TEER-e-ul) CULTURE BACTERIAL CULTURE

In the case of activated sludge, the bacterial culture refers to the group of bacteria classified as AEROBES and FACULTATIVE BACTERIA, which covers a wide range of organisms. Most treatment processes in the United States grow facultative bacteria that use the carbonaceous (carbon compounds) BOD. Facultative bacteria can live when oxygen resources are low. When nitrification is required, the nitrifying organisms are OBLIGATE AEROBES (require oxygen) and must have at least 0.5 mg/L of dissolved oxygen throughout the whole system to function properly.

BALANCED SCORECARD BALANCED SCORECARD

A strategic planning and management system used extensively by water and wastewater facility managers to align business activities to the vision and strategy of the organization, improve communications, and monitor organizational performance against strategic goals. More information is available at www.balancedscorecard.org.

BASE-EXTRA CAPACITY METHOD BASE-EXTRA CAPACITY METHOD

A cost allocation method used by utilities to determine rates for various groups. This method considers base costs (O&M expenses and capital costs), extra capacity costs (additional costs for maximum day and maximum hour demands), customer costs (meter maintenance and reading, billing, collection, accounting), and fire protection costs.

BATCH PROCESS BATCH PROCESS

A treatment process in which a tank or reactor is filled, the wastewater (or other solution) is treated or a chemical solution is prepared, and the tank is emptied. The tank may then be filled and the process repeated. Batch processes are also used to cleanse, stabilize, or condition chemical solutions for use in industrial manufacturing and treatment processes.

BENCHMARK BENCHMARK

A standard or point of reference used to judge or measure quality or value.

BENCHMARKING BENCHMARKING

A process an agency uses to gather and compare information about the productivity and performance of other similar agencies with its own information. The purpose of benchmarking is to identify best practices, set improvement targets, and measure progress.

BIOASSAY (BUY-o-AS-say) BIOASSAY

(1) A way of showing or measuring the effect of biological treatment on a particular substance or waste.

(2) A method of determining the relative toxicity of a test sample of industrial wastes or other wastes by using live test organisms, such as fish.

BIOAUGMENTATION

BIOAUGMENTATION

The addition of bacterial cultures to speed up the breakdown of grease and other organic materials. Bioaugmentation is used to clean sewers and, on a preventive basis, to remove deposits in sewers. This method is also used to prevent grease buildup in lift station wet wells.

BIOCHEMICAL OXYGEN DEMAND (BOD)

BIOCHEMICAL OXYGEN DEMAND (BOD)

See BOD.

BIODEGRADABLE (BUY-o-dee-GRADE-able)

BIODEGRADABLE

Organic matter that can be broken down by bacteria to more stable forms that will not create a nuisance or give off foul odors is considered biodegradable.

BIOFLOCCULATION (BUY-o-flock-yoo-LAY-shun)

BIOFLOCCULATION

The clumping together of fine, dispersed organic particles by the action of certain bacteria and algae. This results in faster and more complete settling of the organic solids in wastewater.

BIOMASS (BUY-o-mass)

BIOMASS

A mass or clump of organic material consisting of living organisms feeding on the wastes in wastewater, dead organisms, and other debris. Also see ZOOGLEAL MASS.

BIOMONITORING

BIOMONITORING

A term used to describe methods of evaluating or measuring the effects of toxic substances in effluents on aquatic organisms in receiving waters. There are two types of biomonitoring, the BIOASSAY and the BIOSURVEY.

BIOSURVEY

BIOSURVEY

A survey of the types and numbers of organisms naturally present in the receiving waters upstream and downstream from plant effluents. Comparisons are made between the aquatic organisms upstream and those organisms downstream of the discharge.

BLANK

BLANK

A bottle containing only dilution water or distilled water; the sample being tested is not added. Tests are frequently run on a sample and a blank and the differences are compared. The procedure helps to eliminate or reduce test result errors that could be caused when the dilution water or distilled water used is contaminated.

BLOCKAGE

BLOCKAGE

Any incident in which a sewer is partially or completely blocked, causing a backup, a service interruption, or an overflow. Also called stoppage.

BOND

BOND

(1) A written promise to pay a specified sum of money (called the face value) at a fixed time in the future (called the date of maturity). A bond also carries interest at a fixed rate, payable periodically. The difference between a note and a bond is that a bond usually runs for a longer period of time and requires greater formality. Utility agencies use bonds as a means of obtaining large amounts of money for capital improvements.

(2) A warranty by an underwriting organization, such as an insurance company, guaranteeing honesty, performance, or payment by a contractor.

BREAKOUT OF CHLORINE (CHLORINE BREAKAWAY)

BREAKOUT OF CHLORINE (CHLORINE BREAKAWAY)

A point at which chlorine leaves solution as a gas because the chlorine feed rate is too high. The solution is saturated and cannot dissolve any more chlorine. The maximum strength a chlorine solution can attain is approximately 3,500 mg/L. Beyond this concentration molecular chlorine will break out of solution and cause off-gassing at the point of application.

BREAKPOINT CHLORINATION

BREAKPOINT CHLORINATION

Addition of chlorine to water or wastewater until the chlorine demand has been satisfied. At this point, further additions of chlorine will result in a free chlorine residual that is directly proportional to the amount of chlorine added beyond the breakpoint.

BUFFER

A solution or liquid whose chemical makeup neutralizes acids or bases without a great change in pH.

BUFFER CAPACITY

A measure of the capacity of a solution or liquid to neutralize acids or bases. This is a measure of the capacity of water or wastewater for offering a resistance to changes in pH.

BULKING

Clouds of billowing sludge that occur throughout secondary clarifiers and sludge thickeners when the sludge does not settle properly. In the activated sludge process, bulking is usually caused by filamentous bacteria or bound water.

BURPING

A term used to describe what happens when billowing solids are swept by the water up and out over the effluent weirs in the discharged effluent. Billowing solids result when the settling tank sludge blanket becomes too deep (occupies too much volume in the bottom of the tank).

C

CERCLA (SIRK-la)

Comprehensive Environmental Response, Compensation, and Liability Act of 1980. This act was passed primarily to correct past mistakes in industrial waste management. The focus of the act is to locate hazardous waste disposal sites that are creating problems through pollution of the environment and, by proper funding and implementation of study and corrective activities, eliminate the problem from these sites. Current users of CERCLA-identified substances must report releases of these substances to the environment when they take place (not just historic ones). This act is also called the Superfund Act. Also see SARA.

CHEMTREC (KEM-trek)

Chemical Transportation Emergency Center. A public service of the American Chemistry Council dedicated to assisting emergency responders deal with incidents involving hazardous materials. Their toll-free 24-hour emergency phone number is (800) 424-9300.

COD (pronounce as separate letters)

Chemical Oxygen Demand. A measure of the oxygen-consuming capacity of organic matter present in wastewater. COD is expressed as the amount of oxygen consumed from a chemical oxidant in mg/L during a specific test. Results are not necessarily related to the biochemical oxygen demand (BOD) because the chemical oxidant may react with substances that bacteria do not stabilize.

CALL DATE

First date a bond can be paid off.

CAPILLARY (KAP-uh-larry) ACTION

The movement of water through very small spaces due to molecular forces.

CAPITAL IMPROVEMENT PLAN (CIP)

A detailed plan that identifies requirements for the repair, replacement, and rehabilitation of facility infrastructure over an extended period, often 20 years or more. A utility usually updates or prepares this plan annually. The plan consists of programs and projects to upgrade and rehabilitate wastewater collection and treatment systems and increase their capacity to allow for future growth.

CATHODIC (kath-ODD-ick) PROTECTION

An electrical system for prevention of rust, corrosion, and pitting of metal surfaces that are in contact with water, wastewater, or soil. A low-voltage current is made to flow through a liquid (water) or a soil in contact with the metal in such a manner that the external electromotive force renders the metal structure cathodic. This concentrates corrosion on auxiliary anodic parts, which are deliberately allowed to corrode instead of letting the structure corrode.

CATION (KAT-EYE-en) EXCHANGE CAPACITY

The ability of a soil or other solid to exchange cations (positive ions such as calcium, Ca^{2+}) with a liquid.

CAVITATION (kav-uh-TAY-shun)

CAVITATION

The formation and collapse of a gas pocket or bubble on the blade of an impeller or the gate of a valve. The collapse of this gas pocket or bubble drives water into the impeller or gate with a terrific force that can knock metal particles off and cause pitting on the impeller or gate surface. Cavitation is accompanied by loud noises that sound like someone is pounding on the impeller or gate with a hammer.

CENTRIFUGE

CENTRIFUGE

A mechanical device that uses centrifugal or rotational forces to separate solids from liquids.

CERTIFICATION EXAMINATION

CERTIFICATION EXAMINATION

An examination administered by a state agency or professional association that operators take to indicate a level of professional competence. In the United States, certification of operators of water treatment plants, wastewater treatment plants, water distribution systems, and small water supply systems is mandatory. In many states, certification of wastewater collection system operators, industrial wastewater treatment plant operators, pretreatment facility inspectors, and small wastewater system operators is voluntary; however, current trends indicate that more states, provinces, and employers will require these operators to be certified in the future. Operator certification is mandatory in the United States for the Chief Operators of water treatment plants, water distribution systems, and wastewater treatment plants.

CERTIFIED OPERATOR

CERTIFIED OPERATOR

A person who has the education and experience required to operate a specific class of treatment facility as indicated by possessing a certificate of professional competence given by a state agency or professional association.

CHAIN OF CUSTODY

CHAIN OF CUSTODY

A record of each person involved in the handling and possession of a sample from the person who collected the sample to the person who analyzed the sample in the laboratory and to the person who witnessed disposal of the sample.

CHEMICAL OXYGEN DEMAND (COD)

CHEMICAL OXYGEN DEMAND (COD)

A measure of the oxygen-consuming capacity of organic matter present in wastewater. COD is expressed as the amount of oxygen consumed from a chemical oxidant in mg/L during a specific test. Results are not necessarily related to the biochemical oxygen demand (BOD) because the chemical oxidant may react with substances that bacteria do not stabilize.

CHLORAMINES (KLOR-uh-means)

CHLORAMINES

Compounds formed by the reaction of hypochlorous acid (or aqueous chlorine) with ammonia.

CHLORINATION (klor-uh-NAY-shun)

CHLORINATION

The application of chlorine to water or wastewater, generally for the purpose of disinfection, but frequently for accomplishing other biological or chemical results.

CHLORINE CONTACT CHAMBER

CHLORINE CONTACT CHAMBER

A baffled basin that provides sufficient detention time of chlorine contact with wastewater for disinfection to occur. The minimum contact time is usually 30 minutes. Also commonly referred to as basin or tank.

CHLORINE BREAKAWAY

CHLORINE BREAKAWAY

See BREAKOUT OF CHLORINE.

CHLORINE DEMAND

CHLORINE DEMAND

Chlorine demand is the difference between the amount of chlorine added to water or wastewater and the amount of chlorine residual remaining after a given contact time. Chlorine demand may change with dosage, time, temperature, pH, and nature and amount of the impurities in the water.

Chlorine Demand, mg/L = Chlorine Applied, mg/L – Chlorine Residual, mg/L

CHLORINE REQUIREMENT

CHLORINE REQUIREMENT

The amount of chlorine that is needed for a particular purpose. Some reasons for adding chlorine are reducing the MPN (Most Probable Number) of coliform bacteria, obtaining a particular chlorine residual, or oxidizing some substance in the water. In each case, a definite dosage of chlorine will be necessary. This dosage is the chlorine requirement.

CHLORINE RESIDUAL
CHLORINE RESIDUAL

The concentration of chlorine present in water after the chlorine demand has been satisfied. The concentration is expressed in terms of the total chlorine residual, which includes both the free and combined or chemically bound chlorine residuals. Also called residual chlorine.

CHLORINE RESIDUAL ANALYZER
CHLORINE RESIDUAL ANALYZER

An instrument used to measure chlorine residual in water or wastewater. This instrument also can be used to control the chlorine dose rate.

CHLORORGANIC (klor-or-GAN-ick)
CHLORORGANIC

Organic compounds combined with chlorine. These compounds generally originate from, or are associated with, living or dead organic materials, including algae in water.

CHRONIC HEALTH EFFECT
CHRONIC HEALTH EFFECT

An adverse effect on a human or animal body with symptoms that develop slowly over a long period of time or that recur frequently.

CLEAR ZONE
CLEAR ZONE

See SUPERNATANT.

COAGULATION (ko-agg-yoo-LAY-shun)
COAGULATION

The clumping together of very fine particles into larger particles (floc) caused by the use of chemicals (coagulants). The chemicals neutralize the electrical charges of the fine particles, allowing them to come closer and form larger clumps. This clumping together makes it easier to separate the solids from the water by settling, skimming, draining, or filtering.

CODE OF FEDERAL REGULATIONS (CFR)
CODE OF FEDERAL REGULATIONS (CFR)

A publication of the US government that contains all of the proposed and finalized federal regulations, including safety and environmental regulations.

COLIFORM (KOAL-i-form)
COLIFORM

A group of bacteria found in the intestines of warm-blooded animals (including humans) and also in plants, soil, air, and water. The presence of coliform bacteria is an indication that the water is polluted and may contain pathogenic (disease-causing) organisms. Fecal coliforms are those coliforms found in the feces of various warm-blooded animals, whereas the term coliform also includes other environmental sources.

COLORIMETRIC MEASUREMENT
COLORIMETRIC MEASUREMENT

A means of measuring unknown chemical concentrations in water by measuring a sample's color intensity. The specific color of the sample, developed by addition of chemical reagents, is measured with a photoelectric colorimeter or is compared with color standards using, or corresponding with, known concentrations of the chemical.

COMBINED AVAILABLE CHLORINE
COMBINED AVAILABLE CHLORINE

The total chlorine, present as chloramine or other derivatives, that is present in a water and is still available for disinfection and for oxidation of organic matter. The combined chlorine compounds are more stable than free chlorine forms, but they are somewhat slower in disinfection action.

COMBINED AVAILABLE CHLORINE RESIDUAL
COMBINED AVAILABLE CHLORINE RESIDUAL

The concentration of chlorine residual that is combined with ammonia, organic nitrogen, or both in water as a chloramine (or other chloro derivative) and yet is still available to oxidize organic matter and help kill bacteria.

COMBINED CHLORINE
COMBINED CHLORINE

The sum of the chlorine species composed of free chlorine and ammonia, including monochloramine, dichloramine, and trichloramine (nitrogen trichloride). Dichloramine is the strongest disinfectant of these chlorine species, but it has less oxidative capacity than free chlorine.

COMMINUTION (kom-mih-NEW-shun)   COMMINUTION

A mechanical treatment process that cuts large pieces of wastes into smaller pieces so they will not plug pipes or damage equipment. Comminution and shredding usually mean the same thing.

COMMINUTOR (kom-mih-NEW-ter)   COMMINUTOR

A device used to reduce the size of the solid materials in wastewater by shredding (comminution). The shredding action is like many scissors cutting to shreds all the large solids in the wastewater.

COMMODITY-DEMAND METHOD   COMMODITY-DEMAND METHOD

A cost allocation method used by water utilities to determine water rates for the various water user groups. This method considers the commodity costs (water, chemicals, power, amount of water use), demand costs (treatment, storage, distribution), customer costs (meter maintenance and reading, billing, collection, accounting), and fire protection costs.

COMMUNITY RIGHT-TO-KNOW   COMMUNITY RIGHT-TO-KNOW

The Superfund Amendments and Reauthorization Act (SARA) of 1986 provides statutory authority for communities to develop right-to-know laws. The act establishes a state and local emergency planning structure, emergency notification procedures, and reporting requirements for facilities. Also see RIGHT-TO-KNOW LAWS and SARA.

COMPOSITE (PROPORTIONAL) SAMPLE   COMPOSITE (PROPORTIONAL) SAMPLE

A composite sample is a collection of individual samples obtained at regular intervals, usually every one or two hours during a 24-hour time span. Each individual sample is combined with the others in proportion to the rate of flow when the sample was collected. Equal volume individual samples also may be collected at intervals after a specific volume of flow passes the sampling point or after equal time intervals and still be referred to as a composite sample. The resulting mixture (composite sample) forms a representative sample and is analyzed to determine the average conditions during the sampling period.

COMPOUND   COMPOUND

A pure substance composed of two or more elements whose composition is constant. For example, table salt (sodium chloride, NaCl) is a compound.

CONFINED SPACE   CONFINED SPACE

Confined space means a space that:

(1) Is large enough and so configured that an employee can bodily enter and perform assigned work; and

(2) Has limited or restricted means for entry or exit (for example, manholes, tanks, vessels, silos, storage bins, hoppers, vaults, and pits are spaces that may have limited means of entry); and

(3) Is not designed for continuous employee occupancy.

Also see DANGEROUS AIR CONTAMINATION and OXYGEN DEFICIENCY.

CONFINED SPACE, PERMIT-REQUIRED   CONFINED SPACE, PERMIT-REQUIRED
     (PERMIT SPACE)   (PERMIT SPACE)

A confined space that has one or more of the following characteristics:

(1) Contains or has a potential to contain a hazardous atmosphere,

(2) Contains a material that has the potential for engulfing an entrant,

(3) Has an internal configuration such that an entrant could be trapped or asphyxiated by inwardly converging walls or by a floor that slopes downward and tapers to a smaller cross section, or

(4) Contains any other recognized serious safety or health hazard.

CONING   CONING

Development of a cone-shaped flow of liquid, like a whirlpool, through sludge. This can occur in a sludge hopper during sludge withdrawal when the sludge becomes too thick. Part of the sludge remains in place while liquid rather than sludge flows out of the hopper. Also called coring.

CONTACT STABILIZATION

CONTACT STABILIZATION

Contact stabilization is a modification of the conventional activated sludge process. In contact stabilization, two aeration tanks are used. One tank is for separate reaeration of the return sludge for at least four hours before it is permitted to flow into the other aeration tank to be mixed with the primary effluent requiring treatment. The process may also occur in one long tank.

CORPORATION STOP

CORPORATION STOP

A water service shutoff valve located at a street water main. This valve cannot be operated from the ground surface because it is buried and there is no valve box. Also called a corporation cock.

COVERAGE RATIO

COVERAGE RATIO

The coverage ratio is a measure of the ability of the utility to pay the principal and interest on loans and bonds (this is known as debt service) in addition to any unexpected expenses.

CURRENT

CURRENT

A movement or flow of electricity. Electric current is measured by the number of coulombs per second flowing past a certain point in a conductor. A coulomb is equal to about 6.25×10^{18} electrons (6,250,000,000,000,000,000 electrons). A flow of one coulomb per second is called one ampere, the unit of the rate of flow of current.

CURVILINEAR (KER-vuh-LYNN-e-ur)

CURVILINEAR

In the shape of a curved line.

CYCLE

CYCLE

A complete alternation of voltage or current in an alternating current (AC) circuit.

D

DNA

DNA

Deoxyribonucleic acid. A chemical that encodes genetic information that is transmitted between generations of cells.

DANGEROUS AIR CONTAMINATION

DANGEROUS AIR CONTAMINATION

An atmosphere presenting a threat of causing death, injury, acute illness, or disablement due to the presence of flammable or explosive, toxic, or otherwise injurious or incapacitating substances.

(1) Dangerous air contamination due to the flammability of a gas, vapor, or mist is defined as an atmosphere containing the gas, vapor, or mist at a concentration greater than 10 percent of its lower explosive (lower flammable) limit (LEL).

(2) Dangerous air contamination due to a combustible particulate is defined as a concentration that meets or exceeds the particulate's lower explosive limit (LEL).

(3) Dangerous air contamination due to the toxicity of a substance is defined as the atmospheric concentration that could result in employee exposure in excess of the substance's permissible exposure limit (PEL).

NOTE: A dangerous situation also occurs when the oxygen level is less than 19.5 percent by volume (OXYGEN DEFICIENCY) or more than 23.5 percent by volume (OXYGEN ENRICHMENT).

DEBT SERVICE

DEBT SERVICE

The amount of money required annually to pay the (1) interest on outstanding debts, or (2) funds due on a maturing bonded debt or the redemption of bonds.

DECHLORINATION (DEE-klor-uh-NAY-shun)

DECHLORINATION

The removal of chlorine from the effluent of a treatment plant. Chlorine needs to be removed because chlorine is toxic to fish and other aquatic life.

DECIBEL (DES-uh-bull)

DECIBEL

A unit for expressing the relative intensity of sounds on a scale from zero for the average least perceptible sound to about 130 for the average level at which sound causes pain to humans. Abbreviated dB.

DECOMPOSITION or DECAY DECOMPOSITION or DECAY

The conversion of chemically unstable materials to more stable forms by chemical or biological action.

DELEGATION DELEGATION

The act in which power is given to another person in the organization to accomplish a specific job.

DENITRIFICATION (dee-NYE-truh-fuh-KAY-shun) DENITRIFICATION

(1) The anoxic biological reduction of nitrate nitrogen to nitrogen gas.

(2) The removal of some nitrogen from a system.

(3) An anoxic process that occurs when nitrite or nitrate ions are reduced to nitrogen gas and nitrogen bubbles are formed as a result of this process. The bubbles attach to the biological floc and float the floc to the surface of the secondary clarifiers. This condition is often the cause of rising sludge observed in secondary clarifiers or gravity thickeners. Also see NITRIFICATION.

DEPRECIATION DEPRECIATION

The gradual loss in service value of a facility or piece of equipment due to all the factors causing the ultimate retirement of the facility or equipment. This loss can be caused by sudden physical damage, wearing out due to age, obsolescence, inadequacy, or availability of a newer, more efficient facility or equipment. The value cannot be restored by maintenance.

DESICCANT (DESS-uh-kant) DESICCANT

A drying agent that is capable of removing or absorbing moisture from the atmosphere in a small enclosure.

DESICCATOR (DESS-uh-kay-tor) DESICCATOR

A closed container into which heated weighing or drying dishes are placed to cool in a dry environment in preparation for weighing. The dishes may be empty or they may contain a sample. Desiccators contain a substance (DESICCANT), such as anhydrous calcium chloride, that absorbs moisture and keeps the relative humidity near zero so that the dish or sample will not gain weight from absorbed moisture.

DETENTION TIME DETENTION TIME

The time required to fill a tank at a given flow or the theoretical time required for a given flow of wastewater to pass through a tank. In septic tanks, this detention time will decrease as the volumes of sludge and scum increase.

DEWATER DEWATER

(1) To remove or separate a portion of the water present in a sludge or slurry. To dry sludge so it can be handled and disposed of.

(2) To remove or drain the water from a tank or a trench. A structure may be dewatered so that it can be inspected or repaired.

DIFFUSED-AIR AERATION DIFFUSED-AIR AERATION

A diffused-air activated sludge plant takes air, compresses it, and then discharges the air below the water surface of the aerator through some type of air diffusion device.

DIFFUSER DIFFUSER

A device (porous plate, tube, bag) used to break the air stream from the blower system into fine bubbles in an aeration tank or reactor.

DISCHARGE HEAD DISCHARGE HEAD

The pressure (in pounds per square inch (psi) or kilopascals (kPa)) measured at the centerline of a pump discharge and very close to the discharge flange, converted into feet or meters. The pressure is measured from the centerline of the pump to the hydraulic grade line of the water in the discharge pipe.

Discharge Head, ft = (Discharge Pressure, psi)(2.31 ft/psi)

or

Discharge Head, m = (Discharge Pressure, kPa)(1 m/9.8 kPa)

DISINFECTION (dis-in-FECT-shun)

The process designed to kill or inactivate most microorganisms in water or wastewater, including essentially all pathogenic (disease-causing) bacteria. There are several ways to disinfect, with chlorination being the most frequently used in water and wastewater treatment plants. Compare with STERILIZATION.

DISSOLVED OXYGEN

Molecular oxygen dissolved in water or wastewater, usually abbreviated DO.

DISTILLATE (DIS-tuh-late)

In the distillation of a sample, a portion is collected by evaporation and recondensation; the part that is recondensed is the distillate.

DRAIN TILE SYSTEM

A system of tile pipes buried under agricultural fields that collect percolated waters and keep the groundwater table below the ground surface to prevent ponding.

DRAINAGE WELLS

Wells that can be pumped to lower the groundwater table and prevent ponding.

DROP JOINT

A sewer pipe joint where one part has dropped out of alignment. Also see VERTICAL OFFSET.

DRY PIT

See DRY WELL.

DRY WELL

A dry room or compartment in a lift station, near or below the water level, where the pumps are located, usually next to the wet well.

DUCKWEED

A small, green, cloverleaf-shaped floating plant, about one-quarter inch (6 mm) across, which appears as a grainy layer on the surface of a pond.

DYNAMIC HEAD

When a pump is operating, the vertical distance (in feet or meters) from a point to the energy grade line. Also see ENERGY GRADE LINE (EGL), STATIC HEAD, and TOTAL DYNAMIC HEAD (TDH).

DYNAMIC PRESSURE

When a pump is operating, pressure resulting from the dynamic head.

Dynamic Pressure, psi = (Dynamic Head, ft)(0.433 psi/ft)

or

Dynamic Pressure, kPa = (Dynamic Head, m)(9.8 kPa/m)

E

EGL

See ENERGY GRADE LINE (EGL).

EASEMENT

Legal right to use the property of others for a specific purpose. For example, a utility company may have a five-foot (1.5 m) easement along the property line of a home. This gives the utility the legal right to install and maintain a sewer line within the easement.

DISINFECTION

DISSOLVED OXYGEN

DISTILLATE

DRAIN TILE SYSTEM

DRAINAGE WELLS

DROP JOINT

DRY PIT

DRY WELL

DUCKWEED

DYNAMIC HEAD

DYNAMIC PRESSURE

EGL

EASEMENT

EDUCTOR (e-DUCK-ter)   EDUCTOR

A hydraulic device used to create a negative pressure (suction) by forcing a liquid through a restriction, such as a Venturi. An eductor or aspirator (the hydraulic device) may be used in the laboratory in place of a vacuum pump. As an injector, it is used to produce vacuum for chlorinators. Sometimes used instead of a suction pump.

EFFECTIVE SOIL DEPTH   EFFECTIVE SOIL DEPTH

The depth of soil in the leach field trench that provides a satisfactory percolation area for the septic tank effluent.

EFFLUENT (EF-loo-ent)   EFFLUENT

Water or other liquid—raw (untreated), partially treated, or completely treated—flowing *FROM* a reservoir, basin, treatment process, or treatment plant.

ELECTROLYSIS (ee-leck-TRAWL-uh-sis)   ELECTROLYSIS

The decomposition of material by an outside electric current.

ELECTROMOTIVE FORCE (EMF)   ELECTROMOTIVE FORCE (EMF)

The electrical pressure available to cause a flow of current (amperage) when an electric circuit is closed. Also called voltage.

ELECTRON   ELECTRON

(1) A very small, negatively charged particle that is practically weightless. According to the electron theory, all electrical and electronic effects are caused either by the movement of electrons from place to place or because there is an excess or lack of electrons at a particular place.

(2) The part of an atom that determines its chemical properties.

ELEMENT   ELEMENT

A substance that cannot be separated into its constituent parts and still retain its chemical identity. For example, sodium (Na) is an element.

ENCLOSED SPACE   ENCLOSED SPACE

See CONFINED SPACE.

END POINT   END POINT

The completion of a desired chemical reaction. Samples of water or wastewater are titrated to the end point. This means that a chemical is added, drop by drop, to a sample until a certain color change (blue to clear, for example) occurs. This is called the end point of the titration. In addition to a color change, an end point may be reached by the formation of a precipitate or the reaching of a specified pH. An end point may be detected by the use of an electronic device, such as a pH meter.

ENDOGENOUS (en-DODGE-en-us) RESPIRATION   ENDOGENOUS RESPIRATION

A situation in which living organisms oxidize some of their own cellular mass instead of new organic matter they adsorb or absorb from their environment.

ENERGY GRADE LINE (EGL)   ENERGY GRADE LINE (EGL)

A line that represents the elevation of energy head (in feet or meters) of water flowing in a pipe, conduit, or channel. The line is drawn above the hydraulic grade line (gradient) a distance equal to the velocity head ($V^2/2g$) of the water flowing at each section or point along the pipe or channel. Also see HYDRAULIC GRADE LINE (HGL).

[SEE DRAWING ON PAGE 445]

ENTRAIN   ENTRAIN

To trap bubbles in water either mechanically through turbulence or chemically through a reaction.

ENZYMES (EN-zimes)   ENZYMES

Organic substances (produced by living organisms) that cause or speed up chemical reactions. Organic catalysts or biochemical catalysts.

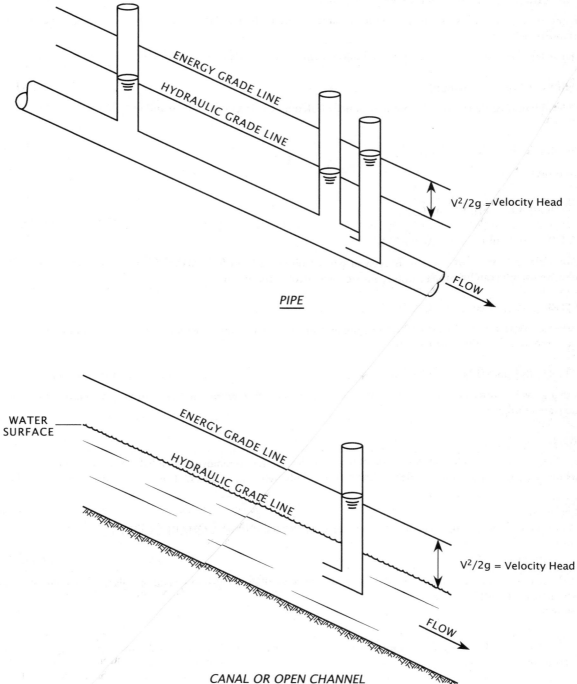

ENERGY GRADE LINE

HYDRAULIC GRADE LINE

$V^2/2g$ = Velocity Head

FLOW

PIPE

WATER SURFACE

ENERGY GRADE LINE

HYDRAULIC GRADE LINE

$V^2/2g$ = Velocity Head

FLOW

CANAL OR OPEN CHANNEL

Energy grade line and hydraulic grade line

EQUITY EQUITY

The value of an investment in a facility.

EVAPOTRANSPIRATION (ee-VAP-o-TRANS-purr-A-shun) EVAPOTRANSPIRATION

(1) The process by which water vapor is released to the atmosphere by living plants. This process is similar to people sweating. Also called transpiration.

(2) The total water removed from an area by transpiration (plants) and by evaporation from soil, snow, and water surfaces.

EXFILTRATION (EX-fill-TRAY-shun) EXFILTRATION

Liquid wastes and liquid-carried wastes that unintentionally leak out of a sewer pipe system and into the environment.

F

F/M RATIO F/M RATIO

See FOOD/MICROORGANISM RATIO.

FACULTATIVE (FACK-ul-tay-tive) BACTERIA FACULTATIVE BACTERIA

Facultative bacteria can use either dissolved oxygen or oxygen obtained from food materials such as sulfate or nitrate ions. In other words, facultative bacteria can live under aerobic, anoxic, or anaerobic conditions.

FACULTATIVE (FACK-ul-tay-tive) POND FACULTATIVE POND

The most common type of pond in current use. The upper portion (supernatant) is aerobic, while the bottom layer is anaerobic. Algae supply most of the oxygen to the supernatant.

FILAMENTOUS (fill-uh-MEN-tuss) ORGANISMS FILAMENTOUS ORGANISMS

Organisms that grow in a thread or filamentous form. Common types are *Thiothrix* and *Actinomycetes*. A common cause of sludge bulking in the activated sludge process.

FILTRATION FILTRATION

The process of passing water through a porous bed of fine granular material to remove suspended matter from the water. The suspended matter is mainly particles of floc, soil, and debris; but it also includes living organisms such as algae, bacteria, viruses, and protozoa.

FIXED COSTS FIXED COSTS

Costs that a utility must cover or pay even if there is no service provided. Also see VARIABLE COSTS.

FLAME POLISHED FLAME POLISHED

Melted by a flame to smooth out irregularities. Sharp or broken edges of glass (such as the end of a glass tube) are rotated in a flame until the edge melts slightly and becomes smooth.

FLIGHTS FLIGHTS

Scraper boards, made from redwood or other rot-resistant woods or plastic, used to collect and move settled sludge or floating scum.

FLOC FLOC

Clumps of bacteria and particles or coagulants and impurities that have come together and formed a cluster. Found in aeration tanks, secondary clarifiers, and chemical precipitation processes.

FLOW LINE FLOW LINE

(1) The top of the wetted line, the water surface, or the hydraulic grade line of water flowing in an open channel or partially full conduit.

(2) The lowest point of the channel inside a pipe, conduit, canal, or manhole. This term is used by some contractors, however the preferred term for this usage is invert.

FOOD/MICROORGANISM (F/M) RATIO

FOOD/MICROORGANISM (F/M) RATIO

Food to microorganism ratio. A measure of food provided to bacteria in an aeration tank.

$$\frac{Food}{Microorganisms} = \frac{BOD,\ lbs/day}{MLVSS,\ lbs}$$

$$= \frac{Flow,\ MGD \times BOD,\ mg/L \times 8.34\ lbs/gal}{Volume,\ MG \times MLVSS,\ mg/L \times 8.34\ lbs/gal}$$

or metric

$$= \frac{BOD,\ kg/day}{MLVSS,\ kg}$$

$$= \frac{Flow,\ ML/day \times BOD,\ mg/L \times 1\ kg/M\ mg}{Volume,\ ML \times MLVSS,\ mg/L \times 1\ kg/M\ mg}$$

FOOT VALVE

FOOT VALVE

A special type of check valve located at the bottom end of the suction pipe on a pump. This valve opens when the pump operates to allow water to enter the suction pipe but closes when the pump shuts off to prevent water from flowing out of the suction pipe.

FREE AVAILABLE CHLORINE

FREE AVAILABLE CHLORINE

The amount of chlorine available in water. This chlorine may be in the form of dissolved gas (Cl_2), hypochlorous acid (HOCl), or hypochlorite ion (OCl^-), but does not include chlorine combined with an amine (ammonia or nitrogen) or other organic compound.

FREE AVAILABLE CHLORINE RESIDUAL

FREE AVAILABLE CHLORINE RESIDUAL

The amount of chlorine available in water at the end of a specified contact period. This chlorine may be in the form of dissolved gas (Cl_2), hypochlorous acid (HOCl), or hypochlorite ion (OCl^-), but does not include chlorine combined with an amine (ammonia or nitrogen) or other organic compound.

FREE OXYGEN

FREE OXYGEN

Molecular oxygen available for respiration by organisms. Molecular oxygen is the oxygen molecule, O_2, that is not combined with another element to form a compound.

FREEBOARD

FREEBOARD

(1) The vertical distance from the normal water surface to the top of the confining wall.

(2) The vertical distance from the sand surface to the underside of a trough in a sand filter. This distance is also called available expansion.

FRICTION LOSS

FRICTION LOSS

The head, pressure, or energy (they are the same) lost by water flowing in a pipe or channel as a result of turbulence caused by the velocity of the flowing water and the roughness of the pipe, channel walls, or restrictions caused by fittings. Water flowing in a pipe loses head, pressure, or energy as a result of friction. Also called head loss.

G

GIS

GIS

See GEOGRAPHIC INFORMATION SYSTEM (GIS).

GARNET

GARNET

A group of hard, reddish, glassy, mineral sands made up of silicates of base metals (calcium, magnesium, iron, and manganese). Garnet has a higher density than sand.

GAS/LIQUID

Gaseous/Liquid. Gaseous/liquid chlorination refers to the fact that free chlorine is delivered to small treatment plants in containers that hold liquid chlorine with a free chlorine gas above the liquid in the container. The release of chlorine gas from the liquid chlorine surface depends on the temperature of the liquid and the pressure of the chlorine gas on the liquid surface. As chlorine gas is removed from the container, the gas pressure drops and more liquid chlorine becomes chlorine gas.

GATE

(1) A movable, watertight barrier for the control of a liquid in a waterway.

(2) A descriptive term used on irrigation distribution piping systems instead of the word valve. Gates cover outlet ports in the pipe segments. Water flows are regulated or distributed by opening the gates by either sliding the gate up or down or by swinging the gate to one side and uncovering an individual port to permit water flow to be discharged or regulated from the pipe at that particular point.

GEOGRAPHIC INFORMATION SYSTEM (GIS)

A computer program that combines mapping with detailed information about the physical locations of structures, such as pipes, valves, and manholes, within geographic areas. The system is used to help operators and maintenance personnel locate utility system features or structures and to assist with the scheduling and performance of maintenance activities.

GRAB SAMPLE

A single sample of water collected at a particular time and place that represents the composition of the water only at that time and place.

GRADE

(1) The elevation of the invert (or bottom) of a pipeline, canal, culvert, sewer, or similar conduit.

(2) The inclination or slope of a pipeline, conduit, stream channel, or natural ground surface; usually expressed in terms of the ratio or percentage of number of units of vertical rise or fall per unit of horizontal distance. A 0.5 percent grade would be a drop of one-half foot per hundred feet (one-half meter per hundred meters) of pipe.

GRAVIMETRIC

A means of measuring unknown concentrations of water quality indicators in a sample by weighing a precipitate or residue of the sample.

GRINDER PUMP

A small, submersible, centrifugal pump with an impeller, designed to grind solids into small pieces before they enter the collection system.

GROUND

An expression representing an electrical connection to earth or a large conductor that is at the earth's potential or neutral voltage.

GROUNDWATER

Subsurface water in the saturation zone from which wells and springs are fed. In a strict sense the term applies only to water below the water table. Also called phreatic water and plerotic water.

GROUNDWATER TABLE

The average depth or elevation of the groundwater over a selected area. Also see ARTIFICIAL GROUNDWATER TABLE, SEASONAL WATER TABLE, and TEMPORARY GROUNDWATER TABLE.

H

HGL   HGL

See HYDRAULIC GRADE LINE (HGL).

HARMFUL PHYSICAL AGENT   HARMFUL PHYSICAL AGENT
  or TOXIC SUBSTANCE   or TOXIC SUBSTANCE

Any chemical substance, biological agent (bacteria, virus, or fungus), or physical stress (noise, heat, cold, vibration, repetitive motion, ionizing and non-ionizing radiation, hypo- or hyperbaric pressure) that:

(1) Is regulated by any state or federal law or rule due to a hazard to health

(2) Is listed in the latest printed edition of the National Institute of Occupational Safety and Health (NIOSH) Registry of Toxic Effects of Chemical Substances (RTECS)

(3) Has yielded positive evidence of an acute or chronic health hazard in human, animal, or other biological testing conducted by, or known to, the employer

(4) Is described by a Material Safety Data Sheet (MSDS) available to the employer that indicates that the material may pose a hazard to human health

Also see ACUTE HEALTH EFFECT and CHRONIC HEALTH EFFECT.

HEAD   HEAD

The vertical distance, height, or energy of water above a reference point. A head of water may be measured in either height (feet or meters) or pressure (pounds per square inch or kilograms per square centimeter). Also see DISCHARGE HEAD, DYNAMIC HEAD, STATIC HEAD, SUCTION HEAD, SUCTION LIFT, and VELOCITY HEAD.

HEAD LOSS   HEAD LOSS

The head, pressure, or energy (they are the same) lost by water flowing in a pipe or channel as a result of turbulence caused by the velocity of the flowing water and the roughness of the pipe, channel walls, or restrictions caused by fittings. Water flowing in a pipe loses head, pressure, or energy as a result of friction. The head loss through a comminutor is due to friction caused by the cutters or shredders as the water passes through them and by the roughness of the comminutor walls conveying the flow through the comminutor. Also called friction loss.

[SEE DRAWING ON PAGE 450]

HEADER   HEADER

A large pipe to which the ends of a series of smaller pipes are connected. Also called a manifold.

HEPATITIS (HEP-uh-TIE-tis)   HEPATITIS

Hepatitis is an inflammation of the liver caused by an acute viral infection. Yellow jaundice is one symptom of hepatitis.

HERTZ (Hz)   HERTZ (Hz)

The number of complete electromagnetic cycles or waves in one second of an electric or electronic circuit. Also called the frequency of the current.

HOT TAP   HOT TAP

Tapping into a sewer line under pressure, such as a force main or a small-diameter sewer under pressure.

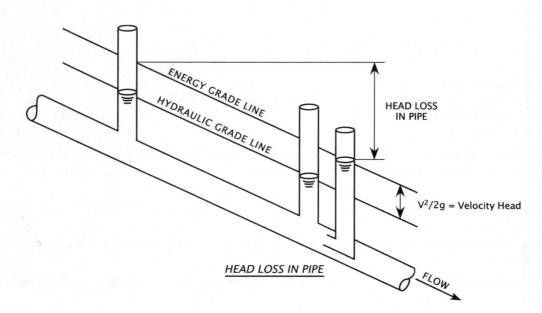

ENERGY GRADE LINE

HYDRAULIC GRADE LINE

HEAD LOSS
IN PIPE

$V^2/2g$ = Velocity Head

HEAD LOSS IN PIPE

FLOW

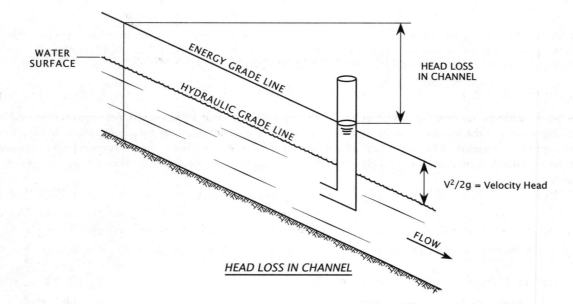

WATER
SURFACE

ENERGY GRADE LINE

HYDRAULIC GRADE LINE

HEAD LOSS
IN CHANNEL

$V^2/2g$ = Velocity Head

FLOW

HEAD LOSS IN CHANNEL

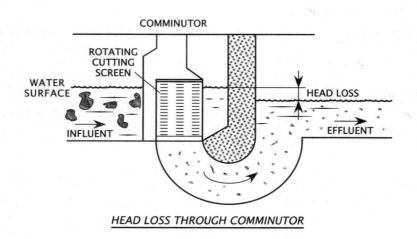

COMMINUTOR

ROTATING
CUTTING
SCREEN

WATER
SURFACE

HEAD LOSS

INFLUENT

EFFLUENT

HEAD LOSS THROUGH COMMINUTOR

Head loss

HYDRAULIC   HYDRAULIC

Referring to water flowing through manmade structures such as pipes or channels or natural environments such as rivers.

HYDRAULIC CONTINUITY   HYDRAULIC CONTINUITY

The smooth flow of wastewater as it moves through sewers, manholes, and pipes.

HYDRAULIC GRADE LINE (HGL)   HYDRAULIC GRADE LINE (HGL)

The surface or profile of water flowing in an open channel or a pipe flowing partially full. If a pipe is under pressure, the hydraulic grade line is that level water would rise to in a small, vertical tube connected to the pipe. Also see ENERGY GRADE LINE (EGL).

[SEE DRAWING ON PAGE 445]

HYDRAULIC JUMP   HYDRAULIC JUMP

The sudden and usually turbulent abrupt rise in water surface in an open channel when water flowing at high velocity is suddenly retarded to a slow velocity.

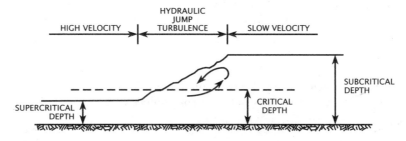

HYDROGEN SULFIDE GAS (H$_2$S)   HYDROGEN SULFIDE GAS (H$_2$S)

Hydrogen sulfide is a gas with a rotten egg odor, produced under anaerobic conditions. Hydrogen sulfide gas is particularly dangerous because it dulls the sense of smell, becoming unnoticeable after you have been around it for a while; in high concentrations, it is only noticeable for a very short time before it dulls the sense of smell. The gas is very poisonous to the respiratory system, explosive, flammable, colorless, and heavier than air.

HYDROLOGIC (HI-dro-LOJ-ick) CYCLE   HYDROLOGIC CYCLE

The process of evaporation of water into the air and its return to earth by precipitation (rain or snow). This process also includes transpiration from plants, groundwater movement, and runoff into rivers, streams, and the ocean. Also called the water cycle.

HYDROPHILIC (hi-dro-FILL-ick)   HYDROPHILIC

Having a strong affinity (liking) for water. The opposite of HYDROPHOBIC.

HYDROPHOBIC (hi-dro-FOE-bick)   HYDROPHOBIC

Having a strong aversion (dislike) for water. The opposite of HYDROPHILIC.

HYPOCHLORINATION (HI-poe-klor-uh-NAY-shun)   HYPOCHLORINATION

The application of hypochlorite compounds to water or wastewater for the purpose of disinfection.

HYPOCHLORINATORS (HI-poe-KLOR-uh-nay-tors)   HYPOCHLORINATORS

Chlorine pumps, chemical feed pumps, or devices used to dispense chlorine solutions made from hypochlorites, such as bleach (sodium hypochlorite) or calcium hypochlorite into the water being treated.

HYPOCHLORITE (HI-poe-KLOR-ite)   HYPOCHLORITE

Chemical compounds containing available chlorine; used for disinfection. They are available as liquids (bleach) or solids (powder, granules, and pellets) in barrels, drums, and cans. Salts of hypochlorous acid.

I

IMHOFF CONE

IMHOFF CONE

A clear, cone-shaped container marked with graduations. The cone is used to measure the volume of settleable solids in a specific volume (usually one liter) of water or wastewater.

IMPELLER

IMPELLER

A rotating set of vanes in a pump or compressor designed to pump or move water or air.

INDICATOR

INDICATOR

(1) (Chemical indicator) A substance that gives a visible change, usually of color, at a desired point in a chemical reaction, generally at a specified end point.

(2) (Instrument indicator) A device that indicates the result of a measurement, usually using either a fixed scale and movable indicator (pointer), such as a pressure gauge, or a moving chart with a movable pen like those used on a circular flow-recording chart. Also called a receiver.

INFILTRATION (in-fill-TRAY-shun)

INFILTRATION

The seepage of groundwater into a sewer system, including service connections. Seepage frequently occurs through defective or cracked pipes, pipe joints and connections, interceptor access risers and covers, or manhole walls.

INFLOW

INFLOW

Water discharged into a sewer system and service connections from sources other than regular connections. This includes flow from yard drains, foundations, and around access and manhole covers. Inflow differs from infiltration in that it is a direct discharge into the sewer rather than a leak in the sewer itself.

INFLUENT

INFLUENT

Water or other liquid—raw (untreated) or partially treated—flowing *INTO* a reservoir, basin, treatment process, or treatment plant.

INORGANIC

INORGANIC

Used to describe material such as sand, salt, iron, calcium salts, and other mineral materials. Inorganic materials are chemical substances of mineral origin, whereas organic substances are usually of animal or plant origin. Also see ORGANIC.

INORGANIC WASTE

INORGANIC WASTE

Waste material such as sand, salt, iron, calcium, and other mineral materials that are only slightly affected by the action of organisms. Inorganic wastes are chemical substances of mineral origin; whereas organic wastes are chemical substances usually of animal or plant origin. Also see NONVOLATILE MATTER, ORGANIC WASTE, and VOLATILE SOLIDS.

INSECTICIDE

INSECTICIDE

Any substance or chemical formulated to kill or control insects.

INTEGRATOR

INTEGRATOR

A device or meter that continuously measures and sums a process rate variable in cumulative fashion over a given time period. For example, total flows displayed in gallons per minute, million gallons per day, cubic feet per second, or some other unit of volume per time period. Also called a totalizer.

INTERCEPTOR

INTERCEPTOR

A septic tank or other holding tank that serves as a temporary wastewater storage reservoir for a septic tank effluent pump (STEP) system. Also see SEPTIC TANK.

INTERCEPTOR (INTERCEPTING) SEWER

A large sewer that receives flow from a number of sewers and conducts the wastewater to a treatment plant. Often called an interceptor. The term interceptor is sometimes used in small communities to describe a septic tank or other holding tank that serves as a temporary wastewater storage reservoir for a septic tank effluent pump (STEP) system.

INTERFACE

The common boundary layer between two substances, such as water and a solid (metal); or between two fluids, such as water and a gas (air); or between a liquid (water) and another liquid (oil).

INTERSTICE (in-TUR-stuhz)

A very small open space in a rock or granular material. Also called a pore, void, or void space. Also see VOID.

INVERT (IN-vert)

The lowest point of the channel inside a pipe, conduit, canal, or manhole. Also called flow line by some contractors, however, the preferred term is invert.

INVERTED SIPHON

A pressure pipeline used to carry wastewater flowing in a gravity collection system under a depression, such as a valley or roadway, or under a structure, such as a building. Also called a depressed sewer.

J

(NO LISTINGS)

K

(NO LISTINGS)

L

LIFT

The vertical distance water is mechanically lifted (usually pumped) from a lower elevation to a higher elevation.

LINEAL (LIN-e-ul)

The length in one direction of a line. For example, a board 12 feet (meters) long has 12 lineal feet (meters) in its length.

LOADING

Quantity of material applied to a device at one time.

LOWER EXPLOSIVE LIMIT (LEL)

The lowest concentration of a gas or vapor (percent by volume in air) that explodes if an ignition source is present at ambient temperature. At temperatures above 250°F (121°C) the LEL decreases because explosibility increases with higher temperature.

M

M or MOLAR　　　　　　　　　　　　　　　　　　　　　　　　　　　　　　　　**M or MOLAR**

A molar solution consists of one gram molecular weight of a compound dissolved in enough water to make one liter of solution. A gram molecular weight is the molecular weight of a compound in grams. For example, the molecular weight of sulfuric acid (H_2SO_4) is 98. A one M solution of sulfuric acid would consist of 98 grams of H_2SO_4 dissolved in enough distilled water to make one liter of solution.

MCRT　　　　　　　　　　　　　　　　　　　　　　　　　　　　　　　　　　　　　**MCRT**

Mean Cell Residence Time. An expression of the average time (days) that a microorganism will spend in the activated sludge process.

$$\text{MCRT, days} = \frac{\text{Total Suspended Solids in Activated Sludge Process, lbs}}{\text{Total Suspended Solids Removed From Process, lbs/day}}$$

or

$$\text{MCRT, days} = \frac{\text{Total Suspended Solids in Activated Sludge Process, kg}}{\text{Total Suspended Solids Removed From Process, kg/day}}$$

NOTE:　Operators at different plants calculate the Total Suspended Solids (TSS) in the Activated Sludge Process, lbs (kg), by three different methods:

1. TSS in the Aeration Basin or Reactor Zone, lbs (kg)

2. TSS in the Aeration Basin and Secondary Clarifier, lbs (kg)

3. TSS in the Aeration Basin and Secondary Clarifier Sludge Blanket, lbs (kg)

These three different methods make it difficult to compare MCRTs in days among different plants unless everyone uses the same method.

mg/L　　　　　　　　　　　　　　　　　　　　　　　　　　　　　　　　　　　　　**mg/L**

See MILLIGRAMS PER LITER, mg/L.

MLSS　　　　　　　　　　　　　　　　　　　　　　　　　　　　　　　　　　　　　**MLSS**

Mixed Liquor Suspended Solids. The amount (mg/L) of suspended solids in the mixed liquor of an aeration tank.

MPN　　　　　　　　　　　　　　　　　　　　　　　　　　　　　　　　　　　　　　**MPN**

MPN is the Most Probable Number of coliform-group organisms per unit volume of sample water. Expressed as a density or population of organisms per 100 mL of sample water.

MSDS　　　　　　　　　　　　　　　　　　　　　　　　　　　　　　　　　　　　　**MSDS**

See MATERIAL SAFETY DATA SHEET (MSDS).

MANIFOLD　　　　　　　　　　　　　　　　　　　　　　　　　　　　　　　　**MANIFOLD**

A large pipe to which the ends of a series of smaller pipes are connected. Also called a header.

MANOMETER (man-NAH-mut-ter)　　　　　　　　　　　　　　　　　　　**MANOMETER**

An instrument for measuring pressure. Usually, a manometer is a glass tube filled with a liquid that is used to measure the difference in pressure across a flow measuring device, such as an orifice or a Venturi meter. The instrument used to measure blood pressure is a type of manometer.

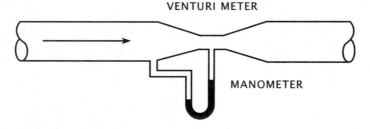

MATERIAL SAFETY DATA SHEET (MSDS)

MATERIAL SAFETY DATA SHEET (MSDS)

A document that provides pertinent information and a profile of a particular hazardous substance or mixture. An MSDS is normally developed by the manufacturer or formulator of the hazardous substance or mixture. The MSDS is required to be made available to employees and operators or inspectors whenever there is the likelihood of the hazardous substance or mixture being introduced into the workplace. Some manufacturers are preparing MSDSs for products that are not considered to be hazardous to show that the product or substance is not hazardous.

MEAN CELL RESIDENCE TIME (MCRT)

MEAN CELL RESIDENCE TIME (MCRT)

See MCRT.

MECHANICAL AERATION

MECHANICAL AERATION

The use of machinery to mix air and water so that oxygen can be absorbed into the water. Some examples are: paddle wheels, mixers, or rotating brushes to agitate the surface of an aeration tank; pumps to create fountains; and pumps to discharge water down a series of steps forming falls or cascades.

MEDIAN

MEDIAN

The middle measurement or value. When several measurements are ranked by magnitude (largest to smallest), half of the measurements will be larger and half will be smaller.

MEGOHM (MEG-ome)

MEGOHM

Millions of ohms. Mega- is a prefix meaning one million, so 5 megohms means 5 million ohms.

MENISCUS (meh-NIS-cuss)

MENISCUS

The curved surface of a column of liquid (water, oil, mercury) in a small tube. When the liquid wets the sides of the container (as with water), the curve forms a valley. When the confining sides are not wetted (as with mercury), the curve forms a hill or upward bulge.

MICRON (MY-kron)

MICRON

µm, Micrometer or Micron. A unit of length. One millionth of a meter or one thousandth of a millimeter. One micron equals 0.00004 of an inch.

MICROORGANISMS (MY-crow-OR-gan-is-ums)

MICROORGANISMS

Very small organisms that can be seen only through a microscope. Some microorganisms use the wastes in wastewater for food and thus remove or alter much of the undesirable matter.

MIL

MIL

A unit of length equal to 0.001 of an inch. The diameter of wires and tubing is measured in mils, as is the thickness of plastic sheeting.

MILLIGRAMS PER LITER, mg/L

MILLIGRAMS PER LITER, mg/L

A measure of the concentration by weight of a substance per unit volume. For practical purposes, one mg/L of a substance in water is equal to one part per million parts (ppm). Thus, a liter of water with a specific gravity of 1.0 weighs one million milligrams. If one liter of water contains 10 milligrams of dissolved oxygen, the concentration is 10 milligrams per million milligrams, or 10 milligrams per liter (10 mg/L), or 10 parts of oxygen per million parts of water, or 10 parts per million (10 ppm), or 10 pounds dissolved oxygen in 1 million pounds of water (10 ppm).

MIXED LIQUOR

MIXED LIQUOR

When the activated sludge in an aeration tank is mixed with primary effluent or the raw wastewater and return sludge, this mixture is then referred to as mixed liquor as long as it is in the aeration tank. Mixed liquor also may refer to the contents of mixed aerobic or anaerobic digesters.

MIXED LIQUOR SUSPENDED SOLIDS (MLSS)

MIXED LIQUOR SUSPENDED SOLIDS (MLSS)

The amount (mg/L) of suspended solids in the mixed liquor of an aeration tank.

MIXED LIQUOR VOLATILE SUSPENDED SOLIDS (MLVSS)

MIXED LIQUOR VOLATILE SUSPENDED SOLIDS (MLVSS)

The amount (mg/L) of organic or volatile suspended solids in the mixed liquor of an aeration tank. This volatile portion is used as a measure or indication of the microorganisms present.

MOLAR

MOLAR

See M or MOLAR.

MOLARITY

MOLARITY

A measure of concentration defined as the number of moles of solute per liter of solution. Also see M or MOLAR.

MOLE

MOLE

The name for a quantity of any chemical substance whose mass in grams is numerically equal to its atomic weight. One mole equals 6.02×10^{23} molecules or atoms. Also see MOLECULAR WEIGHT.

MOLECULAR OXYGEN

MOLECULAR OXYGEN

The oxygen molecule, O_2, that is not combined with another element to form a compound.

MOLECULAR WEIGHT

MOLECULAR WEIGHT

The molecular weight of a compound in grams per mole is the sum of the atomic weights of the elements in the compound. The molecular weight of sulfuric acid (H_2SO_4) in grams is 98.

Element	Atomic Weight	Number of Atoms	Molecular Weight
H	1	2	2
S	32	1	32
O	16	4	64
			98

MOLECULE

MOLECULE

The smallest division of a compound that still retains or exhibits all the properties of the substance.

MOST PROBABLE NUMBER (MPN)

MOST PROBABLE NUMBER (MPN)

See MPN.

N

N or NORMAL

N or NORMAL

A normal solution contains one gram equivalent weight of reactant (compound) per liter of solution. The equivalent weight of an acid is that weight that contains one gram atom of ionizable hydrogen or its chemical equivalent. For example, the equivalent weight of sulfuric acid (H_2SO_4) is 49 (98 divided by 2 because there are two replaceable hydrogen ions). A one N solution of sulfuric acid would consist of 49 grams of H_2SO_4 dissolved in enough water to make one liter of solution.

NPDES PERMIT

NPDES PERMIT

National Pollutant Discharge Elimination System permit is the regulatory agency document issued by either a federal or state agency that is designed to control all discharges of potential pollutants from point sources and stormwater runoff into US waterways. NPDES permits regulate discharges into US waterways from all point sources of pollution, including industries, municipal wastewater treatment plants, sanitary landfills, large animal feedlots, and return irrigation flows.

NTU

Nephelometric Turbidity Units. See TURBIDITY UNITS.

NAMEPLATE

A durable, metal plate found on equipment that lists critical installation and operating conditions for the equipment.

NATURAL CYCLES

Cycles that take place in nature, such as the water or hydrologic cycle where water is transformed or changed from one form to another until the water has returned to the original form, thus completing the cycle. Other natural cycles include the life cycles of aquatic organisms and plants, nutrient cycles, and cycles of self- or natural purification.

NEUTRALIZATION (noo-trull-uh-ZAY-shun)

Addition of an acid or alkali (base) to a liquid to cause the pH of the liquid to move toward a neutral pH of 7.0.

NITRIFICATION (NYE-truh-fuh-KAY-shun)

An aerobic process in which bacteria change the ammonia and organic nitrogen in wastewater into oxidized nitrogen (usually nitrate). The second-stage BOD is sometimes referred to as the nitrogenous BOD (first-stage BOD is called the carbonaceous BOD). Also see DENITRIFICATION.

NITROGENOUS (nye-TRAH-jen-us)

A term used to describe chemical compounds (usually organic) containing nitrogen in combined forms. Proteins and nitrate are nitrogenous compounds.

NONPOINT SOURCE

A runoff or discharge from a field or similar source, in contrast to a point source, which refers to a discharge that comes out the end of a pipe or other clearly identifiable conveyance. Also see POINT SOURCE.

NONVOLATILE MATTER

Material such as sand, salt, iron, calcium, and other mineral materials that are only slightly affected by the actions of organisms and are not lost on ignition of the dry solids at 550°C (1,022°F). Volatile materials are chemical substances usually of animal or plant origin. Also see INORGANIC WASTE and VOLATILE SOLIDS.

NORMAL

See N or NORMAL.

NORMALITY

The number of gram-equivalent weights of solute in one liter of solution. The equivalent weight of any material is the weight that would react with or be produced by the reaction of 8.0 grams of oxygen or 1.0 gram of hydrogen. Normality is used for certain calculations of quantitative analysis. Also see N or NORMAL.

NUTRIENT

Any substance that is assimilated (taken in) by organisms and promotes growth. Nitrogen and phosphorus are nutrients that promote the growth of algae. There are other essential and trace elements that are also considered nutrients. Also see NUTRIENT CYCLE.

NUTRIENT CYCLE

The transformation or change of a nutrient from one form to another until the nutrient has returned to the original form, thus completing the cycle. The cycle may take place under either aerobic or anaerobic conditions.

NTU

NAMEPLATE

NATURAL CYCLES

NEUTRALIZATION

NITRIFICATION

NITROGENOUS

NONPOINT SOURCE

NONVOLATILE MATTER

NORMAL

NORMALITY

NUTRIENT

NUTRIENT CYCLE

O

O&M MANUAL O&M MANUAL

Operation and Maintenance Manual. A manual that describes detailed procedures for operators to follow to operate and maintain a specific treatment plant and the equipment of that plant.

OSHA (O-shuh) OSHA

The Williams-Steiger Occupational Safety and Health Act of 1970 (OSHA) is a federal law designed to protect the health and safety of workers, including collection system and treatment plant operators. The Act regulates the design, construction, operation, and maintenance of industrial plants and wastewater collection systems and treatment plants. The Act does not apply directly to municipalities, *except* in those states that have approved plans and have asserted jurisdiction under Section 18 of the OSHA Act. *However, contract operators and private facilities do have to comply with OSHA requirements.* Wastewater treatment plants have come under stricter regulation in all phases of activity as a result of OSHA standards. OSHA also refers to the federal and state agencies that administer the OSHA regulations.

OBLIGATE AEROBES OBLIGATE AEROBES

Bacteria that must have atmospheric or dissolved molecular oxygen to live and reproduce.

OCCUPATIONAL SAFETY AND
   HEALTH ACT OF 1970 (OSHA)
OCCUPATIONAL SAFETY AND
   HEALTH ACT OF 1970 (OSHA)

See OSHA.

ODOR THRESHOLD ODOR THRESHOLD

The minimum odor of a gas or water sample that can just be detected after successive dilutions with odorless gas or water. Also called threshold odor.

OFFSET JOINT OFFSET JOINT

A pipe joint that is not exactly in line and centered. Also see DROP JOINT and VERTICAL OFFSET.

OHM OHM

The unit of electrical resistance. The resistance of a conductor in which one volt produces a current of one ampere.

OPERATING RATIO OPERATING RATIO

The operating ratio is a measure of the total revenues divided by the total operating expenses.

ORGANIC ORGANIC

Used to describe chemical substances that come from animal or plant sources. Organic substances always contain carbon. (Inorganic materials are chemical substances of mineral origin.) Also see INORGANIC.

ORGANIC WASTE ORGANIC WASTE

Waste material that may come from animal or plant sources. Natural organic wastes generally can be consumed by bacteria and other small organisms. Manufactured or synthetic organic wastes from metal finishing, chemical manufacturing, and petroleum industries may not normally be consumed by bacteria and other organisms. Also see INORGANIC WASTE and VOLATILE SOLIDS.

ORGANIZING ORGANIZING

Deciding who does what work and delegating authority to the appropriate persons.

ORIFICE (OR-uh-fiss) ORIFICE

An opening (hole) in a plate, wall, or partition. An orifice flange or plate placed in a pipe consists of a slot or a calibrated circular hole smaller than the pipe diameter. The difference in pressure in the pipe above and at the orifice may be used to determine the flow in the pipe. In a trickling filter distributor, the wastewater passes through an orifice to the surface of the filter media.

OUTFALL

(1) The point, location, or structure where wastewater or drainage discharges from a sewer, drain, or other conduit.

(2) The conduit leading to the final discharge point or area. Also see OUTFALL SEWER.

OUTFALL SEWER

A sewer that receives wastewater from a collection system or from a wastewater treatment plant and carries it to a point of ultimate or final discharge in the environment. Also see OUTFALL.

OVERHEAD

Indirect costs necessary for a utility to function properly. These costs are not related to the actual collection, treatment, and discharge of wastewater, but include the costs of rent, lights, office supplies, management, and administration.

OVERTURN

The almost spontaneous mixing of all layers of water in a reservoir or lake when the water temperature becomes similar from top to bottom. This may occur in the fall/winter when the surface waters cool to the same temperature as the bottom waters and also in the spring when the surface waters warm after the ice melts. This is also called turnover.

OXIDATION

Oxidation is the addition of oxygen, removal of hydrogen, or the removal of electrons from an element or compound; in the environment and in wastewater treatment processes, organic matter is oxidized to more stable substances. The opposite of REDUCTION.

OXIDATION DITCH

The oxidation ditch is a modified form of the activated sludge process. The ditch consists of two channels placed side by side and connected at the ends to produce one continuous loop of wastewater flow and a brush rotator assembly placed across the channel to provide aeration and circulation.

OXIDATION STATE/OXIDATION NUMBER

In a chemical formula, a number accompanied by a polarity indication (+ or −) that together indicate the charge of an ion as well as the extent to which the ion has been oxidized or reduced in a REDOX REACTION.

Due to the loss of electrons, the charge of an ion that has been oxidized would go from negative toward or to neutral, from neutral to positive, or from positive to more positive. As an example, an oxidation number of 2+ would indicate that an ion has lost two electrons and that its charge has become positive (that it now has an excess of two protons).

Due to the gain of electrons, the charge of the ion that has been reduced would go from positive toward or to neutral, from neutral to negative, or from negative to more negative. As an example, an oxidation number of 2− would indicate that an ion has gained two electrons and that its charge has become negative (that it now has an excess of two electrons). As an ion gains electrons, its oxidation state (or the extent to which it is oxidized) lowers; that is, its oxidation state is reduced. Also see REDOX REACTION.

OXIDATION-REDUCTION POTENTIAL (ORP)

The electrical potential required to transfer electrons from one compound or element (the oxidant) to another compound or element (the reductant); used as a qualitative measure of the state of oxidation in water and wastewater treatment systems. ORP is measured in millivolts, with negative values indicating a tendency to reduce compounds or elements and positive values indicating a tendency to oxidize compounds or elements.

OXIDATION-REDUCTION (REDOX) REACTION

See REDOX REACTION.

OXIDIZING AGENT

Any substance, such as oxygen (O_2) or chlorine (Cl_2), that will readily add (take on) electrons. When oxygen or chlorine is added to water or wastewater, organic substances are oxidized. These oxidized organic substances are more stable and less likely to give off odors or to contain disease-causing bacteria. The opposite is a REDUCING AGENT.

OXYGEN DEFICIENCY
OXYGEN DEFICIENCY

An atmosphere containing oxygen at a concentration of less than 19.5 percent by volume.

OXYGEN ENRICHMENT
OXYGEN ENRICHMENT

An atmosphere containing oxygen at a concentration of more than 23.5 percent by volume.

OZONATION (O-zoe-NAY-shun)
OZONATION

The application of ozone to water, wastewater, or air, generally for the purposes of disinfection or odor control.

P

PEL
PEL

See PERMISSIBLE EXPOSURE LIMIT (PEL).

PARALLEL OPERATION
PARALLEL OPERATION

Wastewater being treated is split and a portion flows to one treatment unit while the remainder flows to another similar treatment unit. Also see SERIES OPERATION.

PARASITIC (pair-uh-SIT-tick) BACTERIA
PARASITIC BACTERIA

Parasitic bacteria are those bacteria that normally live off another living organism, known as the host.

PATHOGENIC (path-o-JEN-ick) ORGANISMS
PATHOGENIC ORGANISMS

Bacteria, viruses, protozoa, or internal parasites that can cause disease (such as giardiasis, cryptosporidiosis, typhoid fever, cholera, or infectious hepatitis) in a host (such as a person). There are many types of organisms that do not cause disease and are not called pathogenic. Many beneficial bacteria are found in wastewater treatment processes actively cleaning up organic wastes.

PATHOGENS (PATH-o-jens)
PATHOGENS

See PATHOGENIC ORGANISMS.

PEAKING FACTOR
PEAKING FACTOR

Ratio of a maximum flow to the average flow, such as maximum hourly flow or maximum daily flow to the average daily flow.

PERCENT SATURATION
PERCENT SATURATION

The amount of a substance that is dissolved in a solution compared with the amount dissolved in the solution at saturation, expressed as a percent.

$$\text{Percent Saturation, \%} = \frac{\text{Amount of Substance That Is Dissolved} \times 100\%}{\text{Amount Dissolved in Solution at Saturation}}$$

PERCOLATION (purr-ko-LAY-shun)
PERCOLATION

The slow passage of water through a filter medium; or, the gradual penetration of soil and rocks by water.

PERFORMANCE INDICATOR
PERFORMANCE INDICATOR

A measurable goal used to determine system performance and level of service provided. Examples of performance indicators include the number of stoppages per 100 miles of sewer per year and the number of lost time accidents per year—measurements of how *well* a utility is doing rather than how *much* a utility is doing. Also see PRODUCTION INDICATOR.

PERMEABILITY (PURR-me-uh-BILL-uh-tee)
PERMEABILITY

The property of a material or soil that permits considerable movement of water through it when it is saturated.

PERMISSIBLE EXPOSURE LIMIT (PEL)

PERMISSIBLE EXPOSURE LIMIT (PEL)

The legal limit in the United States for exposure of a worker to a hazardous substance (such as chemicals, dusts, fumes, mists, gases, or vapors) or agents (such as occupational noise). OSHA sets enforceable permissible exposure limits (PELs) to protect workers against the health effects of excessive exposure. OSHA PELs are based on an 8-hour time-weighted average (TWA) exposure. Permissible exposure limits are listed in the Code of Federal Regulations (CFR) Title 29 Part 1910, Subparts G and Z. Also see TIME-WEIGHTED AVERAGE (TWA).

PERMIT-REQUIRED CONFINED SPACE
   (PERMIT SPACE)

PERMIT-REQUIRED CONFINED SPACE
(PERMIT SPACE)

See CONFINED SPACE, PERMIT-REQUIRED (PERMIT SPACE).

pH (pronounce as separate letters)

pH

pH is an expression of the intensity of the basic or acidic condition of a liquid. Mathematically, pH is the logarithm (base 10) of the reciprocal of the hydrogen ion activity.

$$pH = \text{Log} \frac{1}{\{H^+\}}$$

If $\{H^+\} = 10^{-6.5}$, then pH = 6.5. The pH may range from 0 to 14, where 0 is most acidic, 14 most basic, and 7 neutral.

PHOTOSYNTHESIS (foe-toe-SIN-thuh-sis)

PHOTOSYNTHESIS

A process in which organisms, with the aid of chlorophyll, convert carbon dioxide and inorganic substances into oxygen and additional plant material, using sunlight for energy. All green plants grow by this process.

PIG

PIG

Refers to a polypig, which is a bullet-shaped device made of hard rubber or similar material. This device is used to clean pipes. It is inserted in one end of a pipe, moves through the pipe under pressure, and is removed from the other end of the pipe.

PILLOWS

PILLOWS

Plastic tubes shaped like pillows that contain exact amounts of chemicals or reagents. Cut open the pillow, pour the reagents into the sample being tested, mix thoroughly, and follow test procedures.

PINPOINT FLOC

PINPOINT FLOC

Very small floc (the size of a pin point) that does not settle out of the water in a sedimentation basin or clarifier. Also see FLOC.

PLAN or PLAN VIEW

PLAN or PLAN VIEW

A drawing or photo showing the top view of sewers, manholes, streets, or structures.

PLANNING

PLANNING

Management of utilities to build the resources and financial capability to provide for future needs.

PLUG FLOW

PLUG FLOW

A type of flow that occurs in tanks, basins, or reactors when a slug of water or wastewater moves through a tank without ever dispersing or mixing with the rest of the water or wastewater flowing through the tank.

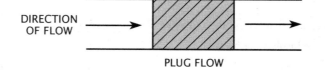

PLUG FLOW

POINT SOURCE

A discharge that comes out the end of a pipe or other clearly identifiable conveyance. Examples of point source conveyances from which pollutants may be discharged include: ditches, channels, tunnels, conduits, wells, containers, rolling stock, concentrated animal feeding operations, landfill leachate collection systems, vessels, or other floating craft. A NONPOINT SOURCE refers to runoff or a discharge from a field or similar source.

POLYELECTROLYTE (POLY-ee-LECK-tro-lite)

A high-molecular-weight (relatively heavy) substance, having points of positive or negative electrical charges, that is formed by either natural or synthetic (manmade) processes. Natural polyelectrolytes may be of biological origin or obtained from starch products or cellulose derivatives. Synthetic polyelectrolytes consist of simple substances that have been made into complex, high-molecular-weight substances. Used with other chemical coagulants to aid in binding small suspended particles to larger chemical flocs for their removal from water. Often called a polymer.

POLYMER (POLY-mer)

A long-chain molecule formed by the union of many monomers (molecules of lower molecular weight). Polymers are used with other chemical coagulants to aid in binding small suspended particles to larger chemical flocs for their removal from water. Also see POLYELECTROLYTE.

POPULATION EQUIVALENT

A means of expressing the strength of organic material in wastewater. In a domestic wastewater system, microorganisms use up about 0.2 pound (90 grams) of oxygen per day for each person using the system (as measured by the standard BOD test). May also be expressed as flow (100 gallons (378 liters)/day/person) or suspended solids (0.2 lb (90 grams) SS/day/person).

$$\text{Population Equivalent, persons} = \frac{\text{Flow, MGD} \times \text{BOD, mg/L} \times 8.34 \text{ lbs/gal}}{0.2 \text{ lb BOD/day/person}}$$

or

$$\text{Population Equivalent, persons} = \frac{\text{Flow, cu m/day} \times \text{BOD, mg/L} \times 10^6 \text{ L/cu m}}{90,000 \text{ mg BOD/day/person}}$$

PORE

A very small open space in a rock or granular material. Also called an interstice, void, or void space. Also see VOID.

POSTCHLORINATION

The addition of chlorine to the plant discharge or effluent, following plant treatment, for disinfection purposes.

POTABLE (POE-tuh-bull) WATER

Water that does not contain objectionable pollution, contamination, minerals, or infective agents and is considered satisfactory for drinking.

PRECHLORINATION

The addition of chlorine in the collection system serving the plant or at the headworks of the plant prior to other treatment processes mainly for odor and corrosion control. Also applied to aid disinfection, to reduce plant BOD load, to aid in settling, to control foaming in Imhoff units, and to help remove oil.

PRECIPITATE (pre-SIP-uh-TATE)

(1) An insoluble, finely divided substance that is a product of a chemical reaction within a liquid.

(2) The separation from solution of an insoluble substance.

PRESENT WORTH

The value of a long-term project expressed in today's dollars. Present worth is calculated by converting (discounting) all future benefits and costs over the life of the project to a single economic value at the start of the project. Calculating the present worth of alternative projects makes it possible to compare them and select the one with the largest positive (beneficial) present worth or minimum present cost.

PRIMARY CLARIFIER

PRIMARY CLARIFIER

A wastewater treatment device that consists of a rectangular or circular tank that allows those substances in wastewater that readily settle or float to be separated from the wastewater being treated.

PRIMARY TREATMENT

PRIMARY TREATMENT

A wastewater treatment process that takes place in a rectangular or circular tank and allows those substances in wastewater that readily settle or float to be separated from the wastewater being treated. A septic tank is also considered primary treatment.

PRIME

PRIME

The action of filling a pump casing with water to remove the air. Most pumps must be primed before start-up or they will not pump any water.

PRODUCTION INDICATOR

PRODUCTION INDICATOR

A measure of a work activity performed by a utility's operators or crews. Examples of production indicators include miles of sewers televised per year and miles of sewers cleaned per year—measurements of how *much* a utility is doing rather than how *well* a utility is doing. Also see PERFORMANCE INDICATOR.

PROFILE

PROFILE

A drawing showing elevation plotted against distance, such as the vertical section or side view of sewers, manholes, or a pipeline.

PROPORTIONAL WEIR (WEER)

PROPORTIONAL WEIR

A specially shaped weir in which the flow through the weir is directly proportional to the head.

PROTOZOA (pro-toe-ZOE-ah)

PROTOZOA

A group of motile, microscopic organisms (usually single-celled and aerobic) that sometimes cluster into colonies and generally consume bacteria as an energy source.

PUTREFACTION (PYOO-truh-FACK-shun)

PUTREFACTION

Biological decomposition of organic matter, with the production of foul-smelling and -tasting products, associated with anaerobic (no oxygen present) conditions.

PYROMETER (pie-ROM-uh-ter)

PYROMETER

An apparatus used to measure high temperatures.

Q

(NO LISTINGS)

R

RNA

RNA

Ribonucleic acid. A chemical that provides the structure for protein synthesis (building up).

RATE OF RETURN

RATE OF RETURN

A value that indicates the return of funds received on the basis of the total equity capital used to finance physical facilities. Similar to the interest rate on savings accounts or loans.

REACTIVE MAINTENANCE

REACTIVE MAINTENANCE

Maintenance activities that are performed in response to problems and emergencies after they occur.

REAGENT (re-A-gent)   REAGENT

A pure, chemical substance that is used to make new products or is used in chemical tests to measure, detect, or examine other substances.

RECEIVER   RECEIVER

A device that indicates the result of a measurement, usually using either a fixed scale and movable indicator (pointer), such as a pressure gauge, or a moving chart with a movable pen like those used on a circular flow-recording chart. Also called an indicator.

RECEIVING WATER   RECEIVING WATER

A stream, river, lake, ocean, or other surface or groundwaters into which treated or untreated wastewater is discharged.

REDOX (REE-docks) REACTION   REDOX REACTION

A two-part reaction between two ions involving a transfer of electrons from one ion to the other. Oxidation is the loss of electrons by one ion, and reduction is the acceptance of electrons by the other ion. Reduction refers to the lowering of the OXIDATION STATE/ OXIDATION NUMBER of the ion accepting the electrons.

In a redox reaction, the ion that gives up the electrons (that is oxidized) is called the reductant because it causes a reduction in the oxidation state or number of the ion that accepts the transferred electrons. The ion that receives the electrons (that is reduced) is called the oxidant because it causes oxidation of the other ion. Oxidation and reduction always occur simultaneously.

REDUCING AGENT   REDUCING AGENT

Any substance, such as base metal (iron) or the sulfide ion (S^{2-}), that will readily donate (give up) electrons. The opposite is an OXIDIZING AGENT.

REDUCTION (re-DUCK-shun)   REDUCTION

Reduction is the addition of hydrogen, removal of oxygen, or the addition of electrons to an element or compound. Under anaerobic conditions (no dissolved oxygen present), sulfur compounds are reduced to odor-producing hydrogen sulfide (H_2S) and other compounds. In the treatment of metal finishing wastewaters, hexavalent chromium (Cr^{6+}) is reduced to the trivalent form (Cr^{3+}). The opposite of OXIDATION.

REFLUX   REFLUX

Flow back. A sample is heated, evaporates, cools, condenses, and flows back to the flask.

REPRESENTATIVE SAMPLE   REPRESENTATIVE SAMPLE

A sample portion of material, water, or wastestream that is as nearly identical in content and consistency as possible to that in the larger body being sampled.

RESIDUAL ANALYZER, CHLORINE   RESIDUAL ANALYZER, CHLORINE
See CHLORINE RESIDUAL ANALYZER.

RESIDUAL CHLORINE   RESIDUAL CHLORINE

The concentration of chlorine present in water after the chlorine demand has been satisfied. The concentration is expressed in terms of the total chlorine residual, which includes both the free and combined or chemically bound chlorine residuals. Also called chlorine residual.

RESISTANCE   RESISTANCE

That property of a conductor or wire that opposes the passage of a current, thus causing electric energy to be transformed into heat.

RESPONSIBILITY   RESPONSIBILITY

Answering to those above in the chain of command to explain how and why you have used your authority.

RIGHT-TO-KNOW LAWS RIGHT-TO-KNOW LAWS

Employee Right-To-Know legislation requires employers to inform employees of the possible health effects resulting from contact with hazardous substances. At locations where this legislation is in force, employers must provide employees with information regarding any hazardous substances they might be exposed to under normal work conditions or reasonably foreseeable emergency conditions resulting from workplace conditions. OSHA's Hazard Communication Standard (HCS) (Title 29 CFR Part 1910.1200) is the federal regulation and state statutes are called Worker Right-To-Know laws. Also see COMMUNITY RIGHT-TO-KNOW and SARA.

RIPRAP RIPRAP

Broken stones, boulders, or other materials placed compactly or irregularly on levees or dikes for the protection of earth surfaces against the erosive action of waves.

RISING SLUDGE RISING SLUDGE

Rising sludge occurs in the secondary clarifiers of activated sludge plants when the sludge settles to the bottom of the clarifier, is compacted, and then starts to rise to the surface, usually as a result of denitrification, or anaerobic biological activity that produces carbon dioxide or methane.

ROD, SEWER ROD, SEWER

A light metal rod, three to five feet long, with a coupling at each end. Rods are joined and pushed into a sewer to dislodge obstructions.

ROTAMETER (ROTE-uh-ME-ter) ROTAMETER

A device used to measure the flow rate of gases and liquids. The gas or liquid being measured flows vertically up a tapered, calibrated tube. Inside the tube is a small ball or bullet-shaped float (it may rotate) that rises or falls depending on the flow rate. The flow rate may be read on a scale behind or on the tube by looking at the middle of the ball or at the widest part or top of the float.

ROTATING BIOLOGICAL CONTACTOR (RBC) ROTATING BIOLOGICAL CONTACTOR (RBC)

A secondary biological treatment process for domestic and biodegradable industrial wastes. Biological contactors have a rotating shaft surrounded by plastic discs called the media. The shaft and media are called the drum. A biological slime grows on the media when conditions are suitable and the microorganisms that make up the slime (biomass) stabilize the waste products by using the organic material for growth and reproduction.

S

SAR SAR

Sodium Adsorption Ratio. This ratio expresses the relative activity of sodium ions in the exchange reactions with soil. The ratio is defined as follows:

$$SAR = \frac{Na}{[\frac{1}{2}(Ca + Mg)]^{\frac{1}{2}}}$$

where Na, Ca, and Mg are concentrations of the respective ions in milliequivalents per liter of water.

$$Na, meq/L = \frac{Na, mg/L}{23.0 \ mg/meq} \qquad Ca, meq/L = \frac{Ca, mg/L}{20.0 \ mg/meq} \qquad Mg, meq/L = \frac{Mg, mg/L}{12.15 \ mg/meq}$$

SARA SARA

Superfund Amendments and Reauthorization Act of 1986. The Comprehensive Environmental Response, Compensation, and Liability Act (CERCLA), commonly known as the Superfund Act, was enacted in 1980. The 1986 amendments increase CERCLA revenues to $8.5 billion and strengthen the EPA's authority to conduct short-term (removal), long-term (remedial), and enforcement actions. The amendments also strengthen state involvements in the cleanup process and the agency's commitments to research and development, training, health assessments, and public participation. A number of new statutory authorities, such as Community Right-To-Know, were also established. Also see CERCLA.

SDGS SYSTEM

See SMALL-DIAMETER GRAVITY SEWER (SDGS) SYSTEM.

STEF SYSTEM

See SEPTIC TANK EFFLUENT FILTER (STEF) SYSTEM.

STEP SYSTEM

See SEPTIC TANK EFFLUENT PUMP (STEP) SYSTEM.

SANITARY SEWER

A pipe or conduit (sewer) intended to carry wastewater or waterborne wastes from homes, businesses, and industries to the treatment works. Stormwater runoff or unpolluted water should be collected and transported in a separate system of pipes or conduits (storm sewers) to natural watercourses.

SAPROPHYTES (SAP-row-fights)

Organisms living on dead or decaying organic matter. They help natural decomposition of organic matter in water or wastewater.

SCUM

A layer or film of foreign matter (such as grease, oil) that has risen to the surface of water or wastewater.

SEASONAL WATER TABLE

A water table that has seasonal changes in depth or elevation.

SECCHI (SECK-key) DISK

A flat, white disk lowered into the water by a rope until it is just barely visible. At this point, the depth of the disk from the water surface is the recorded Secchi disk transparency.

SECONDARY CLARIFIER

A wastewater treatment device consisting of a rectangular or circular tank that allows separation of substances that settle or float not removed by previous treatment processes.

SECONDARY TREATMENT

A wastewater treatment process used to convert dissolved or suspended materials into a form more readily separated from the water being treated. Usually, the process follows primary treatment by sedimentation. The process commonly is a type of biological treatment followed by secondary clarifiers that allow the solids to settle out from the water being treated.

SEDIMENTATION (SED-uh-men-TAY-shun)

The process of settling and depositing of suspended matter carried by water or wastewater. Sedimentation usually occurs by gravity when the velocity of the liquid is reduced below the point at which it can transport the suspended material.

SEIZING or SEIZE UP

Seizing occurs when an engine overheats and a part expands to the point where the engine will not run. Also called freezing.

SEPTAGE (SEPT-age)

The sludge produced in septic tanks.

SEPTIC (SEP-tick)

A condition produced by anaerobic bacteria. If severe, the sludge produces hydrogen sulfide, turns black, gives off foul odors, contains little or no dissolved oxygen, and the wastewater has a high oxygen demand.

SDGS SYSTEM

STEF SYSTEM

STEP SYSTEM

SANITARY SEWER

SAPROPHYTES

SCUM

SEASONAL WATER TABLE

SECCHI DISK

SECONDARY CLARIFIER

SECONDARY TREATMENT

SEDIMENTATION

SEIZING or SEIZE UP

SEPTAGE

SEPTIC

SEPTIC TANK

SEPTIC TANK

A system sometimes used where wastewater collection systems and treatment plants are not available. The system is a settling tank in which settled sludge and floatable scum are in intimate contact with the wastewater flowing through the tank and the organic solids are decomposed by anaerobic bacterial action. Used to treat wastewater and produce an effluent that is usually discharged to subsurface leaching. Also referred to as an interceptor; however, the preferred term is septic tank.

SEPTIC TANK EFFLUENT FILTER (STEF) SYSTEM

SEPTIC TANK EFFLUENT FILTER (STEF) SYSTEM

A facility in which effluent flows from a septic tank into a gravity flow collection system that flows into a gravity sewer, treatment plant, or subsurface leaching system. The gravity flow pipeline is called an effluent drain.

SEPTIC TANK EFFLUENT PUMP (STEP) SYSTEM

SEPTIC TANK EFFLUENT PUMP (STEP) SYSTEM

A facility in which effluent is pumped from a septic tank into a pressurized collection system that may flow into a gravity sewer, treatment plant, or subsurface leaching system.

SEPTICITY (sep-TIS-uh-tee)

SEPTICITY

The condition in which organic matter decomposes to form foul-smelling products associated with the absence of free oxygen. If severe, the wastewater produces hydrogen sulfide, turns black, gives off foul odors, contains little or no dissolved oxygen, and the wastewater has a high oxygen demand.

SEQUENCING BATCH REACTOR (SBR)

SEQUENCING BATCH REACTOR (SBR)

A type of activated sludge system that is specifically designed and automated to mix/aerate untreated wastewater and allow solids flocculation/separation to occur as a batch treatment process.

SERIES OPERATION

SERIES OPERATION

Wastewater being treated flows through one treatment unit and then flows through another similar treatment unit. Also see PARALLEL OPERATION.

SET POINT

SET POINT

The position at which the control or controller is set. This is the same as the desired value of the process variable. For example, a thermostat is set to maintain a desired temperature.

SHOCK LOAD (ACTIVATED SLUDGE)

SHOCK LOAD

The arrival at a plant of a waste that is toxic to organisms in sufficient quantity or strength to cause operating problems. Possible problems include odors and bulking sludge, which will result in a high loss of solids from the secondary clarifiers into the plant effluent and a biological process upset that may require several days to a week to recover. Organic or hydraulic overloads also can cause a shock load.

SHOCK LOAD (TRICKLING FILTERS)

SHOCK LOAD

The arrival at a plant of a waste that is toxic to organisms in sufficient quantity or strength to cause operating problems. Possible problems include odors and sloughing off of the growth or slime on the trickling filter media. Organic or hydraulic overloads also can cause a shock load.

SHORT-CIRCUITING

SHORT-CIRCUITING

A condition that occurs in tanks or basins when some of the flowing water entering a tank or basin flows along a nearly direct pathway from the inlet to the outlet. This is usually undesirable because it may result in shorter contact, reaction, or settling times in comparison with the theoretical (calculated) or presumed detention times.

SHREDDING

SHREDDING

A mechanical treatment process that cuts large pieces of wastes into smaller pieces so they will not plug pipes or damage equipment. Shredding and comminution usually mean the same thing.

SIDESTREAM

SIDESTREAM

Wastewater flows that develop from other storage or treatment facilities. This wastewater may or may not need additional treatment.

SLOPE

SLOPE

The slope or inclination of a trench bottom or a trench side wall is the ratio of the vertical distance to the horizontal distance or rise over run. Also see GRADE (2).

2 VERTICAL

1 HORIZONTAL

2:1 SLOPE (OR 2 IN 1 SLOPE)

SLOUGHED or SLOUGHING (SLUFF-ing)

SLOUGHED or SLOUGHING

The breaking off of biological or biomass growths from the fixed film or rotating biological contactor (RBC) media. The sloughed growth becomes suspended in the effluent and is later removed in the secondary clarifier as sludge.

SLUDGE (SLUJ)

SLUDGE

(1) The settleable solids separated from liquids during processing.

(2) The deposits of foreign materials on the bottoms of streams or other bodies of water or on the bottoms and edges of wastewater collection lines and appurtenances.

SLUDGE AGE

SLUDGE AGE

A measure of the length of time a particle of suspended solids has been retained in the activated sludge process.

$$\text{Sludge Age, days} = \frac{\text{Suspended Solids Under Aeration, lbs or kg}}{\text{Suspended Solids Added, lbs/day or kg/day}}$$

SLUDGE DENSITY INDEX (SDI)

SLUDGE DENSITY INDEX (SDI)

This calculation is used in a way similar to the Sludge Volume Index (SVI) to indicate the settleability of a sludge in a secondary clarifier or effluent. The weight in grams of one milliliter of sludge after settling for 30 minutes. SDI = 100/SVI. Also see SLUDGE VOLUME INDEX (SVI).

SLUDGE VOLUME INDEX (SVI)

SLUDGE VOLUME INDEX (SVI)

A calculation that indicates the tendency of activated sludge solids (aerated solids) to thicken or to become concentrated during the sedimentation/thickening process. SVI is calculated in the following manner: (1) allow a mixed liquor sample from the aeration basin to settle for 30 minutes; (2) determine the suspended solids concentration for a sample of the same mixed liquor; (3) calculate SVI by dividing the measured (or observed) wet volume (mL/L) of the settled sludge by the dry weight concentration of MLSS in grams/L.

$$\text{SVI, mL/gm} = \frac{\text{Settled Sludge Volume/Sample Volume, mL/L}}{\text{Suspended Solids Concentration, mg/L}} \times \frac{1{,}000 \text{ mg}}{\text{gram}}$$

SLUG

SLUG

Intermittent release or discharge of wastewater or industrial wastes.

SLURRY

SLURRY

A watery mixture or suspension of insoluble (not dissolved) matter; a thin, watery mud or any substance resembling it (such as a grit slurry or a lime slurry).

SMALL-DIAMETER GRAVITY SEWER (SDGS) SYSTEM

SMALL-DIAMETER GRAVITY SEWER (SDGS) SYSTEM

A type of collection system in which a series of septic tanks discharge effluent by gravity, pump, or siphon to a small-diameter wastewater collection main. The wastewater flows by gravity to a lift station, a manhole in a conventional gravity collection system, or directly to a wastewater treatment plant.

SOLIDS CONCENTRATION
SOLIDS CONCENTRATION

The solids in the aeration tank that carry microorganisms that feed on wastewater. Expressed as milligrams per liter of mixed liquor volatile suspended solids (MLVSS, mg/L).

SOLUBLE BOD
SOLUBLE BOD

Soluble BOD is the BOD of water that has been filtered in the standard suspended solids test. The soluble BOD is a measure of food for microorganisms that is dissolved in the water being treated.

SOLUTION
SOLUTION

A liquid mixture of dissolved substances. In a solution it is impossible to see all the separate parts.

SPECIFIC GRAVITY
SPECIFIC GRAVITY

(1) Weight of a particle, substance, or chemical solution in relation to the weight of an equal volume of water. Water has a specific gravity of 1.000 at 4°C (39°F). Wastewater particles or substances usually have a specific gravity of 0.5 to 2.5. Particulates with specific gravity less than 1.0 float to the surface and particulates with specific gravity greater than 1.0 sink.

(2) Weight of a particular gas in relation to the weight of an equal volume of air at the same temperature and pressure (air has a specific gravity of 1.0). Chlorine gas has a specific gravity of 2.5.

SPLASH PAD
SPLASH PAD

A structure made of concrete or other durable material to protect bare soil from erosion by splashing or falling water.

SPORE
SPORE

The reproductive body of certain organisms, which is capable of giving rise to a new organism either directly or indirectly. A viable (able to live and grow) body regarded as the resting stage of an organism. A spore is usually more resistant to disinfectants and heat than most organisms. Gangrene and tetanus bacteria are common spore-forming organisms.

STABILIZATION
STABILIZATION

Conversion to a form that resists change. Organic material is stabilized by bacteria that convert the material to gases and other relatively inert substances. Stabilized organic material generally will not give off obnoxious odors.

STABILIZED WASTE
STABILIZED WASTE

A waste that has been treated or decomposed to the extent that, if discharged or released, its rate and state of decomposition would be such that the waste would not cause a nuisance or odors in the receiving water.

STAFF GAUGE
STAFF GAUGE

A ruler or graduated scale used to measure the depth or elevation of water in a channel, tank, or stream.

STANDARD METHODS
STANDARD METHODS

STANDARD METHODS FOR THE EXAMINATION OF WATER AND WASTEWATER, 21st Edition. A joint publication of the American Public Health Association (APHA), American Water Works Association (AWWA), and the Water Environment Federation (WEF) that outlines the accepted laboratory procedures used to analyze the impurities in water and wastewater. Available from: American Water Works Association, Bookstore, 6666 West Quincy Avenue, Denver, CO 80235. Order No. 10084. Price to members, $198.50; nonmembers, $266.00; price includes cost of shipping and handling. Also available from Water Environment Federation, Publications Order Department, PO Box 18044, Merrifield, VA 22118-0045. Order No. S82011. Price to members, $203.00; nonmembers, $268.00; price includes cost of shipping and handling.

STANDARD SOLUTION
STANDARD SOLUTION

A solution in which the exact concentration of a chemical or compound is known.

STANDARDIZE

STANDARDIZE

To compare with a standard.

(1) In wet chemistry, to find out the exact strength of a solution by comparing it with a standard of known strength. This information is used to adjust the strength by adding more water or more of the substance dissolved.

(2) To set up an instrument or device to read a standard. This allows you to adjust the instrument so that it reads accurately, or enables you to apply a correction factor to the readings.

STATIC HEAD

STATIC HEAD

When water is not moving, the vertical distance (in feet or meters) from a reference point to the water surface is the static head. Also see DYNAMIC HEAD, DYNAMIC PRESSURE, and STATIC PRESSURE.

STATIC LIFT

STATIC LIFT

Vertical distance water is lifted from upstream water surface to downstream water surface (which is at a higher elevation) when no water is being pumped.

STATIC PRESSURE

STATIC PRESSURE

When water is not moving, the vertical distance (in feet or meters) from a specific point to the water surface is the static head. The static pressure in psi (or kPa) is the static head in feet times 0.433 psi/ft (or meters × 9.81 kPa/m). Also see DYNAMIC HEAD, DYNAMIC PRESSURE, and STATIC HEAD.

STATIC WATER HEAD

STATIC WATER HEAD

Elevation or surface of water that is not being pumped.

STEP-FEED AERATION

STEP-FEED AERATION

Step-feed aeration is a modification of the conventional activated sludge process. In step-feed aeration, primary effluent enters the aeration tank at several points along the length of the tank, rather than at the beginning or head of the tank and flowing through the entire tank in a plug flow mode.

STERILIZATION (STAIR-uh-luh-ZAY-shun)

STERILIZATION

The removal or destruction of all microorganisms, including pathogens and other bacteria, vegetative forms, and spores. Compare with DISINFECTION.

STOP LOG

STOP LOG

A log or board in an outlet box or device used to control the water level in ponds and also the flow from one pond to another pond or system.

STOPPAGE

STOPPAGE

Any incident in which a sewer is partially or completely blocked, causing a backup, a service interruption, or an overflow. Also called blockage.

SUBSURFACE LEACHING SYSTEM

SUBSURFACE LEACHING SYSTEM

A method of treatment and discharge of septic tank effluent, sand filter effluent, or other treated wastewater. The effluent is applied to soil below the ground surface through open-jointed pipes or drains or through perforated pipes (holes in the pipes). The effluent is treated as it passes through porous soil or rock strata (layers). Newer subsurface leaching systems include chamber and gravelless systems, and also gravel trenches without pipe the full length of the trench.

SUCTION HEAD

SUCTION HEAD

The positive pressure [in feet (meters) of water or pounds per square inch (kilograms per square centimeter) of mercury vacuum] on the suction side of a pump. The pressure can be measured from the centerline of the pump up to the elevation of the hydraulic grade line on the suction side of the pump.

SUCTION LIFT SUCTION LIFT

The negative pressure [in feet (meters) of water or inches (centimeters) of mercury vacuum] on the suction side of a pump. The pressure can be measured from the centerline of the pump down to (lift) the elevation of the hydraulic grade line on the suction side of the pump.

SUMP SUMP

This term refers to a facility or structure that connects an industrial discharger to a public sewer. The sump could be a sample box, a clarifier, or an intercepting sewer.

SUPERFUND ACT SUPERFUND ACT

See CERCLA.

SUPERNATANT (soo-per-NAY-tent) SUPERNATANT

The relatively clear water layer between the sludge on the bottom and the scum on the surface of an anaerobic digester or septic tank (interceptor).

(1)   From an anaerobic digester, this water is usually returned to the influent wet well or to the primary clarifier.

(2)   From a septic tank, this water is discharged by gravity or by a pump to a leaching system or a wastewater collection system.

Also called clear zone.

SURCHARGE SURCHARGE

Sewers are surcharged when the supply of water to be carried is greater than the capacity of the pipes to carry the flow. The surface of the wastewater in manholes rises above the top of the sewer pipe, and the sewer is under pressure or a head, rather than at atmospheric pressure.

SURFACTANT (sir-FAC-tent) SURFACTANT

Abbreviation for surface-active agent. The active agent in detergents that possesses a high cleaning ability.

SUSPENDED SOLIDS SUSPENDED SOLIDS

(1)   Solids that either float on the surface or are suspended in water, wastewater, or other liquids, and that are largely removable by laboratory filtering.

(2)   The quantity of material removed from water or wastewater in a laboratory test, as prescribed in *STANDARD METHODS FOR THE EXAMINATION OF WATER AND WASTEWATER,* and referred to as Total Suspended Solids Dried at 103–105°C.

T

TOC (pronounce as separate letters) TOC
Total Organic Carbon. TOC measures the amount of organic carbon in water.

TWA TWA

See TIME-WEIGHTED AVERAGE (TWA).

TAILGATE SAFETY MEETING TAILGATE SAFETY MEETING

Brief (10 to 20 minutes) safety meetings held every 7 to 10 working days. The term comes from the safety meetings regularly held by the construction industry around the tailgate of a truck.

TELEMETERING EQUIPMENT TELEMETERING EQUIPMENT

Equipment that translates physical measurements into electrical impulses that are transmitted to dials or recorders.

TEMPORARY GROUNDWATER TABLE

TEMPORARY GROUNDWATER TABLE

(1) During and for a period following heavy rainfall or snow melt, the soil is saturated at elevations above the normal, stabilized, or seasonal groundwater table, often from the surface of the soil downward. This is referred to as a temporary condition and thus is a temporary groundwater table.

(2) When a collection system serves agricultural areas in its vicinity, irrigation of these areas can cause a temporary rise in the elevation of the groundwater table.

TERTIARY (TER-she-air-ee) TREATMENT

TERTIARY TREATMENT

Any process of water renovation that upgrades treated wastewater to meet specific reuse requirements. May include general cleanup of water or removal of specific parts of wastes insufficiently removed by conventional treatment processes. Typical processes include chemical treatment and pressure filtration. Also called advanced waste treatment.

THIEF HOLE

THIEF HOLE

A digester sampling well that allows sampling of the digester contents without venting digester gas.

THRESHOLD ODOR

THRESHOLD ODOR

The minimum odor of a gas or water sample that can just be detected after successive dilutions with odorless gas or water. Also called odor threshold.

THRUST BLOCK

THRUST BLOCK

A mass of concrete or similar material appropriately placed around a pipe to prevent movement when the pipe is carrying water. Usually placed at bends and valve structures.

TIME-WEIGHTED AVERAGE (TWA)

TIME-WEIGHTED AVERAGE (TWA)

A time-weighted average is used to calculate a worker's daily exposure to a hazardous substance (such as chemicals, dusts, fumes, mists, gases, or vapors) or agent (such as occupational noise), averaged to an 8-hour workday, taking into account the average levels of the substance or agent and the time spent in the area. This is the guideline OSHA uses to determine permissible exposure limits (PELs) and is essential in assessing a worker's exposure and determining what protective measures should be taken. A time-weighted average is equal to the sum of the portion of each time period (as a decimal, such as 0.25 hour) multiplied by the levels of the substance or agent during the time period divided by the hours in the workday (usually 8 hours). Also see PERMISSIBLE EXPOSURE LIMIT (PEL).

TITRATE (TIE-trate)

TITRATE

To titrate a sample, a chemical solution of known strength is added drop by drop until a certain color change, precipitate, or pH change in the sample is observed (end point). Titration is the process of adding the chemical reagent in small increments (0.1–1.0 milliliter) until completion of the reaction, as signaled by the end point.

TOPOGRAPHY (toe-PAH-gruh-fee)

TOPOGRAPHY

The arrangement of hills and valleys in a geographic area.

TOTAL CHLORINE

TOTAL CHLORINE

The total concentration of chlorine in water, including the combined chlorine (such as inorganic and organic chloramines) and the free available chlorine.

TOTAL CHLORINE RESIDUAL

TOTAL CHLORINE RESIDUAL

The total amount of chlorine residual (including both free chlorine and chemically bound chlorine) present in a water sample after a given contact time.

TOTAL DYNAMIC HEAD (TDH)

TOTAL DYNAMIC HEAD (TDH)

When a pump is lifting or pumping water, the vertical distance (in feet or meters) from the elevation of the energy grade line on the suction side of the pump to the elevation of the energy grade line on the discharge side of the pump. The total dynamic head is the static head plus pipe friction losses.

TOTALIZER

A device or meter that continuously measures and sums a process rate variable in cumulative fashion over a given time period. For example, total flows displayed in gallons per minute, million gallons per day, cubic feet per second, or some other unit of volume per time period. Also called an integrator.

TOXIC

A substance that is poisonous to a living organism. Toxic substances may be classified in terms of their physiological action, such as irritants, asphyxiants, systemic poisons, and anesthetics and narcotics. Irritants are corrosive substances that attack the mucous membrane surfaces of the body. Asphyxiants interfere with breathing. Systemic poisons are hazardous substances that injure or destroy internal organs of the body. Anesthetics and narcotics are hazardous substances that depress the central nervous system and lead to unconsciousness.

TOXIC SUBSTANCE

See HARMFUL PHYSICAL AGENT and TOXIC.

TOXICITY (tox-IS-it-tee)

The relative degree of being poisonous or toxic. A condition that may exist in wastes and will inhibit or destroy the growth or function of certain organisms.

TRANSPIRATION (TRAN-spur-RAY-shun)

The process by which water vapor is released to the atmosphere by living plants. This process is similar to people sweating. Also called evapotranspiration.

TRICKLING FILTER

A treatment process in which wastewater trickling over media enables the formation of slimes or biomass, which contain organisms that feed upon and remove wastes from the water being treated.

TURBIDIMETER

See TURBIDITY METER.

TURBIDITY (ter-BID-it-tee)

The cloudy appearance of water caused by the presence of suspended and colloidal matter. In the waterworks field, a turbidity measurement is used to indicate the clarity of water. Technically, turbidity is an optical property of the water based on the amount of light reflected by suspended particles. Turbidity cannot be directly equated to suspended solids because white particles reflect more light than dark-colored particles and many small particles will reflect more light than an equivalent large particle.

TURBIDITY (ter-BID-it-tee) METER

An instrument for measuring and comparing the turbidity of liquids by passing light through them and determining how much light is reflected by the particles in the liquid. The normal measuring range is 0 to 100 and is expressed as nephelometric turbidity units (NTUs). Also called a turbidimeter.

TURBIDITY (ter-BID-it-tee) UNITS (TU)

Turbidity units are a measure of the cloudiness of water. If measured by a nephelometric (deflected light) instrumental procedure, turbidity units are expressed in nephelometric turbidity units (NTU) or simply TU. Those turbidity units obtained by visual methods are expressed in Jackson turbidity units (JTU), which are a measure of the cloudiness of water; they are used to indicate the clarity of water. There is no real connection between NTUs and JTUs. The Jackson turbidimeter is a visual method and the nephelometer is an instrumental method based on deflected light.

U

UNIFORMITY COEFFICIENT (UC) UNIFORMITY COEFFICIENT (UC)

The ratio of (1) the diameter of a grain (particle) of a size that is barely too large to pass through a sieve that allows 60 percent of the material (by weight) to pass through, to (2) the diameter of a grain (particle) of a size that is barely too large to pass through a sieve that allows 10 percent of the material (by weight) to pass through. The resulting ratio is a measure of the degree of uniformity in a granular material, such as filter media.

$$\text{Uniformity Coefficient} = \frac{\text{Particle Diameter}_{60\%}}{\text{Particle Diameter}_{10\%}}$$

V

V-NOTCH WEIR V-NOTCH WEIR

A triangular weir with a V-shaped notch calibrated in gallons (liters) per minute readings. The weir can be placed in a pipe or open channel. As the flow passes through the V-notch, the depth of water flowing over the weir can be measured and converted to a flow in gallons (liters) per minute.

VARIABLE COSTS VARIABLE COSTS

Costs that a utility must cover or pay that are associated with the actual collection, treatment, and discharge of wastewater. These costs vary or fluctuate on the basis of the volume of wastewater collected, treated, and discharged or reused. Also see FIXED COSTS.

VAULT VAULT

A small, box-like structure that contains valves used to regulate flows.

VECTOR VECTOR

An insect or other organism capable of transmitting germs or other agents of disease.

VELOCITY HEAD VELOCITY HEAD

The energy in flowing water as determined by a vertical height (in feet or meters) equal to the square of the velocity of flowing water divided by twice the acceleration due to gravity ($V^2/2g$).

VERTICAL OFFSET VERTICAL OFFSET

A pipe joint in which one section is connected to another at a different elevation, such as a DROP JOINT.

VOID VOID

A pore or open space in rock, soil, or other granular material, not occupied by solid matter. The pore or open space may be occupied by air, water, or other gaseous or liquid material. Also called an interstice, pore, or void space.

VOLATILE (VOL-uh-tull) VOLATILE

(1) A volatile substance is one that is capable of being evaporated or changed to a vapor at relatively low temperatures. Volatile substances can be partially removed from water or wastewater by the air stripping process.

(2) In terms of solids analysis, volatile refers to materials lost (including most organic matter) upon ignition in a muffle furnace for 60 minutes at 550°C (1,022°F). Natural volatile materials are chemical substances usually of animal or plant origin. Manufactured or synthetic volatile materials, such as plastics, ether, acetone, and carbon tetrachloride, are highly volatile and not of plant or animal origin. Also see NONVOLATILE MATTER.

VOLATILE ACIDS VOLATILE ACIDS

Fatty acids produced during digestion that are soluble in water and can be steam-distilled at atmospheric pressure. Also called organic acids. Volatile acids are commonly reported as equivalent to acetic acid.

VOLATILE LIQUIDS VOLATILE LIQUIDS

Liquids that easily vaporize or evaporate at room temperature.

VOLATILE SOLIDS VOLATILE SOLIDS

Those solids in water, wastewater, or other liquids that are lost on ignition of the dry solids at 550°C (1,022°F). Also called organic solids and volatile matter.

VOLTAGE VOLTAGE

The electrical pressure available to cause a flow of current (amperage) when an electric circuit is closed. Also called electromotive force (EMF).

VOLUMETRIC VOLUMETRIC

A measurement based on the volume of some factor. Volumetric titration is a means of measuring unknown concentrations of water quality indicators in a sample by determining the volume of titrant or liquid reagent needed to complete particular reactions.

VOLUTE (vol-LOOT) VOLUTE

The spiral-shaped casing that surrounds a pump, blower, or turbine impeller and collects the liquid or gas discharged by the impeller.

W

WASTEWATER WASTEWATER

A community's used water and water-carried solids (including used water from industrial processes) that flow to a treatment plant. Stormwater, surface water, and groundwater infiltration also may be included in the wastewater that enters a wastewater treatment plant. The term sewage usually refers to household wastes, but this word is being replaced by the term wastewater.

WATER CYCLE WATER CYCLE

The process of evaporation of water into the air and its return to earth by precipitation (rain or snow). This process also includes transpiration from plants, groundwater movement, and runoff into rivers, streams, and the ocean. Also called the hydrologic cycle.

WATT WATT

A unit of power equal to one joule per second. The power of a current of one ampere flowing across a potential difference of one volt.

WEIR (WEER) WEIR

(1) A wall or plate placed in an open channel and used to measure the flow of water. The depth of the flow over the weir can be used to calculate the flow rate, or a chart or conversion table may be used to convert depth to flow. Also see PROPORTIONAL WEIR.

(2) A wall or obstruction used to control flow (from settling tanks and clarifiers) to ensure a uniform flow rate and avoid short-circuiting.

WET CHEMISTRY WET CHEMISTRY

Laboratory procedures used to analyze a sample of water using liquid chemical solutions (wet) instead of, or in addition to, laboratory instruments.

WET PIT WET PIT

See WET WELL.

WET WELL WET WELL

A compartment or tank in which wastewater is collected. The suction pipe of a pump may be connected to the wet well or a submersible pump may be located in the wet well.

X

(NO LISTINGS)

Y

(NO LISTINGS)

Z

ZOOGLEAL (ZOE-uh-glee-ul) MASS

Jelly-like masses of bacteria found in both the trickling filter and activated sludge processes. These masses may be formed for or function as the protection against predators and for storage of food supplies. Also see BIOMASS.

ZOOGLEAL (ZOE-uh-glee-al) MAT

A complex population of organisms that form a slime growth on the sand filter media and break down the organic matter in wastewater. These slimes consist of living organisms feeding on the wastes in wastewater, dead organisms, silt, and other debris. On a properly loaded and operating sand filter, these mats are so thin as to be invisible to the naked eye. Slime growth is a more common term.

SUBJECT INDEX

A

NOTES

NOTES

NOTES

NOTES

NOTES